Capitalism and Environmental Collapse

Luiz Marques

Capitalism and Environmental Collapse

 Springer

Luiz Marques
Department of History
Univerity of Campinas
Campinas, São Paulo, Brazil

Translated from the Portuguese language edition: "Capitalismo e Colapso Ambiental", third edition, by Luiz Marques. Copyright (c) Editora da Unicamp, 2018. All rights reserved. Translated by Rebecca de Faria Slenes with the support of the grant 2019/14233-3 from the São Paulo Research Foundation (Fundação de Amparo à Pesquisa do Estado de São Paulo - FAPESP). The opinions, hypotheses and conclusions expressed in this material are the author's responsibility and don't necessarily reflect FAPESP's vision. Translation reviewed by William Shelton.

ISBN 978-3-030-47526-0   ISBN 978-3-030-47527-7   (eBook)
https://doi.org/10.1007/978-3-030-47527-7

This Springer imprint is published by the registered company Springer Nature Switzerland AG
The registered company address is: Gewerbestrasse 11, 6330 Cham, Switzerland

Foreword

You are about to read a book that should be required reading at universities all over the world. In it, Luiz Marques lays out in clear detail the trajectory that has been forced upon humanity by a dysfunctional economic system designed and installed to benefit the banks, corporations, and wealthy elites of the world.

This economic model, actually known as NeoClassical or "growth economics," posits at the axiomatic level that the human economy can grow forever. The growth model, and those who teach and promote it, goes on to functionally implement economic growth at an exponential rate. Nothing can grow forever within a limited container, no less grow at an exponential rate. Only in the reductionist human imagination is such a thing possible. But this is of little concern to those who have for the past hundred years learned, taught, and practiced this catastrophically destructive model of how the human economy should work.

> "Anyone who believes that exponential growth can go on forever in a finite world is either a madman or an economist." – Kenneth Boulding

To say that humanity is at a crossroads is an understatement. In fact, we have already been forced to choose the "road to ruin" simply because it benefits the wealthy interests who control today's political-economy. There is much talk of "tipping points" in today's conversations about climate change and our assault on the ecosystems of Earth, which in fact support humanity and all of life, and without which we would perish. I prefer to call them "thresholds" of an irreversible decline.

The point here is that we must all become critical thinkers, able to acclimate to new paradigms, and the new terms which must be used to communicate those paradigms. We must see clearly and think differently than what has come before. In the famous words of Albert Einstein, "We cannot solve our problems with the same thinking we used when we created them." This book will help you think differently.

Lest the picture I paint be considered too bleak, it is imperative for you to read on. While it is unlikely that we can completely "solve" the problems we've created, in the normal sense of the word "solve," we can still "intervene" to avert the very worst outcomes. It is this imperative for "interventions" that should motivate you, the reader, to delve deeply into this remarkably clear exposition and the recommendations it contains.

Chief among those interventions is that we must reconceive how a human economy could, would, and should function. The model Luiz and I regard as most appropriate and beneficial is known as **ecological economics**. While it is not named explicitly, its imprint is detectable throughout this scholarly work.

In the words of Yogi Berra, the legendary catcher for the New York Yankees, and oft-quoted, humorous philosopher, "It ain't over 'til it's over." As humans, it is our moral and ethical obligation to do whatever we can to avert an ecological disaster. Personally, I feel it is the only thing really worth doing anymore. This book makes our errors of the past unmistakably clear and points the way to how we might blunt the impact of our ongoing collusion with Nature. So read on.

Stuart Scott
Eco-Social Strategist
Executive Director, http://scientistswarning.org/
Executive Producer, http://scientistswarning.tv/

Preface to the English Edition

The English edition of this book is quite different from its three editions in Portuguese (2015–2018). With the exception of the final three chapters, all the others have been partially or totally rewritten and enriched with new data, analyses, and projections. Since 2018, fresh scientific knowledge has further reinforced the perception of how much faster capitalism's inherently expansive mode of functioning is destroying our bountiful biosphere and changing the stable climate of the last twelve millennia that has made mankind's material and spiritual journey during the Holocene possible. A global environmental collapse is slowly unfolding and its first unequivocal signs are already here. Since the beginning of the twenty-first century, we have experienced the most lethal heatwaves, droughts, hurricanes, floods, and wildfires ever recorded. We are witnessing accelerating defaunation and species extinctions in all taxonomic groups, less fertile soils, desertification, growing food insecurity, sea level rise at an unexpected speed, and the emergence of millions of people who fall into a new category of refugees – climate refugees. We are also suffering unprecedented sanitary crises, with increasing antibiotic resistance and the cancer burden jumping to 18.1 million new cases in 2018, up from 12.6 million cases in 2012 (+44%), more threatening pandemics such as Covid-19, more devastating plagues such as Africa and Asia and South America's recent locust outbreak, more widespread pollution, intoxication, and hormonal disturbances. This new world is accurately portrayed not only by the flourishing dystopian fiction of the last decades, but also, and with even darker tones, by a growing sum of converging scientific data and projections.

Acceleration is maybe the most defining feature of our time. In 2018, I drew attention to the fact that what had been previously expected to occur only at the end of the century was then expected by mid-century. These imbalances include a global warming of at least 2 °C above the pre-industrial period, the disappearance of several tropical forests, the transition of the eastern and southern parts of the Amazon forest to a savanna vegetation, the first iceless September in the Arctic, extensive thawing of land and seabed permafrost, global mean sea level rise of over 50 cm (above the 2000 level), and large-scale extinctions of species. This English edition shows that the possibility of these occurrences is now being projected for the second quarter of the century.

The historical cycle of twentieth-century capitalism's relative material success pertains to… the twentieth century. The world of which our parents and grandparents were beneficiaries has ended. GDP growth rates are now steadily decreasing and, most importantly, economic growth no longer translates into a better quality of life. No significant social improvement is possible in a rapidly deteriorating environment. The so-called "emerging" economies, such as those of India and China, are typically trapped in a devastating machine that is, in reality, making them sink more deeply, and before other countries, into extreme levels of pollution, very serious public health crises, lethal climactic anomalies, and acute water scarcity. Their high growth rates have, in short, put them at the forefront of the environmental nightmare with which our planetary society is contending.

The truth is that the curve measuring environmental costs against the economic benefits of capitalism has irreversibly entered into a negative phase. Despite this, there are still those who argue that global warming, deforestation, biodiversity loss, pollution, and increasing intoxication of organisms are the price to pay for economic growth. They persist in the belief that states and corporations can be "educated" to have a more environmentally sustainable behavior. *L'éducation fait tout* hails the long-living enlightenment tenet... Most are also deluded by the empty words of the Kyoto Protocol, the Paris Agreement, the Sustainable Development Goals, the Aichi Biodiversity Targets, and the New York Declaration on Forests (NYDF), among other texts that delight diplomats who are experts at negotiating new targets to reduce devastation, as long as they do not undermine the growth plans of corporations and their governments. Proceed on the path of GDP growth and fit back into the biosphere, thanks to the miraculous formulas of "sustainable capitalism"! Not many escape this magical thinking. But the few who do, know that any given quantity that grows at 2% rate per year doubles every 35 years, or every 23 years if it grows at 3% yearly. They also know that there is no available or future technology capable of allowing for more indiscriminate economic growth without having a further impact on the biosphere and on a climate that is already transitioning towards an additional 1 °C above the current averages. In short, these few people know that there is no longer a chance of escaping a cascade of socioenvironmental disasters unless we radically redefine the meaning and purpose of economic activity, reduce social inequality, and enforce democratic control of strategic investments. This means putting a new economy at the service of reducing environmental impact. This also means investing in widespread sanitation infrastructure, in local and organic agriculture, and in public policies of demographic degrowth, beginning with the right to free and state-assisted access to all forms of contraception and termination of unwanted pregnancies.

In the history of mankind, there is no record of a rupture in the growth paradigm, nor of changes of this magnitude and at such speed. There is no record because such rupture and such changes have never been necessary. Now, they are. It is not a question of underestimating the extreme difficulty of putting this program into practice. But nothing can be considered impossible when what is at stake is the fate of our societies, if not our species. The few who perceive the causal relationship, in essence a very simple one, between corporate expansion and environmental collapse in a finite world are growing in number and in persuasion. And they are beginning to have a weight on the balance. Not yet, it is true, on the balance of political power, but at least on the balance of ideas, and especially in the growing conviction that capitalism's obsolete economic and political recipe will not offer a way out of the impasse and challenges posed by the Anthropocene. A new radicalism of philosophical thought and political action is required in our times. We need a recovered confidence, reborn from the ashes, that we, as a society, are capable of overcoming capitalism in the direction of a new social contract founded on a natural contract, as proposed by Michel Serres, among others. No social justice project is politically conceivable in a scorched Earth.

This book seeks to raise a more acute awareness of the extreme gravity of our situation. Its aim is also to add more evidence to the fact, already highlighted by two or three generations of scholars, that "sustainable capitalism" is a contradiction in terms. Capitalism is the disease. It cannot offer, therefore, any credible therapy to mitigate the ongoing environmental collapse. We are, of course, guided by science, but the question of whether societies will be able to significantly mitigate this collapse is not a scientific one. "Rather it is a question," as pointed out by Damon Matthews and colleagues (2018), "of what we believe human societies to be capable of achieving." The ultimate hope of this book is to contribute in some way so that this belief does not falter among its readers.

São Paulo, Brazil   Luiz Marques

Reference

MATTHEWS, Damon *et al.*, "Focus on cumulative emissions, global carbon budgets and the implications for climate mitigation targets". *Environmental Research Letters*, 13, 1, 12/I/2018.

Acknowledgments

It is with renewed pleasure that I express my gratitude to my friends, colleagues, and institutions cited in the acknowledgments section of this book's first three editions in Brazil. A special word of thanks goes to Francisco Foot Hardman, who read and discussed the book's Introduction with me at length and wrote the generous text that has graced the back cover of its Brazilian editions. I also wish to restate my gratitude to my friends and to my undergraduate and graduate students in the History Department at the State University of Campinas – UNICAMP, the former, for having embraced my proposals for courses about the climate and environmental crises, and to the latter for discussions both in and outside class. Unicamp's Rector Marcelo Knobel, Gabriela Castellano, and Marco Aurélio Pinheiro Lima, three cherished colleagues at the Gleb Wataghin Physics Institute; Marcelo Cunha, Lázlo Nagy, and Enrique Ortega, my friends at the Laboratory for Ecological Engineering and Applied Computer Sciences (LEIA – acronym in Portuguese); and Néri de Barros Almeida, of our University's Board of Directors for Human Rights have provided many opportunities to broaden discussion about contemporary environmental crises, both on and off campus. And finally, my gratitude goes to the young people of the São Paulo Climate Coalition, to my friends and colleagues in the Brazilian section of the International Society for Ecological Economics (ISEE), in particular to Clóvis Cavalcanti, José Eustáquio Diniz Alves, and Ademar Romeiro, as well as to my new friends from Coletivo 660.

In this English edition, my first words of appreciation go to Márcia Abreu, director of my University's publishing house, whose support has been extremely important for this current edition of my book to become a reality. Rebecca de Faria Slenes translated it with great proficiency, William Shelton reviewed her translation with no less accuracy, and Fernando Chaves prepared the graphics, with his customary competence. Throughout this journey, I have received valuable assistance and orientation from Bruno Fiuza, Editor at Springer's office in São Paulo, Brazil. Now is the time to thank them once again for his tremendous work and dedication.

Stuart Scott, Eco-Social Strategist, Executive Director of ScientistsWarning.org, and Executive Producer of ScientistsWarning.TV, drafted the text presenting this book in terms that again demonstrate his immense generosity and unswerving commitment to our common cause. David Lapola, Cristiana Simão Seixas, and José Eustáquio Diniz Alves read, corrected, and, with many observations, enriched the chapters about the decline in forests, the decline in land-based biodiversity, and the demographic question, respectively. I am immensely grateful to them. Their careful reading made me feel more secure about the data and analyses proposed in these chapters. Obviously, any potential errors remaining are my sole responsibility. I received indispensable financial support for the Brazilian edition and for its English translation from the São Paulo Foundation to Assist Research (FAPESP – acronym in Portuguese).

This book would be different or, more probably, would not exist at all without the tremendous amount and quality of critical remarks and contributions by Sabine Pompeia, my wife. Even more so, I owe her for motivating and encouraging me to carry out the unpleasant task of scrutinizing the socioenvironmental collapse that is manifesting itself more evidently in these times. I also dedicate this book to her and to our children, Elena and Leon.

São Paulo,
March 2020

Contents

Part I
The Convergence of Environmental Crises

Chapter 1
Introduction

In 1856, Alexis de Tocqueville opens his reflections on the French Revolution with a warning: "Philosophers and statesmen may learn a valuable lesson of modesty from the history of our Revolution, for there never were greater events, better prepared, longer matured, and yet so little foreseen."[1] Since Tocqueville, the very principle of historical predictability, dear to the eighteenth and nineteenth centuries,[2] was gradually undermined. In 1928, before Karl Popper, Paul Valéry signed its death certificate: "Nothing was more destroyed by the last war than the pretension of foresight."[3]

Since unpredictability is embedded in history, it comes as no surprise that the most decisive historical processes and events of the past century were not predicted: the carnage of the World War I and the rise of modern chemical and nuclear warfare, the 1929 crisis, totalitarianism, racism and its genocides, the Cold War, the 1968 protests that raged throughout the world, the 1973 oil crisis, the Berlin Wall and its fall, the disintegration of the Soviet Union, the implosion of the Western Communist Parties, China's emergence as an imperialist power, the advent of computers and of the Internet, the resurgence of modern fundamentalism (which is now deeply imbedded in all of the three main monotheistic religions), the creationist assault on science

---

[1] Cf. A. de Tocqueville (1856, p. 13): "Il n'y a rien de plus propre à rappeler les philosophes et les hommes d'État à la modestie que l'histoire de notre Révolution; car il n'y eut jamais d'événements plus grands, conduits de plus loin, mieux préparés et moins prévus."

[2] See, for example, Condorcet (1793/1993, p. 189): "man can predict almost with total certainty those phenomenon which he knows the laws of; (...) even when these are unknown he can, based on past experience, foresee the events of the future with great probability [that they will occur]."

[3] "Rien n'a été plus ruiné par la dernière guerre que la prétention de prévoir." *De l'Histoire* (1928), Cf. Popper (1936/1957) and (1963), Chapter 16: Prediction and Prophecy in Social Sciences. In his *Prison notebooks* (1929–1935), Gramsci distanced himself from the Marxism of manuals, such as that of Bukharin, regarding the idea of historical predictability. As pointed out by Ives (2004, p. 141): "Gramsci contends that it is absurd to speak of objective prediction if one means that the one making such predictions is impartial or does not connect the predicted outcome with his or her own political desire and conception of the world."

© Springer Nature Switzerland AG 2020
L. Marques, *Capitalism and Environmental Collapse*,
https://doi.org/10.1007/978-3-030-47527-7_1

education, the rise in public debt in industrialized countries, the Asian financial crisis of 1997, civil wars and the waves of migration from Afghanistan and Arab countries (triggered, in part, by the chaos that Western invasions sowed in these countries), the financial crises of 2007–2009,[4] and, finally, the pandemic of Covid-19, with its unpredictable developments. Among the most relevant and equally unpredictable events in the last years are the instability of oil prices and the indecipherable incognito of their future prices; the rise of extreme right-wing parties and movements in the United States, Brazil, Germany, Hungary, Poland, Austria, Italy, and other European countries; the revival of nationalisms; and the rejection of immigrants, fundamental ingredients in the rapid ideological disintegration of the European project, reaching its climax with Britain's final decision to leave the EU. All this was fueled by the emergence of a new type of interaction between technology and politics, with the use of virtual reality, tweets, and fake and hate news (disseminated through social media networks by replicating algorithms) as weapons for disinformation and mass emotional and intellectual debasement.[5] The few scholars who did predict these vicissitudes of the historical drama did not win a general audience except ex post facto and precisely because of such exploits.[6]

One aspect of history, once considered peripheral, has, however, proven to be less unpredictable: the impacts of industrial societies on nature and its backlashes, the object of this book. Since 1820, Lamarck (1744–1829), one of the first naturalists to introduce the term "biology," foresaw the causal link between industrial civilization and environmental collapse:[7]

---

[4] The opening words of the *World Economic Outlook* published by the IMF in April 2007 sound today rather comical: "Notwithstanding the recent bout of financial volatility, the world economy still looks well set for continued robust growth in 2007 and 2008."

[5] Although the use of such weapons has become commonplace among political rulers, Donald Trump occupies a prominent position here. The systematic use of lies by the US president has raised concerns that his behavior patterns are typical of a psychopath. Cf. Baynes (2017); Blow (2017); Kessler et al. (2017); Lee (2017); Kentish (2017). The same diagnosis applies to the current president of Brazil, Jair Bolsonaro (Lichterbeck 2020).

[6] Some works were celebrated for predicting these great historical *catastrophae*: in *Impossibilités techniques et économiques d'une guerre entre grandes puissances*. Paris, Paul Dupont, 1899, Jan de Bloch warned of the dire consequences of a war between industrialized countries ("Les guerres ne pourront donc se terminer," he concluded, "autrement que par l'épuisement entier des deux adversaires ou par un cataclisme social"). Concerning the end of the Soviet Union, cf. Todd (1976). David Levy and Nouriel Roubini, among very few other economists, warned about the coming of the 2007–2008 financial crisis. Cf. "8 who saw the crisis coming..." CNNMoney/Fortune, August 2008.

[7] Cf. Lamarck (1820, p. 154). English translation by Burkhardt (1977/1995, p. 214): "L'homme, par son égoïsme trop peu clairvoyant pour ses propres intérêts, par son penchant à jouir de tout ce qui est à disposition, en un mot, par son insouciance pour l'avenir et pour ses semblables, semble travailler à l'anéantissement de ses moyens de conservation et à la destruction même de sa propre espèce. Em détruisant partout les grands végétaux qui protégeaient le sol, pour des objets qui satisfont son avidité du moment, il amène rapidement à la stérilité ce sol qu'il habite, donne lieu au tarissement des sources, em écarte les animaux qui y trouvaient leur subsistance et fait que de grandes parties du globe, autrefois très fertiles et très peuplées à tous égards, sont maintenant nues, stériles, inhabitables et desertes [...] On dirait que l'homme est destine à s'exterminer lui-même

> By his egoism too short-sighted for his own good, by his tendency to revel in all that is at his disposal, in short, by his lack of concern for the future and for his fellow man, man seems to work for the annihilation of his means of conservation and for the destruction of his own species. In destroying everywhere the large plants that protect the soil in order to secure things to satisfy his greediness of the moment, man rapidly brings about the sterility of the ground on which he leaves, dries up the springs, and chases away the animals that once found their subsistence there. He causes large parts of the globe that were once very fertile and well-populated in all respects to become dead, sterile, uninhabitable, and deserted. (…) One could say that he is destined to exterminate himself, after having rendered the globe uninhabitable.

The terms under which the famous French naturalist formulates the problem, that of the "human egoism," belong more to his century than to ours. But Lamarck and other philosophers and naturalists of the nineteenth century—Alexander von Humboldt (Wulf 2015), Ernst Haeckel, George Perkins Marsh, and John Muir, to name only the most celebrated—accurately foresaw the tendency for environmental collapse caused by the increasing anthropogenic interference in the Earth system, and their perception of this process does not substantially differ from that of the current scientific consensus. Matthias Aengenheyster and colleagues (2018) echo this consensus when they remind us that "the Earth system is currently in a state of rapid warming that is unprecedented even in geological records." This rapidity manifests itself distinguishably in the absence, in the last half century, of decennial oscillations in the rhythm of global warming, oscillations that had been occurring until the seventh decade of the twentieth century. In its 2018 Special Report, the IPCC showed that (1) the impacts of a warming of 1.5 °C put mankind clearly beyond the safe zone and (2) any warming above this threshold is terribly threatening to the elementary functioning of contemporary societies. According to this report, to stick to 1.5 °C, however, would require reducing GHG emissions by 2030 to the levels of 1977, that is, reduce the emissions of GHG per capita to the levels of 1955 (Marland et al. 2019), something implausible if we maintain the existing socioeconomic and political structures.

The chances of not exceeding an average global warming of 1.5 °C above the pre-industrial period have, therefore, become insignificant, if not to say nil, and in the second quarter of this century, the Earth will see an average temperature never faced by our species. In other words, the future predicted by Lamarck in 1820 coincides roughly with what was proposed by Erik M. Conway and Naomi Oreskes in their remarkable book manifest, *The Collapse of Western Civilization* (2014). Actually, current environmental imbalances as a whole—not only climate change but the so-called nine planetary boundaries as a whole (Rockström et al. 2009; Lynas 2007, 2011; Rockström and Wijkman 2012; Steffen et al. 2015; Raworth 2017),[8] four of which have already been exceeded[9]—are already producing a radical change in the coordinates that enable life on this planet as we know it.

---

après avoir rendu le globe inhabitable."

[8] See also the editorial of *Nature*, "Earth's boundaries?", 461, 24/9/2009, pp. 447–448.

[9] The four limits are land use change (mainly deforestation), biodiversity loss, high atmospheric

Science and politics are much more intertwined than we have portrayed them in the past. In his speech at the Rio+20 in 2012, José Mujica, former President of Uruguay, affirmed: "The big crisis is not ecological; it is political." Without misunderstanding the specific environmental nature of these crises, Mujica stresses that societies and other ecosystems now share a common fate. What will decide the evolution of the environmental crises will be, above all, the ability of societies, informed by the scientific consensuses, to take on radically democratic forms of government, without which it will not be possible to significantly attenuate the ongoing environmental collapse. In the conclusion, I return briefly to the crucial issue of these new forms of democracy; their analysis, however, is beyond the scope of this book.

1.1   The Great Inversion and the Limits of Environmental Awareness

Throughout millenniums, the security of societies in the face of scarcity and other adversities depended fundamentally on the capacity to accumulate surplus from the continuous increase in soil use, technology, work productivity, production, and consumer goods. The historical situation today suddenly became not only diverse but also *inverse* in its relationship to this long past: the environmental crises of our time, unleashed precisely by the success of industrial societies in the endless multiplication of surplus, not only imposes new forms of scarcity but also systemic threats to our security. The aim of this book is to demonstrate that the equation "more surplus = more security," which throughout millenniums was made part of our *forma mentis*, has today been converted into the equation "more surplus = less security."

The difficulty in understanding this inversion, its seriousness, and the extent of its implications is the main cognitive obstacle to widespread awareness of the environmental deadlock that threatens us. Next to the totem of GDP growth, which has taken the shape of a religious dogma and of a "social pathology" (Boyd 2013), the destruction of ecosystems (when recognized) is still considered a "cost" or an inevitable collateral effect, and a problem that can be bypassed, thanks to continuous technological innovation, gains in efficiency, improvement in security protocols, and better risk management. Although illusionary, this belief in the possibility of continuous economic growth is easy to understand for 90% of the worldwide population, those in need of a minimum amount of material comfort. But certainly, the main problem does not lie here: satisfying the basic needs of 90% of humanity would not increase, and might even decrease, the human impact over the ecosystems. "Some 3 in 10 people worldwide, or 2.1 billion, lack access to safe, readily available water at home, and 6 in 10, or 4.45 billion, lack safely managed

---

concentrations of CO_2 (above 350 ppm), and water eutrophication due to sewage and input of agricultural nutrients (phosphorous and nitrogen).

sanitation."[10] Providing them with this infrastructure would bring a decrease, not an increase, in the extent of human environmental impact. Figure 1.1 captures, therefore, where the problem lies.

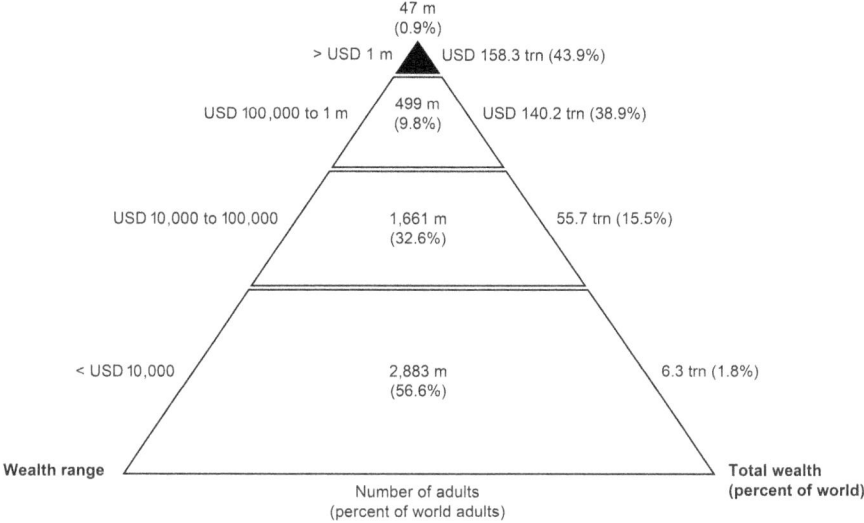

Fig. 1.1 The Global Wealth Pyramid in 2019. Based on A. Shorrocks, J. Davies & R. Lluberas, *The Crédit Suisse Global Wealth Report 2019: The year in review.* The Crédit Suisse Research Institute. Observation: Wealth here is understood as the combined assets of an adult individual

In 2019, the wealth of the adult human population (5090 million people) reached 360.5 trillion dollars. At the apex of the pyramid, 47 million people (0.9% of the adult population), with assets of more than US$ 1 million, owned 43.9% of the global wealth (US$ 158.3 trillion). Altogether, the two upper strata of the pyramid—546 million individuals, or 10.7% of the adult population—owned about 82.8% of the world's wealth (US$ 298.5 trillion out of a total of US$ 360.5 trillion in 2019). At the bottom of the pyramid, 56.6% of adult individuals owned only 1.8% (US$ 6.3 trillion) of global wealth, with assets of less than US$ 10,000 each.

In Chap. 13, I discuss the anatomy of the small pyramid formed by the vertex of this pyramid. It is only important to note here that this unequal concentration of global wealth has been deepening in the last 9 years. Table 1.1 compares the global wealth pyramids published by the Crédit Suisse Research Institute in 2010 and 2019:

In 2010, the wealth of the world adult population (4400 million people) amounted to US$ 194.5 trillion. In 2019, this wealth jumped to US$ 360.5 trillion (5090 million people). To whom did the US$ 166 trillion, accrued in these 9 years, go to?

---

[10] Cf. World Health Organization, "2.1 billion people lack safe drinking water at home, more than twice as many lack safe sanitation." *Public Health, Environmental and Social Determinants of Health* ((PHE), 93, June–July 2017.

Table 1.1 Comparison between the Global Wealth Pyramids (2010 and 2019)

	Population (millions of adults = m.a.)	<10 thousands (US$)	10 th.–100 th. (US$)	100 th.–1 million (US$)	>1 mil. (US$)
2010	4400	3038 m.a. (68.4%/4.2%)	1045 m.a. (23.5%/16.5%)	334 m.a. (7.5%/43.7%)	24.2 m.a. (0.5%/35.6%)
2019	5090	2883 m.a. (56.6%/1.8%)	1661 m.a. (32.6%/15.5%)	499 m.a. (9.8%/38.9%)	47 m.a. (0.9%/43.9%)

Source: A. Shorrocks & J. Davies, *The Crédit Suisse Global Wealth Report* 2010 and A. Shorrocks, J. Davies & R. Lluberas, *The Crédit Suisse Global Wealth Report 2019: The year in review.* Observation: m.a. = millions of adults. The percentages in parenthesis represent the percentage of the total adult population, to the left; the percentage of global wealth owned by each extract of the pyramid, to the right)

Mostly, to the owners of more than US$ one million located at the apex of the global wealth pyramid. They pocketed US$ 89.1 trillion or 53.7% of the accrued world wealth during these 9 years. The wealthiest have nearly doubled in number in these 9 years, going from 24.2 million in 2010 to 47 million people in 2019. They represented 0.5% of the adult world population in 2010 and 0.9% in 2019. Their wealth increased nearly 128%, going from US$ 69.2 trillion in 2010 to US$ 158.3 trillion. Most importantly, in 2010, they accumulated 35% of the global wealth; in 2019, this percentage reached 43.9%.

In 2010 the second upper segment of the pyramid (individual assets in the range US$ 100,000 – US$ one million) owned US$ 85 trillion or 43.7% of the world wealth. In 2019, this figure jumped to US$ 140.2 trillion, which represented 38.9% of the world wealth. So they pocketed US$ 55.2 trillion or 34% of the increase in this wealth (US$ 166 trillion) during these 9 years. In 2019, the wealth of this second group of 499 million people was four times their share of the adult population. The middle class in developed nations typically belongs to this group (Shorrocks et al. 2019). Together, the two upper brackets of the world wealth pyramid appropriated US$ 144.3 trillion or nearly 87% of the increase in the world wealth during this period.

Thus, only US$ 21.7 trillion or 13% of the global wealth accrued between 2010 and 2019 went to the pockets of the 4554 million people, or 89.2% of the 2019 world adult population, representing the two bottom strata of the wealth pyramid. In the bottom bracket of the global wealth pyramid (2883 million people, or 56.6%, with wealth below US$ 10,000), there has been impoverishment in every sense of the word. In 2010, these 3038 million people (68.4% of the adult population) owned US$ 8.2 trillion. In 2019, they were 2883 million people. The poorest people of the world have only marginally decreased since 2010 (155 million people) but owned only US$ 6.3 trillion in 2019. They lost US$ 1.9 trillion in 9 years! In 2010, their per capita wealth was US$ 2699; in 2019, it declined to US$ 2185, an impoverishment of US$ 514. In the last 9 years, their share in the world wealth fell from 4.2% to 1.8%.

In 2018 and 2019, this global inequality widened further. According to Oxfam (Lawson et al. 2019), "between 2017 and 2018, a new billionaire was created every two days." In 2018 alone, "the wealth of the world's billionaires increased by US$ 900 billion, or US$ 2.5 billion a day. Meanwhile the wealth of the poorest half of humanity, 3.8 billion people, fell by 11%." Also, according to Oxfam (Coffey et al. 2020), "in 2019, the world's billionaires, only 2,153 people, had more wealth than 4.6 billion people."

It is important to note here that there is increasing inequality not only in assets but also in income. Today, 46% of the world, or 3.4 billion people, live on less than US$ 5.50 a day (World Bank 2018). In the United States, for instance, according to Paul Krugman (2014), "since the late 1970s real wages for the bottom half of the work force have stagnated or fallen, while the incomes of the top 1% have nearly quadrupled (and the incomes of the top 0.1% have risen even more)" (see also Lauer 2014). The *2013 Survey of Consumer Finances* (SCF), a triennial cross-sectional survey of US families conducted by the Federal Reserve Board, indicates that 30.5% of the total income in the United States was concentrated in the hands of the 3% richest North Americans in 2013, compared to 27.7% in 2010. According to the same survey, in 2013, 90% of those in the base of the income pyramid had their economic participation reduced to 24.7%, against 33.2% in 2010. This phenomenon is generalized. Oxfam (*Working for the Few* 2014) showed that in 2014 seven in ten people lived in countries where the inequality of income increased.

The economy of the planet is propelled to satisfy the appetite of 546 million people (the 10.7% of the world's adult population who owned US$ 298.5 trillion, or 82.8% of the world wealth in 2019), leading to environmental crises, starting with climate change caused by anthropogenic carbon emissions. In a paper titled "Carbon and Inequality: From Kyoto to Paris," Lucas Chancel and Thomas Piketty (2015) showed that income and CO_2e emissions inequalities increased within countries over the 1998–2013 period: "Global CO_2e emissions remain highly concentrated today: the top 10% emitters contribute to 45% of global emissions, while the bottom 50% contribute to 13% of global emissions." According to the authors, in 2015, among the top 10% global emitters, 40% of CO_2e emissions come from the United States, 20% from the EU, and 10% from China. More recently, Nicholas Beuret (2019) endorses the same conclusion:

> Most of the world's population produces very little in the way of either carbon emissions or broader environmental impacts. We can go further here by also looking at imported carbon emissions – that is, the emissions that come from the production of goods and services in countries such as China that are then consumed in the wealthy countries of the global north. (…) When we approach carbon emissions this way, it's clear the problem isn't overpopulation or China, but the richest people on earth. After all, being rich, especially ultra-rich, means being directly responsible, either through consumption or control, for the majority of the world's carbon emissions.

This structure of wealth and income and the tendency for its concentration manifest a mechanism that is at the heart of the economic system and that propels a tiny parcel of the population to control the investment decisions and to accumulate wealth as an end in itself. This mechanism, which is nothing other than the accumulation of capital, is self-renewing, not only in action but also ideologically. The belief that societies' security and prosperity depend on the accumulation of capital is, as stated above, a great cognitive obstacle to the understanding that this accumulative mechanism is actually pushing us toward a social and environmental collapse.

In Antiquity, the absence of limits is discussed in a saying attributed to Epicurus: "wealth, if limits are not set for it, is great poverty."[11] In our days, the truth of this motto has not only reached a new extreme—never was the economic system conceived so perfectly so as to satisfy the anxieties of the rich to become richer—but it also assumes a new dimension: although it is a fact that environmental crises affect the poor most severely, the intensification of these crises will end up throwing the rich and the poor in the same precarious situation. Contrary to the hidden garden that protected the *onesta brigata* of ten youth from the Black Plague in Boccaccio's *Decameron*, there is no wall capable of sheltering the rich from the effects of environmental crises, given their systemic character: global warming, decline of biodiversity, droughts, water shortage, desertification, devastating fires, more extreme meteorological events, floods, heat and cold waves capable of threatening our lives and energy security, rising sea levels, air, soil and water pollution, food poisoning, cities obstructed by cars and trash, and with increasing levels of insalubrious sanitary conditions.

Of course, this perspective is not of great concern to those who control flows of investments. In 2017 alone, the top five US banks—JPMorgan, Morgan Stanley, Goldman Sachs, Bank of America, and Citi Group—lent $ 1.5 billion to the major coal corporations, Peabody Energy, Arch Coal, and Alpha Natural Resources (Flitter 2018). The Davos 2018 *Global Risk Report* shows a bleak picture of corporate leaders' disinterest in environmental crises. The survey, titled *Global Risk of Highest Concern for Doing Business*, interviewed 14,375 executives from 148 economies and received 12,775 responses from 133 economies. It classifies risks into five categories: (a) economic (eight risks); (b) geopolitical (five risks); (c) environmental (five risks); (d) social (six risks); and (e) technological (five risks). Among the 29 proposed risks pertaining to these five categories, the Davos respondents were asked to select the five global risks that are of greatest concern for doing business in their country in the next 10 years. The first environmental risk selected by the interviewees (extreme weather events) appears only in 18th place, since only 12.9% of respondents mentioned it among their top concerns. The risk of climate change appears in 21st place, since only 11.4% of managers included it among their top

---

[11] *Porphyrius ad Marcellam*, 27, p. 207, 31 Nauck, in Hermann Usener, *Epicurea*, p. 161, translated/ed. Ilaria Ramelli, Milan, Bompiani, 2002, p. 367. Seneca also attributes the sentence to Epicurus and writes in the Ep. II to Lucilius: *non qui parum habet, sed qui plus cupit, pauper est* (poor is not the one who has little, but the one who desires more). In modern times, see, for example, Leonardo da Vinci, *Scritti letterati*, Milan, Rizzoli, 1987, p. 222: "De' non m'avere a vil ch'i non son povero; povero è quel che assai cose desidera" (Don't take me for a village man, as I am not poor; poor is the one who desires many things).

concerns, and the risk of a collapse of biodiversity appears in 26th place, indicated by only 6.6% of respondents as being one of their top concerns. Let's not be fooled: for those who control the levers of the world economy, environmental crises remain among the last of their concerns, despite their verbiage on "sustainability."

Moreover, as stressed by Ilona Otto et al. (2019), "any form of policy targeted at the superrich is bound to meet with strong resistance. The rich are over-represented in national governments and there are strong ties between the wealthy and the political elites."

What do we say, however, of the near indifference of the majority of the world's population, those most immediately vulnerable to the global environmental crises? Political, economic, and social marginalization, as well as the excruciating fight for survival, explains this near indifference. But one cannot underestimate the explanatory power of another factor: the promises of consumer society. As Ivan Illich (1973/2003) wrote almost 50 years ago, the consumer society is comprised of "two types of slaves: those intoxicated and those who want to become intoxicated; the initiated and the neophytes." Since Ivan Illich, many have meditated on this voluntary new servitude to consumerism; yet, it is necessary to return to this issue, even at the risk of redundancy.

Capitalism presents its legitimacy in tangible, and previously unimaginable, comforts that were brought to important sectors of industrialized and "emerging" contemporary societies. But even these countries are now finding themselves in a process of impoverishment. This process does not restrict itself only to the generation born between 1980 and 2000, the "unlucky millennials" (Davies et al. 2017, p. 27). It reaches significant segments of the population of all age brackets in OECD countries. In 2018, the World Bank included high-income countries in its global estimates of people living in poverty and deep poverty. There were then 13.8 million deeply poor people, living on US$ 4 a day or less, in the United States (5.3 million people), in selected countries of the European Union (6.9 million people), in Canada, Japan, and Australia (Deaton 2018). Among white non-Hispanic Americans with no more than a high school education, life expectancy is falling, and mortality rates from drugs, alcohol, and suicide are rising (Case and Deaton 2015). Philip Alston (2017), UN special rapporteur on extreme poverty and human rights, affirmed that one in four young people in the United States lived below the poverty line. Furthermore, the United States occupies 36th position in the world in terms of access to drinkable water and sanitation infrastructure, and it has the highest demographic incarceration rate on the planet: 2.16 million people, or 655 out of every 100 thousand people in 2016, a rate much above that of Cuba, China, Turkmenistan, Thailand, the Philippines, Russia, and Brazil (Kaeble and Cowhig 2016; Kann 2018). One political consequence of these levels of poverty and social exclusion is the further limiting of social representation in US democracy: only 70% of US citizens of voting age were registered to vote and only 55.7% of these took part in the 2016 presidential election (Alston 2017).

This consensus on continuous economic growth being a condition for a secure and prosperous society obviously caters to corporations and to their "classical" political spectrum. But leftist groups, with few exceptions, not only endorse this consensus but also claim they are better equipped than the right-wing political parties at guaranteeing robust economic growth. They remain anchored in an ideological automatism: a historical conception, taken from Marx, centered on the protagonism of productive forces, almost identifying the development of these productive forces with historical "progress." The *locus classicus* of this idea in Marx's thought is the passage from the *A Contribution to the Critique of Political Economy*:

> At a certain stage of development, the material productive forces of society come into conflict with the existing relations of production or – this merely expresses the same thing in legal terms – with the property relations within the framework of which they have operated hitherto. From forms of development of the productive forces these relations turn into their fetters. Then begins an era of social revolution.

The left's resistance to relegate this type of Mechanics of history "without nature" to the nineteenth century has blinded them from noticing that, throughout the twentieth century, capitalist relations of production did not hinder the development of productive forces (in fact, the contrary happened) and that, *precisely because of this*, the distinctive trait of global capitalism in the twenty-first century is the tendency for environmental collapse. Faced with this defining tendency of our century, conserving what is left of the biosphere has become the first condition not only for any possibility of social advance (which will be increasingly unlikely and ephemeral if we keep the religion of economic growth) but also for the maintenance of any organized society. Not realizing the radical novelty of our current situation, and especially its gravity, almost all leftist groups dissociate the social agenda from the ecological agenda, reserving the latter to a secondary status in their ideals and programs. Seeing the planet as a stock of resources (and even more grave, as an infinite stock of resources), the left subscribes to the premise that takes the capitalist point of view as universal, one that deems the continuous accumulation of surplus and energy as positive and even as necessary. The left doesn't notice that the only criticism to reach the root of the capitalist system is the criticism of this premise and of the suicidal type of society that it implies. As stated by Cornelius Castoriadis (2005): "Ecology is subversive since it puts into question the capitalist imagination that dominates the planet. It refuses [capitalism's] central motive which states that it is our destiny to continuously increase production and consumption." They do not notice, therefore, that this dismissal—tragic like their previous contempt for "bourgeois" freedoms[12]—allows conservative sectors to mollify and neutralize the environmental movement's potential for criticism. The delaying of an *aggiornamento—svecchiamento* might be a better word—by the majority of those in the

---

[12] The historical error of the left, tolerating tyranny in the name of socialism, is being repeated, giving the right the opportunity to sell itself as the guardians of civil liberties, in itself an absurdity.

political left has led to the current paucity of political alternatives to the socioenvironmental crises.

Reinforcing this cognitive obstacle, common across the ideological spectrum, there are at least three psychological mechanisms that keep people from becoming aware of the seriousness of these environmental crises and, a fortiori, from taking political action.

(1) The first, as stated by George Marshall (2014),[13] is loss aversion which, in decision-making theory, is a tendency to strongly prefer avoiding losses to acquiring gains. The problem with environmental crises is their perfect formula for inaction, even among those who do not deny their reality. Let us take the conclusion of the 2006 Stern Report, reinforced by Nicholas Stern in 2010, and transformed today into the mantra of ecological economists: "ignoring climate change is not a viable option – inaction will be far more costly than adaptation" (p. 616). Similarly, Simon Dietz et al. (2016) state that:

> The expected 'climate value at risk' (climate VaR) of global financial assets today is 1.8% along a business-as-usual emissions path. Taking a representative estimate of global financial assets, this amounts to US$ 2.5 trillion. However, much of the risk is in the tail. For example, the 99th percentile climate VaR is 16.9%, or US$ 24.2 trillion.

According to UNEP's *Green Economy Report* (2008/2012), it would be necessary to invest 1.3 trillion dollars per year until 2050 to finance the transition to a "green" economy. The same document mentions a similar estimate made by IEA: it would be necessary to invest 750 billion dollars every year from 2030 to 2050 to reduce by half only those CO_2 emissions linked to the production of energy. More recently, the Global Commission on Adaptation (2019), chaired by Ban Ki-moon, Bill Gates, and Kristalina Georgieva, estimates that investing US$ 1.8 trillion globally in five areas from 2020 to 2030 could generate US$ 7.1 trillion in total net benefits. Even if the future costs of inaction are admittedly much greater, mitigating emissions and adapting to climate change require accepting loss *here* and *now*: renouncing fossil fuels and major investments in low-carbon renewables and in climate-resilient infrastructure, in addition to other sacrifices for the upper- and middle-income classes which should be concrete and quantifiable. In this context, it is easy to understand the inefficiency of the environmental community's appeals to an immediate and vigorous reaction to these crises and the relative success of denialists and "merchants of doubt" (Oreskes and Conway 2010), since their rhetoric reinforces what everyone wants to hear. According to Marshall, the information that veers away from this consensus is unconsciously selected or framed in a way that distorts it so that it does not seriously conflict with the worldview of the receiver. These mechanisms of aversion to loss and neutralization of dissonance led Daniel Kahneman, Noble Prize in Economic Sciences for his research on the psychological biases that distort decision-making, to affirm: "I am very sorry, but I am deeply pessimistic.

---

[13] See also "Hear no climate evil." *New Scientist*, 16/VIII/2014, p. 24.

I really see no path to success on climate change" (quoted in Marshall 2014, p. 56). Similarly, referring to this collective resistance in making rational decisions, even when faced with mounting evidence of the acceleration of environmental crises, Daniel Gilbert, a specialist in cognitive psychology at Harvard, made the following affirmation: "It really has everything going against it. A psychologist could barely dream up a better scenario for paralysis" (quoted in Marshall (2014, p. 91).

(2) The second psychological mechanism in action is the process of habituation, a type of adaptive behavior that consists of a reduction in the response to a repetitive stimulus, but without immediate consequences. Aesop's fable of the wolf that never comes is a good illustration of this process. The reiteration of scientific prognosis about the aggravation of the environmental crisis tends to have a decreasing impact on people's consciousness, as these crises are not manifested in the form of an immediate risk. As Clive Hamilton (2010) mentions, natural selection has reinforced our capacity to react in a visceral and instantaneous way to immediate risks. But "we are at a loss when confronted with global warming which requires us to rely heavily on cognitive processing." The repetition of warnings without these being followed by immediate consequences leads to a progressive decrease in the notion of risk and in the energy needed to respond to it. Decades of living with the threat of nuclear war and of catastrophes in nuclear plants has had an anesthetic function. This explains, in part, our inertia toward the threats inherent in the environmental crises. Inspired by the catastrophe of Union Carbide in Bhopal, in 1984, and by the impact of the Chernobyl disaster in 1986, Ulrich Beck wrote *Risk Society: Towards a New Modernity*. Jean-Pierre Dupuy published *Pour un catastrophisme éclairé. Quand l'impossible est certain* (or "Enlightened Doomsaying") in 2002, inspired by Hans Jonas' criticism that we do not attribute a sufficient *weight of reality* to the future catastrophe: "We are neither cognitively nor emotionally affected by the anticipation of the misfortune to come" (p. 199). Since the warnings issued by Jonas, Beck, and Dupuy, the aging of nuclear plants[14] and the revival of the military nuclear threat (Krepon 2019) have become two among many environmental catastrophes waiting to happen, while the emotional perception of these risks tends to not correspond to what is at stake.

(3) The third psychological mechanism that reinforces this cognitive obstacle is dissociation (Worthy 2013; Humes 2013) between structural causes and specific effects. The difficulty to recognize the wolf persists, even when the signs of his arrival are multiplying. Tendencies evolve in spatial and temporal scales that are not accessible to everyday radars, which are much more sensitive to events. Events do not come, however, with a label of the tendency that they

---

[14]The 15 nuclear reactors in Ukraine were built over 40 years ago by the Soviet Union but conceived to last only 30 years. Cf. Le Hir (2014). The accident in February 2014 in the US Air Force Waste Isolation Pilot Plant received very little media attention, despite its catastrophic potential. See *Nature* editorial 7500, 509, 15/V/2014, p. 259: "An accident waiting to happen."

express. It is true that a new body of research, known as Probabilistic Event Attribution (PEA), is now examining to what extent wildfires or extreme weather events can be associated with past anthropogenic emissions and long-term climate tendencies. According to Schiermeier (2018) and a *Nature* editorial of July 2018, climate science "has finally generated the tools to attribute heatwaves and downpours to global warming."[15] In short, we can affirm that the intensity of certain events would be extremely unlikely in the absence of the ongoing anthropogenic climate change (Herring et al. 2019). The relative uncertainty between isolated facts and structural causes is being reduced. Nevertheless, an extreme meteorological event, a heatwave, an exceptional drought, or a new plague are still perceived by the public in an fragmented way, only as one more piece of "news" that the press puts out like instant shots of a clip, aligned with other news bits and diluted next to sports chronicles, criminal reports, corruption scandals, as if "environmental" were just one more adjective among other adjectives—economic, financial, moral, educational, etc.—linked to the noun "crisis."

Outside the realm of psychology, the most important factor to reinforce this cognitive obstacle may be the belief that our governments are capable of "saving" us from environmental collapse, or at least of assuming a bigger share of responsibility in implementing policies to revert the environmental degradation of the biosphere. This belief does not take into account the new and deep relationship that is being formed between the state and corporations, as we will demonstrate next.

1.2 An Ongoing Change in the Nature of the State: The "State-Corporation"

As we will see in detail in Chap. 7, between 1990 (the reference year of the Kyoto Protocol) and 2017, global CO_2 emissions increased by 63% (GCP, Global Carbon Budget 2018). And they will continue to increase in the foreseeable future (although probably at lower rates), given the ongoing increases in deforestation and wildfires, mostly in Brazil, Australia, Russia, and the United States, as well as new investments in thermoelectric power plants. It seems, therefore, more and more remote that the policies advocated by the IPCC to contain global warming will be implemented. The verdicts of the *Climate Change Performance Index 2020* (CCPI) are peremptory in this respect:

> None of the countries assessed is already on a path compatible with the Paris climate targets. (…) Eight G20 countries are remaining in the worst category of the index ("very low"). Australia (56th out of 61), Saudi Arabia and above all the USA perform particularly

---

[15] Cf. "Pinning extreme weather on climate change is now routine and reliable Science." *Nature*, editorial, 560, 5, 30/VII/2018. But see some *caveats* by Rowan Sutton (2018).

poor – the USA is the worst performer for the first time. Under the Trump administration, the USA is rated "low" or "very low" in almost all categories; in the category climate policy only Australia performed worse, which received 0 out of 100 possible points based on the assessment of climate experts in the country. 'This science based assessment shows again that in particular the large climate polluters do hardly anything for the transformational shift we need to deep emissions reductions to curtail the run to potentially irreversible climate change' (Stephan Singer).

The CCPI 2020 only reiterated the previous CCPI reports. Two years ago, for instance, its authors wrote: "we still see a huge ambition gap in the countries' greenhouse gas reduction targets and their progress regarding a sufficient implementation of the Paris Agreement in national legislation" (Burck et al. 2018).

The CCPI measures countries' efforts in terms of state initiatives, which, however, many are unable to take on. This incapacity lies in what we consider the central tendency in economic and political history of our time: the ongoing change in the nature of the state. In fact, a difference between the nature of the twenty-first century state and the state generated by the so-called Second Industrial Revolution begins to take shape. The latter was characterized by the emergence of financial and industrial conglomerates with much greater technological and capital density, one that—as is well known—implied a new interaction between capital and the state. This second phase of the Industrial Revolution, the prototype of which is seen in the cohabitation between the German state and industrial conglomerates such as Krupp and IG Farben (which in 1945 split, leading to Agfa, BASF, Hoechst and Bayer), gave rise to what we refer to as state capitalism, a term apparently coined by Wilhelm Liebknecht (1896), one of the founders of the Social Democratic Party (SPD) of Germany.

Throughout the twentieth century, the complementary relations between national states and corporations deepened and became more widespread. From the end of the century, they have acquired forms sufficiently typical to justify the hypothesis of a new phase of state capitalism, or even, as I suggest here, of a true change in the nature of the state, with the emergence of what might be called the "state-corporation," a new model of symbioses between the state and corporations. This new model is brought about by the conversion to capitalism of China, the former Soviet Union, and Eastern European countries but also by state leverage of the economies of the "Asian tigers" and of less industrialized countries, such as India and Brazil (at least until 2016).

Everyone knows that the Thatcher and Reagan administrations meant the beginning of the dismantling of social democracy, the globalization of commerce, the deregulation of the financial market, and privatization of state assets. But the privatizations that hit countries like Brazil (1995–2002), Russia (1991–1999), and India after the abolition of the Licence Raj in 1990 (Sibal 2012) did not necessarily involve state retreat from the energy and financial sectors and meant, moreover, greater state involvement in other sectors of corporate capital. According to the MSCI World Index,[16] state-owned enterprises (SOE) account for 80% of the value

---

[16] The MSCI World Index (formerly Morgan Stanley Capital International) is a stock market index

of the stock market in China, 62% in Russia, and 38% in Brazil. The proportion of SOE among the Fortune Global 500 (which ranks the largest companies in the world by revenue) has grown from 9% in 2005 to 23% in 2014.[17] Also according to the OECD's report, *Ownership and Governance of State-Owned Enterprises* (2018), "they account for over a fifth of the world's largest enterprises as opposed to ten years ago where only one or two SOEs could be found at the top of the league table." In Russia, the state's share in GDP expanded from about 40% in 2006 to 46% in 2016 (Di Bella et al. 2019). In 2019, for the first time since the first Fortune Global 500 (1990), there were more Chinese companies than American ones. Only 15% of China's 109 corporations listed on the Fortune Global 500 are privately owned (Guluzade 2019). In fact, the top 12 Chinese companies are all state-owned. They include massive banks and oil companies that the central government controls through the State-Owned Assets Supervision and Administration Commission of the State Council (SASAC) (Cendrowski 2015). Furthermore, of the ten most valuable corporations in the world, four are state-owned, three in China and one in Japan.[18]

Consider now another fact, revealed by Richard Heede (2014): the historical records of global CO_2 and methane gas released into the atmosphere (914 $GtCO_2$_eq) between 1854 and 2010 show that 63% of the global emissions that occurred between 1751 and 2010 originated from the activities of 90 corporations in the fossil fuel and cement industries and that half of these emissions were released into the atmosphere after 1986. Of these 90 carbon majors, 50 are private corporations (investor-owned) and 40 are state-owned or nation-state. In terms of numbers, state and private companies are almost equal (40–50), but in terms of the quantity of gigatons of CO_2 released, the responsibility of the 40 state-owned or nation-state companies (600 $GtCO_2e$) is much greater—almost double—than that of the 50 private companies (315 $GtCO_2e$). In addition, the state-owned companies are the biggest from this group: the ten biggest gas and petroleum companies in the world, measured by their reserves, are nation-state corporations, while the thirteen biggest ones, which together account for three-fourths of the world petroleum reserves, have some form of state participation.[19] The International Energy Agency (IEA) estimates that 74% of all the coal, petroleum, and gas reserves are state-owned companies, while Exxon and Shell, the two first private petroleum majors, measured by profit, own less than 10% of the world's reserves (Carrington 2015; Bezat 2015).

These two facts—the increasing participation of state capital in key sectors of the economy and its decisive weight in the fossil fuel industry—explain why states are reluctant to reduce greenhouse gas emissions. Despite solemn declarations about greenhouse gas reduction strategies, their policies, investments, and subsidies are being slowly shaped by the interests of their own assets, following the logic of

---

of 1631 world stocks. It is used as a common benchmark for global stock funds.

[17] See the PwC Report, *State-Owned Enterprises. Catalysts for Public Value Creation?* 2015.

[18] See "The Visible Hand." *The Economist*, 26/I/2013.

[19] See "The Rise of State Capitalism" and "The Visible Hand." *The Economist*, 26/I/2013.

profitability through which the state promotes selective support, participates in shareholders' agreements, and encourages or inhibits market trends, following a reasoning that is not essentially different from that of a large corporation.

These two facts are linked to two others: (1) the financial survival of a good number of states depends on the dividends of state-owned enterprises; (2) states are becoming—in general—more dependent on corporations. In *Development in an Era of Capital Control* (2017), Clara Hackett rightly observes that "the government of more dependent states may be resigned to the whim of the corporation and, as such, can be described as being as a hostage to capital." This second point is of crucial importance. In the previous political order, the state legitimized itself to the extent that it was able to put itself, or appear to put itself, above the social conflicts at stake. Its relative financial autonomy made it more capable of assuming environmental and social responsibilities, which sometimes contradicted the immediate interests of corporations. All the labor and environmental legislation imposed on businesses in the nineteenth and twentieth centuries, thanks to the pressure of environmental and social movements, but also to the "notion of the state" in public power, gives proof to the mediation capacity that the state once exercised in the conflicting dynamics of society. In short, there was an irreducible difference of identity between the state and corporations until the 1980s.

From the 1980s on, this difference in identity begins to disappear as a result of the factors mentioned above, but also due to two other trends: (1) the great transnational mobility of goods and capital makes the environmental and social movements less able to interfere in public policy; (2) states' fiscal deficit and growing public debt stall their investment capacity and force them to subordinate their environmental and social policies to the logic of the market.

The entire more or less democratic framework of political representation created by a long history of nation states thus loses its relative effectiveness. Increasingly destitute of real sovereignty, while also functioning as creditors and debtors, partners, and competitors of big capital, states are absorbed in a logic of national or transnational corporate networks; they tend to function and, above all, to *think of themselves*, as constituent of this dynamic. This new condition of the state calls for an update of the historical constant formulated by Marx, according to which, in a capitalist system, the state ultimately represents the interests of capital. The contemporary state does not represent capital because representation supposes a relationship between two distinct entities, while what happens today is a *continuum* between both, always functioning, of course, according to the logic of capital. It would often be more accurate today to replace the term "corporate network" with "corporate-state network" and to assert that in contemporary capitalism the state not only represents corporations but also intertwines its assets with theirs while at the same time depending existentially on them to manage its structural debt. It is both these conditions—of partner and of debtor—that lead the state to no longer distinguish the *raison d'être* of public management from that of private management. A concept such as "crony capitalism" (Zingales 2012) tries to capture these new forms of overlap between the state and corporations, which no longer only affect economic policy, but, again, the very identity of the state itself. The phenomenon was well described by Sheldon Wolin (2008), who coined the terms "democracy incorporated" and "inverted totalitarianism." According to Wolin, in the "classic" totalitarianism:

> The state was conceived as the main center of power (…) Inverted totalitarianism, by contrast, while exploiting the authority and resources of the state, gains its dynamic by combining with other forms of power, such as evangelical religions, and most notably by encouraging a symbiotic relationship between traditional government and the system of 'private' governance represented by the modern business corporation. The result is not a system of codetermination by equal partners who retain their distinctive identities, but rather a system that represents the political coming-of-age of corporate power.

This does not mean that there is no longer any tension between the state and corporations. But the nature of this tension has changed. Previously, the state was the mirror of a given relationship between forces, that is, the ability of each class to be present and to have an impact—through social struggle, unions, parties, parliamentary representations, etc.—on state socioeconomic policies and fundamental political directives. Today, other factors predominate in the tension between the state and corporations. Among them are (1) fiscal legislation, its implementation and monitoring, manipulating accounts and tax evasion in fiscal paradises (see Chap. 13); (2) importing into the state conflicts from different groups in the corporate world due to an alignment of interests and alliances of the state with this or that entrepreneurial group; (3) dysfunctional interactions between the state and corporations, such as corruption and bureaucracy; and (4) finally, but *less importantly*, civil society pressures so that the state assumes its historical identity as a promoter of environmental policies and social well-being, less importantly because social movements are very weak and fragmented, at least hitherto, and because the capacity and availability of states to meet these pressures are increasingly conditioned by the corporate pact which governs this emerging nature of the state. In summary, the tensions between the state and corporations result from the metabolizing, in fieri, of one organism into another in symbiotic digestion, a metabolizing that should remain imperfect, as it is only through conserving a residue of identity and autonomy in relation to corporations that this new hybrid entity, the state-corporation, can legitimize itself in the eyes of society and perceive itself as functional.

Therefore, when we asked above, why, according to the *Climate Change Performance Index 2020* (CCPI), "none of the countries assessed is already on a path compatible with the Paris climate targets," the answer begins to take shape. States no longer have the power, interest, or even the perception that they should act, as a public power, in the name of preserving the most universal of goods—their natural heritage. The interests of states and of corporations now fundamentally coincide: to increase production and consumption and guarantee the international flow of natural resources at prices that guarantee the maximum rate of profit for private and state companies, in short, for the state-corporation.

1.3   The Regression of Multilateralism

In the 4 years after the military victory over Nazi-fascism and the Bretton-Woods agreement (1944), the allied countries, under US hegemony, remodeled the international institutional framework, still partially in force today, to their image and likeness with the creation of the IMF (1944), IBRD (1944), the World Bank (1945), GATT (1947, WTO since 1995), OAS (1948), NATO (1949), and the Marshall Plan, which resulted in the 1948 Organisation for European Economic Co-operation (and later OECD), among others.

But alongside this new instrument designed to consolidate and legitimize the *Pax Americana*, other institutions, movements, agreements, and treaties emerged whose vocation was to strengthen an embryo of multilateralism. In this way, together with the decolonization process, a series of initiatives came forth in the 45 years after the war, which, taken altogether, can be considered a cornerstone of global governance building. Let us remember a few of them: the UN (1945), the International Court of Justice (1946), the Antarctic Treaty System (ATS, 1959–1961), the Movement of Non-Aligned Countries (1961–1963), the Organization of African Unity (1963), the International Covenant on Civil and Political Rights (1966), the Stockholm Conference on the Human Environment (1972), the Convention on International Trade in Endangered Species of Wild Fauna and Flora—CITES (1973), the UN Convention on the Law of the Sea (UNCLOS) of 1982, the Summit of Reykjavik and Washington on nuclear disarmament (1986–1987),[20] the creation of the Brundtland Committee (1983–1987), the Montreal Protocol on Substances that Deplete the Ozone Layer (1987), the IPCC (1988), the Basel Convention on Motion Control of Transboundary Hazardous Waste (1989), and the formation of regional blocs, such as the European Union, which was born as an ingenious and generous political project. In 1987, at the opening of the Brundtland Committee, *Our Common Future*, Gro Harlem Brundtland wrote: "Perhaps our most urgent task today is to persuade nations of the need to return to multilateralism."

Still in the 1990s, the hope of a slow evolution in the direction of effective international governance was fueled by the fall of the Berlin Wall (1989), the first IPCC Assessment Report (1990), aimed at governments, and by the Earth Summit, or ECO-92, which led to seven major agreements,[21] as well as to important documents and protocols, such as the Agenda 21, the Rio Declaration on Environment and Development, and the Earth Charter (2000). Throughout the decade, there were other important meetings: in 1993, the Vienna Conference on Human Rights, which resulted in the Vienna Declaration and Program of Action (VDPA) and the Office of the United Nations High Commissioner for Human Rights (OHCHR); in 1994, the

---

[20] As a result of this treaty, the number of nuclear weapons in the world is actually down from 70,000 in 1986 to around 14,000 in 2020 (see "Nuclear Weapons." BBC, 14/I/2020). But a new nuclear arms race is now being sparked by Trump, Putin, and Xi Jinping.

[21] Cf. Ronald B. Mitchell, *2002–2012, International Environmental Database Project.* University of Oregon.

Cairo Conference on Population and Development; in 1995, the accession of 38
nations to the Treaty on the Non-Proliferation of Nuclear Weapons (NPT-UNODA);
and in 1996, the World Food Summit (WFS) and the Fourth Beijing Conference on
Equality of the Sexes. Outside the official circuits, there was, in 1998, the creation
of ATTAC (*Association pour la taxation des transactions financières et pour l'action
citoyenne*) in Paris, present today in 28 countries, and, in 1999, the Battle of Seattle
against the corporate establishment which showed the momentum of several socio-
environmental movements in favor of a then emerging alter-globalization. These
movements have tried to consolidate themselves, since 2001, in the World
Social Forum.

But already at the end of the 1990s, diplomatic multilateralism had lost the
momentum that the ECO-92 had given it, and was reduced, with the G8 (1997) and
G20 (1999), to mere concerted efforts to manage crises in the financial markets.
And with the invasion of Afghanistan and Iraq by a heteroclite military commission
forged by the US in 2001 and 2003, the very principle of international law was definitely invalidated. In this way, the Rio+10 in Johannesburg remained incapable of
implementing the agreements set in 1992, and this inability confirmed itself in the
Rio+20. In addition, the goal set in Rio+10 of restoring fish stocks by 2015 remained
absolutely ineffective. Today, beyond the unilateralism of the Trump "doctrine," it
is the degradation of natural resources, the prospect of scarcity, or the complete
transformation of these resources into commodities that make state-corporations
less inclined to respect multilateral bodies, to include clauses of environmental sustainability in their trade agreements, and to enter into binding international agreements or to ratify them and stick to them.

The Kyoto Protocol (1997–2012) offers an emblematic example of non-
compliance with multilateral agreements. It compelled signatories to reduce, by
2012, greenhouse gas emissions (GHG) by 5.2% in relation to 1990 levels.
Figure 1.2 shows what has happened between 1997 and 2012.

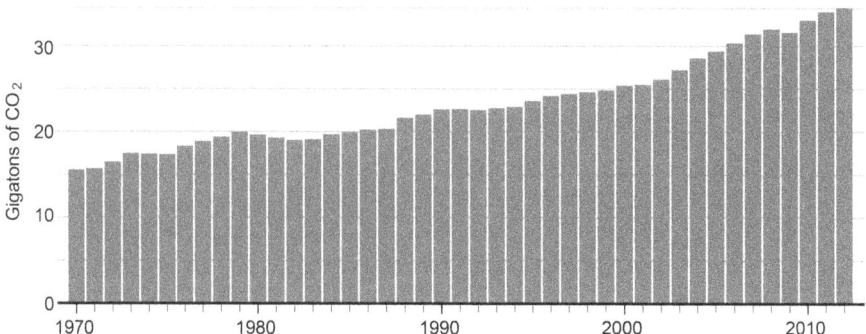

Fig. 1.2 CO_2 Global Emissions between 1970 and 2012 (in Gt). Based on Jeff Tollefson, *Nature*,
503, 14/XI/2013, p. 175, from PBL Netherlands Environ. Assessment Agency/UNEP

Instead of falling 5.2%, CO_2 atmospheric emissions increased by more than 50% between 1992 and 2012, causing an 11% increase in its atmospheric concentration. According to Jeff Tollefson and Natasha Gilbert (2012):

> The climate numbers are downright discouraging. The world pumped 22.7 billion tonnes of carbon dioxide into the atmosphere in 1990 (…). By 2010 that amount had increased roughly 45% to 33 billion tonnes. Carbon dioxide emissions skyrocketed by more than 5% in 2010 alone, marking the fastest growth in more than two decades as the global economy recovered from its slump.

The Global Carbon Project (GCP) calculates an increase of 61% in greenhouse gas emissions between 1990 and 2013. These estimates only account for CO_2 emissions from both industrial activity and the burning of fossil fuels and do not include fuels used in maritime and air transport (or *bunker fuels*), as well as emissions from deforestation, fires, melting of the permafrost, agriculture and livestock, hydroelectric dams, etc.

1.4   From Rio+20 to the Present

Kumi Naidoo, Executive Director of Greenpeace, dubbed the UN Conference Rio+20 "an epic failure." Ban Ki-moon, former UN Secretary-General, admirably summed up its results: "Let me be frank. Our efforts have not lived up to the measure of the challenge. Nature does not wait. Nature does not negotiate with human beings."[22] An important reason for the regression of multilateralism is the loss of the surprise factor. In 1992, the corporate universe was neutralized by immense media exposure and by general enthusiasm. Twenty years later, company lobbies and state-corporations no longer had the surprise factor against them. They returned to Rio committed to obstruct any agreement on global governance. This resistance, already evident in the Copenhagen Summit failure (2009), continued in Lima, in December 2013 (COP 20). An International Monetary Fund Working Paper, uploaded in May 2015, puts the cost of subsidizing fossil fuels at US$ 5.3 trillion a year, which is more than double the projected 2014 estimate that the IMF released in April 2014 (Coady et al. 2014; Vorrath 2014). According to Nicholas Stern[23]:

> This very important analysis shatters the myth that fossil fuels are cheap by showing just how huge their real costs are. There is no justification for these enormous subsidies for fossil fuels, which distort markets and damages economies, particularly in poorer countries.

The Paris Agreement was hailed by former President Barack Obama with these words: "Today, we can be more confident that this planet is going to be in better shape for the next generation." But this Agreement does not even mention the issue of such huge subsidies to fossil fuels. Not even the most foolish optimist expects

---

[22] Cf. "Rio+20 Has Become the Summit of Futility." *Der Spiegel*, 21/VI/2012.

[23] Quoted by Damian Carrington, "Fossil fuel subsidised by $10 m a minute, says IMF." *The Guardian*, 18/V/2015.

that our governments will go far enough as to phase out these subsidies. Legally non-binding and without any clauses on penalties in case of non-compliance, the Paris Agreement has all the ingredients to prove what James Hansen stated (Milman 2015):

> It's a fraud really, a fake. It's just bullshit for them to say: 'We'll have a 2°C warming target and then try to do a little better every five years.' It's just worthless words. There is no action, just promises. As long as fossil fuels appear to be the cheapest fuels out there, they will be continued to be burned.

The facts confirm James Hansen's perception: in the fourth quarter of 2015, the world consumed 96 million barrels of oil per day, and in the fourth quarter of 2019, the world increased its daily consumption to almost 101 million barrels of oil (EIA, Short-Term Energy Outlook, 14/I/2020). Despite the United Kingdom signing up to the Paris agreement, its government (through overseas aid and lines of credit) gave 4.6 billion pounds to overseas fossil fuel projects between 2010 and 2017, and there is no downward trend in the fossil fuel funding. When we know that JSW, a coal corporation, sponsored the COP24 in 2018, that the United States already gave a formal notice of intention to withdraw from the Paris Agreement, and that Brazil withdrew its offer to stage the COP25, deemed by the way the worst climate summit ever (Harvey 2019), what remains of the credibility of this "historic" climate agreement?

1.5   Natural Reserves and Horror Vacui

The increase in GHG emissions is only one example of a greater context of regression. The Convention on Biological Diversity and the UN Convention to Combat Desertification (UNCCD), two agreements made by the ECO-92, present an equally negative balance. The 2010–2020 UN *Decade for Deserts and the Fight against Desertification* did not meet its goals: today, "land degradation and desertification affect one third of the land used for agriculture" (FAO 2019). Multilateralism has also struck a negative balance with respect to stemming the collapse of biodiversity, deforestation, pollution of the soil, atmosphere, and hydrosphere, overfishing, the increase in waste and international trafficking of waste, wood, and animals, eutrophication of water by industrial fertilizers, intoxication from pesticides and other substances, etc. The conclusion is that in the last 20 years, there has not been a reduction in the rate of degradation of ecosystems. Vandana Shiva, Director of the Research Foundation for Science, Technology and Natural Resource Policy, stated in 2010:[24]

> When we think of wars in our times, our minds turn to Iraq and Afghanistan. But the bigger war is the war against the planet. This war has its roots in an economy that fails to respect

---

[24] See *Time to End War Against the Earth*. Speech upon receiving the Sydney Peace Prize (4/XI/2010).

ecological and ethical limits - limits to inequality, limits to injustice, limits to greed and economic concentration. A handful of corporations and of powerful countries seeks to control the earth's resources and transform the planet into a supermarket in which everything is for sale. They want to sell our water, genes, cells, organs, knowledge, cultures and future.

The idea of establishing natural "reserves" is symptomatic of this war, because reserves are thought of as demilitarized zones. Nevertheless, just as in the physics of Aristotle, *natura abhorret a vacuo*,[25] so too the logic that sees nature as a raw material is characterized by the horror vacui. Nature reserves themselves do not escape this logic. The data is unequivocal. In 2013, the vast majority of the 621 biosphere reserves recognized by UNESCO in 117 countries[26] were in a process of degradation.[27] In Brazil, according to Manuela Carneiro da Cunha, "indigenous lands and conservation units, lands that are kept outside of the market, are being threatened more than ever before." This statement dates back to 2014.[28] Today, Brazilian indigenous and their lands are the targets of a tragic *Blitzkrieg*. On an international scale, 10% of the 183 UNESCO natural heritage sites, strongholds and symbols of the concept of ecological sanctuaries, are already at risk or in the process of being degraded (Landrin 2012). Global marine reserves (marine protected areas or MPA) do not surpass 3% of oceans, and MPAs "are ineffective or only partially effective" (Toropova et al. 2010). The destruction of 50% of the Great Barrier Reef in Australia, protected by UNESCO since 1981, presents an exemplary case of the vulnerability of the planet's most important nature "reserves." In November 2014, the World Parks Congress of the IUCN took place in Sydney; it was their decennial meeting which establishes the protection agenda for "natural reserves" for the next 10 years. Commenting on this event, which is at the top of the conservationist calendar, the editorial of *Nature* (6/XII/2014) posits: "At the next World Parks Congress, around 2024, what will attendees discuss? Will there be any truly wild rhinos left? Will the Great Barrier Reef be in terminal decline?" On March 19, 2018, the last male Northern White Rhino died. Indeed, the Great Barrier Reef decline is now probably terminal because, according to Terry Hughes and colleagues (2019), "as a consequence of mass mortality of adult brood stock in 2016 and 2017 owing to heat stress, the amount of larval recruitment declined in 2018 by 89% compared to historical levels." How do we not fear also for mangrove biodiversity, as more than 35% of these tropical ecosystems have already been destroyed worldwide?

---

[25] Cf. Aristotle, *Physics*, IV, section 8. The expression *natura abhorret vacum* was apparently introduced by Rabelais in *Gargantua et Pantagruel* (1534), book I, chapter 5.

[26] See the program, "Man and the Biosphere" (MAB), in cooperation with UNEP and IUCN.

[27] On reserves in Africa, see Amadou Boureima, *Réserves de la biosphère en Afrique de l'Ouest*, 2008.

[28] Cf. D. Chiaretti, "Cresce disputa pelas terras dos índios no país." *Valor econômico*, 17/IV/2014, p. 4.

1.6   Unsustainability and the Increasing Severity of Environmental Crises

Given this brief assessment of societies' inability—subject to alleged economic *Diktats* and, above all, victims of their own mental blocks—to respond in a concerted way to the degradation of the biosphere, it is easy to understand the growing appeal of the idea of "sustainable development." Coined about 30 years ago, this idea has become, over time, a blah-blah-blah, an advertising slogan, and a synonym for "green" economic growth. In the final text of Rio+20, "The Future We Want," for example, the word "sustainable" is repeated 115 times, without being linked to a single concrete action to make it effective. To restore its significance, it is necessary to look at what the term denotes. A socioeconomic system is sustainable if and only if: (1) the economic activity does not destroy biodiversity and does not change the environmental parameters in a faster rate than their ability for restoration and adaptation; (2) the economic activity "meets the needs of the present without compromising the ability of future generations to meet their own needs" (see the 1987 Brundtland Report, *Our Common Future*). The words of Herman Daly (1990a/1993) on this topic are enlightening: "the term 'sustainable development' (…) makes sense for the economy, but only if it is understood as 'development without growth.'" Also emblematic is the verdict of James Lovelock (2006):

> The error that they [the acolytes of market *laissez-faire* and those who desire the so-called sustainable development] share is the belief that further development is possible and that the Earth will continue, more or less as now, for at least the first half of the century.

At this advanced stage of environmental crises, all growth clashes with a physical impossibility: the entropy generated by growth itself, as was described almost half a century ago by the *opus magnum* of Nicholas Georgescu-Roegen (1971). It also clashes with a basic principle of economic theory, that is, Herman Daly's theorem of impossibility, discussed in Chap. 13. In short, the current scale of economic activities and the steady increase in industries and agribusiness with high environmental impact have come to be *incompatible* not only with the available supply of natural resources but also with the equilibrium of the Earth's system itself which has allowed for the rapid development of civilizations since the end of the last glacial period about 11.7 millennia ago. Kevin Anderson (2015) rightly maintains that: "A 2°C emission pathway cannot be reconciled with the repeated high-level claims that in transitioning to a low-carbon energy system 'global economic growth would not be strongly affected' [IPCC]."[29]

We can test this hypothesis on the incompatibility between increased economic growth and the planet's limits by examining two questions: (1) To what extent can the difficulty in resuming the average economic growth rates of 1945–1973 be attributed to the already unsustainable economic system? (2) What will the impact of this economic unsustainability be on future economic crises?

---

[29] For the IPCC quote, see "The Concluding Instalment of the Fifth Assessment Report," 2/XI/2014.

The crisis of 2008 broke out by the culmination of the bursting of a housing bubble, an excessive expansion of credit, and a resale of subprime derivatives, which generated a domino effect of defaults, a liquidity crisis, and a violent contraction of credit. Economists claim that this mechanism, itself a part of the financial system, bears some similarities with other previous credit crises and other speculative and real-estate bubbles, such as the long crisis that began with the "Panic of 1873" and the one that has affected Japan's economy since the 1990s.

More than 10 years after the financial panic of 2008, its effects remain so persistently that the global economy has not yet completely overcome them. Almost certainly it will never regain its previous performance (and all the more so because of the long-lasting effects of the Covid-19 pandemic). Furthermore, according to the former Bank of England governor Lord Mervyn King and to Kristalina Georgieva, managing director of the IMF, the global economy is sleepwalking toward a new massive financial disaster (Elliot 2019; Inman 2020). Joe Davis, head of investment strategy at Vanguard, believes there is 50% chance that a new epic stock market crash could come as early as 2020 (Hoy 2019). Financial cyclical crises are part of the game in a capitalist economy. But on a deeper level, the next crises may be different from the previous ones. It may not be simply another "classical" crisis of capitalism because environmental difficulties can override its perennial cycle. As put by Patrick Bolton et al. (2020) in a book report published by the Bank for International Settlements in Basel (BIS) and the European Central Bank (ECB), climate change could lead to "green swan" events and be the cause of the next systemic financial crisis:

> Climate-related physical and transition risks involve interacting, nonlinear and fundamentally unpredictable environmental, social, economic, and geopolitical dynamics that are irreversibly transformed by the growing concentration of greenhouse gases in the atmosphere.

This conjunction of cyclical capitalist crises with environmental crises will produce "eco-crises" in the near future which are much more profound, prolonged, and difficult to resolve than the previous ones. In fact, the weight of environmental conditions on the performance of the world economy is already considerable (see Chap. 7, Sect. 7.4 Suffering and Greater Lethality Due to the Current Warming). A report by the United Nations Office for Disaster Risk Reduction (UNISDR) warns that:[30]

> In 1998–2017 disaster-hit countries (…) reported direct economic losses valued at US$ 2,908 billion, of which climate-related disasters caused US$ 2,245 billion or 77% of the total. (…) Overall, reported losses from extreme weather events rose by 151% between these two 20-year periods" (1978–1997 vs. 1998–2017).

Ultimately and above all, we must take into account the economic losses caused by the destruction of natural resources that must be valued for their non-use, this "invisible" value consisting of the essential services that the existence of forests and

---

[30]Cf. UNISDR, *Economic Losses, Poverty & Disasters, 1998–2017*, based on the Centre for Research on the Epidemiology of Disasters (CRED) data from the Université Catholique de Louvain.

biodiversity provides to life and, therefore, to man's survival. According to Pavan Sukhdev (2010) and his Economics of Ecosystems and Biodiversity (TEEB) report, in 2050 the value of destroyed ecosystems will correspond to 18% of total world production. This projection is modest when compared to the US\$ 60 trillion loss caused by a possible massive release of methane from the Arctic in the coming decades, as calculated by Gail Whiteman, Chris Hope, and Peter Wadhams (2013).

As we can see, in the next crises, the relative weight of economic and environmental factors in the cause of crises will tend to be reversed, relegating the classic cycle of capitalism to a supporting role, while the rising costs of environmental crises will increasingly take on a protagonist role. This is what James Leape, International Director-General of WWF, affirmed in the foreword to the World Wildlife Fund's Living Planet 2008 Report: "The world is currently struggling with the consequences of over-valuing its financial assets. However, a more fundamental crisis looms ahead – an ecological credit crunch caused by under-valuing the environmental assets that are the basis of all life and prosperity." Environmental degradation is becoming, in short, the structural component of the crisis of global capitalism.

1.7   The Phoenix That Turned into a Chicken

Discussions about how to heal the anemia that has pervaded economic activity on a global scale since 2008 will prolong themselves ad nauseam (or *ad bellum*) until we recognize the exhaustion of the capitalist pharmacopeia. Deficits, default on loan payments, and declining economic growth are chronic and structural problems. Gone indeed are the times of the elegant mathematical models of long-term growth, such as Robert Solow's famous "Contribution to the Theory of Economic Growth" (1956). These theories flourished in the postwar period and worked well in a vacuum when environmental deficits were still manageable, the climate still stable, and cheap oil and other natural resources still abundant. The defenders of capitalism will continue to insist on the same two things: the salvific prowess of technological innovation and the exceptional resilience of this economic system. To a certain extent, they are right: the couplet written by Mayakovsky in 1917 now has an archaeological flavor:

> Eat pineapples, chew grouse
> Your last day is approaching, bourgeois!

The Dies irae of the *bourgeois* did not arrive. The prediction did not take place neither in Russia nor elsewhere. Capitalism was able to provide institutional legality, to manage social pressure, or, when pressured, to eliminate it by Nazi-fascism and similar regimes. Its worst economic crises and most extreme near-death experiences spurred mechanisms of partial autophagy, as well as the survival of the fittest, the re-concentration of capital, and technological innovation. These reactions allowed the system to be successively reset, thus being reborn stronger and more vigorous from its ashes.

Since the 2008 crisis, the phoenix, however, is slow to be reborn, or rather, it is being reborn with the flight range of a chicken, not because it has unlearned how to fly, but because the height of its roof is constantly being lowered: the limits of nature. These limits form an iron ring that begins to close itself around the global economy and no economic policy seems capable of breaking it. In fact, doping the economy with steroids (subsidies, facilitation of credit, monetary easing, technological innovation for greater productivity, etc.) in view of reestablishing its past performance will only increase the pressure on natural resources and further erode what remains of the pillars that sustain life on this planet. And, in the same proportion, it will further decrease the chances that the accumulation paradigm will function. What seems to be just another crisis *in* capitalism is, in reality, a crisis *of* capitalism, or, more precisely, of the relations between the economic system and its physical boundaries. Thus, although the business cycle of capitalism continues to repeat itself, Fig. 1.3, elaborated by Gail Tverberg, shows how growth in each of these cycles is decreasing:

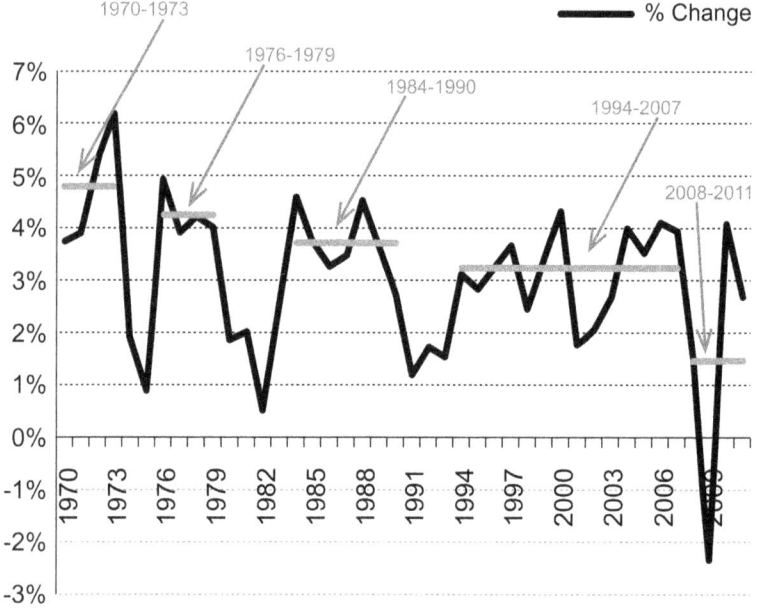

Fig. 1.3 World GDP. % change between 1970 and 2011 (year by year and average by time period). (Source: Based on Gail Tverberg, *Our Finite World*, 18/VII/2012)

The average growth of real-world GDP (discounted for inflation) from 1970 to 1973 was almost 5%; from 1976 to 1979 it was just over 4%; from 1984 to 1990 it was less than 4%; from 1994 to 2007 it was just over 3%; and from 2008 to 2011 it was about 1.5%, with a moment of negative growth for the first time since 1945. According to the unfailingly optimistic IMF (WEO, January 2020), "across advanced economies, growth is projected to stabilize at 1.6% in 2020–2021". Euro area GDP is forecast to expand by 1.2% in 2020 and 2021. These projections of anemic growth were proposed on the eve of the outbreak of the current pandemic, which will condemn the global economy to a depression of unpredictable duration. Reversing this downward trend would require stability in the climate system and an abundance of energy and natural resources at low cost, things that no longer exist.

This is the trap that ensnares the global economic system: the higher the scale of exploration of energy, minerals, soil, water, etc., the scarcer these resources become and, hence, the more polluting is their exploitation and the more intense the techno-logical innovation required to maintain this scale. This leads the economic system to resort to more invasive, costly, and destructive activities, which, in turn, cause the economy to generate higher entropy in itself and in the environment, unbalancing the biophysical parameters that prevailed in the mild Holocene. In this context, a new law arises in contemporary global capitalism: scarcity and/or pollution of natu-ral resources, climate change, biodiversity loss, and other environmental imbal-ances are now increasingly the decisive factors in determining the profit rate of capital.

1.8 "What Were We Thinking?", Denial, and Self-Deception

Echoing Paul Gilding's book, *The Great Disruption*, in his *The New York Times* column, Thomas L. Friedman (2011) wrote:

> You really do have to wonder whether a few years from now we'll look back at the first decade of the 21st century — when food prices spiked, energy prices soared, world popula-tion surged, tornados plowed through cities, floods and droughts set records, populations were displaced and governments were threatened by the confluence of it all — and ask ourselves: What were we thinking? How did we not panic when the evidence was so obvi-ous that we'd crossed some growth/climate /natural resource/population redlines all at once?

Ten years later, how can we continue to fool ourselves with cultivated and useless discussions on the 17 Sustainable Development Goals and the Paris Agreement? What are we thinking of in 2020 as "the world now faces climate-change existential crisis which may result in 'outright chaos' and an end to human civilization as we know it"? (Dunlop and Spratt 2017). In other words, we find ourselves in the ante-room of an environmental collapse of global dimensions, triggered, to start with, by the inevitability of an average global warming of 1.5 °C or more above the pre-industrial period by 2030. "Human kind cannot bear very much reality," wrote T.S. Eliot around 1935. Today, this verdict applies with all the more force. With the intensification of environmental crises in these final years of the second decade of this century, the transgression of all the red lines alluded to by Thomas Friedman becomes even more evident. This intensification also makes the efficiency of the three psychological mechanisms analyzed above (loss aversion, habituation, and dissociation) even more apparent. They result in a behavior of paralysis that is similar to panic. Animals run in the face of danger. But when danger is upon them, they tend to freeze, in the hope that they will go unnoticed by the predator. This animal behavior takes, in us, the form of a fourth psychological mechanism: that fear will not reach us if we are capable of creating a tranquilizing auto-narrative.

This type of narrative, also called self-deception, or more precisely "straight self-deception" (Mele 2001), consists in accepting as real information that at another level of consciousness we admit being dubious or false. It should not be confused with denial, which is characterized by a childish refusal of evidence. Both behaviors were well examined by Clive Hamilton (2010). Self-deception does not deny the evidence of the anthropogenic environmental crises. Although we admit, simply through reading the newspapers, that what we are doing or promising to do for sustainability is not enough to deter the ongoing environmental collapse, self-deception induces us to see reality through transfigured lenses which magnify the positive factors and minimize the negative ones. This is how self-deception blocks the perception of reality: when evaluating positive and negative factors, one does not take into account the fact that the negative factors exceed the positive ones in scale, speed, and acceleration. This is shown in all areas: greater concentrations of greenhouse gases in the atmosphere; climate change; deforestation; quick decline in water resources; more severe and prolonged droughts; soil erosion and desertification; stronger and more frequent wildfires; destruction of habitats; biodiversity loss; pollution of the soil and ocean by sewage, municipal, and industrial waste; chemical intoxication of organisms; increased risk of pandemics from the abuse of antibiotics and from more intense contact with other species; heating, acidification, and deoxygenation of the oceans; increase in dead zones in rivers, lakes, and oceans; bleaching and demise of coral reefs; increasing concentrations of ozone and other toxic particles in the troposphere; melting of sea ice in the Arctic, increased melting in Greenland, in Antarctica, in the so-called Third Pole, and in the permafrost; increased and potentially catastrophic liberation of CO_2 and methane into the atmosphere; rising of sea levels; and intensification of extreme meteorological events: bigger hurricanes, torrential rains, flooding, torrid summers, rigorous winters with more intense snowstorms, paradoxically in a hotter world. The list is far from complete.

These are facts that are cumulative, synergetic, and convergent. But there are still a few who, as Tacitus says, *fingunt simul credunque* (*Ann*, V, 10), that is, who believe in what they fantasize. Some believe, in fact, against all the evidence that the advance of "clean" energy will soon imply a reduction in the consumption of fossil fuels, GHG emissions, and other anthropic impacts on the Earth system; others that people in advanced economies are starting to consume less; still others think that decisions made at COPs will be enforced. And they all disqualify—as catastrophic and apocalyptic—the views of those who doubt that market forces will, ultimately, end up prioritizing global sustainability in detriment to their own priorities, or that both priorities will end up coinciding.

No silver bullet technology will arrive to our rescue at the last minute and solve even the problems that technology itself created or intensified, as in the deus ex machina of classical theater. The problem of technology is not its greater or lesser advance. The problem is its appropriation by a logic that converts it into an amplifier of the crisis. Self-deception is, in reality, the most seductive and insidious case of denial of capitalism's unsustainability. Without it, it would be impossible for us to stay in our precarious zones of material and psychological comfort and lull our belief that, as bad as the everyday environmental news is, in the end "all will be well."

1.9 The Goal and the Two Central Arguments of This Book

Every line of this book has the objective of arguing for the opposite (of the "all will be well" vision). It argues that, if we are not able to react *now* and act appropriately to the challenges that confront us, everything will end badly—and soon—for an uncountable number of species, including our own.

The book is divided into two parts. The first (The Convergence of Environmental Crises) has the goal of gathering and analyzing what science has put forth, not as hypotheses, but as findings: a global warming of more than 2 °C compared to the pre-industrial period during the second quarter of the century already seems inevitable and the increasing anthropic interference on the Earth system is producing ruptures in the physical, chemical, and biological equilibrium on which the web of life is based on. This web of life, therefore, is falling apart. We do not know how close we are from tipping points, and even if we have already crossed some of them (see Chap. 8). But we know that, if the same trajectory is kept, the chances of not crossing these thresholds become smaller and smaller.

We also know what the nature of our agenda is. In the late 1960s, when the growth paradigm began to be questioned, the precautionary principle developed by Hans Jonas' made sense, as it was still possible to prevent the outbreak of destructive dynamics that characterize our present-day situation. In 2020, the time of prophylaxis is far behind us. The aggravation of all the environmental crises mentioned above is no longer a possibility; it is a reality whose effects are already being felt in virtually all parts of the globe. Today, in short, the agenda is not precaution, but mitigation of the ongoing environmental collapse, because the sooner we act, the more we will be able to attenuate its destructive impacts. This is exactly what Erik M. Conway and Naomi Oreskes posit (2014, p. 78):

> The precautionary principle deals with what one should do when there is evidence that something may be a problem, but we're not quite sure, or not sure of the extent of it. We *are* sure that climate change is happening – we already see damage – and we know beyond a reasonable doubt that business as usual will lead to more damage, possibly devastating damage. (…) It's way too late for precaution. Now we are talking about damage control.

The definite characteristic of this moment is, therefore, a race against the clock, a race that we are undoubtedly losing.

The second part of the book (Three Concentric Illusions) continues through the last three chapters. Its ambition is to contribute to the recognition of the fact that all contemporary environmental crises, due to their scale, ubiquity, and acceleration, redefine the themes and priorities of the socioeconomic and political debates that polarize our societies today. Maybe it is not within his power for *Homo sapiens* to disassemble the trap that his ingenuity made him fall into. But the first condition to face the present environmental crises is to see it, without subterfuge, as the central and unavoidable issue of humanity. In his much-quoted Preface to *A Contribution to the Critique of Political Economy* (1859), Marx hypothesized that "mankind inevitably sets itself only such tasks as it is able to solve." This hypothesis is as encouraging as it is "unfalsifiable." But there is a prior point worthy of attention: humanity is incapable of solving a problem if it does not recognize it as such. What needs to be recognized is expressed in the two central arguments discussed in the second part of this book.

Thesis 1. Capitalism is environmentally unsustainable and the hope to make it sustainable can be considered the most misleading illusion in contemporary political, social, and economic thought. The global socioeconomic system that we call capitalism is defined by two characteristics: (1) an institutional order granting capital owners or state managers the right to make investment decisions based solely on economic grounds and (2) an economic logic according to which the natural resources and the productive forces in society are allocated and organized in order to increase production and maximize profit. If capitalism cannot envision the mythical stationary state described by John Stuart Mill in his *Principles of Political Economy*, it cannot, a fortiori, conceive of itself as a system of monitored degrowth, a key point discussed in Chap. 13. In 1844 John Stuart Mill himself admitted that expansion was a goal of the capitalist system itself, rooting it in human nature, since "man is a being who is determined, by the necessity of his nature, to prefer a greater portion of wealth to a smaller." The discussion about the existence of a human nature or about the final cause of human actions is fruitless, but it is a widely acknowledged fact that the final cause of the cycle of capital in the capitalist system is its own amplified reproduction. No one has analyzed this final cause of capitalism better than Marx (1867/1887, I, Part 4, section V, chapter 14, p. 251):

> Political Economy, which, as an independent science, first sprang into being during the period of manufacture, views the social division of labour only from the standpoint of manufacture, and sees in it only the means of producing more commodities with a given quantity of labour, and, consequently, of cheapening commodities and hurrying the accumulation of capital. In a most striking contrast to this accentuation of quantity and exchange-value, is the attitude of the writers of classical antiquity, who hold exclusively to quality and use-value.

In capitalism, to be *is* to expand. Being and expanding are the same thing in the cellular metabolism of this system. In a biosphere that is being annihilated and in a planet with increasing thermal energy imbalance and with finite natural resources, the expression "sustainable capitalism" therefore expresses a contradiction in terms. Capitalism's environmental unsustainability, far from being a "childhood illness," is a congenital, chronic, and degenerative disease of the socioeconomic system. Herman Daly (1990b) famously maintained that "since the human economy is a subsystem of a finite global ecosystem which does not grow, even though it does develop, it is clear that growth of the economy cannot be sustainable over long periods of time. The term sustainable growth should be rejected as a bad oxymoron." In more blunt terms, one can, thus, formulate the first central thesis of this book: capitalism is an intrinsically expansive system and the more difficult it is to expand, the more environmentally destructive it becomes. Under such a socioeconomic system, mankind will no longer "fit" into the biosphere, implying that society in the future will be post-capitalist or will not be a complex society, and there might even be no society at all.

Thesis 2. The second central thesis of this book, discussed in Chaps. 14 and 15, is that this first illusion of capitalism becoming sustainable emerges from a second and third illusion, both deeply rooted in European history, the matrix of contemporary hegemonic societies. The second illusion is that the more excess material and energy we can produce, the more secure will be our existence as a species in the face of scarcity and the adversity of nature. In physics, energy is often defined as the ability of a physical system to do work on other physical systems or everything that allows the alteration of a given state of a system. From the human point of view, energy is everything that allows for greater appropriation of environmental resources. Until the eighteenth century, the equation "more surplus = more security" was anything but illusory since man's ability to enhance energy in an exosomatic way allowed him to capture only energy flows and recent energy stocks: domesticated animal work, mechanical systems, forces from wind, water, and fire, as well as solar radiation, mainly through photosynthesis. The various stages of the industrial revolution gave us access not only to these physical flows and recent stocks of energy but also to the immense primary energy stocks stored in other geological eras through fossil fuels, and, since the postwar period, to energy trapped in the nucleus of the atom. The industrial use of these new energy sources which allowed for an explosive increase not only in population but also in production and consumption of per capita goods increased the impact of human action on all ecosystems—forests and other native vegetation coverings, soil, water, minerals, etc. Today, we begin to realize that the *more* surplus and energy we accumulate, the *less* secure we become. The ability to multiply surplus, a supreme good until the eighteenth century, has become, with the global capitalism of the twentieth century, an evil that threatens to disrupt the climate system, to kill the biosphere and, not least, the human species.

The third illusion—from which these first two and, indeed, all other illusions of consumer society are anchored—is the anthropocentric illusion of metaphysical and religious nature that Lucretius, in the first half of the first century BC, already called by its real name when he exclaimed: "to say that for the sake of men they [the gods] willed to prepare this world's magnificence, (…), Memmius, is madness."[31] This madness is the belief that the biosphere is a means to an end and that the right to reduce it to an energy apparatus for human benefit is rooted in the singularity of our

---

[31] Lucrécio, *De rerum natura*, V, 156–165: *Dicere porro hominum causa voluisse parare/praeclaram mundi naturam (…), Memmi, desiperest.*

species or in a radical discontinuity between us and the web of life. The flight of Icarus is a good image of this anthropocentric illusion, one that will be discussed in the last chapter of this book.

These two central theses of the book can be summarized into one. We cannot avoid environmental collapse if we fail to overcome capitalism, but capitalism will not be overcome until we overcome both these illusions that nourish, naturalize, and even sacralize it: the illusion that the growth of surplus is still an asset to our societies and the anthropocentric illusion.

Without breaking the mental frame which confines us into these three illusions—the illusion of a sustainable capitalism, of unlimited growth, and of our exceptionality in the web of life—man will not depart from capitalism. The state appropriation of economic surplus does not eliminate capitalism, as the various revolutions of the twentieth century believed in. We will only overcome capitalism—*assuming that it is possible to overcome it*—when it is no longer conceivable to destroy habitats for money, when accumulation of surplus ceases to be an end in itself to become a variable dependent on the possibilities of the biosphere, and when the biosphere is conceived as a subject with rights or, if we refuse to give it such status, as an insurmountable physical limit at the risk of collapse.

The strength of capitalism lies in the fact that it projects into our conscience an inverted image of itself so that the disorder that is produced emerges as the natural order of things. This naturalization of a historical social order impedes the perception that it is possible, at least in theory, to transcend these fossilized behavior patterns. If we are not capable of becoming, socioeconomically and politically, more than what these patterns have made us, if capitalism is the best that our global society, and ultimately our species—also endowed with reason, prudence, aesthetic sense, and morals—can do, then we deserve a somber future, or maybe the non-future to which we are condemning ourselves.

* * *

This book will have achieved its objective if the arguments in favor of the aforementioned theses are convincing. It is not within the scope of this work—quite far from it actually—to attack the bigger problem which we point to, namely, the collective creation of a post-capitalist society after the failure of the socialist experiment of the twentieth century.

We cannot, however, ignore this greater problem. In the conclusion we will suggest, without pretense of offering any political prescription, the conditions for the possibility of an environmentally sustainable society. There is no reason to articulate them here, but it is prudent to lay out the three cornerstones of my argument. The first is that, even though environmental sustainability requires an alternative society to capitalism, this does not exclude gradualisms and political mediations that are part of our contemporary reality: all the actions of the state, of political parties, of NGOs, of companies, of different civil society institutions, and of individuals in the direction of environmental sustainability, even if this sustainability is unattainable under capitalism, are precious. Aiming to address climate change and economic inequality, initiatives such as the Green New Deal Group in United Kingdom (2008) and the Green New Deal, sponsored by Rep. Alexandria Ocasio-Cortez and Sen. Ed Markey in the United States (2018), are highly valuable attempts to radically transform the UK and the US economies, and ultimately our global economy, away from fossil fuels and from the corporative food system.

The second cornerstone is that, if it is not possible to imagine effective solutions to sustainability without overcoming the logic of economic growth, this is not conceivable without a strengthening of democracy. Science and technology are fundamental allies in making these solutions viable, but the solutions are in the hands of society through strategic policy decisions. The third one is that a post-capitalist society will also be a post-socialist society, since, environmentally speaking, socialism was as catastrophic as is capitalism. Socialism is in itself a failed experience, and even more so if evaluated by its program to overcome capitalism. The book, *Écosocialisme. L'alternative radicale à la catastrofe écologique capitaliste* (2013), by Michael Löwy, offers a flawless diagnosis of capitalism, which implies the necessity, in the words of the author, for "a new society and a new mode of production, but *also a new paradigm of civilization*" (p. 101). Exactly. But it is impossible to reconcile this new paradigm of civilization with socialism, even the socialism imagined by Marx and Engels. As Löwy himself states (p. 98): "Marx and Engels lack a full ecological perspective." We owe to Marx nothing less than the understanding of the genesis and the logic of the world in which we live in. But we do not owe to him the understanding of its final developments and of its end. After all, why should the tendency for socioenvironmental collapse—a historical problem that has increased in magnitude and reached our awareness only in the 1960s—be a topic of discussion in the middle of the nineteenth century? The problem only became fully visible much later, as pointed out, in extremis, by Arnold Toynbee, in 1976 (p. 566):

> It was not till the Industrial Revolution had been in progress for two centuries that mankind realized that the effects of mechanization were threatening to make the biosphere uninhabitable for all species of life by polluting it, not locally, but globally, and uninhabitable for Man in particular by using non-replaceable natural resources that had become indispensable for him.

Michael Löwy is also right when he perceives Marx's contradictions: "The first of these contradictions is, of course, that between the productivist creed of certain texts and the intuition that progress can be a source of irreversible destruction of the natural environment." Well formulated: Marx and Engels came to *intuit* the possibility of the devastation of ecosystems by the productive forces, but their fundamental *creed* was productivity and, as such, it inclined them to imagine a future flatly denied by current evidence. It is necessary, therefore, to differentiate socialism—a system whose accumulative metabolism, similar to that of capitalism, secreted insoluble environmental crises—from eco-socialism, a term that refers to a "new paradigm of civilization" or, similarly, to a "revolution of degrowth," as Serge Latouche (2014) proposes:

> In short, we somehow arrived at the 'moment of truth', at a historical turning point, a real 'crisis of civilization'. It is the crisis of Western civilization, from which a revolution will

come in the true sense of the word (that is, a total change, even on a cultural level, which I call the 'degrowth revolution' or even 'eco-socialism'), or barbarism. For the moment, it seems to me that we are rather well on the road to barbarism.

Despite the current fervor about post-capitalist ideas, a concrete political path to overcome capitalism still does not exist. Stéphane Hessel and Edgar Morin (2011) are on point in affirming that "those who denounce capitalism are unable to articulate any credible alternative to it." But they are even more on point when they criticize the opposite attitude: "those who consider capitalism immortal resign themselves to it." If overcoming capitalism seems like an unrealistic political program, the environmental collapse that this system is leading us toward has shown that it is unrealistic for us not to overcome it. Those who prefer "realism," remembering the sinister dystopias generated by the socialism of the twentieth century, do not notice that this resignation is the open door through which we see new dystopias emerging, more sinister than the ones which we lived or imagined.

Cornered by its environmental unsustainability, the contemporary world finds itself in the contingency of choosing between two unrealistic agendas: the unrealism of self-deception according to which capitalism would have the power to metamorphose—like a caterpillar into a butterfly—into an environmentally sustainable system and the unrealism that consists of affirming the possibility of redefining the position of *Homo sapiens* in the biosphere, an unprecedented redefinition in which, certainly, there would be no place for societies entrenched in nation-states and subject to the imperative of continuous increase in power, production, and consumption. Given that environmental catastrophe is increasingly inescapable and that the ideas of the new generation are different from that of the old one, it is possible to imagine an "unrealistic" mutation of youth in time to mitigate the ultimate consequences of the current storm. Aware of the ongoing environmental collapse, the global ecological movement is now growing exponentially and requires radical changes in humanity's relationship with the Earth system. If there is room for hope that the worst will be avoided, it is because hope is born from the certainty that the future of societies is not known in the present and is, therefore, uncertain. More than ever we need to understand the words of Tocqueville and Valéry that open the Introduction to this book, about the unpredictability of history.

References

AENGENHEYSTER, Matthias et al., "The point of no return for climate action: effects of climate uncertainty and risk tolerance", *Earth System Dynamics*, 9, 2018, pp. 1085–1095.

ALSTON, Philip, "Statement on Visit to the USA", United Nations Special Rapporteur on Extreme Poverty and Human Rights, Washington D.C., 15/XII/2017.

ANDERSON, Kevin, "Duality in climate science". *Nature Geoscience*, 8, December 2015, pp. 898–900.

BAYNES, Chris, "Anne Frank Centre warns of 'alarming parallels' between Trump's America and Hitler's Germany". *The Independent*, 9/VIII/2017.

BEURET, Nicholas, "Emissions inequality: there is a gulf between global rich and poor". *World Economic Forum*, 4/IV/2019.

BEZAT, Jean-Michel, "Pétrole-gaz: la bataille de l'accès aux réserves". *Le Monde,* 10/IV/2015.

BLOW, Charles M., "Trump Isn't Hitler. But the Lying…". *The New York Times*, 19/X/2017.

BOLTON Patrick et al., The green swan: central banking and financial stability in the age of climate change. Bank of International Settlements (BIS), European Central Bank, Banque de France, January 2020.

BOYD, Roger, "Economic Growth: A Social Pathology". *Resilience*, 8/XI/2013.

BURCK, Jan, MARTEN, Franziska, BALS, Christoph, HÖHNE, Niklas, *Climate Change Performance Index*, Results 2018.

CARRINGTON, Damian, "10 myths about fossil fuel divestment put to the sword". *The Guardian*, 9/III/2015.

CASE, Anne & DEATON, Angus, "Rising morbidity and mortality in midlife among white non-Hispanic Americans in the 21st century". *PNAS*, 112, 49, 2/XI/2015, pp. 15078–15083.

CASTORIADIS, Cornelius. *Une Société à la derive* (1974–1997). Paris, Seuil, 2005.

CENDROWSKI, Scott, "China's Global 500 companies are bigger than ever—and mostly state-owned". *Fortune*, 22/VII/2015.

CHANCEL Lucas & PIKETTY Thomas, "Carbon and Inequality: From Kyoto to Paris". Paris, 3/XI/2015.

COADY David, PARRY Ian, SEARS Louis, & SHANG Baoping, "How Large Are Global Energy Subsidies?" A IMF Working Paper, 2014.

COFFEY Clare et al., *Time to care. Unpaid and underpaid care work and the global inequality crisis.* Oxfam, 2020.

CONDORCET. *Esquisse d'un tableau historique des progrès de l'esprit humain* (1793). Paris, 1989.

CONWAY, Erik M. & ORESKES, Naomi. *Merchants of doubt. How a Handful of Scientists Obscured the Truth on Issues from Tobacco Smoke to Global Warming.* New York, 2010.

———. *The collapse of Western Civilization. A view from the future.* New York, Columbia University Press, 2014.

DALY, Herman E. "Sustainable Growth. An Impossibility Theorem" (1990a). *In*: DALY, Herman E. & TOWNSEND, Kenneth (eds.). *Valuing the Earth: Economics, Ecology, Ethics.* Cambridge (Mass.), MIT Press, 1993, pp. 267–285.

———. "Toward some operational principles of sustainable development". *Ecological Economics*, 2, 1990b, pp. 1–6.

DAVIES, J., LLUBERAS, R. & SHORROCKS, A., *The Crédit Suisse Global Wealth Report*, 2017.

DEATON, Angus, "The U.S. Can No Longer Hide from Its Deep Poverty Problem". *The New York Times*, 24/I/2018.

DI BELLA Gabriel, DYNNIKOVA Oksana & SLAVOV Slavi, "The Russian State's Size and its Footprint: Have they Increased?". IMF Working Paper, 2/III/2019.

DIETZ Simon, BOWEN Alex, DIXON Charlie & GRADWEL Philip. "'Climate value at risk' of global financial assets". *Nature Climate Change* (*Letters*), 4/IV/2016.

DUNLOP Ian & SPRATT David, *Disaster Alley. Climate Risk, Conflict and Risk.* Breakthrough. National Centre for Climate Restoration. Melbourne, June 2017.

ELLIOT Larry, "World economy is sleepwalking into a new financial crisis, warns Mervyn King". *The Guardian*, 20/X/2019.

FAO 2019, *Land Degradation & Restoration. Background Paper.* Rome, 2019.

FLITTER Emily, 'Think the Big Banks have abandoned coal? Think again". *The New York Times*, 28/V/2018.

FRIEDMAN, Thomas L., "The Earth is Full". *The New York Times*, 7/VI/2011.

GEORGESCU-ROEGEN Nicholas, *The Entropy Law and the Economic Process.* Harvard University Press, 1971.

GULUZADE Amir, "Explained, the role of China's state-owned companies". *World Economic Forum*, 7/V/2019.

HACKETT Clara, *Development in an era of capital control*: *corporate social responsibility within a transnational regulatory framework*, London, McMillan Education, 2017.

HAMILTON Clive. *Requiem for a species. Why we resist the truth about Climate Change*, Earthscan, 2010.

HARVEY Fiona, "UN climate talks end with limited progress on emissions targets". *The Guardian*, 15/XII/2019.

HEEDE Richard, "Tracing anthropogenic carbon dioxide and methane emissions to fossil fuel and cement producers, 1854–2010". *Climatic Change*, 122, 1–2, January 2014, pp. 229–241.

HERRING S. C. et al., Explaining Extreme Events of 2017 from a Climate Perspective. *Bulletin of the American Meteorological Society*, 100, 1, January 2019.

HESSEL Stéphane & MORIN Edgar. *Le chemin de l'espérance*. Paris, Fayard, 2011.

HOY Laura, "An Epic Stock Market Crash Is Looming, Analysts Warn". *CNN*, 29/XII/2019.

HUGHES Terry P. et al., Global warming impairs stock–recruitment dynamics of corals. *Nature*, 3/IV/2019.

HUMES Edward, "Blanking out the mess". *Nature*, 500, 7460, 1/VIII/2013, pp. 26–27.

ILLICH Ivan. *La convivialité. Oeuvres completes*, vol. I. Paris, 1973/2003, pp. 451–580.

IMF (International Monetary Fund), "World Economic Outlook (WEO). Tentative Stabilization, Sluggish Recovery?" January 2020.

INMAN Phillip, "IMF boss says global economy risks return of Great Depression". *The Guardian*, 17/I/2020.

IVES Peter R. *Gramsci's politics of language: engaging the Bathlin Circle and the Frankfurt School*. University of Toronto Press, 2004.

KAEBLE Danielle & COWHIG Mary, "Correctional Population in the United States, 2016. U.S. Department of Justice.

KANN Drew, "5 Facts Behind America's High Incarceration Rate". *CNN*, 10/VII/2018

KENTISH Ben, "Donald Trump is a psychopath, suffers psychosis and is an 'enormous present danger', says psychiatrist". *The Independent*, 30/IX/2017.

KESSLER Glenn, KELLY Meg, LEWIS Nicole, "President Trump has made 1,628 false or misleading claims over 298 days". *The Washington Post*, 14/XI/2017.

KREPON Michael, "The New Age of Nuclear Confrontation Will Not End Well". *The New York Times*, 3/III/2019.

KRUGMAN Paul, "The Undeserving Rich". *The New York Times*, 19/I/2014.

LAMARCK, Jean-Baptiste de. *Système analytique des connaissances positives de l'homme restreintes à celles qui proviennent directement ou indirectement de l'observation*. Paris, A. Belin, 1820.

LANDRIN Sophie, "Les périls se multiplient sur les sites naturels du Patrimoine mondial". *Le Monde*, 28/VI/2012.

LAUER Stéphane, "Les inégalités continuent de se creuser aux États-Unis". *Le Monde*, 5/IX/2014.

LAWSON Max et al., *Public good or private wealth?* Oxfam, January 2019.

LE HIR, Pierre, "Inquiétudes sur la sûreté nucléaire en Ukraine". *Le Monde*, 5/XII/2014.

LEE Bandy X. (ed.), *The Dangerous Case of Donald Trump: 27 Psychiatrists and Mental Health Experts Assess a President*, New York, Thomas Dunes Books, 2017.

LICHTERBECK, Philipp, "Com um 'bullshiter' no Palácio, Covid-19 pode virar a peste negra no Brasil". *Deutsche Welle*, 17/III/2020.

LIEBKNECHT, Wilhelm, "Our recent Congress". *Justice*, 15/VIII/1896.

LOVELOCK, James. *The Revenge of Gaia*. London, Penguin Books, 2006.

LYNAS, Mark, *Six Degrees: Our Future on a Hotter Planet*. London, HarperCollins, 2007.

———. *The God Species: How the planet can survive the age of humans*. London, Fourth State, 2011.

MARLAND, Gregg, ODA, Tom & BODEN, Thomas A., "Cut emissions per capita to 1955 levels". *Nature*, 565, 31/I/2019, p. 567.

MARSHALL, George, *Don't even think about it. Why our brains are wired to ignore climate change*. London, Bloomsbury, 2014.

MELE, Alfred, *Self-deception unmasked*, Princeton University Press, 2001.

MILMAN, Oliver, "James Hansen, father of climate change awareness, calls Paris talks 'a fraud.'" *The Guardian*, 12/XII/2015.

OECD, *Ownership and Governance of State-Owned Enterprises: A Compendium of National Practices*, 2018.

OTTO Ilona M. et al., "Shift the focus from the super-poor to the super-rich". *Nature Climate Change*, 9, February 2019, pp. 82–87.

OXFAM, *Working for the Few*. London, 2014.

POPPER, Karl. *The Poverty of Historicism* (1936). New York, Routledge, 1957.

———. *Conjectures and refutations. The Growth of Scientific Knowledge* (1963). New York, Routledge, 2002.

RAWORTH, Kate, *Doughnut Economics. Seven Ways to Think Like a 21ˢᵗ-Century Economist*, Chelsea Green, 2017.

ROCKSTRÖM, Johan & WIJKMAN, Anders. *Bankrupting Nature. Denying our Planetary Boundaries. A Report to the Club of Rome*. London, Routledge, 2012.

ROCKSTRÖM, Johan et al. "A safe operating space for humanity". *Nature*, 2009, 461, 24/IX/2009, pp. 472–475.

SCHIERMEIER, Quirin, "Droughts, heatwaves and floods: How to tell when climate change is to blame". *Nature*, 560, 5, 30/VII/2018.

SHORROCKS, Anthony & DAVIES, James, *Global Wealth Report 2010*. Crédit Suisse Research Institute. Geneva October 2010.

SHORROCKS, Anthony, DAVIES, James & LLUBERAS, Rodrigo, *Global Wealth Report 2019: The year in review*. Crédit Suisse Research Institute. Geneva 2019.

SIBAL, D. Rajeev, "The Untold Story of India's Economy". *LSE, The London School of Economics and Political Science*, March 2012.

SOLOW Robert M. "A Contribution to the Theory of Economic Growth". *The Quarterly Journal of Economics*, 1956, 70, 1, pp. 65–94.

STEFFEN, Will et al., "Planetary boundaries: Guiding human development on a changing planet". *Science*, 15/I/2015.

SUTTON Rowan, "Attributing extreme weather to climate change is not a done deal". *Nature*, 561, 12/IX/2018.

TOCQUEVILLE Alexis de. *L'Ancien Régime et la Révolution*. Paris, Michel Lévy, 1856. English edition, *The Old Regime and the Revolution*, translated by John Bonner, New York, Harpes & Brother, 1856.

TODD, Emmanuel. *La chute finale: Essais sur la décomposition de la sphère soviétique*. Paris, 1976.

TOLLEFSON Jeff & GILBERT Natasha, "Earth Summit: Rio report card". *Nature*, 6/VI/2012.

TOROPOVA Caithlin et al., *Global Ocean Protection. Present Status and Future Possibilities*. Gland, IUCN, 2010.

TOYNBEE, Arnold J., *Mankind and Mother Earth*. Oxford University Press, 1976.

VALÉRY, Paul, "De l'histoire" (1928). *Regards sur le monde actuel* (1945). *Oeuvres*, vol. II. Paris, Gallimard, 1960, pp. 935–938.

VORRATH, Sophie, "Fossil fuel subsidies costing global economy $ 2 trillion: IMF". *REnew economy*, 29/IV/2014,

WHITEMAN, Gail, HOPE, Chris, WADHAMS, Peter, "Vast costs of Arctic change", *Nature*, 499, 25/VII/2013.

WOLIN, Sheldon S. *Democracy Incorporated. Managed Democracy and the Specter of the Inverted Totalitarianism*. Princeton University Press, 2008.

WORTHY, Kenneth. *Invisible Nature: Healing the Destructive Divide Between People and the Environment*. New York, Prometheus, 2013.

WULF, Andrea, *The Invention of Nature. Alexander von Humboldt's New World*. New York, Vintage Books, 2015.

ZINGALES, Luigi, "Crony Capitalism and the Crisis in the West", *The Wall Street Journal*, 6/VI/2012.

Chapter 2
Decrease and Degradation of Forests

"Time is running out for the world's forests, whose total area is shrinking by the day." This is the key message of FAO's *State of the World's Forests 2018* (SOFO). An assessment of the global extent and distribution of forest trees proposed by Thomas Crowther and 37 colleagues from 15 countries (2015) offered "the first spatially continuous map of forest tree density at a global scale." The authors concluded that, until 2014, human societies had already suppressed almost half of the planet's trees:

> The global number of trees is approximately 3.04 trillion, an order of magnitude higher than the previous estimate. Of these trees, approximately 1.30 trillion exist in tropical and subtropical forests, with 0.74 trillion in boreal regions and 0.66 trillion in temperate regions. (…) Based on our projected tree densities, we estimate that over 15 billion trees are cut down each year, and the global number of trees has fallen by approximately 46% since the start of human civilization.

In 2014, we were facing the destruction of 28,500 trees per minute, a rate that is obviously suicidal for the human species and deadly to countless other species. But the pace of deforestation has worsened dramatically over the past 6 years. Since the much celebrated New York Declaration on Forests (NYDF), endorsed in September 2014 by a coalition of governments, companies, civil society, and indigenous peoples' organizations (over 200 signatories in 2019), the global rate of gross tree cover loss has increased by 43% (and average annual tropical primary forest loss has accelerated by 44%), according to a 5-year assessment report published in September 2019 by Climate Focus and other organizations (Streck et al. 2019). Between 2014 and 2018, 260,000 km^2, an area of tree cover the size of the UK, have been lost every year. These figures do not include the impact of the 2019 and 2020 forest fires in the Arctic, the Amazon, Indonesia, Democratic Republic of Congo, Siberia, California, Alberta, Canada, Australia, etc. In the Brazilian Amazon rainforest, "the number of active fires in August 2019 was nearly three times higher than in August 2018 and the highest since 2010. There is strong evidence this increase in fire was linked to deforestation" (Barlow et al. 2019). In Australia, fires have been starting earlier in the 2019–2020 season. As of January 14, 2020, fires have burned an

© Springer Nature Switzerland AG 2020
L. Marques, *Capitalism and Environmental Collapse*,
https://doi.org/10.1007/978-3-030-47527-7_2

estimated 186,000 km^2, a globally unprecedented percentage of any continental forest biome: 21% of the Australian forest biome has burned in a single season (Boer et al. 2020). According to this analysis, "these unprecedented fires may indicate that the more flammable future projected to eventuate under climate change has arrived earlier than anticipated." Indeed, in 2019, Global Forest Watch Fires (GFW Fires) counted over 4.5 million fires worldwide that were larger than 1 km^2. That's a total of 400,000 more fires than 2018 (Schauenberg 2020).

2.1   We Can't Live Without Forests

Michael Jenkins and David Kaimowitz began a preface to a study on forest conservation with these words: "The future of the world's forests and the future of millions of the world's poorest people are inextricably linked" (Scherr et al. 2003). This is correct. But the most accurate statement is the one expressed in the title and central message of a document launched in 2014 by FAO: *We can't live without forests. Forests are key to supporting life on Earth.* As living communities in which trees and other plants, animals, fungi, and microorganisms reproduce and interact in various ways, forests are recognized as crucial for soil preservation, water regulation, nutrient cycling, trade balance of gases in the atmosphere, and global climatic stability. Furthermore, "forests are home to more than 80% of all terrestrial species of animals, plants, and insects" (UNDP 2015; FAO 2012), including the vast majority of threatened species. Thus, it is impossible to exaggerate the importance of two facts:

1. FAO's *State of the World's Forests 2012* estimates that, by the end of the last Ice Age, forests covered 60 million km^2; today, "30 percent of the global forest cover has been cleared, while another 20 percent has been degraded. Most of the rest has been fragmented, leaving only about 15 percent intact."
2. Contrary to FAO's *Global Forest Resources Assessment 2015* (FRA 2015), the pace of deforestation has accelerated in the first two decades of the twenty-first century (Tollefson 2015).

2.2   The Upward Curve of Deforestation (1800–2019)

Deforestation is driven by a combination of several factors: logging, land use change to industrial-scale agriculture and cattle ranching, human-caused wildfires, mining, hydroelectric dams, urban encroachment, land grabbing, and new roads in the forest. In the Amazon rainforest, for instance, nearly 95% of all deforestation occurred within 5.5 km of roads (Barber et al. 2014). Regarding logging, the figures leave no room for doubt:

1. Data from the International Tropical Timber Organization (ITTO) shows that "of the 400 million hectares (more than half of the world's total) of tropical forests used for timber production today, less than 8% is sustainably managed" (Tollefson 2013).
2. According to "The Global Initiative against Transnational Organized Crime" (2018), "Illegal logging is the most lucrative crime pertaining to natural resources and constitutes US $52–$157 billion in profits. Organized crime groups, as well as terrorist networks, are reaping in these profits with illegal logging, creating lower risk, but very high return." This flourishing business would obviously be unfeasible without the indulgence or complicity of governments, such as that of Brazil.

2.3 Acceleration

"The global consequences of human activity are not something to face in the future. They are with us now. All of these changes are ongoing, and in many cases accelerating," wrote Peter Vitousek and colleagues already in 1997. Deforestation is one of the most accurately measurable cases of acceleration. "During the past 10,000 years, vegetation across about half of Earth's ice-free land surface has been cleared or become otherwise dominated by human activity" (Boakes et al. 2010). The deforestation carried out by pre-industrial societies, partially offset by the recovery of secondary forests, was a gradual and multi-millennial process that goes back to the beginning of agriculture and the use of wood for construction, boats, furniture, energy, etc. Drawing on Michael Williams' (2003) extensive research on deforestation over the last few millennia, FAO's *State of the World's Forests 2012* estimates that 8 million km^2 of forests were lost globally before 1800. During the nineteenth and twentieth centuries, the synergy between capitalist expansion and population explosion generated an exponential growth in the supply and demand of commodities and raw materials, giving the phenomenon a new scale, one that Fig. 2.1 captures to perfection.

In only 210 years (1800–2010), the world lost 10 million km^2 of forests, 2 million more than during the preceding millennia following the end of the last Ice Age.

Millennia before 1800   = 8 million km^2
1800–2010   = 10 million km^2

This total 18 million km^2 of destroyed forests corresponds precisely to the aforementioned 30% of the total of 60 million km^2 of forests existing at the end of the last

Deforestation (billion hectares)   Population (billion)

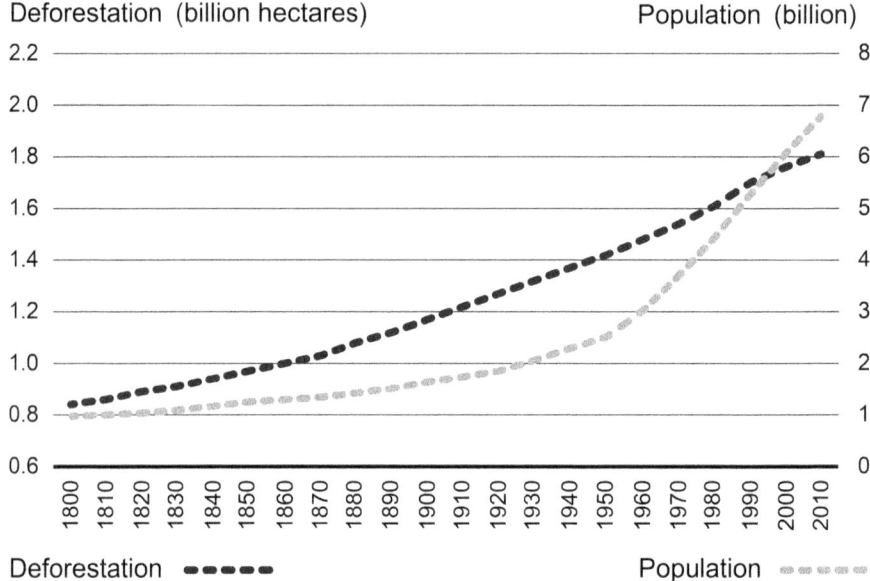

Deforestation ●●●●●   Population ●●●●●

Fig. 2.1 Global deforestation 1800–2010. (Based on FAO—*State of the World's Forests*, 2012, p. 28)

Ice Age. Since 1950, demographic expansion and global capitalism have depleted and destroyed our planet's forests at an unprecedented rate, as shown by the 2005 *Millennium Ecosystem Assessment* (p. 2):

> More land was converted to cropland in the 30 years after 1950 than in the 150 years between 1700 and 1850. Cultivated systems (areas where at least 30% of the landscape is in croplands, shifting cultivation, confined livestock production, or freshwater aquaculture) now cover one quarter of Earth's terrestrial surface.

Since the 1980s, global deforestation seems to have reached a paroxysm of damage, as recent estimates suggest that a quarter of global forest loss over the last ten millennia has occurred over the past 30 years (Boakes et al. 2010; Middleton 1995/2013). According to the *Global Forest Resources Assessment 2015* (FRA 2015), a report published by FAO every 5 years, while in 1990 forests covered 31.6% of the planet's land, or about 41.28 million km², this percentage fell to 30.6% in 2015, or about 39.99 million km², a net loss of 1.29 million km² in only 25 years. This loss is even greater in relative terms, as in 1990 there were 0.8 hectares of forest per capita, while in 2015 this number dropped to 0.6 ha, a decrease of 25%. According to FAO's *State of the World's Forests* 2016, "there was a net forest loss of 7 million hectares per year in tropical countries in 2000–2010 and a net gain in agricultural land of 6 million hectares per year."

2.4   Global Forest Watch (GFW)

No matter how alarming, the FAO data does not reflect the problem in its entirety. In 2008, NASA and USGS began to make available the archives of Earth images obtained by their Landsat satellites. And in 2014, the World Resources Institute (WRI), together with Google, the University of Maryland, and about 70 partners, made accessible the Global Forest Watch (GFW), an interactive forest monitoring system based on satellite images with a resolution of 30 × 30 m which enables it to provide much more speedy and accurate information on the death or removal of trees with canopies reaching at least 5 m in height. In a first set of results, Matthew Hansen et al. (2013) showed that:

> Globally, 2.3 million square kilometers of forest were lost during the 12-year study period and 0.8 million square kilometers of new forest were gained. The tropics exhibited both the greatest losses and the greatest gains (through regrowth and plantation), with losses outstripping gains.

As pointed out in 2013 by the World Resources Institute, "the world lost the equivalent of 68,000 soccer fields of forest every day over the past 13 years – 50 soccer fields per minute" (Sizer et al. 2013). And the pace of gross forest lost has increased even more between 2011 and 2016, as shown in Fig. 2.2.

On a larger scale (2001–2018), Fig. 2.3 shows a tree cover loss area of 3.61 million km², equivalent to a 9% decrease in tree cover, since 2000.

The two graphs below (Figs. 2.2 and 2.3) show that throughout this century, we have gone from an annual loss of 133,000 km² in 2001 to 248,000 km² in 2018. Thus, we can assume with a high degree of confidence that FAO's claim of a slowdown in deforestation is incorrect and that GFW offers a more accurate picture of

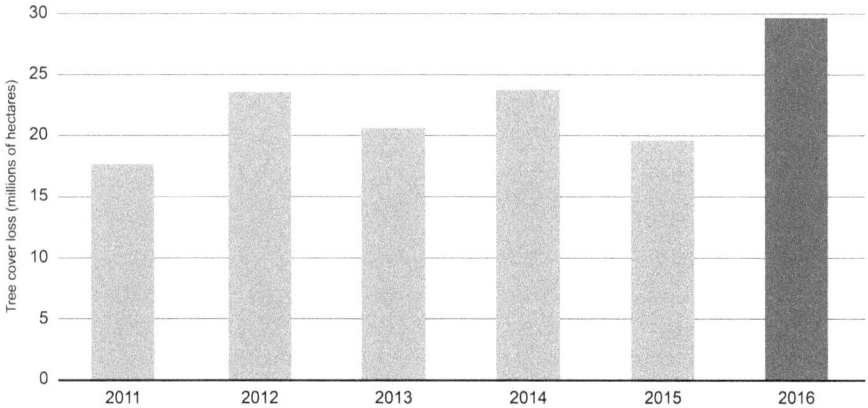

Fig. 2.2   Global tree cover loss between 2011 and 2016 (millions of hectares). (Source: Mikaela Weisse & Liz Goldman, "Global Tree Cover Loss Rose 51% in 2016", 18/X/2017, GFW)

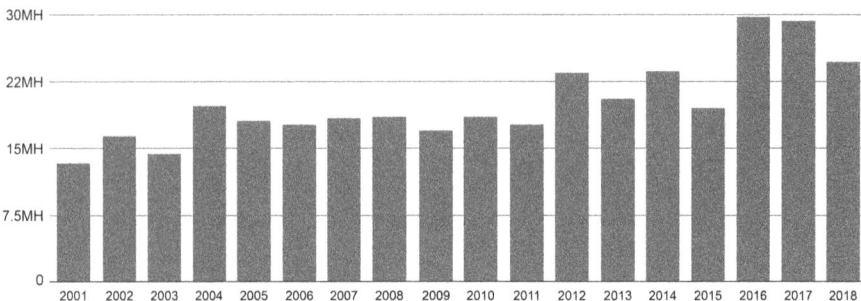

Fig. 2.3 Global annual tree cover loss between 2001 and 2018 (millions of hectares). (Source: Global Forest Watch. Forest Change)

the rate of progression of global deforestation than FAO. Indeed, for Matthew Hansen, FAO assessments (FAO 2015) should not be treated as a benchmark by anyone seeking to understand changes in forest cover. But in the end, these differences in the total area of forest loss, although very important, must give way to two crucial points that we cannot lose sight of:

1. Although greatly underestimated, FAO figures show that, in the span of just one generation (1990–2015), our planet suffered a loss of 1.29 million km^2 of forests, an area larger than South Africa (1.22 million km^2). The difference between the FAO and GFW numbers, however large, indicates only two different speeds in our trajectory of environmental collapse.
2. Between 2000 and 2018, tree cover loss released 98 Gt of CO_2 into the atmosphere (GFW). Deforestation is the second leading cause of climate change and accounts for nearly 20% of all GHG emissions (FAO 2018). And yet the problem received almost no attention in the Paris Agreement (COP 21) of December 2015. Article 5.2 of this agreement states that "Parties are encouraged to take action to implement and support (...) activities relating to reducing emissions from deforestation and forest degradation." This is all that is said on deforestation in this highly acclaimed text.

2.5   Boreal Forests

Since 1950, deforestation increased in the boreal forests. From 2001 to 2017, Canada, for instance, lost nearly 40 million hectares of forest (Berman 2019). Russia's taiga, the largest forested region on Earth (approximately 12 million km^2), is being consumed by fires and logging. In 2019 alone, the fires spread across at least 2.5 million hectares and were possibly started deliberately to cover up illegal logging (WWF, "Russia's Boreal forest").

"The boreal forest is breaking apart. The question is what will replace it?", asks Dennis Murray, a professor of ecology at Trent University in Peterborough, Ontario.

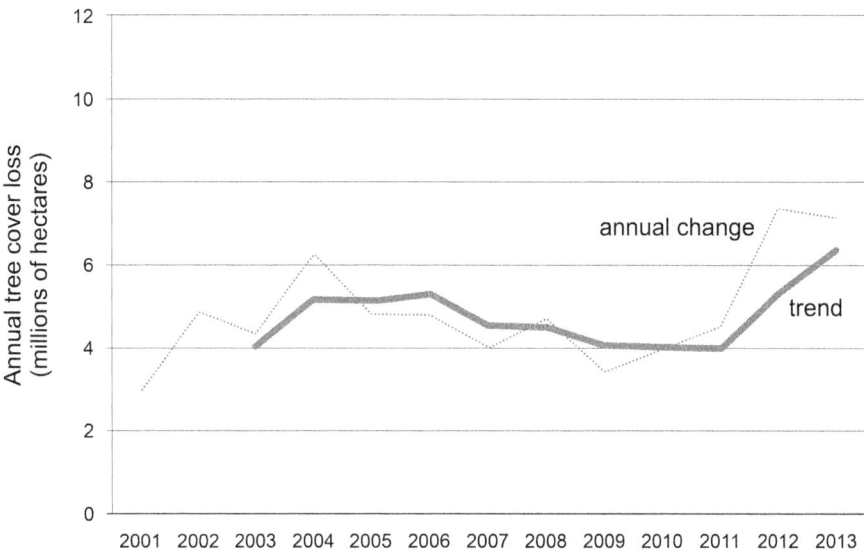

Fig. 2.4 Annual loss of forest cover in the boreal region, in millions of hectares. The dotted line indicates annual change and the continuous line indicates average triennial change. All figures are calculated with a minimum of 30% of tree cover canopy density. (Source: Based on Nigel Sizer; Rachel Peterson, James Anderson; Matt Hansen, Peter Potapov & David Thau, "Tree Cover Loss Spikes in Russia and Canada, Remains High Globally." WRI, 2/IV/2015)

Jim Robbins (2015) lists the main reasons for the decline of this giant ecosystem that extends across North America and Eurasia and makes up 30% of the globe's forest cover:

> The Arctic and the boreal region are warming twice as fast as other parts of the world. Permafrost is thawing and even burning, fires are burning unprecedented acres of forest, and insect outbreaks have gobbled up increasing numbers of trees. Climate zones are moving north ten times faster than forests can migrate. And this comes on top of increased industrial development of the boreal, from logging to oil and gas. The same phenomena are seen in Russia, Scandinavia, and Finland.

Indeed, starting in 2011, the boreal forests begin to decline at a much faster rate than forests in other regions, surpassing a loss of 60,000 km^2, as shown in Fig. 2.4 (Sizer et al. 2015).

2.6 Deforestation Accelerates in the Tropics

In 1952, in his classic handbook, *The Tropical Rain Forest*, Paul W. Richards drew attention to the fact that "the destruction in modern times of a forest that is millions of years old is a major event in the world's history. It is larger in scale than the clearing of the forests in temperate Eurasia and America, and it will be accomplished in a much shorter time." Over the last 70 years, this ominous prognosis has been restated in a formidable library of measurements, studies, analyses, appeals, and warnings on the acceleration of deforestation in the tropics published by forest and conservation specialists, from Richards to Norman Myers (Hance 2008), who declared in 2008 with evident exasperation:

> I'm going to give you my bottom-line message right now, up front, this is a super crisis that we are facing, it's an appalling crisis, it's one of the worst crises since we came out of our caves 10,000 years ago. I'm referring of course to elimination of tropical forests and of their millions of species.

Before and after this declaration, Adrian Sommer (1976), Michael Williams, Thomas Lovejoy, Paulo Nogueira Neto, Claude Martin (2015), Antônio Donato Nobre (2014), Carlos Nobre (2019), José Marengo (2011), Philip Fearnside (2001, 2002, 2005), and Matthew Hansen, among many others, have been expressing similar warning cries. And we must not forget the past and current resistance of forest communities, silenced by armed evictions, assassinations, or massacres. But to no avail. In 2016, around three quarters of the most biologically diverse habitat on Earth, still home to more than half of the world's terrestrial flora and fauna (Sudarshana et al. 2012), has already disappeared or is being degraded and disappearing. In *Forests and Grasslands*, an educational book published by the Encyclopedia Britannica, John P. Rafferty (2011) says that "as recently as the 19th century tropical forests covered approximately 20% of the dry land area on Earth. By the end of the 20th century this figure had dropped to less than 7%." A collective survey published by Hans ter Steege and 157 colleagues from 21 countries shows that "most of the world's 40,000 tropical tree species now qualify as globally threatened" (Steege et al. 2015).

As mentioned above, FAO's *State of the World's Forests 2012* estimates that at the end of the last great Ice Age, forests covered 60 million km². Tropical rainforests then represented a bit more than one quarter of this area, or 16 million km², an estimate proposed by Adrian Sommer (1976), author of the first attempted assessment of tropical deforestation. Sommer assumed that by the 1970s, this area had shrunk to 9.35 million km², a regression of 6.65 million km², or 41.6% of the world's total area of tropical rainforests. In 2015, Claude Martin, former director general of the WWF International, published *On the Edge*, the 34th Report to the Club of Rome and the most comprehensive assessment to this date of the state and the fate of tropical forests. According to him, "today less than half this area [16 million km²] remains as undisturbed forest – nobody knows exactly how much – and about another quarter survives as fragmented and degraded forest." These figures are widely accepted. For Chaitanya Iyyer (2009), for instance:

> Global deforestation sharply accelerated around 1852. It has been estimated that about half of the Earth's mature tropical forests – between 7.5 and 8 million km² (2.9 to 3 million sq mi) of the original 15 to 16 million km² (5.8 to 6.2 sq mi) that until 1947 covered the planet – have now been cleared.

Since 1920, the pace of tropical deforestation has been increasing, and from 1950 to 1980, it has increased even further. According to Michael Williams (2003), "it is

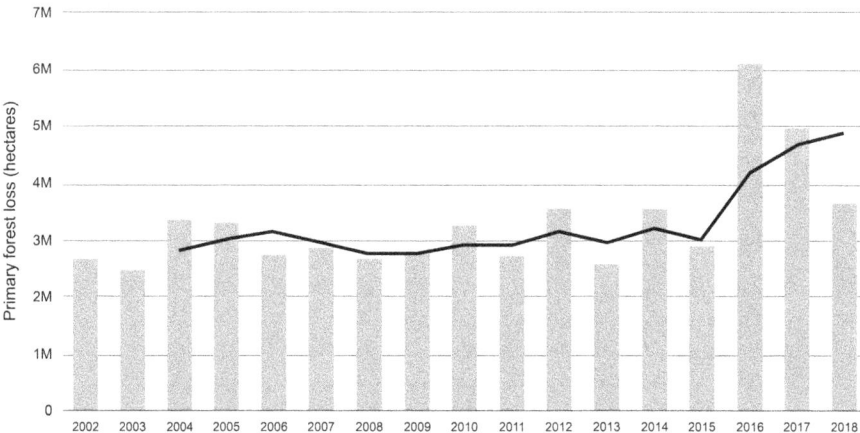

Fig. 2.5 Tree cover losses in tropical countries, 2002–2018 (millions of hectares). (Source: World Resources Institute. See Mikaela Weisse & Liz Goldman, "The World Lost a Belgium-sized Area of Primary Rainforest Last Year" Global Forest Watch, 25/IV/2019)

probably safe to say that between 1950 and 1980 a total of 318 million ha [of forests] disappeared in the tropical world." Just to have a parameter, this loss of tropical forests is almost equivalent to the area of India (3,287,590 km²).

FAO's claim (FRA 2015) that the rate of net forest loss has declined sharply since peaking in the 1990s has been contested by several studies, as pointed out by Do-Hyung Kim, Joseph O. Sexton, and John R. Townshend (2015):

> The Forest Resources Assessment (FRA) has been criticized for inconsistencies in the definition of forest among countries and over time, as well as its dependence on national self-reporting. (...) In the tropics especially, the FRA reported a declining rate of deforestation from the 1980s to the 1990s, while studies based on satellite data observed opposite trends.

Estimates proposed by Kim, Sexton, and Townshend (2015) "indicate a 62% acceleration in net deforestation in the humid tropics from the 1990s to the 2000s." Figure 2.5 shows that there was a tremendous leap in tropical deforestation in the years 2016–2018.

In 2017 alone, the tropics experienced 15.8 million ha (39 million acres) of tree cover loss, and in 2018, 12 million ha, of which 3.6 million ha in primary rainforest (Weisse and Goldman 2019) were lost. In Colombia, especially on the border of the Amazon biome, tree cover loss in 2017 reached 2200 km², more than double the rate of loss from 2001 to 2015. With the demobilization of the Revolutionary Armed Forces of Colombia (FARC) in 2016, a power vacuum emerged, leading to much more extensive deforestation (Weisse and Goldman 2018).

In Africa, an analysis of deforestation (Ordway et al. 2017) shows that:

> Four Congo Basin countries, Sierra Leone, Liberia, and Côte d'Ivoire were most at risk in terms of exposure, vulnerability and pressures from agricultural expansion. These countries averaged the highest percent forest cover (58% ± 17.93) and lowest proportions of potentially available cropland outside forest areas (1% ± 0.89). Foreign investment in these countries was concentrated in oil palm production (81%), with a median investment area of 41,582 thousand ha.

Under the impact of the colonial extraction of wood, especially since 1900, "West and East African forests have undergone almost complete decline (83.3% and 93%, respectively)" (Aleman et al. 2017). In 2011, during the COP17 in Durban, Helen Gichohi, President of the African Wildlife Foundation, warned[1]:

> Deforestation rates in Africa are accelerating. (…) 9% of forest cover has been lost between 1995 and 2005 across sub-Saharan Africa, representing an average loss of 40,000 km² of forest per year. For example, Kenya has lost the majority of its forest cover to settlements and agriculture, leaving only 1.7% of its land still forested.

In 2015, Liberia and Sierra Leone, respectively, experienced a 6- and 12-fold increase in tree cover loss rates compared to 2001 (Weisse et al. 2017). In Asia, Indonesian primary forest loss totaled over 60,000 km² from 2000 to 2012 and increased on average by 476 km² per year (Margono et al. 2014).

2.7 Intact Forest Landscapes (IFL)[2]

According to the World Resources Institute (2014):

> Almost 95% of the world's remaining Intact Forest Landscapes are in the tropical and boreal regions. (...) Just three countries – Canada, Russia and Brazil – together contain 65% of the world's remaining Intact Forest Landscapes. These countries also accounted for over half of all IFL degradation with road building, often linked to logging and extractive industries, being a key driver. Other drivers vary significantly in different regions, from human-caused fires in Russia to agricultural conversion in Brazil.

In 2000, IFL covered 12.8 million km², or 9.7%, of the Earth's land area (excluding Antarctica and Greenland). By year 2013, the total IFL area had decreased by 8.1% and made up 11.8 million km²; "that means human activities disturbed 20,000 hectares of pristine forest every day for the past 13 years" (Harris, Petersen, Minnemeyer 2014). Very close to this estimate, the IPCC *Climate Change and Land* Report (2019) considers that "forests (intact or primary) with minimal human use amounts to 9% of the global ice-free land surface (130 Million km²)."

---

[1] See Forest Day 5, "New Waves of Deforestation Threatens Africa's Climate Resilience", 4/XI/2011.

[2] See *Intact Forests Landscapes* (http://www.intactforests.org/concept.html): "Technically, an IFL is defined as a territory within today's global extent of forest cover which contains forest and non-forest ecosystems minimally influenced by human economic activity, with an area of at least 500 km² (50,000 ha) and a minimal width of 10 km (measured as the diameter of a circle that is entirely inscribed within the boundaries of the territory)."

2.8   Brazil (1970–2019): The Most Fulminating Ecocide Ever Perpetrated by the Human Species

The vegetation cover of Brazil is divided into six biomes: Pampas, Atlantic Forest, Caatinga, Cerrado, Pantanal, and Amazon. During the five centuries of its history, the fundamental socioeconomic structures of Brazilian society were built upon the massacre and expulsion of Indians and the predatory occupation of their territory, in an east–west direction. The Atlantic Forest that once bordered the coastal strip of the country was the first victim of predation. Of its original area of about 1.3 million km^2 (15% of the Brazilian territory), only 8.5% remain in stretches over 100 hectares (an area that represents a minimum conservation of biodiversity), and only 1% remain of primary forest (Oliveira 2019). Deforestation of this native vegetation cover remains; from 1985 to 2016, more than 19,000 km^2 were deforested, with a slight acceleration in the years 2015–2016.[3] In the 2017–2018 biennium, the biome lost an additional 239 km^2 of forest (INPE 24/V/2019).

2.9   The Cerrado Might Disappear by 2030

The same process of destruction occurs with the second largest Brazilian vegetation area, the Cerrado, which encompasses three physiognomies (tropical fields, savanna, and cerradão). Before its systematic destruction, the Cerrado occupied nearly 2 million km^2, almost one fourth of the Brazilian territory. Until the 1970s, there were about 10,000 species of plants (almost half of them found only there) and almost 300 species of mammals. Moreover, "the Cerrado harbors a high concentration of bird species, with almost half of the species recorded in Brazil occurring in this biome. One of the first major reviews of the birds in the Cerrado still indicated the existence of 837 species in this area, with a good part of them (82%) having some degree of dependence on forest environments" (Machado et al. 2009). In this biome, 800 species of fish and 14,425 species of insects have been cataloged (Brawn et al. 2012). But in 2017, only 19.8% of its vegetation cover remained untouched, and strictly protected areas represent only 3% of the Cerrado (Françoso et al. 2015). As Bernardo Strassburg et al. (2017) affirm, "of what is left of the Cerrado, 88.4% is suitable land for the cultivation of soybeans and 68.7% of sugarcane, crops for which a strong increase in demand is foreseen in the coming decades." The picture painted by the authors is a gloomy one regarding the expected extinction of species; it contains a warning that the main actor responsible for this ecocide—the agribusiness sector—will soon be the victim of its crimes:

> Our calculations based on the species–area relationship suggest that this projected deforestation will drive ~480 endemic plant species to extinction—over three times all documented

---

[3] Cf. INPE, "Deforestation of the Atlantic Forest grows almost 60% in one year," 29/V/2017.

plant extinctions since the year 1500. This will in turn have profound consequences for
Brazil's environmental standing and damaging repercussions for its agribusiness sector.

Among the observed consequences of this *Blitzkrieg* against the Cerrado are reduced
rainfall levels, an increase in fires, loss of biodiversity, dryness and erosion of soils,
and reduction of the three large aquifers (Guarani, Bambuí, and Urucuia) on which
the three watersheds of the region (Tocantins–Araguaia, Paraná–Prata, and São
Francisco) depend. These watersheds provide 43% of Brazilian surface waters, outside the Amazon River. As Tiago Reis, from the Institute of Amazon Environmental
Research (IPAM), recalls: "the loss of native vegetation in the Cerrado compromises the formation of rainfall by evapotranspiration and the infiltration of groundwater to recharge the aquifers and rivers in the region."[4] Added to this is the pollution
of the aquifers by agrichemicals and a growing demand for intensive irrigation agriculture, typical of the agribusiness industry that exports water in the form of soybeans, beef, and other products, which has led to completely unsustainable water
utilization.

2.10 Amazon: The Other Ecocide

In the historical process of deforestation from east to west, global capitalism finally
arrived at the Amazon, a region believed to be home to 10% of known species on
Earth, nearly 500 species of mammals and more than 1000 species of birds; 75% of
the plant species found there are unique to this region (see WWF, "Amazon deforestation"). The largest tropical forest in the world is also, with its 3000 species of
fish, the greatest freshwater fish nursery in the world. Although immense, this combination of ecosystems is very vulnerable. Undoubtedly immense, "the Brazilian
Amazon hosts approximately one third of the rainforests in the world, an area that
covers 4.1 million square kilometers" (Barreto et al. 2005). But no less vulnerable.
An inventory of the forest as a whole—spanning nine countries and about 6 million
km²—was carried out by a team of 120 researchers. They calculate, by extrapolating
from 1170 observation points, that the Amazon forest as a whole is composed of
about 390 billion trees with trunks of at least 10 cm in diameter, belonging to 16,000
different species. Of this total, 227 species are considered "hyperdominant," that is,
they are so recurrent that together they account for more than half of all the Amazon's
trees, while the 11,000 rarer species account for only 0.12% of the total number of
trees (Steege et al. 2013). This disproportion is an important indicator of the fragil-
ity of the forest, since a change in the climatic and/or precipitation parameters of
this ecosystem could destroy these 227 species that are specially adapted to survive
and reproduce in these precise environmental coordinates. On the other hand, the
11,000 rarer species can be eradicated by deforestation, since they represent only

---

[4] Quoted in "Desmatamento do Cerrado supera o da Amazônia, indica dado oficial". *Observatório
do Clima*, 25/VII/2017.

0.12% of all trees. In fact, the team of 158 researchers from 21 countries, managed by Hans ter Steege (2015), published an evaluation of the current risks of forest extinction. They show that:

> At least 36% and up to 57% of all Amazonian tree species are likely to qualify as globally threatened under the International Union for Conservation of Nature (IUCN) Red List criteria. If confirmed, these results would increase the number of threatened plant species on Earth by 22%.

2.11   The Military Catastrophe

According to National Institute of Amazonian Research (INPA), until the mid-twentieth century, human occupation of the Amazon, constant and intense during millennia, did not decrease the area and the density of its forest cover. The military unleashed the dynamic of its destruction. The consequences of the military coup (1964) were tragic for Brazilian democracy, as evidenced by the National Truth Commission, and they were no less so for the Amazon, its forest, and its peoples. The militarist ideology of national "integration" of the Amazon resulted in the opposite: its disintegration and its insertion into the international circuit of commodities, leading to the destruction of the greatest natural heritage in the country and one of the greatest on the planet.

The aggression against the forest waged by tyrants with fire and sword ripped the forest fabric with highways and burned it down for the expansion of agriculture and livestock frontiers and the settlement of population groups from other regions of the country. The result was the destabilization of the region's socio-environmental balance, with impacts on watersheds as a consequence of floods and hydroelectric dams, mining activities, gold rushes, mercury pollution, clear cutting, fragmentation of the forest canopy, and loss of biodiversity. The military opened a Pandora box of destruction of the Amazon. The militarist slogan, "The Amazon is ours," now stupidly repeated by President Jair Bolsonaro, left as its mark the depletion, fragmentation, amputation, and demise—in the near future—of one of the most important foundations of life on Earth.

2.12   The End of the World Machine

Between 1970 and 2019, Brazil's military and civilian governments—instruments of the expansion of mining companies and large hydroelectric plants and, especially, of traders and of the international agrichemical industry—destroyed over 800,000 km^2 of the Amazon forest, or approximately 20% of its original area, through clear cutting. Let us pause for a moment to reflect on what is undoubtedly the most monstrous ecocide ever-perpetrated by the human species. As shown by Antônio Donato Nobre (2014):

One must imagine a tractor with a three-meter blade moving at 756 km/h without interruption for forty years: a type of end of the world machine. According to the estimates, this represents 42 billion trees destroyed, that is, two thousand trees cut down per minute, or 2 million trees per day, a number difficult to imagine due to its monstrosity. And here we are speaking only of clear cutting. Rarely do we evoke the forests that suffer human degradation, the zones that satellite images don't distinguish, where the only thing left are a few trees that hide a more gradual deforestation; this includes whole regions in which the forest is no longer functional as an ecosystem. According to deforestation data from 2007 and 2010, these zones of degradation cover 1.3 million km^2, while the total measurement of areas affected by clear cutting and by degradation represent about two million km^2, or 40% of the Brazilian Amazon forest.

According to data from the MapBiomas project (http://mapbiomas.org/), coordinated by Tasso Azevedo, at the end of the brutal period of the dictatorship (1964–1985), the Brazilian Amazon's vegetated area, although already largely destroyed, still covered an area of 4,756,845 km^2. But in 2018, this area had been reduced to 4,151,763 km^2. An additional 600,000 km^2 of the forest have been cleared since 1986, mainly by agribusiness, to make way for livestock and soybean plantations. Between 2000 and 2014, there was an increase in the extension of Brazilian cropland, which went from 260,000 km^2 to 461,000 km^2 (Zalles et al. 2018). Of this total expansion (about 200,000 km^2) of agriculture and livestock, induced mainly by the international market, and especially by China and Europe, "79% occurred on repurposed pasture lands and 20% was from the conversion of natural vegetation." According to the official data published by Brazil's space research agency (INPE), from August 2018 to July 2019, clear-cut deforestation of the Amazon primary forest reached almost 10,000 km^2 (9762 km^2), nearly 30% up from the previous year. This was by far the highest level of devastation in the second decade of the century and also the biggest annual increase in more than two decades. These last 12 months (August 2018–July 2019) ended a period of relatively stable losses, as the average between 2012 and 2018 was 6727 km^2. We should add that 95% of this destruction is being done illegally, with explicit complicity of far-right President Jair Bolsonaro, a mad man who should stand trial for ecocide at The Hague (Londoño 2019). According to previous estimates by the WWF ("Amazon deforestation"), nothing less than "27% of the Amazon biome will be without trees by 2030 if the current rate of deforestation continues." In fact, under Bolsonaro, the situation may turn out to be much worse than these projections indicate.

2.13 Tipping Point: Forest Dieback

For millions of years, forests have evolved while adapting to disturbances. But the admirable resilience of these ecosystems is now succumbing to diebacks,[5] due to the combined impacts of deforestation, fragmentation, droughts and floods, extreme heat

---

[5] A forest dieback is a condition characterized by progressive or sudden death of twigs, branches, or roots, in a large number of trees or even entire forests, starting at the tips.

waves, more destructive fires, invasive species, and the proliferation of pathogens. Many forests are now facing environmental conditions outside the ranges in which they evolved. With the intensification of these stressors, and especially of deforestation on a global scale during the first two decades of the century, there emerges the central and ever more recurring question of our time: how much disturbance or stress is too much for the forests? Put in other words, "how to define where the transition between 'normal' and 'too much' stress takes place and how to determine whether this transition is an abrupt threshold or a linear decline" (Trumbore et al. 2015):

> Increased disturbance intensity, disturbance frequency, or even the introduction of new kinds of disturbances can trigger abrupt nonlinear declines in the ability of forests to perform intrinsic functions. [...] Of particular concern is the coupling of direct, local, human-related disturbances with ongoing, more diffuse changes in climate and atmospheric composition.

This topic of forest tipping points (see also Chap. 8) is discussed in the scientific literature, presenting results that are convergent, yet not always identical, given that forests can react in different ways to such pressures. Some semi-arid forests have already crossed this tipping point, as stressed by Romà Ogaya et al. (2020):

> Hotter and drier conditions are driving an increase in forest dieback and stem mortality in many water-limited ecosystems, such as Mediterranean forests, with a progressive substitution of the current dominant species by species of tall shrubs better adapted to future climate.

But not only semi-arid forests. There are now several lines of evidence to suggest that large regions of the Amazon rainforest, for example, are very close to reaching a critical point beyond which are triggered positive feedbacks and, ultimately, non-linear changes that lead to its irreversible decline and its rapid conversion into a savanna-type vegetation (although without Brazilian savanna's rich biodiversity). For Thomas Lovejoy and Carlos Nobre (2018), what they call the "Amazon Tipping Point" is imminent:

> We believe that negative synergies between deforestation, climate change, and widespread use of fire indicate a tipping point for the Amazon system to flip to non-forest ecosystems in eastern, southern, and central Amazonia at 20–25% deforestation. The severity of the droughts of 2005, 2010, and 2015–16 could well represent the first flickers of this ecological tipping point. These events, together with the severe floods of 2009, 2012 (and 2014 over SW Amazonia), suggest that the whole system is oscillating. For the last two decades the dry season over the southern and eastern Amazon has been increasing. Large scale factors such as warmer sea surface temperatures over the tropical North Atlantic also seem to be associated with the changes on land.

Given the recent acceleration of deforestation in the Amazon, Thomas Lovejoy and Carlos Nobre (2019) wrote still another warning, which took the form of an ultimatum, or a "last chance for action." They declared in December 2019: "Today, we stand exactly in a moment of destiny: The tipping point is here, it is now." Antônio Donato Nobre reinforces this perception (as cited by Rawnsley 2020):

> Half of the Amazon rainforest to the east is gone – it's losing the battle, going in the direction of a savannah. When you clear land in a healthy system, it bounces back. But once you

cross a certain threshold, a tipping point, it turns into a different kind of equilibrium. It becomes drier, there's less rain. It's no longer a forest.

The consequences of crossing this tipping point were well summarized by Ignacio Amigo (2020):

> If that happens, it would not only affect the millions of people and animals in the region. It could also mean billions of tonnes of carbon dioxide will be emitted into the atmosphere as trees die and vegetation burns; less rainfall throughout central and southern South America; and altered climate patterns farther afield.

Although some uncertainty still remains about Amazon rainforest's resilience (Amigo 2020), the destruction of growing parts of the forest has been predicted for at least 10 years. A report titled *Assessment of the Risk of the Amazon Dieback* (Vergara and Scholz 2011) evaluated the risk of collapse of part of the Amazon forest due to the combination of climate change with deforestation and fires caused by agribusiness. The results of this report are overwhelming: by 2075, only 5% of the eastern Amazon forests would be left. A work published in 2017 paints a similar, yet more nuanced, picture (Zemp et al. 2017):

> Although our findings do not indicate that the projected rainfall changes for the end of the twenty-first century will lead to complete Amazon dieback, they suggest that frequent extreme drought events have the potential to destabilize large parts of the Amazon forest.

According to Monica de Bolle (2019), a senior fellow at the Peterson Institute for International Economics (PIIE):

> maintaining the current rate of deforestation [between January and August 2019] through the rest of 2019 and over the next few years would bring the Amazon dangerously close to the estimated "tipping point" as soon as 2021, beyond which the rainforest can no longer generate enough rain to sustain itself.

Reacting to this new projection, Carlos Nobre declared to Dom Phillips (2019), "I hope she is wrong. If she is right, it is the end of the world." Thomas Lovejoy believes that de Bolle's projection could come true because global heating, soaring deforestation, and an increase in Amazon fires have created a "negative synergy" that is accelerating its destruction: "We are seeing the first flickering of that tipping. It's sort of like a seal trying to balance a rubber ball on its nose … the only sensible thing to do is to do some reforestation and build back that margin of safety."

2.14 Cavitation or Plant Embolism: The Threshold of Hydraulic Failure

Higher temperatures (which increase tree transpiration) and/or increased water shortage in the soil leads the roots of trees to pump water more intensely through their vascular system. An important consequence of this more intense pumping is the formation of air bubbles in the tree xylem (the tissue through which the sap circulates), a phenomenon that can cause its hydraulic failure. "Once a tree has too many bubbles, it's usually done for," says Atticus Stovall, a forest ecologist at NASA's Goddard Space Flight Center (Pennisi 2019). The conclusion of a 2010 study on the increased vulnerability of forests in 88 zones of the planet is thus described (Allen et al. 2010):

> Although episodic mortality occurs in the absence of climate change, studies compiled here suggest that at least some of the world's forested ecosystems already may be responding to climate change and raise concern that forests may become increasingly vulnerable to higher background tree mortality rates and die-off in response to future warming and drought, even in environments that are not normally considered water-limited. This further suggests risks to ecosystem services, including the loss of sequestered forest carbon and associated atmospheric feedbacks.

According to Michel Vennetier, co-author of this study, "in 20 years, the areas affected by the demise of forests multiplied by four" (quoted by Larousserie 2012). Another study led by Brendon Choat et al. (2012) shows that 70% of 226 tree species in different types of forests located in 81 different latitudes of the planet already operate with narrow safety margins in relation to decreasing humidity, in such a way that the intensification of droughts in various regions of the globe, predicted by climate models, can lead them to succumb to these processes of cavitation or plant embolism. For Hervé Cochard, one of the co-authors of this study, not only Mediterranean forests but also tropical forests have little room for maneuver: "all the trees and all the forests of the globe are living on the edge of their hydraulic rupture. There is, therefore, a global functional convergence of the response of these ecosystems to droughts" (Larousserie 2012). Commenting on this study, Bettina Engelbrecht (2012) confirms these results and declares: "The suggestion that all forests are on the brink of succumbing to drought, and may already be responding to climate change, is supported by observations of increased drought-induced forest die-offs and tree mortality in many ecosystems." And she adds: "the majority of species appear to be right on the edge. Just a little more drought will push them over" (quoted by Lemonick 2012). The phenomenon of the Sudden Aspen Decline (SAD), that is, of the decline of the poplar (*Populus tremuloides*) woods in western North America since 2004, was associated with this phenomena of cavitation or plant embolism, as a consequence of the droughts from 2000 to 2003 in that region: "we find substantial evidence of hydraulic failure of roots and branches linked to landscape patterns of canopy and root mortality in this species" (Anderegg et al. 2011).

There are other causes of weakening of forests as a consequence of droughts or of excessive CO_2 atmospheric concentrations. An experiment conducted in temperate forests, titled FACE (Free-Air Carbon dioxide Enrichment), shows that more CO_2 atmospheric concentrations can promote plant growth, a process called CO_2 fertilization (Grossman and Lapola 2018; Fleischer et al. 2019). This experiment was recently launched in the Amazon forest led by Brazil's National Institute of Amazonian Research (INPA/AmazonFACE https://amazonface.inpa.gov.br/). Its preliminary results were published by Daniel Grossman and David Lapola (2018). Compared with temperate forests, soil of the central Amazon contains little phosphorus, and maybe "the dearth of this important plant nutrient will impede any benefit to the forest from increased CO_2 in the atmosphere." In any case, they add, the benefits plants get from increased CO_2 "might not scale up to whole forests." In 2019, Katrin Fleischer and a team of 27 researchers of the AmazonFACE project published a study on the potential interactions between elevated CO_2 and nutrient (N and P) feedbacks in a mature Amazonian rainforest. They suggest that Amazon forest response to CO_2 fertilization is highly dependent on plant phosphorus acquisition. Given phosphorus scarcity in the Amazon soil, "the resilience of the region to climate change may be much less than previously assumed." In fact, in a situation of excessive CO_2 atmospheric concentration, or of hydric or temperature stress, trees could react by closing their stomata (which is what enables them to undergo transpiration and other exchanges with the atmosphere, among them the emission of water vapor into the atmosphere and the absorption of CO_2), thereby impairing or interrupting photosynthesis. Thus, according to Grossman and Lapola (2018), "with stomata constricted, trees might undermine their own source of rain." It goes without mentioning that the weakening of trees makes them more vulnerable to infestations by microorganisms and/or insects that colonize and wear them out, many times to the point of death. It is the case, for example, with the beetle infestation of mountain pines; the beetles survive milder winters, and they have affected hundreds of thousands of forest hectares in six states of the USA and in Canada, British Columbia (Gillis 2011).

2.15 Last Century of the Rainforests

A survey released in 2015 by the WWF identified 11 global deforestation fronts, "where the largest concentrations of forest loss or severe degradation are projected between 2010 and 2030." Collectively, these places will account for over 80% of the forest loss projected globally by 2030, i.e., up to 170 million ha. If we maintain the current trajectory, 289 million ha of tropical forests will be depleted until 2050, causing emissions of about 169 $GtCO_2$, equivalent to more than 4 years of the current global anthropogenic emissions of this gas (Busch and Engelmann 2017). Here, as in other processes of environmental degradation, the worst projection is, as a rule, the most recent. This tendency has been establishing itself as a general rule in the history of the acceleration of our path toward socio-environmental collapse.

Overall, there is a growing risk that the twenty-first century will be the last century of tropical forests. According to NASA's Earth Observatory, "if the current rate of deforestation continues, the world's rain forests will vanish within 100 years – causing unknown effects on global climate and eliminating the majority of plant and animal species on the planet" (Urquhart et al. 2001). Twenty years later, the prognostics for the disappearance of tropical forests in Africa, Asia, and Oceania are now very short term. In Ghana, the last great forests might disappear in less than 25 years (Ogutu 2014). In addition to deforestation, some of the most precious and symbolic species of African forests are succumbing to the pressures of droughts and climatic change. This is the case of three of the nine species of the 1000-year-old baobab (*Adansonia digitata*, *Adansonia perrieri*, and *A. suarezensis*), already placed on the Red List of the IUCN and now dying in Madagascar and Senegal. Five countries bathed by the Mekong River—Cambodia, Laos, Burma, Thailand, and Vietnam—could be left with little more than 10–20% of their original cover by 2030 (Vidal 2013). According to FAO estimates, 98% of original rainforest in Indonesia will disappear by 2022 (Prokurat 2013). Also, in Papua New Guinea, deforestation has been on the rise since 2013 and saw a 70% jump between 2014 and 2015, reaching 18,000 km² in this last year At this pace, its forests will be reduced to a small fraction of their original area in the near future (Weisse et al. 2017). As pointed out by Andrés Viña et al. (2016), China has been importing wood and other forest products from several countries of Asia, as well as from Africa and Northern Eurasia, thereby exacerbating forest degradation in these regions.

In Brazil, the agriculture and livestock sector has always dominated the political and economic guidelines that are leading to the rapid disappearance and degradation of forests. This is the case today more than ever. Emboldened by President Bolsonaro's lax enforcement policies, land grabbers take over protected land, often from terrified indigenous people, extract commercially valuable wood, and set fire to the forest. Claiming the land, once cleared, as theirs, they sell it to Brazilian agribusiness "developers," mostly cattle ranchers and soybean conglomerates. In August 2019, during the dry season, land grabbers and farmers celebrated a "day of fires" to honor Bolsonaro (Rawnsley 2020; Matias 2019). In one day alone, 400 fires were lit along the highway BR-163, near the city of Altamira, in Amazonia.

Man-made wildfires in Brazil will be exacerbated by climate change. A study suggestively titled "The gathering firestorm in southern Amazonia," by Paulo Brando et al. (2020), indicates that "Amazon fire regimes will intensify under both low- and high-emission scenarios. Our results indicate that projected climatic changes will double the area burned by wildfires, affecting up to 16% of the region's forests by 2050." There is no longer a remote chance that such a process will stop at the initiative of the international agriculture and livestock sector, a grand deforestation coalition that includes large-scale farmers and their State representatives and lobbies, the military, megacorporations of agrichemical products, agricultural equipment, and Big Food, big traders, the financial system, and university technocrats. The final elimination of tropical forests, with its terminal impacts on the biosphere and on humanity, is imminent. It will only be avoided in extremis if social movements in exporting countries have active support from importing countries, namely, China, the USA, the European Union, Russia, and some Middle Eastern countries (Kehoe et al. 2019; Levis et al. 2020). It is in the existential interest of these countries to refuse agriculture and livestock commodities produced at the cost of the disappearance of tropical forests (and of its indigenous peoples), for they will not be the last ones to suffer the global consequences of the tropical ecocide in progress. Indeed, as stressed by Daniel Grossman and David Lapola (2018), "the forest regulates the entire planet's temperature by abating the amount of CO_2 building up in the atmosphere."

2.16   The Socio-environmental Cancer of Deforestation

In September 2014, Kátia Abreu, Agriculture Minister during Brazilian President Dilma Rousseff's mandate (2011–2016), wrote[6]: "A pejorative meaning was linked to the word deforestation, as if it meant a voluntary and arbitrary act of destruction of nature." Deforestation is, in fact, a voluntary and arbitrary act of destroying nature. The most atrocious. It is the most direct and immediate way of killing the greatest number of life forms on a planetary scale. Deforestation invades, like a cancer, the social and natural organism. As a social cancer, it is the realm of brutality and organized crime, led by agribusiness, in Brazil and other soft commodity exporting countries. The G8, Interpol, the European Union, UNEP (UN Environment Programme), and UNICRI (United Nations Interregional Crime and Justice Research Institute) view deforestation as the fifth big area of environmental crimes. In Brazil and elsewhere, deforestation is at the center of violence against traditional forest populations. The Global Witness reports of 2014 and 2017 affirm that Brazil continues to be the most dangerous place in the world for those who try to defend the forest. Of the 908 assassinations documented by the NGO between 2002 and 2013 worldwide, 448 happened in Brazil (49%). Between 2010 and 2016, there were 200 documented assassinations of peasants, indigenous peoples, and activists, perpetrated at the behest of agribusiness, loggers, and other corporate interests.[7] And this data reveals only the tip of the iceberg. As a natural cancer, deforestation is a lethal blow to the biosphere. Among the 17 types of threats to biodiversity cited by a study on the decline of mammals in Brazil, deforestation appears as the biggest cause (Chiarello et al. 2008):

> Seventeen types of threats were cited as the main causes of the decline of species found in the national list. The majority of species (88.4%) is threatened by *habitat* destruction and deforestation (73.9%), factors that are more intense in the Cerrado, Atlantic forest, and Caatinga, but are obviously not restricted to these biomes.

Deforestation may have aggravated not only the droughts of 2005, 2010, and 2016 (Erfanian et al. 2017) but also the 2015 floods in the Brazilian and Peruvian Amazon. In reality, all of the most serious imbalances in the biosphere have deforestation as a departure point or as a crucial factor for its escalation. The crises that plague the biosphere are, largely, metastases of the cancer of deforestation.

---

[6] Cf. "Desmatamento eleitoreiro". *Folha de São Paulo*, 27/IX/2014

[7] Cf. Global Witness, *Defenders of the Earth. Global killings of land and environmental defenders*, July 2017

References

ALEMAN, Julie C., JARZYNA, Marta A. & STAVER, A. Carla, "Forest extent and deforestation in tropical Africa since 1900". *Nature Ecology & Evolution*, 2, 26–33, 11/XII/2017.

ALLEN, Craig D. *et al.* "A global overview of drought and heat-induced tree mortality reveals emerging climate change risks for forests". *Forest Ecology and Management*, 259, 2010, pp. 660–684.

AMIGO, Ignacio, "The Amazon's Fragile Future". *Nature*, 578, 7796, 27/II/2020, pp. 505–507.

ANDEREGG, William *et al.*, "The roles of hydraulic and carbon stress in a widespread climate-induced forest die-off". *PNAS*, 109, 13/XII/2011.

BARBER, Christopher P. *et al.*, "Roads, deforestation, and the mitigating effect of protected areas in the Amazon". *Biological Conservation*, 177, September 2014, pp. 203–209.

BARLOW, Jos, BERENGER, Erika, CARMENTA, Rachel & FRANÇA, Filipe, "Clarifying Amazonia's burning crisis". *Global Change Biology*, 26, 2, 15/XI/2019.

BARRETO, Paulo *et al. Pressão humana na Floresta Amazônica brasileira*. Belém, WRI, Imazon, 2005.

BERMAN, Tzeporah, "Canada clearcuts one million acres of boreal forest every year … a lot of it for toilet paper". *The Narwhal*, 21/III/2019.

BOAKES, Elizabeth H.; MACE, Georgina M. & McGOWAN, Richard A. Fuller, "Extreme conta-gion in global habitat clearance". *Proceedings of the Royal Society B*, 24/II/2010.

BOER, Matthias M., RESCO DE DIOS, Víctor, BRADSTOCK, Ross A., "Unprecedented burn área of Australian mega forest fires". *Nature Climate Change*, 24/II/2020.

BRANDO, Paulo *et al.*, "The gathering firestorm in southern Amazonia". *Science Advances*, 6, 2, 10/I/2020.

BRAWN, Jeffrey, WARD, Michael & KENT, Angelat, "Biodiversity, Species Loss, and Ecosystem Function". In Tom Theis & Jonathan Tomkin (eds.), *Sustainability: A Comprehensive Foundation Collection*. Rice University, 2012.

BUSCH, Jonah, ENGELMANN, Jens, "Cost-effectiveness of reducing emissions from tropical deforestation, 2016–2050". *Environmental Research Letters*, 20/XII/2017.

CHIARELLO, Adriano G. *et al.* "Mamíferos ameaçados de extinção no Brasil". *Livro Vermelho da Fauna Brasileira Ameaçada de Extinção*, MMA, Brasilia, 2008.

CHOAT, Brendon *et al.* "Global convergence in the vulnerability of forests to drought". *Nature*, 21/XI/2012.

CROWTHER, Thomas W. *et al.* (2015). "Mapping tree density at a global scale" *Nature,* 2/IX/2015.

DE BOLLE, Monica, "The Amazon Is a Carbon Bomb: How Can Brazil and the World Work Together to Avoid Setting It Off?". Peterson Institute for International Economics (PIIE), October 2019.

ENGELBRECHT, Bettina M., "Plant ecology: Forests on the brink". *Nature,* 21/XI/2012.

ERFANIAN, Amir, WANG, Guiling, FOMENKO, Lori, "Unprecedented drought over tropical South America in 2016: significantly under-predicted by tropical SST". *Scientific Reports* 7, 5811, 2017.

FAO, *Global Forest Resources Assessment*, 2010.

FAO, *Global Forest Resources Assessment. How are the world's forests changing?* Rome, 2015.

FAO, *State of the World's Forests.* Rome, 2012.

FAO's *State of the World's Forests.* Rome, 2016.

FAO, *State of the World's Forests.* Rome, 2018.

FEARNSIDE, Philip M. "Soybean cultivation as a threat to the environmental in Brazil". *Environmental Conservation*, 28, 2001, pp. 23–38.

_____. "Greenhouse Gas Emissions from a Hydroelectric Reservoir (Brazil's Tucuruí Dam) and the Energy Policy Implications". *Water, Air, and Soil Pollution*, January 2002, 133, 1–4, pp. 69–96.

_____. "Desmatamento na Amazônia brasileira: História, índices e consequências" (INPA). *Megadiversidade*, I, 1, July 2005, pp. 113–123.

FLEISCHER, Katrin *et al.*, "Amazon forest response to CO_2 fertilization dependent on plant phosphorus acquisition". *Nature Geoscience*, 12, 5/VIII/2019, pp. 736–741.

FRANÇOSO, Renata et al. "Habitat loss and effectiveness of protected areas in the Cerrado Biodiversity Hotspot". *Natureza & Conservação*, Volume 13, issue 1, January-June 2015, pp. 35–40.

GILLIS, Justin, "With Deaths of Forests, a Loss of Key Climate Protectors". *The New York Times*, 1/X/2011.

GROSSMAN, Daniel & LAPOLA, David M., *Floresta em risco. As mudanças climáticas destruirão a floresta amazônica?*, AmazonFACE, Campinas, Biblioteca/Unicamp, 2018.

HANCE, Jeremy, "Tropical deforestation is 'one of the worst crises since we came out of our caves." *Mongabay*, 15/V/2008.

HANSEN, Matthew C. *et al.*, "High-Resolution Global Maps of 21st-Century Forest Cover Change". *Science*, 342, 6160, 15/XI/2013, pp. 850–853.

HARRIS, Nancy, PETERSEN, Rachael, MINNEMEYER, Susan, "World lost 8 percent of its remaining pristine forests since 2000". *Global Forest Watch*, 4/IX/2014.

INPE (2019), SOS Mata Atlântica e INPE lançam novos dados do Atlas do bioma, 24/V/2019.

IYYER, Chaitanya, *Land Management. Challenges and Strategies*. Delhi, 2009.

KEHOE, Laura *et al.* (604 signatories), "Make EU trade with Brazil sustainable". *Science*, 364, 6438, 26/IV/2019.

KIM, Do-Hyung; SEXTON, Joseph O. & TOWNSHEND, John R. "Accelerated deforestation in the humid tropics from the 1990s to the 2000s". *Geophysical Research Letters*, 7/V/2015.

LAROUSSERIE, David, "Les deux tiers des arbres dans le monde sont menacés de dépérissement". *Le Monde*, 24/XI/2012.

LEMONICK, Michael D., "Drought Puts Trees the World Over 'At the Edge'". *Climate Central*, 21/XI/2012.

LEVIS, Carolina *et al.*, "Help restore Brazil's governance of globally important ecosystem services". *Nature Ecology & Evolution*, 4, 3/II/2020, pp. 172–173.

LONDOÑO, Ernesto, "Imagine Jair Bolsonaro Standing Trial for Ecocide at The Hague. A group of activists already has". *The New York Times*, 21/IX/2019.

LOVEJOY, Thomas & NOBRE, Carlos, "Amazon Tipping Point". *Science Advances*, 4, 2, 21/II/2018.

LOVEJOY, Thomas & NOBRE, Carlos, "Amazon Tipping Point. Last chance for action". *Science Advances,* 5, 12, 20/XII/2019.

MACHADO, Ricardo B. *et al.*, "Caracterização da fauna e da flora do Cerrado". In EMBRAPA, *Savanas – desafios e estratégias para o equilíbrio entre sociedade, agronegócio e recursos naturais*, Brasília, 2009, chapter 9.

MARENGO, José, *et al.*, *Riscos de mudanças climáticas no Brasil e limites à adaptação. Análise conjunta Brasil-Reino Unido sobre os Impactos das Mudanças Climáticas e do Desmatamento na Amazônia*, INPE, May 2011.

MARGONO, Belinda Arunawati *et al.* "Primary forest cover loss in Indonesia over 2000–2012". *Nature Climate Change*, 29/VI/2014.

MARTIN, Claude. *On the Edge. The State and Fate of World's Tropical Rainforests. A Report to the Club of Rome*. Preface by Thomas E. Lovejoy. Vancouver, Berkeley, Greystone Books, 2015.

MATIAS, Ivaci, "Dia do Fogo Foi Organizado por Três Grupos de Whatsapp". *Globo Rural*, 23/X/2019.

MIDDLETON, Nick. *The Global Casino* (1995). London, Routledge, 5th edition, 2013.

NOBRE, Antônio Donato. "O futuro climático da Amazônia". Relatório de Avaliação para a Articulación Regional Amazônica (ARA), 2014.

_____. "Il faut un effort de guerre pour reboiser l'Amazonie". *Le Monde*, 24/XI/2014.

NOBRE, Carlos A., MARENGO, José A., SOARES, W.R. (eds.), *Climate Change Risks in Brazil.* Amsterdam, Springer, 2019.

OGAYA, Romà *et al.*, "Stem Mortality and Forest Dieback in a 20-Years Experimental Drought in a Mediterranean Holm Oak Forest". *Frontiers in Forest and Global Change*, 8/I/2020.

OGUTU, Judy, "Ghana's forests could completely disappear in less than 25 years". *Mongabay*, 25/VIII/2014.

OLIVEIRA, Elida, "Desmatamento da Mata Atlântica cresce em cinco estados do país". *G1*, 27/V/2019.

ORDWAY, Elsa M., ASNER, Gregory P. & LAMBIN, Eric F., "Deforestation risk due to commodity crop expansion in sub-Saharan Africa". *Environmental Research Letters*, 4/IV/2017.

PENNISI, Elizabeth, "Sturdy as they are, giant trees are particularly susceptible to these three killers". *Science*, 4/IX/2019.

PHILLIPS, Dom, "Amazon rainforest 'close to irreversible tipping point'". *The Guardian*, 23/X/2019.

PROKURAT, Sergiusz, "Palm oil - strategic source of renewable energy in Indonesia and Malaysia". *Journal of Modern Sciences*, 3/X/2013, pp. 425–443.

RAWNSLEY, Jessica, "Amazon rainforest reaches point of no return". *Climate News Network*, 16/III/2020.

RICHARDS, Paul W. *The Tropical Rain Forest. An Ecological Study.* Cambridge University Press, 1952, 2nd edition, 1996.

ROBBINS, Jim, "The Rapid and Startling Decline of World's Vast Boreal Forests". *Yale environment 360*, 12/X/2015.

SCHAUENBERG, Tim, "Wildfires: Climate Change and deforestation increase the global risk". *DW*, 8/I/2020.

SCHERR, Sara; WHITE, Andy & KAIMOWITZ, David. *A new agenda for forest conservation and poverty reduction: making markets work for low-income producers.* Washington, DC, Forest Trends and Cifor, 2003.

SIZER, *et al.*, "Tree Cover Loss Spikes in Russia and Canada, Remains High Globally". World Resources Institute, 2/IV/2015.

SIZER, Nigel, HANSEN, Matt, & MOORE, Rebecca, "New High-Resolution Forest Maps Reveal World Losses 50 Soccer Fields of Trees Per Minute". World Resources Institute, 14/XI/2013.

SOMMER, Adrian, "Attempt at an assessment of the world tropical moist forests". FAO, Committee on Forest Development in the Tropics, Rome, 1976.

STEEGE, Hans ter *et al.* "Hyperdominance in the Amazonian Tree Flora". *Science, 342, 6156,* 18/X/2013.

_____. "Estimating the global conservation status of more than 15,000 Amazonian tree species". *Science Advances*, 1, 10, 20/XI/2015a.

STRASSBURG, Bernardo *et al.*, "Moment of truth for the Cerrado hotspot". *Nature. Ecology and Evolution*, 1, 23/III/2017.

STRECK, Charlotte *et al.*, Progress on the New York Declaration on Forests. Protecting and Restoring Forests: A Story of Large Commitments yet Limited Progress. Five-Year Assessment Report. September 2019

SUDARSHANA, Padmini; NAGESWARA-RAO, Madhugiri & SONEJI, Jaya R. Soneji. *Tropical Forests*. Croatia, InTech, 2012.

TOLLEFSON, Jeff. "Plastic wood is no green guarantee". *Nature*, 498, 6/VI/2013.

_____, "Tropical forest losses outpace UN estimates". *Nature*, 26/II/2015.

TRUMBORE, S.; BRANDO, P. & HARTMANN, H. "Forest health and global change". *Science*, 349, 6.250, 21/VIII/2015, pp. 814–818.

UNDP, 2015. *Transforming our world: the 2030 Agenda for Sustainable Development*. "Goal 15: Sustainably manage forests, combat desertification, halt and reverse land degradation, halt biodiversity loss."

URQUHART, Gerald *et al.*, "Tropical deforestation". Earth Observatory, NASA, 2001.

VERGARA, Walter & SCHOLZ, Sebastian M. (eds.). *Assessment of the Risk of Amazon Dieback. A World Bank Study*, 2011.

VIDAL, John, "Greater Mekong countries lost one-third of forest cover in 40 years". *The Guardian*, 2/V/2013.

VIÑA, Andres *et al.*, "Effects of conservation policy on China's forest recovery". *Science Advances*, 18/III/2016.

VITOUSEK, Peter M. *et al.*, "Human Domination of Earth's Ecosystems". *Science*, 277, 5325, 25/VII/1997, pp. 494–499.

WEISSE, Mikaela & GOLDMAN, Liz, "Global Tree Cover Loss Rose 51% in 2016", 18/X/2017, Global Forest Watch http://blog.globalforestwatch.org/data/global-tree-cover-loss-rose-51-percent-in-2016.html

WEISSE, Mikaela & GOLDMAN, Elizabeth Dow, "2017 was the second-worst year on record for tropical tree-cover loss". Global Forest Watch, 27/VI/2018.

WEISSE, Mikaela & GOLDMAN, Elizabeth Dow, "The World Lost a Belgium-sized Area of Primary Rainforests Last Year". World Resources Institute, 25/IV/2019.

WILLIAMS, Michael. *Deforesting the earth: from prehistory to global crisis*. University of Chicago Press, 2003.

World Resources Institute (WRI), "New Analysis Finds Over 100 Million Hectares of Intact Forest Area Degraded Since 2000". Washington, D.C., 4/IX/2014.

WWF, *Living Forests Report*. Chapter 5: "Saving Forests at Risk" "Saving Forests at Risk", 2015

ZALLES, Viviane *et al.*, "Near doubling of Brazil's intensive row crop area since 2000". *PNAS*, 17/XII/2018.

ZEMP, Delphine Clara *et al.*, "Self-amplified Amazon forest loss due to vegetation-atmosphere feedbacks". *Nature Communications* 8, 14681, 13/III/2017.

Chapter 3
Water and Soil

Quantitative decline and qualitative degradation of water resources form two inseparable aspects of one of the largest environmental crises on the planet. For the sake
of clarity, the first aspect—scarcity—will be analyzed in this chapter. The second
aspect—the degradation of these resources—will be the subject of Chap. 4 (item
4.1, Sewers) and 11 (item 11.2, Eutrophication, hypoxia, and anoxia).

3.1 Decline of Water Resources

About 70% of the planet's freshwater is conserved in glaciers, which are in rapid
decline. The seasonal thawing of some of these glaciers supplies water to various
populous regions, including those dependent on the Himalayas, the Andes, and
other mountain ranges (Marks 2013; Pritchard 2019). According to data cited by
UNEP, "the total usable freshwater supply for ecosystems and humans is ~
200,000 km³ of water, which is <1% of all freshwater resources, and only 0.01% of
all water on Earth." Usable fresh water is therefore relatively scarce. And this scarcity is increasing.[1] This increase can be explained by the synergy between deforestation and climate change, but also by an increase in per capita water consumption.
The world population increased by about 3.6 times in the twentieth century (from
1.65 billion in 1900 to about six billion in 2000), while in the same period global
human consumption of water multiplied by eight, with a clear preponderance of
intensively irrigated agriculture (Kirby 2004). The gap between population and
world consumption only increases: between 1990 and 2010, the population rose

---

[1] The Falkenmark Water Stress Indicator establishes three levels of water scarcity: (1) water stress,
when the renewable water supply is below the threshold of 1700 m³ per capita per year; (2) water
scarcity, when this level is less than 1000 m³ per capita per year; and (3) absolute scarcity, when
this level is less than 500 m³ per capita per year. This parameter includes domestic, agricultural,
industrial, energy, and environmental needs.

© Springer Nature Switzerland AG 2020
L. Marques, *Capitalism and Environmental Collapse*,
https://doi.org/10.1007/978-3-030-47527-7_3

from 5.3 billion to 6.8 billion, a growth of less than 20%, while human consumption of water increased 100% (Spring and Cohen 2011). This increase in water consumption is marked by tremendous inequality. According to the World Water Council, while residential use of water in North America is around 400 l per person a day, in many sub-Saharan countries, the average use is 10–20 l per person a day.

3.1.1   Intensification of Water Scarcity on a Global Scale

On the eve of this century's second decade, Charles J. Vörösmarty et al. (2010) published what they called the first worldwide synthesis to jointly consider human and biodiversity perspectives on water security. They found that "nearly 80% (4.8 billion) of the world's population (for 2000) lives in areas where either incident human water security or biodiversity threat exceeds the 75th percentile." In 2013, *The Bonn Declaration on Global Water Security*, a document issued by the Conference, "Water in the Anthropocene, " projected this estimate for the near future and defined the economic causes of this process with surgical precision:

> In the short span of one or two generations, the majority of the nine billion people on Earth will be living under the handicap of severe pressure on fresh water. […] Faced with a choice of water for short-term economic gain or for the more general health of aquatic ecosystems, society overwhelmingly chooses development, often with deleterious consequences on the very water systems that provide the resource.

The pace of intensification of water scarcity on a global scale has been overwhelming the scientific projections. Thus, in 2006, the report of the International Water Management Institute (IWMI) stated: "One-third of the world's population is short of water—a situation we were not predicted to arrive at until 2025" (Coghlan 2006). But in 2016, water scarcity already threatens two-thirds of humanity. As pointed out by Arjen Hoekstra, "if you look at environmental problems, [water scarcity] is certainly the top problem" (Carrington 2016). In fact, Mesfin Mekonnen and Arjen Hoekstra (2016) show that the overall water situation is much worse than previous studies suggest. Those studies estimated that such scarcity would reach "only" between 1.7 and 3.1 billion people:

> Previous global water scarcity assessments, measuring water scarcity annually, have underestimated experienced water scarcity by failing to capture the seasonal fluctuations in water consumption and availability. […] We find that about 71% of the global population (4.3 billion people) lives under conditions of moderate to severe water scarcity (WS > 1.0) at least 1 month of the year. About 66% (4.0 billion people) lives under severe water scarcity (WS > 2.0) at least 1 month of the year. Of these 4.0 billion people, 1.0 billion live in India and another 0.9 billion live in China. Significant populations facing severe water scarcity during at least part of the year further live in Bangladesh (130 million), the United States (130 million, mostly in western states such as California and southern states such as Texas and Florida), Pakistan (120 million, of which 85% are in the Indus basin), Nigeria (110 million), and Mexico (90 million).

Table 3.1 reports data from the World Resources Institute (WRI—National Water Stress Rankings). As of 2019, the following 44 countries are already suffering high and extremely high baseline water stress (BWS).

Table 3.1 Countries suffering high and extremely high baseline water stress (BWS)

Extremely high baseline water stress (nearly 25% of the world's population. In these countries, more of 80% of available supply is withdrawn every year by agriculture, industry, and municipalities)			
1. Qatar	6. Libya	10. United Arab Emirates	14. Pakistan
2. Israel	7. Kuwait	11. San Marino	15. Turkmenistan
3. Lebanon	8. Saudi Arabia	12. Bahrain	16. Oman
4. Iran	9. Eritrea	13. India	17. Botswana
5. Jordan			
High baseline water stress (40–80% of available supply is withdrawn every year)			
18. Chile	25. Uzbekistan	32. Turkey	39. Niger
19. Cyprus	26. Greece	33. Albania	40. Nepal
20. Yemen	27. Afghanistan	34. Armenia	41. Portugal
21. Andorra	28. Spain	35. Burkina Faso	42. Iraq
22. Morocco	29. Algeria	36. Djibouti	43. Egypt
23. Belgium 30.	Tunisia	37. Namibia	44. Italy
24. Mexico	31. Syria	38. Kyrgyzstan	

Source: World Resources Institute, Aqueduct Water Risk Atlas, 2019

3.1.2   The Impact of Climate Change

Christel Prudhomme et al. (2014) analyzed the impact of climate change on hydrological droughts:

> Using an ensemble of 35 simulations, we show a likely increase in the overall severity of drought by the end of the 21st century, with regional hotspots including South America and Central and Western Europe in which the frequency of drought increases by more than 20%.

The droughts plaguing São Paulo (2014), Cape Town (2015–2018), and Chennai (India's sixth largest city, 2019) illustrate this trend worldwide. In Cape Town, drought drove five of the city's six reservoirs to near-absolute collapse, and the rains that saved the city, in extremis, from Day Zero brought the dams back only to the 2016 level. Six years ago, Jacob Schewe et al. (2014) made the following dire projection:

[…] a global warming of 2°C above present (approximately 2.7°C above preindustrial) will confront an additional approximate 15% of the global population with a severe decrease in water resources and will increase the number of people living under absolute water scarcity (< 500 m³ per capita per year) by another 40% (…) compared with the effect of population growth alone.

Commenting on these results, Quirin Schiermeier (2014) considers that, in fact, "even modest climate change might drastically affect the living conditions of billions of people, whether through water scarcity, crop shortages or extremes of weather." On regional impacts, several important findings were put forward in the IPCC's Fourth Assessment Report (AR4 2007). With regard to Africa, for instance, it was stated that by 2020, between 75 and 250 million people are projected to be exposed to increased water stress due to climate change.[2] In fact, according to the World Health Organization (Regional Office for Africa), in 2020 water scarcity already affects nearly 400 million people or 1 in 3 people in the African Region.

3.2 Rivers, Lakes, and Reservoirs

In *Ecclesiastes* (I,4-9), the most famous of the Sapiential Books of the Old Testament, one reads: "All streams flow into the sea, yet the sea is never full. To the place the streams come from, there they return again." As pointed out by Kader Asmal (2000), these "words are beautiful, haunting and, suddenly, anachronistic." Arjen Hoekstra et al. (2012) show that there is indeed something new under the sun:

We analyzed 405 river basins for the period 1996–2005. In 201 basins with 2.67 billion inhabitants there was severe water scarcity during at least one month of the year. The ecological and economic consequences of increasing degrees of water scarcity – as evidenced by the Rio Grande (Rio Bravo), Indus, and Murray-Darling River Basins – can include complete desiccation during dry seasons, decimation of aquatic biodiversity, and substantial economic disruption.

3.2.1 Pakistan, India, and Bangladesh

The estimated 95,000 glaciers in the high mountains of Asia play a very important role in the hydrological cycle of the Indus, Ganges, Brahmaputra, and many other glacier-fed river basins in this region. About 800 million people rely on these glaciers on seasonal and longer timescales. But they are shrinking faster and faster. Josh Maurer et al. (2019) have calculated the acceleration of ice loss across the Himalayas during the intervals 1975–2000 and 2000–2016:

---

[2] See "Statement of Dr. RK Pachauri, Chairman (IPCC) at the 18th Conference of the Parties, Doha, 28/XI/2012.

We observe consistent ice loss along the entire 2000-km transect for both intervals and find a doubling of the average loss rate during 2000–2016 (…) compared to 1975–2000. The similar magnitude and acceleration of ice loss across the Himalayas suggests a regionally coherent climate forcing, consistent with atmospheric warming and associated energy fluxes as the dominant drivers of glacier change.

In a report published in 2016, the Pakistan Council of Research in Water Resources (PCRWR) warned that the country could reach the level of "absolute water scarcity" by 2025 (Roberts 2017). Pakistan is one of the most water-stressed countries in the world (see above WRI's Aqueduct Water Risk Atlas, 2019). It is, in fact, undergoing an undeniable water collapse. According to the Asian Development Bank (ADB 2013): "At present, Pakistan's storage capacity is limited to a 30-day supply, well below the recommended 1000 days for countries with a similar climate. Climate change is affecting snowmelt and reducing flows into the Indus River, the main supply source." Forced to use brackish groundwater, the poor population of Karachi revolted, demanding that water be supplied by trucks at least once a week. The Indus River and the Hub Dam are no longer capable of supplying the city, and the increasing water scarcity has attracted organized crime, as might be expected (Jawad 2014).

In 2018, India was plagued by the biggest water crisis in its history, according to a government report, titled *Composite Water Management Index* (June 2018), issued by the National Institute for Transforming India (NITI Aayog). Nearly half of the population—600 million people—face water shortages "from high to extreme," with thousands of people dying each year due to lack of access to safe water. According to this document, by 2030 water supply will be half of the demand in this country. "There are around 90 cities in India which are water stressed. They face crisis today, tomorrow and the day after," declared Rajendra Singh, founder and Chairman of Tarun Bharat Sangh, a conservation group (Abi-Habib and Kumar 2018). The number of rivers defined as "polluted" more than doubled in India in the last 5 years, jumping from 121 to 275. "In view of population increase, demand for freshwater for all uses will be unmanageable" (Burke 2015). In Bangladesh, more than 15 rivers, including the Teesta, Brahmaputra, Dharla, Dudhkumor, Phulkumor, Gangadhar, Kalzani, and Zinziram, have declined significantly in recent years, and many of them are on the verge of extinction. The part of the Teesta River that flows through Bangladesh (115 km) has dried up because of India's withdrawal of water. Moreover, many other rivers that are also fed by the Himalayan glaciers and tributaries of the Ganges-Brahmaputra Delta are compromised.

3.2.2 Brazil, Mexico, and the USA

Despite its gigantic rivers and aquifers, Brazil suffers from different levels of hydric stress. The hydrographic regions of the Atlantic, where almost half of the urban population of the country live, contain only 3% of the country's hydric capacity, which is in decline. The 2011 *Atlas of Urban Water Supply* informs us that 55% of

Brazilian municipalities (73% of the demand) will be subject to water scarcity in the next decade. One of the causes of this future hydric scarcity is the removal of about 20% of the Brazilian Amazon forest and more than half of the vegetation cover of the Cerrado. This biome is home to three great aquifers (Guarani, Bambuí, and Urucuia) and three great watersheds (Tocantins-Araguaia, Paraná-Prata, and São Francisco) are born in it. According to Tiago Reis of the Amazon Environmental Research Institute (IPAM), "the loss of native vegetation in the Cerrado jeopardizes the formation of rains by evapotranspiration and the infiltration of water in the soil to recharge aquifers and rivers of the region."[3] The pollution of aquifers by agro-chemicals and the growing demand for irrigation-intensive agriculture, typical of agrobusinesses that export water in the form of soy, cow meat, and other products, has led to the unsustainable use of water (Marques 2017). At least since 2011, 49 rivers in 11 Brazilian states are threatened; 24.5% of them are facing a high degree of pollution from pesticides, fertilizers, and sewage, and none are in a situation considered great or good. No longer protected by forests, these rivers tend to lose volume. One of the most emblematic declines in volume is that of the São Francisco river which irrigates 7.5% of the Brazilian territory. Many of its perennial tributaries have become dry or very empty. Already in 2012, the flow rate of one of its tributaries, the Rio Doce (853 km)—today destroyed by mining corporations—had been reduced to one-third of its original rate. Deforestation, excessive use of water for irrigation, hydroelectric plants, increased demand by the population, and the biggest drought in its history, begun in 2012, have led to a huge decrease in the flow rate of the São Francisco river, from 2000 to 2500 m³ per second to only 600 m³/s in 2017 (Iandoli 2017). With a decrease in its flow rate, the mouth of the São Francisco river is being invaded by the ocean. In São Paulo state, the most populous state in the country, out of 54 river sources evaluated in 2003, 34 were found with less than half the volume of water and 29 had dried up by 2014. Furthermore, in 81% of the rivers evaluated, there was a decrease in water quality. In 2014–2015, the level of the main reservoirs that supply the metropolitan region of São Paulo (21.5 million inhabitants) came very close to collapse.

The availability of water in Mexico is declining (Spring and Cohen 2011). The northern region of the country now uses more than 40% of the average natural water availability, a percentage defined by the UN as "strong pressure on water resources." The Yaqui River Basin, the largest river in northwest Mexico, once a habitat of the American crocodile (*Crocodylus acutus*), is drying up due to intensive use for irrigation, a series of impoundments, and urban growth. The Rio Grande (Rio Bravo), which separates Mexico from Texas, is now reduced to a fifth of its flow when it arrives at the Gulf of Mexico. In 2001, for the first time in history, the Rio Grande dried up before reaching its mouth and this has occurred several times since then.

Many of the more than 250 thousand rivers in the USA are threatened. Since 1984, the American Rivers, a nonprofit conservation organization, has published

---

[3] Cited in "Desmatamento do Cerrado supera o da Amazônia, indica dado oficial." *Observatório do Clima*, 25/VII/2017.

annual reports, titled *America's Most Endangered Rivers*. The picture provided is bleak and too complex to be summarized here. But one must at least point out the calamitous state of the Colorado River, victim of a "slower-burning catastrophe" (Lustgarten et al. 2015). Today, it rarely reaches its delta in the Gulf of California. An essential artery in the Southwest of the USA, it provides water to 40 million people in seven states, accounts for 15% of the country's food production, and is the source of two of the largest water reserves in the country. Meteorological drought and warming have been shrinking this crucial water resource that supports more than USD 1 trillion per year of economic activity, according to Paul Christopher Milly and Krista Dunne (2020). They estimate that its annual-mean discharge has been decreasing by 9.3% per °C of warming due to increased evapotranspiration. Also according to Bradley Udall and Jonathan Overpeck (2017), "between 2000 and 2014, annual Colorado River flows averaged 19% below the 1906–1999 average, the worst 15-year drought on record. At least one-sixth to one-half (average at one-third) of this loss is due to unprecedented temperatures (0.9°C above the 1906–1999 average)" (see also Seager et al. 2012). Many other rivers are seriously threatened in the USA—White Rivers in the state of Washington, Flint (Georgia), San Saba (Texas), Catawba (North Carolina and South Carolina), and so on—either due to flow rate loss or to a deterioration in the biological quality of its waters.

3.2.3   Middle East and Central Asia

In the sixteenth century, Luís de Camões, in his poem *The Lusíads* (I, 8), laments that other than Christians "still drink the liquor of the Holy River" ("inda bebe o licor do Santo Rio"). Today, the Jordan River has become a stagnant stream of polluted water from which no one can drink. Until the 1960s, the river's flow rate into the Dead Sea was 1.3 billion m^3 per year. By 2017 it was reduced to 200–300 million m^3. Its basin, measuring 251 km, borders Syria, Lebanon, the West Bank, Israel, and Jordan. These last two countries, which are in strong demographic expansion and more directly dependent on the river, are the main ones responsible for its decline. The project of a 180 km channel that allows the Dead Sea to receive 100 million m^3 of water per year from the Red Sea (the Red Sea-Dead Sea Water Conveyance Project) is considered to be of high environmental risk, with potential formation of red algae, gypsum, and chemical alteration of this unique ecosystem. This channel would represent, in any case, only a small fraction of the input of water once received from the Jordan River.

In Iran, the Zayandeh-rud ("life giver") river, crossed by beautiful bridges and largely responsible for the prehistoric and historical civilizations of the Iranian central plateau, was substantially intact until the 1960s. It dried up completely in 2010, after years of partial drying. With the drying up of its largest river and with the rapid decline of its dams, Iran is at risk of desertification. The Amu Darya River (Darya means sea), or Jayhoun (Gihon, in Hebrew), considered to be one of the four rivers of the Garden of Eden in the *Book of Genesis*, flows through Central Asia, defining the borders of Afghanistan with Tajikistan and then with Turkmenistan and Uzbekistan. Today, it dies about 110 km before reaching the now-defunct Aral Sea.

3.2.4 "Water: China's Achilles Heel"

This is the title of an article, published on September 9, 2019, in *The Japan Times*, by Yoichi Funabashi, Asia Pacific Initiative's chairman. The situation of China's rivers is critical, especially in the north, a region that suffers from "absolute water scarcity" (consumption of less than 500 m^3 per capita per year). "Eight of China's northern provinces are currently experiencing absolute water scarcity, while another 11 are at the water scarcity level," writes Funabashi. The number of rivers with significant areas of influence fell from more than 50,000 in the 1950s to only 23,000 in 2013 (Toor 2013). Therefore, more than half of China's rivers had simply disappeared in 2013, driven by irrigation and industrial water use. And much of the rest of its water resources is affected by pollution. Song Lanhe, chief engineer of the urban water quality monitoring center of China's Ministry of Interior, said that only half of the water sources in the cities are drinkable.

The case of the Yellow River (Huang He), the "Mother River", cradle of Chinese civilization, is emblematic. It irrigates 15% of China's land and feeds 12% of its population. In 1972 it completely dried up before reaching the sea, and in 1997, it dried up for 236 days to an extent of about 700 km, severing the river's connection to the Bohai Sea. Each time it dries up, it affects more than 140 million people in an area of 74 thousand km^2 of arable land. In 2019, the water resources provided by the Yellow River were a tenth of what they had been in the 1940s (Funabashi 2019). In 2016, Yongnan Zhu and colleagues evaluated projections of the impacts of climate change on the Yellow River according to three scenarios (RCP 2.6, 4.5, and 8.5 W/m^2) of the IPCC-AR5 (2013). They concluded that all results indicate a reduction in the water resources of the Yellow River throughout this century (Zhu et al. 2016). In addition to water shortage, pollution levels of the Yellow River are extreme. In 2007, the Yellow River Conservancy Commission, a government agency, inspected 13,000 km of the river's course and of its tributaries and concluded that one-third of the water is inappropriate even for agricultural irrigation. This is because 4000 petrochemical industries were built on its banks and only about 40% of the water used by Chinese industry is recycled. Moreover, the river is a victim of pollution from fertilizers and agrochemicals. In the end of 2012, the Chinese press reported that they had discovered 300 human corpses floating in the Yellow River, in the area of the city of Lanzhou. They are the latest victims of the 10,000 or so corpses (mostly suicidal, according to police) found on the river since the 1960s.

The Hai, Huai, Tarim, and Jiapingtang rivers in China also present varying degrees of pollution and waning. The Huai River in central China contains high levels of arsenic. In March 2013, the Jiapingtang River (and its tributary, Huangpu), which supplies Shanghai's 23 million inhabitants, was poisoned by the corpses of 16,000 pigs and 1000 ducks. About 500 pig carcasses are found per month in these waters. In four of the ten summers from 2003 to 2012, more than half of the 1200 km of the Tarim River in northwest China dried up. According to Niels Thevs of the University of Greifswald, cotton farmers who irrigate plantations with the river water react by multiplying and deepening well drilling, which accelerates the depletion of the region's fossil aquifers. Furthermore, the Tarim River receives 40% of its waters from seasonal thawing and, in the last 40 years, an 8% decrease in the volume and a 7% decrease in the area of the glacial covers that feed it further aggravates the river's situation (Ives 2012). For the Hai River basin, a World Bank study shows a 40-billion-ton decrease in water per year.

3.2.5 Degradation and Death of Lakes

"Lakes are generally facing rapid decline in the arid and semi-arid regions of the world" (Chen et al. 2019). The 13th World Lakes Conference (Wuhan, China, 2009) warns that "the ecological state of the lakes worldwide has deteriorated alarmingly during the last decades." In fact, an increasing number of the more than 50,000 natural and man-made lakes around the world are drying up and/or becoming degraded. The overall picture is so pervasive and catastrophic in all continents, except Antarctica, that, once again, it cannot be reproduced here, not even in a general form. In addition to the well-known direct human impacts, there is a new factor of imbalance: global warming. In 167 of the world's largest lakes—including the Great Lakes, Tahoe (California), Baikal (Siberia), and Tanganyika—higher temperatures of up to 2.2 °C have been registered between 1985 and 2009; in some cases, this constitutes a rate of warming of up to seven times the atmospheric rate in the same region and period (Schneider and Hook 2010).

3.2.6 Lakes of Central Asia: Aral, Balkhash, Urmia, Hamoun, Baikal, etc.

The Aral Sea, between Kazakhstan and Uzbekistan, was once the fourth largest lake in the world (68,000 km^2). It provided one-sixth of the Soviet Union's fish consumption. From the 1960s onward, the excessive collection of water from its tributaries (the Amu Daria and Syr Daria rivers) for the irrigation of cotton fields progressively transformed it into little more than a mire of toxic substances, given the indiscriminate use of insecticides in these fields. Could Kazakhstan's Lake Balkhash (16,000 km^2) be the next Aral? Scientists believe so, as it is becoming more and more shallow and saline. Water levels in Lake Balkhash have been declining, mostly due to evaporation and increased water usage for irrigation along the Ili River, which flows from China to Kazakhstan: "the gradual degradation of the lake ecosystems is being hastened by the construction of hydroelectric installations in

China" (Le Sourd and Rizzolio 2004). On the Irano-Afghan border, the Hamoun-e Puzak and its two sister lakes (Hamoun-e Sabari and Hamoun-e Helmand) "are now seas of sand, and the marshes they supported have withered away" (Stone 2018).

Finally, Lake Baikal in Southeastern Siberia, the oldest and deepest lake in the world, home to 3600 species of animals and plants, many of which exist only there (hence, its name, Galapagos of Russia), is under attack. The omul (*Coregonus migratorius*), for example, a whitefish species of the salmon family endemic to Lake Baikal, was included in the Russian list of endangered species in 2004 and its total biomass diminished from 25 million tons to just 10 million. Overfishing is admittedly a big problem, but the omul crisis is also tied to drought, pollution, and climate as Baikal's waters are warming at a rate omul might not be able to adapt to. "The beautiful Baikal," as Piotr Kropotkin called it in his autobiography (1899), is at its lowest level in 60 years.[4] This natural heritage site into which 300 rivers flow is also the destination of increasing amounts of waste from local populations and tourist boats, leading to the extinction of sponges and the proliferation of putrid algae. Moreover, Baikal will soon be the dumping site for 90% of the radioactive material rejected by the Angarsk Enriched Uranium Center (see UNESCO, "Saving Lake Baikal").

3.2.7 China's Lakes

"The 24,800 Chinese lakes cover an area of over 80,000 km²—and with a few exceptions nearly all lakes are heavily polluted or on the verge of drying up" (Cribb 2017). Some of the lakes in the semi-arid areas of the Inner Mongolian Plateau have degraded or even disappeared during the past three decades (Chen et al. 2019). Hongjiannao Lake, about 500 km west of Beijing, makes the Muus Desert habitable in Shaanxi province, yet the lake has been disappearing since the 1970s. Its level declines 60 cm per year. In 1969, the lake stretched for 67 km²; in 2000, it averaged 50 km²; in 2009, 46 km²; and in 2014, 32 km²; this amounts to a retraction of more than 50% in 45 years (Ying 2016). According to Kang Liang and Guozhen Yan (2017), the lake area ranged from 55.02 km² in 1997 to 30.90 km² in 2015, with an overall downward trend of −0.94 km²/year. The remaining waters, in an advanced deterioration process, have become more alkaline, reaching a pH of 9.6, when the maximum that most fish are able to bear is a pH of 8.5. As a result, the lake, where more than 300 tons of fish were fished each year, has been emptied of animal life. Ren Leijie, a local authority, says that "the Hongjiannao could vanish in only 10 years."[5] A second example should be remembered, this time in Southwest China. Poyang Lake is located in northern Jiangxi Province and the southern bank of the middle reach of the Yangtze River. It is the largest freshwater lake of the country and receives water from five major rivers upstream: the Gan, Fu, Xin, Rao, and Xiu riv-

---

[4] Cf. "World's Largest Freshwater lake, Lake Baykal under threat." *Euronews*, 10/I/2015

[5] Quoted in "China's largest desert lake could vanish in 10 years." *Want China Times*, 24/XII/2011; "China's largest desert freshwater lake shrinking." *News Xinhuanet*, 28/XI/2013.

ers. After regulation and storage, the water of Poyang Lake flows into the Yangtze River. According to Yao Wu et al. (2019), "due to human activities in modern times (e.g., Three Gorges hydraulic projects in the Yangtze River and other hydraulic projects in Poyang Lake Basin), flood and drought disasters have occurred frequently in Poyang Lake in recent years." In fact, Poyang Lake is facing a collapse comparable to that of the Aral Sea. According to Wang Hao, a scientist from the China Institute of Water Resources and Hydropower Research, its area has gone from almost 5200 km^2 in 1950 to just over 3600 km^2 in 2003 and to only 200 km^2 in 2012.[6] In 2016, the lake nearly dried up completely.

3.2.8   Africa: Chad, Songor, Faguibine, etc.

In 2006, UNEP published *Africa's Lakes: Atlas of Our Changing Environment.* Its conclusions were presented by Maria Mutagamba, President of the African Ministers' Council on Water: "Satellite images of Africa confirm that dramatic environmental changes are affecting the continent's 677 natural and human made reservoirs, which contain an estimated 30,000 km^3 of fresh water—the world's largest volume." Here are some of the most catastrophic cases. Lake Chad, surrounded by four countries (Nigeria, Niger, Chad, and Cameroon), used to be the main source of livelihood for more than 20 million people. Between the 1960s and 1990s, its volume fell by 90%. Climate change is perhaps accelerating the decline of the lake as temperatures in the region are rising 1.5 times faster than the global average (Brown and Vivekananda 2019), although there is no general agreement about this. In any case, human water use certainly accounts for roughly 50% of the observed decrease in lake area since the 1960s and 1970s (Coe and Foley 2012). Since 2007, there have been small improvements, but in the dry months, Lake Chad is little more than a swamp, and in the months of maximum rainfall, it does not exceed seven meters of depth. According to the United Nations, 10.7 million people in the Lake Chad basin now need humanitarian relief to survive (Ross 2018). A similar decline occurs with Songor Lagoon in Ghana and with the Lake Faguibine in Mali, the last of five interconnected lakes in the interior delta of the Niger River, once one of the largest lacustrine complexes in West Africa, with 860 km^2. NASA satellite photos from 1974 and 2006 show almost complete drying.

3.2.9   The Americas

In the USA, many lakes are in decline: the Salton Sea, Owens, Tulare, and Mono lakes in California, Walker Lake in Nevada, Lake Mead (Nevada and Arizona), Lake Meredith in Texas, etc. Lake Owens once stretched out for 240 km^2 and had depths

---

[6] See *China English News*, 27/VIII/2012.

of 7–16 m. As the rivers that fed it were diverted to supply Los Angeles, the lake dried up almost completely, with the water that remains reaching a depth of no more than 1 m. A source of sandstorms, it is now considered the largest single cause of dust pollution in the USA (Siegler 2013). Lakes Powell and Mead, both on the Colorado River and separated by the Grand Canyon, are the largest man-made reservoirs in the USA. Their fates are closely linked. The consumption of water from Lake Mead exceeds the volume of water that flows into it from the Colorado River, threatening its equilibrium. Its capacity fell from 90% in 2000 to less than 40% in 2015. By April 2015, its level fell to its lowest since 1937, when it was still being filled after the construction of the Hoover Dam. Lake Mead has not yet fallen into critical condition because it is being fed unsustainably by Lake Powell. But one lake cannot rescue the other forever. Indeed, between 2000 and 2018, Lake Powell, the second largest man-made reservoir in the USA, dropped 94 feet (Davis 2018). In 2015 it was already at only 45% of its capacity (McGreal 2015). Lakes Superior, Michigan, Huron, Erie, and Ontario make up the Great Lakes system, which contain about one-fifth of the world's surface fresh water. Over 30 million people and more than 3500 species of plants and animals, including at least 170 species of fish, inhabit the Great Lakes basin. As recently as 2013 Lakes Huron and Michigan hit record low levels, and Lakes Superior, Ontario, and Erie were below their historical averages (Sutherland 2013). Lake Superior was also one of the most rapidly warming lakes in the world (Borre 2012). But in 2014 there was abrupt turnaround, and the Great Lakes are now reaching record highs. This poses just as many threats to people and biodiversity in the region, including shoreline erosion, floods, and delays in planting spring crops. Polar vortex disruptions caused by climate change could be the culprit for this trend reversal; hence rapid and unpredictable transitions between extreme high and low water levels in the North American Great Lakes may represent the "new normal" (Gronewold and Rood 2019).

In Latin America, Lake Chapala, located near the city of Guadalajara (Mexico) and considered a national treasure, is the largest lake in Mexico. Since the late 1970s, the lake has experienced long-term decline and has lost about 80% of its water. In 2004, the Global Nature Fund labeled it as "Threatened Lake of the Year". Lake Atescatempa in southwest Guatemala dried up almost completely in 2018, and Lake Izabal (590 km^2) in the northeast of the country is completely polluted by a nickel mine owned by CGN-Pronico and controlled by Solway, a Swiss corporation (Michel 2019). In the Bolivian highlands, at almost 3700 m, Lake Poopó, the second largest lake in Bolivia and former habitat of 75 bird species that flew over its 1340–2500 km^2, was officially declared dry in January 2016. Its end was an effect of climate change, droughts, the melting of the Andes glaciers, as well as the diversion of water from the Desaguadero River and its tributaries for agriculture and mining. More than a hundred mines served themselves with the waters of this lacustrine ecosystem, polluting its tributaries; among them is Huanuni, the largest tin mine in the country.

3.3 Aquifers

This is the central message of the United Nations World Water Development Report (2014, p. 26): "There is clear evidence that groundwater supplies are diminishing, with an estimated 20% of the world's aquifers being over-exploited, some massively so. Globally, the rate of groundwater abstraction is increasing by 1% to 2% per year." The scale of the global aquifer crisis was equally well summed up by Margaret Catley-Carlson (2017), and it is best to turn to her without delay:

> It is astonishing to many that lakes and rivers account for less than one-third of 1% of global fresh water. Some 95% of unfrozen fresh water resides unsung and underground, dimly visible at the bottom of a well or gushing from a pump. Big cities such as Buenos Aires and entire countries, including Germany, depend hugely on groundwater. About 70% of it goes into irrigation, accounting for more than half of irrigated agriculture—which in turn provides nearly half of the global food basket. (…) Everywhere, aquifers are poorly measured and managed. As a result, no scientific consensus exists on the details of this vast and vital source of fresh water—although there is consensus on the fact that we face a worldwide problem.

Contrary to the decline of rivers and lakes, the decline of aquifers is a surreptitious process that takes its users by surprise (Brown 2004, 2013). But lakes and rivers often depend on groundwater and when its levels drop, discharges from groundwater to streams, lakes, and rivers also decline, with potentially devastating effects on aquatic ecosystems (De Graaf et al. 2019). In 2000, the countries with the highest rates of use of renewable and non-renewable aquifers were, in descending order, India, Pakistan, the USA, Iran, China, Mexico, and Saudi Arabia, whose populations add up to 3 billion and 300 million people, representing almost 50% of the world's population (Wada et al. 2012). Between 1960 and 2000, the use of fossil aquifers has tripled, and new measurements have now elicited the gravity of the situation. Tom Gleeson et al. (2012) evaluated the global groundwater footprint for the first time. The authors estimated that "the size of the global groundwater footprint is currently about 3.5 times the actual area of aquifers and that about 1.7 billion people live in areas where groundwater resources and/or groundwater-dependent ecosystems are under threat." Two years later, James Famiglietti (2014) summed up what is at stake:

> Most of the major aquifers in the world's arid and semi-arid zones […] are experiencing rapid rates of groundwater depletion. […] These include the North China Plain, Australia's Canning Basin, the Northwest Sahara Aquifer System, the Guarani Aquifer in South America, the High Plains and Central Valley aquifers of the United States, and the aquifers beneath northwestern India and the Middle East.

Two papers published by Alexandra Richey et al. (2015a) showed that 21 of the 37 largest aquifers in the world were in a state of imbalance, as more water was being withdrawn from them than their capacity to recharge. The most overused aquifers in the world were in Saudi Arabia, under the Indus basin supplying Pakistan and northwest India and under the Murzuk basin in North Africa. Measurements done between 2003 and 2013 allow us to affirm that 13 of these 21 aquifers were being used while receiving almost no replenishment, or none at all; 8 were classified as overstressed because they were not being compensated by a natural recharge and five were defined as "extremely" or "highly" stressed. Currently, about 70% of the pumped groundwater worldwide is used to sustain irrigation. Inge de Graaf et al. (2019) estimate that, by 2050, environmental flow limits will be reached for approximately 42–79% of the watersheds in which there is groundwater pumping worldwide. As pointed out by Richey et al. (2015a), "the combination of population growth and increased food demand with the potential for increased hydrologic extremes due to climate change may further increase the rate of use in groundwater systems as surface supplies become less accessible."[7]

3.3.1 India

India's current water precariousness—at quantitative and qualitative levels—is unprecedented. Currently, the country has a population of 1.36 billion that is estimated to reach 1.7 billion by 2050; the current enrichment of its middle class will imply, at least for this growing stratum of the population, greater demand for water per capita. In an assessment done by the World Bank and confirmed by Asit Biswas, Cecilia Tortajada, and Udisha Saklani (2017), more than 60% of India's irrigated agriculture and 85% of its domestic water use now comes from aquifers. India now uses more aquifers than China and the USA combined. If the current trend is maintained, by 2030 about 30% of the country's aquifers will be in critical condition, putting at risk about 25% of its agricultural production. According to India's Ministry of Agriculture, 22% of the territory and 17% of the population of the country will suffer a "total shortage of water" by 2050. The western, northwestern, and southern states of the country suffer from absolute water scarcity (Bouissou 2013; Johari 2013). In the northern region of Gujarat, the aquifer levels are falling at the rate of six meters per year. In the state of Maharashtra in western India alone, there are two million wells; this is more than double the number found in 1985. State governor Prithviraj Chavan claimed in March 2013 that "in recorded history the reservoirs have never been so low in central Maharashtra" (O'Brien 2013).

3.3.2 China

"China is facing two prominent challenges: water shortages and pollution," affirms Ma Jun, Director of the Institute of Public and Environmental Affairs and author of *China's Water Crisis* (1999), the first book to warn about the issue. About 60% of

---

[7] See also "Study: Third of Big Groundwater Basins in Distress." NASA, Jet Propulsion Laboratory. California Institute of Technology, 16/VI/2015.

the country's aquifers are polluted, according to a survey by the Ministry of Land and Resources. Tests in 4778 points of these aquifers in 203 cities show that the water quality of 44% of them is "relatively poor," which means that it only becomes drinkable after treatment and that 15.7% is "very poor," which means it is no longer drinkable. Only 3% of urban aquifers can be classified as "clean" in a country where one-third of the water resources come from aquifers (Kaiman 2014). Beijing's case is well-known. The Probe International Beijing Group Report titled *Beijing's Water Crisis, 1949–2008 Olympics* (June 2008), drafted by a group of Chinese experts who, for security reasons, remain anonymous, states[8]:

> Beijing […] is running out of water. Although more than 200 rivers and streams can still be found on official maps of Beijing, the sad reality is that little or no water flows there anymore. […] Dozens of reservoirs built since the 1950s have dried up. Finding a clean source of water anywhere in the city has become impossible. As recently as 30 years ago, Beijing residents regarded groundwater as an inexhaustible resource. Now hydrogeologists warn it too is running out. Beijing's groundwater table is dropping, water is being pumped out faster than it can be replenished, and more and more groundwater is becoming polluted. Today, more than two-thirds of the municipality's total water supply comes from groundwater. The rest is surface water coming from Beijing's dwindling reservoirs and rivers. The municipality's two largest reservoirs, Miyun and Guanting, now hold less than ten percent of their original storage capacity and Guanting is so polluted it hasn't been used as a drinking water source since 1997.

He Qingcheng affirms that in order to supply Beijing with water, it is necessary to resort to fossil aquifers 1000 m below the surface, a depth five times greater than 25 years ago. A study commissioned by the Ministry of Land and Resources concluded that "the North China Plain suffers from severe groundwater pollution with over 70% of overall groundwater classified as Grade IV+, in other words, unfit for human touch" (China Water Risk 2013). Today, thanks to the South-North Water Transfer Project, water travels from Southern China in a 1432-km canal to get to Beijing. This additional water provides about two-thirds of the city's drinking water. This is, of course, only a short-term solution, as water sources from the south are also shrinking (Carney 2018).

3.3.3   United States

There is currently a trend of drilling ever-deeper for groundwater in the USA, according to an analysis of 11.8 million groundwater-well locations, depths, and purposes across the country (Perrone and Jasechko 2019). The authors conclude that "widespread deeper well drilling represents an unsustainable stopgap to groundwater depletion that is limited by socioeconomic conditions, hydrogeology and groundwater quality." A rise in agricultural demand, combined with recent and last-

---

[8] The report was funded by the Open Society Institute, edited by Dai Qing and translated from Chinese by Probe International of Canada, a division of the Energy Probe Research Foundation.

ing droughts due to climate change, has accelerated a decline in the level of the High Plains aquifer, as pumping was increased to compensate for lack of rain. Michon Scott (2019) sums up the past and present situation of this region admirably well:

> In the early twentieth century, farmers converted large stretches of the Great Plains from grassland to cropland. Drought and stress on the soils led to the 1930s Dust Bowl. Better soil conservation and irrigation techniques tamed the dust and boosted the regional economy. (…) However, well outputs in the central and southern parts of the aquifer are declining due to excessive pumping, and prolonged droughts have parched the area, bringing back Dust Bowl-style storms, according to the Fourth National Climate Assessment (NCA4, 2018).

In October 2015, the Central Valley aquifer in California was at its lowest historical level. With its more than 450 thousand km^2 stretching through eight states of the country, the shallow Ogallala aquifer provides water for 170,000 wells that irrigate about 1.3 million km^2 of farms. The Ogallala has been dropping by an average of six feet per year. Meanwhile, the annual recharge rate is thought to be around half an inch, but no more than one inch. Data from 2012 showed that 30% of the groundwater had already been pumped (Steward et al. 2013). According to Leonard Konokow (2013), the decline between 2001 and 2008 accounts for 32% of the decline accumulated throughout the twentieth century. Bridget Scanlon and colleagues (2012) state that:

> Extrapolation of the current depletion rate suggests that 35% of the southern High Plains will be unable to support irrigation within the next 30 years. Reducing irrigation withdrawals could extend the lifespan of the aquifer but would not result in sustainable management of this fossil groundwater.

Furthermore, because of its shallow depth, the High Plains contains high concentrations of sodium, nitrates, and triazine herbicides, such as atrazine, one of the most common pesticides in the world. Atrazine has been banned in Europe since 2004 due to its environmental impacts and the fact that it is in an endocrine disruptor in invertebrates (Vogel et al. 2015) and vertebrates, especially in amphibians. This was demonstrated by Tyrone Hayes who became victim of a smear campaign, orchestrated by Syngenta, to discredit him (Hayes et al. 2010). Warming temperatures represent an important complementary factor in the decline of US shallow aquifers. Laura Condon, Adam Atchley, and Reed Maxwell (2020) showed that in the USA, even a moderate warming of 1.5 °C above the pre-industrial period "can shift groundwater surface water exchanges and lead to substantial and persistent storage losses."

3.3.4 Middle East

In 2002, two-thirds of the 1.6 trillion liters of water destined for Saudi Arabia's agriculture came from fossil aquifers. These aquifers are not only running out (which led to a halving of the country's wheat crop in 2002), but the water left behind is increasingly saline and must be filtered to remove its metals even before use in agriculture, with costs that exceed, in some cases, those of producing the same amount of oil (Smith 2003).

A quantity of freshwater equivalent to that of the Black Sea was lost in several parts of the Middle East in underground areas of Turkey, Syria, Iraq, and Iran, along the Tigris and the Euphrates rivers. Between 2003 and 2009, these reservoirs lost 144 km^3, the second largest loss of aquifers after India. About 60% of the total loss is due to the pumping of these underground reservoirs for irrigation, including for use in 1000 wells in Iraq, and 20% is due to the prolonged impact of the 2007 drought, a decline in glaciers, and desertification. The remaining 20% is attributed to a reduction in surface waters (rivers, lakes, and dams).[9]

3.4 More Severe and Widespread Droughts

"Over the past decades, drought episodes have become more widespread and prolonged in many parts of the world, with increased socio-economic and environmental impacts" (FAO 2017). The Palmer Drought Severity Index (PDSI), created by Wayne Palmer (1965), measures the cumulative departures, relative to local average conditions, in atmospheric moisture at the surface. "The global very dry areas (PDSI < −3.0) have more than doubled since the 1970s" (Dai et al. 2004), and drought losses have significantly increased in recent years around the world (Su et al. 2018). According to Aiguo Dai (2011), "climate models project increased aridity in the 21st century over most of Africa, southern Europe, and the Middle East, most of the Americas, Australia, and Southeast Asia." Dai's projections for the decades after 2030 point to a PDSI of −4 to −6 for many zones of the globe, reaching −8 in some areas of the Mediterranean. These projections began dramatically to be confirmed in 2015, as seen by a report of the United Nations International Strategy for Disaster Reduction (UNISDR). Robert Glasser, head of the UNISDR, declared in 2016: "the most obvious impact was the 32 major droughts recorded which was more than double the ten-year annual average. These affected 50.5 million people and many have continued into this year particularly in Africa." More recently, Benjamin Cook (2018) confirmed that "the most recent research shows climate change is already making many parts of the world drier and droughts are likely to pack more punch as the climate warms further." According to Cook, the relationship between warming and bigger droughts already seems undeniable, at least in the Pacific Northwest, the Western USA, and the Mediterranean. David Wallace-Wells (2019) reports projections that "at 2°C of warming, droughts will wallop the Mediterranean and much of India, and corn and sorghum all around the world will suffer, straining food supply." Based on estimates from two IPCC scenarios (RCP4.5 and RCP8.5 W/m^2), Chang-Eui Park et al. (2018) project that "substantial aridification occurs over 42% and 49% of the total land surface under the RCP4.5 and RCP8.5 scenarios, respectively, by 2100." Let us briefly describe how this global problem manifests itself in some particularly vulnerable regions of the planet.

---

[9] See NASA's Gravity Recovery and Climate Experiment. *Water Resources Research. American Geophysical Union*, 15/II/2013.

3.4.1   The Amazon Region

From 2012 to 2017, the semi-arid region of northeastern Brazil experienced the longest of a historical series of droughts that began in 1845. The Amazon region, once extremely humid, is no longer free of droughts; in its southeastern region, some rivers have been subject to drying up completely since, at least, 2009 (Marengo et al. 2009). There is a strong synergy between droughts and tree cover loss, which is why this issue was briefly addressed in the previous chapter (section 2.4 Tipping Point: Forest Dieback). Largely a result of deforestation and climate change, droughts in the Amazon are now not only more intense but also much more frequent. In fact, the Amazon suffered droughts in 1997, 2005–2006, 2010, and 2016, with three "100-year" droughts happening in the space of just 10 years (2005–2015). Several studies have described the increase in droughts in the Amazon region and the consequent greater vulnerability of the forest.[10] Ten years ago, José Marengo et al. (2009) had already made the following prediction:

> […] an increase in dry (or drought) periods in the eastern Amazon and part of the northeast, while the number of consecutive days with high humidity will decrease in most of the northeastern and midwestern regions of Brazil, as well as in the west and south of the Amazon. […] By 2030, the dominant pattern will be a reduction in the total amount of rainfall and in the number of wet days in tropical South America.

According to Simon Lewis et al. (2011), while in 2005 the drought reached 37% of the forest, in 2010 it affected 57% of it. In a 2014 interview granted to the newspaper *Le Monde*, Antônio Donato Nobre once again drew attention to the fact that "the system is in the process of becoming deregulated." This phenomenon was stressed by Thomas Hilker et al. (2014) in the same year.

Since the year 2000, the Amazon forest has declined across an area of 5.4 million km² as a result of well-described reductions in rainfall. […] If drying continues across Amazonia, which is predicted by several global climate models, this drying may accelerate global climate change through associated feedbacks in carbon and hydrological cycles.

It was noted that the 2016 drought (which was stronger than the droughts of 2005 and 2010) cannot be explained solely by the El Niño effect and that deforestation also probably played a role in it (Erfanian et al. 2017):

> Warmer-than-usual SSTs [surface sea temperatures] in the Tropical Pacific (including El Niño events) and Atlantic were the main drivers of extreme droughts in South America, but are unable to explain the severity of the 2016 observed rainfall deficits for a substantial portion of the Amazonia and Nordeste [Northeastern] regions. This strongly suggests potential contribution of non-oceanic factors (e.g., land cover change and CO_2-induced warming) to the 2016 drought.

---

[10] Only the most important references are given here: Nepstad (1999), Salati (2001), Fearnside (2005), Levine et al. (4/II/2016), Vergara and Scholz (2011), Hilker et al. (2014), A.D. Nobre (2014), Steege (2015), C. Nobre (2016), Zemp et al. (2017), Lovejoy and C. Nobre (21/II/2018).

As stated above, the combination of deforestation, climate change, and drought—three accelerating phenomena—is putting the Amazon rainforest on the threshold of a tipping point that will trigger an accelerated transition toward a non-forest biome, with likely catastrophic consequences for the whole world. In the previous chapter, we have cited at length the editorial of *Science Advances*, signed by Thomas Lovejoy and Carlos Nobre (2018), and appropriately titled "Amazon Tipping Point." Also, Clara Delphine Zemp et al. (2017) warned that "reduced rainfall increases the risk of forest dieback, while in return forest loss might intensify regional droughts (…). The risk of self-amplified Amazon forest loss increases nonlinearly with dry-season intensification."

3.4.2   South Europe and the Mediterranean Basin

In southern Europe, the rate of warming has been higher than the average world rate. With regard to the regions of the Tagus, Douro, and Guadiana river basins, covering 40% of the Iberian Peninsula, all 15 climate models analyzed by Guerreiro, Kilsby, and Fowler (2017) project:

> […] an intensification of drought conditions for the three basins. (…) Some project small increases in drought conditions but most project multi-year droughts reaching up to a Drought Severity Index -12 (DSI-12) of 800% (i.e. 8 years of mean annual rainfall missing), and an average of 80% of each basin area experiencing extreme drought, by the end of the century.

In the Mediterranean Basin, the frequency and intensity of droughts have increased since 1950 (Cramer et al. 2018). In 2016, Benjamin Cook and colleagues estimated that "the recent 15-year drought in the Levant (1998–2012) is the driest ever recorded (…). There is an 89% likelihood that this drought is drier than any comparable period of the last 900 years and a 98% likelihood that it is drier than the last 500 years."

3.4.3   China, Central Asia, and Iran

In 2007, China's southeast region experienced a drought that hurt more than 13 million hectares of farmland and that extended to Vietnam. In 2009 and 2011, the country experienced the worst droughts since 1950. And in 2014 and 2015, and even more in the 2017–2018 winter, several provinces in the north of China suffered from droughts that were among the worst on record (Wang et al. 2019). Since the late twentieth century, droughts in these provinces have become more serious and frequent, and the areas affected by them have been increasing at a rate of 3.72% a decade for the past 50 years. This has particularly serious consequences for China because most of the wheat and corn production in the country comes from these northern provinces which are home to nearly half of its population. Scientists attribute these extreme weather patterns, especially in northern China, to climate change (Wong 2017). "Most projections agree that the warming rate of China will be faster than the global mean and the country might be seriously threatened by global warming-induced disasters" (Su et al. 2018). According to Buda Su et al. (2018), droughts affected about a sixth of China's arable land from 1949 to 2017 (2,090,000 km² per year), with corn and wheat among the worst-hit crops. Limiting the temperature to a 1.5 °C increase (compared to a 2 °C increase) can considerably reduce the drought-related losses in China: a global warming of 1.5 °C could cause economic losses of US\$ 47 billion annually, while an average global warming of 2 °C, which is probably already inevitable, could almost double those losses (US\$ 84 billion/year).

Coupled with climate change and growing anthropogenic pressures, drought is becoming a significant threat to Central Asia's water security. As stressed by FAO (2017), in this region "climate change predictions indicate a likely increase in the frequency and intensity of drought (IPCC 2014)." A distinguishing characteristic of Central Asia is that the majority of precipitation falls as snow. In countries such as Pakistan, Afghanistan, Tajikistan, Turkmenistan, Uzbekistan, and Kyrgyzstan, Himalayan glaciers still serve as hydrological buffers, given the high interannual rainfall variability typical of these countries. But for more than half of the region, water stress is already classed as medium to extremely high and is increasing. According to Hamish Pritchard (2019):

> This is particularly the case in semi-arid Pakistan, which has the most water-intensive economy (water use per unit gross domestic product (GDP)) of any country, a population that is growing by more than three million annually, a power demand that is increasing by 8% annually, and a per-capita water supply that has fallen almost to the scarcity threshold (1000 m³ year⁻¹, below which normal demand can no longer be satisfied), with demand expected to exceed supply by 44% (103 km³) by 2025.

In February 1, 2019, the newspaper *Tehran Times* reported that, based on data collected over the past decade, approximately, 97% of Iran's territory is affected by long-term drought: "some 12.7% of the country is hit by extremely severe drought, 53.6% of the country is affected by severe drought while 24.8% of the country is facing moderate drought and 5.9 percent of the country is withstanding mild drought." A 2019 report released by the Iranian parliament predicts that by 2022 "more than 50 million Iranians [total population 83 million people] will be affected by water shortages for drinking purposes." The country's future is looking bleak. "Iran will experience an increase of 2.6°C in mean temperatures and a 35% decline in precipitation in the next decades" (Daneshvar et al. 2019).

3.4.4   Australia and Africa

Since 2004, Australia has been experiencing the worst droughts of the last 117 years (when records began), with increasing incidence and severity of fires. A 2011 study predicts that, by the year 2050, the days where the risk of fire in the country is "very high" or "extremely high" will increase by 70% (Slezak 2013). The drought of 2019 in Australia is being regarded as the worst in its modern history. Furthermore, heat waves and bush fires in 2019 broke records across the country. The re-election of Prime Minister Scott Morrison (2019), a politician advocating for increased coal production, shows a lack of awareness among the Australian electorate of the causal link between rising greenhouse gas emissions and more severe droughts and bush fires. Denying scientific evidence, Australian voters have clearly opted for still more severe heat waves, more desertification, and the irreversible demise of the Great Barrier Reef.

Much of Africa is characterized by a semi-arid climate, with extreme annual variations in rainfall. Despite this great variability, a significant downward trend in rainfall was observed from 1900 to 2009 across the Sahel (Thomson and Mantilla 2013). Not only in the Sahel, but in Kenya and in the countries of the Horn of Africa (Somalia, Ethiopia, Djibouti, and Eritrea), the worst droughts in the last 60 years, occurring in 2011 and 2012, have threatened 130 million people and reduced eight million to hunger. Ethiopia, in particular, faced a drought in 2016 that is considered to be equivalent to or worse than the drought of the 1980s, which, in combination with the civil war, killed about one million people.

Mmatlou Kalaba (2019) analyzes how, in recent years, droughts have become more commonplace in the south of the African continent, especially in South Africa and Namibia. In the southern and western regions of South Africa, "in the past two decades since 1990, 12 of those years were defined as drier years, compared to only seven years in the previous 20 years." In February 2018, the drought was reclassified as a "national disaster." Between 2015 and 2017, the Cape Town region experienced the lowest rainfall years on record. Like São Paulo in southeastern Brazil during the summer of 2014–2015, the city of Cape Town was forced in the summer of 2017–2018 to set severe water restrictions in order to avoid "Day Zero," as dam levels dropped to below 20% (below 10% in the case of São Paulo). Mark New, Friederike Otto, and Piotr Wolski (2018) calculated that the risk of drought in this region of South Africa "has increased by a factor of just over three" because of global warming. Namibia too has been afflicted by a succession of droughts since 2013, and in 2019 its government declared a state of emergency, the third in 3 years. In 2018 and in the first semester of 2019, nearly 130,000 domestic animals died, and 500,000 people were left without access to enough food.

3.4.5 Southwest and Central Plains of the USA

In the 1930s, the Dust Bowl, which decimated 250,000 square miles of land in the United States, served as inspiration for the famous *Deserts on the March* (1935) by Paul Sears, *The Grapes of Wrath* (1939) by John Steinbeck, and the movie by the same name by John Ford (1940). The drought that has hit southwestern USA for years has been called the *New Dust Bowl,*[11] and the pumping of the aquifers can no longer mask the "hydric bubble" which had allowed for intensive irrigation. According to John Laird, secretary of California's Natural Resources Agency, the agribusiness sector in the state lost 160,000 hectares of agricultural land and laid off 17,000 workers.[12] In February 2016, the US Drought Monitor reported that 99.5% of California was abnormally dry and 38% of the state was facing an exceptionally severe drought.

Most of the states west of the Mississippi River have been hit by a megadrought which, according to a survey by the American Geophysical Union (Magil 2013), has no parallel in the country's history. Eleven years between 2000 and 2014 have been drought years in parts of Arizona, California, Nevada, New Mexico, Oklahoma, and Texas (Goldenberg 2015). According to Benjamin Cook, Toby Ault, and Jason Smerdon (2015):

> In the Southwest and Central Plains of Western North America, climate change is expected to increase drought severity in the coming decades. [...] Our results point to a remarkably drier future that falls far outside the contemporary experience of natural and human systems in Western North America, conditions that may present a substantial challenge to adaptation. [...] Notably future drought risk will likely exceed even the driest centuries of the Medieval Climate Anomaly (1100–1300 CE) in both moderate (RCP 4.5) and high (RCP 8.5) future emissions scenarios, leading to unprecedented drought conditions during the last millennium.

3.5 Soil Degradation and Desertification

"Twelve million hectares of productive land become barren every year due to desertification and drought alone" (UNCCD report *Desertification. The Invisible frontline* 2014). Given this critical situation, the United Nations Food and Agriculture Organization (FAO) decreed 2015 the International Year of Soils, an appropriate occasion for the launch of the *Status of the World's Soil Resources* (SWSR), a monumental report produced with the support of FAO and announced as "the first major global assessment ever on soils and related issues." Its message is, in short, encapsulated in the statement of FAO director-general, José Graziano da Silva:

> Today, 33 percent of land is moderately to highly degraded due to the erosion, salinization, compaction, acidification and chemical pollution of soils. Further loss of productive soils would severely damage food production and food security, amplify food-price volatility, and potentially plunge millions of people into hunger and poverty.

If current rates of soil degradation persist, "the global amount of arable and productive land per person in 2050 will be only a quarter of the level in 1960," said Maria-Helena Semedo, Deputy Director of FAO (Arsenault 2014). According to FAO data,

---

[11] The term has been widely employed. See, for example, "The New Dust Bowl. High Plains Aquifer Pumped Dry". *Daily Kos,* 20/V/2013; Bryan Walsh, "Rising Temperatures and Drought Create Fears of a New Dust Bowl". *Times*, 5/VII/2012.

[12] Quoted by Chris Megerian, "California faces 'Dust Bowl'-like conditions amid drought, says climate tracker." *The Los Angeles Times*, 9/IV/2015.

reported in a *Nature* editorial (22/I/2015), "an area of soil the size of Costa Rica [51,000 km^2] is lost every year to factors such as erosion, compaction and saliniza-tion. (…) When soil becomes sicker, so too do the people who rely on it. Contamination soars and crop yields and human health decline." Each year, an esti-mated 24 billion tons of fertile soil are lost due to erosion (Weigelt 2015). The unsustainability of land use is irreversible if we maintain the current paradigm of our food system, one that is conditioned by agribusiness. This is because, according to Jes Weigelt, it takes an average of 500 years for two centimeters of fertile soil to be formed. Soil erosion from agricultural fields is estimated to be currently 10–20 times (no tillage) to more than 100 times (conventional tillage) higher than the soil formation rate (IPCC 2019). David Wallace-Wells (2019) reports that in the USA "the rate of erosion is ten times as high as the natural replenishment rate; in China and India, it is thirty to forty times as fast."

Pedodiversity (the variety of soils in a region) is collapsing at a rate comparable to that of biodiversity, especially in dryland areas (Howgego 2015). Drylands cover one-third of the world's land area and account for 44% of cultivated areas, two-thirds of which are occupied by livestock. About 73% of pastureland is being degraded. Table 3.2 reports data from the *2010–2020 UN Decade for Deserts and the Fight against Desertification* (resolution 62/195). Deserts and drylands occu-pied 41.3% of lands above sea level in 2010, according to the following proportions[13]:

Table 3.2 Deserts and drylands in area and percentage of lands above sea level in 2010

	Area (millions km^2)	(%)
Hyper-arid (desert)	9.8	6.6
Arid	15.7	10.6
Semi-arid	22.6	15.2
Dry sub-humid	12.8	8.7
Total	**60.9**	**41.3**

Source: 2010–2020 UN Decade for Deserts and the Fight against Desertification

The document adds that, globally, 24% of the land is being degraded, in the fol-lowing proportions: cropland 20%, rangeland 20–25%, and forests 42%. This is occurring due to the causes pointed out in Fig. 3.1.

---

[13] The data reported here are slightly different from that published by Qi Feng et al. (2015).

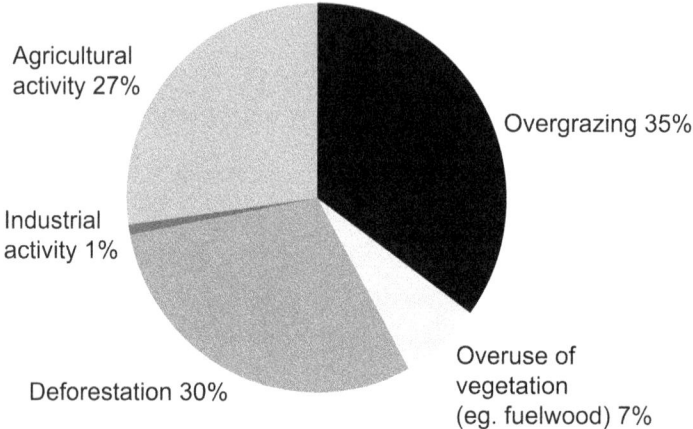

Fig. 3.1 Main causes of soil degradation on a global scale (%). (Source: FAO/UNEP, *Food Energy and Water Resources* (http://www.fewresources.org/))

3.5.1 Desertification

Desertification is the degradation of soils in dryland areas, resulting from several factors, including climatic variations and human activity, especially deforestation (Kirby and Landmark 2011).[14] In 2015, about 500 (380–620) million people lived within areas which experienced desertification between the 1980s and 2000s (IPCC 2019). According to the UN document cited above, "the livelihoods of more than one billion people in 100 countries are threatened by desertification." The *Millennium Ecosystem Assessment* warns that desertification is "potentially the most threatening ecosystem change in terms of its impact on the livelihoods of the poor." With regard to the next three decades, a press release titled "Sustainability. Stability. Security," published in 2016 by the UN Convention to Combat Desertification (UNCCD), reminds us that more than 90% of Africa's economy depends on a climate-sensitive natural resource like rain-fed, subsistence agriculture and issues a very clear warning: "Unless we change the way we manage our land, in the next 30 years we may leave a billion or more vulnerable poor people with little choice but to fight or flee." The phenomenon does not spare Europe and the USA. The UNCCD report *Desertification: The Invisible Frontline* (2014) states that "169 of the 194 parties declared that they were affected by desertification."

---

[14] See also *Report of the United Nations Conference on Environment and Development*. Chap. 12 – *Managing Fragile Ecosystems: Combating Desertification and Drought*. Rio de Janeiro, 1992.

3.5.2 Latin America

It is estimated that two million km^2 in Latin America have already been degraded by overgrazing and intensive agriculture. In Brazil, and particularly in the Cerrado, responsible for 55% of the country's meat production, 320 thousand km^2 of pastures (60% of the 530 thousand km^2 of the Cerrado pasture) have already been degraded. The semi-arid region of Brazil, covering almost one million square kilometers, is considered one of the largest areas in the world that is susceptible to desertification. According to the *Atlas of Areas Susceptible to Desertification in Brazil* (Santana 2007), the ongoing desertification process in the semi-arid and adjacent areas affects the nine Northeastern states, as well as the north of Minas Gerais and Espírito Santo. The *Atlas* states that, out of a total area of 1340,000 km^2, "180 thousand km^2 are already in a situation of serious or very serious desertification, concentrated primarily in the states of the Northeast, where 55.25% of the territory is affected by different degrees of environmental deterioration." In 2013 in the northeast of the country, 230,000 km^2 of land (against 180 thousand km^2 in 2007) was severely or very severely affected by desertification. Areas where soil and vegetation no longer respond to rainfall have increased. Under normal conditions, it takes between 11 and 15 days after rainfall for the Caatinga vegetation to sprout. But, in these areas, the vegetation does not respond; it does not grow, no matter how much it rains (Carvalho 2013).

3.5.3 China and Mongolia

By 2014, China's deserts accounted for a total of 2.6 million km^2, approximately 28% of the country's surface area (Cao et al. 2018). According to the China Ministry of Water Resources, reported by Fengshi Wu and Richard Edmonds (2017):

> Close to one-third (31.12%) of China's total land area is affected by soil erosion. (…) Arid (less than 250 mm of precipitation per annum) and semi-arid (between 250 mm and 500 mm per annum) lands make up approximately another one-third of China's territory. And, over 40% of the arid, semi-arid, and sub-humid tillage is in the process of desertification.

Sand and dust storms have always been part of China's history, with records of this dating back to 300 B.C. In this long history, the aggravation of the problem is known to have happened during five periods, the last one occurring between 1820 and 1890. But throughout the second half of the twentieth century, sandstorms in the north of the country have increased in number, intensity, size of the area affected, and duration. In recent years, they have been occurring earlier and increasing in frequency, duration, and intensity. Serious storms of this type increased from 5 in the 1950s, to 8 in the 1960s, then to 13 in the 1970s, to 14 in the 1980s, and to 23 in the 1990s. Beijing has been hit by unusually strong storms. On average, 5 to 6 sand and dust storms choke Beijing each year. But this rate is increasing, with ten storms hitting Beijing in 2006. In addition to this, global warming produces milder, drier winters, allowing the spring winds to carry more soil (Hays 2014). Lester Brown, Janet Larsen, and Bernie Fischlowitz-Roberts (2002) point out that:

> The process of desertification itself directly affects 40% of China's landmass, including Sinkiang Province and Tibet in the far west and Qinghai, Gansu, Ningxia, and Inner Mongolia Provinces in the north-central region. Although desertification is concentrated in these six provinces, it is now spreading into Sichuan, Shaanxi, Shanxi, and Hebei Provinces as well.

Under the Three-North Shelter Forest Program (also called the Great Green Wall) and the Combating of Desertification Program, China has planted more than 66 billion trees across 13 provinces, mostly in its north, since 1978. The results seem very positive, not least because "far from just planting trees in arid areas, China's re-vegetation codes vary for different regions and greening programmes" (Shu and Xu 2019). In any case, in 2000, deserts across the country were expanding by 10,400 km^2 a year; in 2017, they were shrinking by more than 2400 km^2 a year, according to China's State Forestry Administration. Notwithstanding the above, some scientists doubt that China will eventually win the war against desertification. Mark Zastrow (2019) reports that:

> Large parts of China—including some areas where trees are being planted—are becoming drier. (…) A study published online in July found that semi-arid areas in the country grew by 33% between 1994 and 2008 compared to between 1948 and 1962. Another paper, co-authored by Sternberg, found that arid areas in China had increased by roughly 1.6 million km^2, about the size of Iran, since 1980—probably due in large part to anthropogenic climate change.

In fact, since 1975, China's deserts have expanded by 54,000 km^2 (Haner et al. 2016). As Qi Feng et al. (2015) show, "the total desertified area has decreased in many areas, but in others, desertification has continued to expand," mainly because afforestation in drylands requires irrigation to ensure tree survival. In the northwest of the country, the Taklamakan and Kumtag deserts are merging. Also, there is a tendency for the Badain Jaran (49,000 km^2) and Tengger (36,700 km^2) deserts to come together. Today, the Gobi Desert covers almost 1,300,000 km^2. Since 1950, an estimated 24,000 villages in northwestern China, as well as thousands of kilometers of roads, have been totally or partially covered by sand.

Mongolia also suffers from desertification. Between 2002 and 2012, the soil in its southern region, a transition area between the steppes and the Gobi Desert, lost 40% of its biomass, while the country as a whole (1.6 million km^2) lost 12% of its soil biomass. No less than 70% of its prairies are considered degraded, either because the soil was covered by sand or because it was depleted by overgrazing. In fact, 80% of the loss of vegetation in that decade (2002–2012) is due to the huge increase (from 26 million heads in 1990 to 45 million in 2012) of cattle, sheep, goats, and yaks (Hilker et al. 2013). With the end of the Soviet Union, Mongolia became a major exporter of wool, at the price of rapid desertification. Now, added to this phenomenon, is further devastation caused by mining, as the country holds coal reserves valued at seven billion tons, in addition to copper, gold, and uranium reserves.

3.5.4 Soil Impoverishment by Industrial Agriculture

The problems of desertification and soil loss are compounded by the impoverishment and industrial pollution of soils. Most of the recycling of nutrients that maintain the biosphere takes place in soils. The diversity of microbial biomass is crucial for the process of the formation and balance of ecosystems. The illusion that agricultural productivity and profitability can be maximized indefinitely through monoculture, the systematic use of industrial pesticides and fertilizers, and the transformation of food into commodities is leading to a vicious cycle of diminishing microbial diversity in the soil and its consequent impoverishment. As stated by Fernando Dini Andreote (as cited in Freire 2014):

> The greatest source of genetic and metabolic biodiversity on the planet is found in soil and plants: about 1 billion live cells for each gram of soil; there are 30 thousand different species. [...] The plant selects the microorganisms that will associate with it. If the biodiversity of the environment is high, the selection is more efficient. If this diversity is reduced, there is a greater chance of colonization by opportunistic organisms, or pathogens; this explains the greater occurrence of root diseases in areas of monoculture, since biodiversity is restricted.

Also, use of agrochemicals reduces microbial diversity in the soil. As will be seen in the next chapter, monoculture and the progressive increase in pesticide doses per hectare have led to a loss of soil fertility in many countries, caused, among other things, by the disappearance of the microorganisms that revitalize it.

References

ABI-HABIB, Maria & KUMAR, Hari, "Deadly Tensions Rise as India's Water Supply Runs Dangerously Low". *The New York Times*, 17/VI/2018.

ADB (2013), Asian Development Bank, Asian Development Outlook 2013 Asia's Energy Challenge, 2013.

ARSENAULT, Chris, "Only 60 Years of Farming Left If Soil Degradation Continues". *Scientific American*, 5/XII/2014.

ASMAL, Kader, "Globalization from below". Preface to *Dams and Development. A new framework for decision-making*, World Commission on Dams, New York, 2000.

BISWAS, A., TORJATADA, C. & SAKLANI, U., "India is facing its worst water crisis in generations. *Quartz*, 15/III/2017.

BORRE, Lisa, "Warming Lakes: Climate Change and Variability Drive Low Water Levels on the Great Lakes". *National Geographic*, 20/XI/2012.

BOUISSOU, Julien, "L'ouest de l'Inde fait à sa plus grave sécheresse depuis plus de 40 ans". *Le Monde*, 15/III/2013.

BROWN, Lester, *Outgrowing the Earth: The Food Security Challenge in an Age of Falling Water Tables and Rising Temperatures*. Washington, Earth Policy Institute; New York, W. W. Norton & Company, 2004.

———. "Aquifer Depletion". *Encyclopedia of Earth* (2010). Ed. Cutler J. Cleveland, Washington, 2013.

BROWN, Lester, LARSEN, Janet & FISCHLOWITZ-ROBERT, Bernie, *The Earth Policy Reader*, 2002.

BROWN, Oli & VIVEKANANDA, Janani, "Lake Chad shrinking? It's a story that masks serious failures of governance". *The Guardian*, 22/X/2019.

BURKE, Jason, "Half of India's rivers are polluted, says government report". *The Guardian*, 7/IV/2015.

CAO, Hui et al., "Characterizing Sand and Dust Storms (SDS) Intensity in China Based on Meteorological Data". *Sustainability*, 9/VII/2018.

CARNEY, Matthew, "Forget geopolitics, water scarcity shapes up as the biggest threat to China's rise". *ABC* (Australia), 26/XI/2018.

CARRINGTON, Damian, "Four billion people face severe water scarcity, new research finds". *The Guardian*, 12/II/2016.

CARVALHO, Cleide, "Desertificação já atinge uma área de 230 mil km² no Nordeste." *The globe*, 9/VII/2013.

CATLEY-CARLSON, Margareth, "Water supply: The emptying well". *Nature*, 542, 23/II/2017, pp. 412–413.

CHEN, Jiaqi et al., "External Groundwater Alleviates the Degradation of Closed Lakes in Semi-Arid Regions of China". *Remote Sensing*, 20/XII/2019.

China Water Risk (CWR), North China Plain Groundwater: >70% Unfit for Human Touch, 26/II/2013.

COE, Michael T. & FOLEY, Jonathan A., "Human and Natural Impacts on the water resources of the Lake Chad basin". *Journal of Geophysical Research*, 21/IX/2012.

COGHLAN, Andy, "Global water crisis looms larger". *New Scientist*, 28/VIII/2006.

CONDON, Laura, ATCHLEY, Adam & MAXWELL, Reed, "Evapotranspiration depletes groundwater under warming over the contiguous United States". *Nature Communications*, 11, 873, 13/II/2020.

COOK, Benjamin I., AULT, Toby R. & SMERDON, Jason, "Unprecedented 21st century drought risk in the American Southwest and Central Plains". *Science Advances*, 1/II/2015.

COOK, Benjamin et al., "Spatiotemporal drought variability in the Mediterranean over the last 900 years". *Journal of Geophysical Research*, 4/III/2016.

COOK, Benjamin, "Climate change is already making droughts worse". *Carbon Brief*, 14/V/2018.

CRAMER, Wolfgang et al., "Climate change and interconnected risks to sustainable development in the Mediterranean". *Nature Climate Change*, 8, 22/X/2018, pp. 972–980.

CRIBB, Julian, *Surviving the 21th Century. Humanity's Ten Great Challenges and How We Can Overcome Them*. Springer 2017.

DAI, Aiguo, TRENBERTH, Kevin E., QIAN, Taotao, "A Global Dataset of Palmer Drought Severity Index for 1870–2002: Relationship with Soil Moisture and Effects of Surface Warming". *Journal of Hydrometeorology*, December 2004.

DAI, Aiguo, "Drought under global warming: a review". *WIREs Climate Change*, 2, 2011, pp. 45–65.

DANESHVAR, Mohammad Reza M., EBRAHIMI, Majid & NEJADSOLEYMANI, Hamid, "An overview of climate change in Iran: facts and statistics". *Environmental Systems Research*, 1/IX/2019.

DAVIS, Tony, "Feds hurting Lake Powell to prop up Lake Mead, scientists warn". *Tucson.com*, 5/IX/2018.

DE GRAAF, Inge E.M. et al. "Environmental flow limits to global groundwater pumping". *Nature*, 574, 2019, pp. 90–94.

ERFANIAN, Amir, WANG, Guiling, FOMENKO, Lori, "Unprecedented drought over tropical South America in 2016: significantly under-predicted by tropical SST". *Scientific Reports* 7, 5811, 2017.

FAMIGLIETTI, James S., "The global groundwater crisis". *Nature Climate Change*, 4, 29/X/2014, pp. 945–948.

FAO, *Drought characteristics and management in Central Asia and Turkey*, FAO, Rome, 2017.

FEARNSIDE, Philip. M., "Desmatamento na Amazônia brasileira: História, índices e consequências" (INPA). *Megadiversidade*, I, 1, July 2005, pp. 113–123.

FENG, Qi et al., "What Has Caused Desertification in China?". *Nature Scientific Reports*, 5, 3/XI/2015.

FREIRE, Diogo, "Scientists suggest sustainable solutions to agricultural challenges". *Agência FAPESP*, 26/IX/2014.

FUNABASHI, Yoichi, "Water: China's Achilles heel". *The Japan Times*, 9/IX/2019.

GLEESON, Tom et al., "Water balance of global aquifers revealed by groundwater footprint". *Nature*, 488, 9/VIII/2012, pp. 197–200.

GOLDENBERG, Suzanne, "US faces worst droughts in 1,000 years, predict scientists". *The Guardian*, 12/II/2015.

GRONEWOLD, Drew & ROOD, Richard, "Climate change is driving rapid shifts between high and low water levels on the Great Lakes". *The Conversation*, 4/VI/2019.

GUERREIRO, Selma B., KILSBY, Chris & FOWLER, Hayley J., "Assessing the threat of future megadrought in Iberia". *International Journal of Climatology*, 25/V/2017.

HANER, Josh et al., "Living in China Expanding Deserts". *The New York Times*, 24/X/2016.

HAYES, Tyrone B. et al., "Atrazine induces complete feminization and chemical castration in male African clawed frogs". *PNAS*, 107, 10, 9/III/2010, pp. 4.612–4.617.

HAYS, Jeffrey, "Sand, dust, rain and ice storms in China". Facts and Details, April 2014. <http://factsanddetails.com/china/cat10/sub64/item1149.html>.

HILKER, Thomas et al., "Satellite observed widespread decline in Mongolian grasslands largely due to overgrazing". Global Change Biology, 20, 2, 22/VIII/2013.

HILKER, Thomas et al., "Vegetation dynamics and rainfall sensitivity of the Amazon". *PNAS*, 11, 45, 2014, pp. 16.041–16.046.

HOEKSTRA, Arjen, Y. et al., "Global Monthly Water Scarcity: Blue Water Footprints versus Blue Water Availability". *PLOS One*, 29/II/2012.

HOWGEGO, Joshua, "Save our soils". *New Scientist*, 10/X/2015, pp. 42–45.

IANDOLI, Rafael, "De onde vem a crise hídrica que seca a bacia do rio São Francisco". *Nexo*, 22/X/2017.

IPCC (2014), Fifth Assessment Report – Climate Change 2014.

IPCC (2019), Climate Change and Land. An IPCC Special Report on climate change, desertification, land degradation, sustainable land management, food security, and greenhouse gas fluxes in terrestrial ecosystems. Summary for Policymakers, Approved Draft, 7/VIII/2019.

IVES, Mike, "Melting Glaciers May Worsen Northwest China's Water Woes". *Yale Environment 360*, 26/VII/2012.

JAWAD, Adil, "Pakistan's largest city thirsts for water supply". *Philly.com. The Associated Press*, 24/VIII/2014.

JOHARI, Aarefa, "Maharashtra: State of despair". *Hindustan Times*, 30/III/2013.

KAIMAN, Jonathan, "China says more than half of its groundwater is polluted." *The Guardian*, 23/IV/2014.

KALABA, Mmatlou, "How droughts will affect South Africa's broader economy". *The Conversation*, 6/III/2019.

KIRBY, Alex, "Water scarcity: A looming crisis?" BBC, 19/X/2004.

KIRBY, Alex & LANDMARK, Karen, *Global Drylands*, UNCCD, 2011.

KONOKOW, Leonard, "Groundwater depletion in the United States (1900–2008)". *Scientific Investigations Report*, Virginia, U.S. Geological Survey, 2013.

LE SOURD, Guillaume, RIZZOLIO, Diana, *Global Resource Information Database* (GRID), UNEP, 2004. <http://www.grid.unep.ch/activities/sustainable/balkhash/index.php>.

LEVINE, Naomi M. et al., "Ecosystem heterogeneity determines the ecological resilience of the Amazon to climate change". *PNAS*, 113, 3, 19/I/2016.

LEWIS, Simon L. et al., "The 2010 Amazon Drought". *Science*, 331, 6.017, 4/II/2011, pp. 554–556.

LIANG, Kang Liang & YAN, Guozhen, "Application of Landsat Imagery to Investigate Lake Area Variations and Relict Gull Habitat in Hongjian Lake, Ordos Plateau, China". *Remote Sensing*, 9, 2017.

LOVEJOY, Thomas & NOBRE, Carlos, "Amazon Tipping Point". *Science Advances*, 4, 2, 21/II/2018.

LUSTGARTEN, Abraham, KIRCHNER, Lauren & ZAMORA, Amanda, "California's Drought Is Part of a Much Bigger Water Crisis". *Scientific American*, 26/VI/2015.

McGREAL, Chris, "Disappearing Lake Powell underlies drought crisis facing Colorado River". *The Guardian*, 17/V/2015.

MAGIL, Bobby, "Is the West's Dry Spell Really a Megadrought?" *Climate Central*, 12/XII/2013.

MARENGO, José A. et al., "Mudanças Climáticas e Eventos Extremos no Brasil". Fundação Brasileira para o Desenvolvimento Sustentável (FBDS), 2009.

MARKS, P., "Fly-bys warn of water shortages". *New Scientist,* 15/VI/2013.

MARQUES, Luiz, "Em defesa da Amazônia e do Cerrado". *Jornal da Unicamp*, 28/VIII/2017.

MAURER, Josh M. et al., "Acceleration of ice loss across the Himalayas over the past 40 years". *Science Advances*, 5, 6, 19/VI/2019.

MEKONNEN, Mesfin M. & HOEKSTRA, Arjen Y. "Four billion people facing severe water scarcity". *Science Advances, 2, 2,* 12/II/2016.

MICHEL, Anne, "Au Guatemala, les morts du lac Izabal". *Le Monde*, 21/VI/2019.

MILLY, Paul C.D. & DUNNE, Krista E., "Colorado River flow dwindles as warming-driven loss of reflective snow energizes evaporation". *Science*, 20/II/2020.

NEPSTAD, Daniel et al., "Large-scale impoverishment of Amazonian forests by logging and fire". *Nature*, 398, 1999, pp. 505–508.

NEW, Mark, OTTO, Friederike & WOLSKI, Piotr, "Climate Scientists are warning Cape Town to brace for more severe droughts than it had in 2018". *Quartz Africa*, 19/XII/2018.

NOBRE, Antônio Donato, "O futuro climático da Amazônia". Relatório de Avaliação para a Articulación Regional Amazônica (ARA), 2014.

NOBRE, Carlos A. et al., *Riscos de mudanças climáticas no Brasil e limites à adaptação*, March 2016.

O'BRIEN, Rachel, "Millions of Indians facing worst drought in decades". *Phys.org*, 6/III/2013.

PARK, Chang-Eui et al., "Keeping global warming within 1.5°C constrains emergence of aridification". *Nature Climate Change*, 1/I/2018.

PERRONE, Debra & JASECHKO, Scott, "Deeper well drilling an unsustainable stopgap to groundwater depletion". *Nature Sustainability*, 22/VII/2019.

PRITCHARD, Hamish, "Asia's shrinking glaciers protect large populations from drought stress". *Nature*, 569, 30/V/2019.

PRUDHOMME, Christel et al., "Hydrological droughts in the 21st century, hotspots and uncertainties from a global multimodel ensemble experiment". *PNAS*, 111, 9, 4/III/2014.

RICHEY, Alexandra S. et al., "Uncertainty in Global Groundwater Storage Estimates in a Total Groundwater Stress Framework". *Water Resources Research*, 16/VI/2015a.

———. "Quantifying renewable groundwater stress with GRACE". *Water Resources Research*, 17/VI/2015b.

ROBERTS, Rachel, "Pakistan could face mass droughts by 2025 as water level nears 'absolute scarcity'". *The Independent*, 15/IX/2017.

ROSS, Will, "Lake Chad: Can the vanishing lake be saved?". *BBC*, 31/III/2018.

SALATI, Eneas, "Mudanças climáticas e o ciclo hidrológico na Amazônia". *In*: Fleischresser, V. (ed.). *Causas e dinâmica do desmatamento na Amazônia*. Brasilia, MMA, 2001, p. 153–172.

SANTANA, Marcos Oliveira (ed.), *Atlas das áreas susceptíveis à desertificação do Brasil.* Ministério do Meio Ambiente, Secretaria de Recursos Hídricos, Universidade Federal da Paraíba, 2007.

São Paulo (2014) is not a bibliographical reference. It is the date of a very severe drought in the city of São Paulo, in Brazil.

SCANLON, Bridget R. et al., "Groundwater depletion and sustainability of irrigation in the US High Plains and Central Valley". *PNAS*, 29/V/2012.

SCHEWE, Jacob et al., "Multimodel assessment of water scarcity under climate change". *PNAS*, 111, 93245-250, 4/III/2014.

SCHIERMEIER, Quirin, "Water risk as world warms". *Nature*, 7492, 505, 2/I/2014, pp. 10–11.

SCHNEIDER, Philip & HOOK, Simon J., "Space observations of inland water bodies show rapid surface warming since 1985". *Geophysical Research Letters*, 24/XI/2010.

SCOTT, Michon, "Study shows changes in Great Plains' Ogallala Aquifer". *Weather Nation*, 21/II/2019.

SEAGER, R. et al., "Projections of declining surface-water availability for the southwestern United States". *Nature Climate Change*, 23/XII/2012.

SHU, Lele & XU, Zexuan, "China's different shades of greening". *Nature*, 577, 29, 31/XII/2019.

SIEGLER, Kirk, "Owens Valley Salty as Los Angeles Water Battle Flows into Court". *NPR*, 11/III/2013.

SLEZAK, Michael, "Australian inferno previews fire-prone future". *New Scientist*, 17/I/2013.

SMITH, Craig S., "Saudis Worry as They Waste Their Scarce Water". *The New York Times*, 26/I/2003.

SPRING, Ursula Oswald & COHEN, Ignácio Sanchez, "Water resources in Mexico. A Conceptual Introduction". *In*: SPRING, Ursula Oswald (ed.), *Water Resources in Mexico: Scarcity, Degradation, Stress, Conflicts, Management, and Policy*. Heildelberg, Springer Verlag, 2011.

STEEGE, Hans ter et al., "Hyperdominance in the Amazonian Tree Flora". *Science, 342, 6156*, 18/X/2013.

―――. "Estimating the global conservation status of more than 15,000 Amazonian tree species". *Science Advances*, 1, 10, 20/XI/2015.

STEWARD, David R. et al., "Tapping unsustainable groundwater stores for agricultural production in the High Plains Aquifer of Kansas, projections to 2110". *PNAS*, 26/VIII/2013.

STONE, Richard, "Can Iran and Afghanistan cooperate to bring an oasis back from the dead?". *Science*, 21/II/2018.

SU, Buda et al., "Drought losses in China might double between the 1.5°C and 2.0°C warming", *PNAS*, 16/X/2018.

SUTHERLAND, Scott, "Lake Michigan, Lake Huron now at lowest levels on record". *Geekquinox. Science and Weather,* 6/II/2013.

THOMSON, Madeleine C. & MANTILLA, Gilma C., "EPID: Focus on Surveillance. Integrating Climate Information into Surveillance Systems for Infectious Diseases: New Opportunities for Improved Public Health Outcomes in a Changing Climate". Institute on Science for Global Policy (ISGP), 2013.

TOOR, Amar, "Why did 28,000 rivers in China suddenly disappear?" *The Verge*, 3/IV/2013

UDALL, Bradley & OVERPECK, Jonathan, "The twenty-first century Colorado River drought and the implications for the future". *Water Resources Research*, 17/II/2017.

VERGARA, Walter & SCHOLZ, Sebastian M. (eds.). *Assessment of the Risk of Amazon Dieback. A World Bank Study*, 2011.

VOGEL, Andrea et al., "Effects of atrazine exposure on male reproductive performance in *Drosophila melangaster*". *Journal of Insect Physiology*, 72, January 2015, pp. 14–21.

VÖRÖSMARTY, Charles et al., "Global threats to human water security and river biodiversity". *Nature*, 467, 30/IX/2010, pp. 555–561.

WADA, Y., VAN BEEK, L. P. H., BIERKENS, M. F. P., "Nonsustainable groundwater sustaining irrigation: A global assessment". *Water Resources Research*, 48, 2012.

WALLACE-WELLS, David, *The Unhabitable Earth. Life After Warming.* New York, 2019.

WANG, Lijuan et al., "The 2017–2018 Winter Drought in North China and Its Causes". *Atmosphere*, 10, 2, 60, 2019.

WEIGELT, Jes (coord.), *Soil. The Substance of Transformation*. Global Soil Week 2015. Institute for Advanced Sustainability Studies (IASS). Potsdam, 2015.

WONG, Edward, "Drought in Northern China is Worst on Record". *The New York Times*, 29/VI/2017.

WU, Fengshi & EDMONDS, Richard L., "China's Three-Fold Environmental Degradation". In C. Tubilewicz (ed.), *Critical Issues in Contemporary China*, 2nd edition, Routledge, 2017 [early draft].

WU, Yao et al., "Reviewing the Poyang Lake Hydraulic Project Based on Humans' Changing Cognition of Water Conservancy Projects". *Sustainability*, 2019.

YING, Liu, "Analysis of Hongjiannao Lake Area Based on SMMI". CNKI, 2016.

ZASTROW, Mark, "China's tree-planting drive could falter in a warming world". *Nature*, 573, 23/
    IX/2019.
ZEMP, Delphine Clara et al., "Self-amplified Amazon forest loss due to vegetation-atmosphere
    feedbacks". *Nature Communications* 8, 14681, 13/III/2017.
ZHU, Yongnan *et al.*, "Impacts of Climate Changes on Water Resources in Yellow River Basin".
    Procedia Engineering, 154, 2016, pp. 687–695.

Chapter 4
Waste and Industrial Intoxication

The metabolic residues of living creatures pertain to phases in a process of recomposition of matter and of interaction between the worlds of minerals, plants, and animals. Nature does not produce trash; it produces metamorphoses and nutrients. It is only the secretions of man in the Industrial Age that do not reintegrate into the cycle of recomposition of matter. This is due to their scale, the rate at which they multiply, and the fact that they are, in large part, materials that are chemically more stable.

In an effort to assert our distinctiveness in the chain of life (see Chap. 15), the human species has exalted itself with the claim of exclusivity of our attributes, such as the capacity for symbolization, use of language, self-awareness, tool making, use of clothing, cooking, aesthetic sense, and moral sense. It is superfluous to say that, in terms of cognitive and aesthetic symbolization, man has attained an astonishing complexity and sophistication in the last millennia. But science has recently shown how other species—not just those gifted with a neocortex—share with us, albeit to a much lesser degree, deductive, cognitive, and even aesthetic capacities previously credited exclusively to humans. What would be *qualitatively* exclusive to our species is a type of "historical" angst, that is, our consciousness of an origin and an end: as individuals, as civilizations, and as a species. As Michel Serres (1998) states: "Certainly, we become the men we are by learning – will we ever know how? – that we were going to die." In the twentieth century, with the discovery of the destructive potential of our psychic drives and of the technological means of realizing this potential, this angst began to define us in a more essential way, as indicated by an immense collection of fictional, philosophical, and scientific works on the topic.

Furthermore, with the new hegemony of the chemical and petrochemical industry, the *Homo sapiens* has acquired a peculiar behavior which, in turn, has gradually emerged as one of his foremost attributes: his sweeping appropriation of the ecosystems surrounding him generates, *inevitably,* an increasing amount of waste that has a weak passive interaction with these ecosystems and a strong toxic-active interaction with them. In other words, this human appropriation of the world generates industrial waste at an industrial scale, with toxic discharges at an equally industrial scale. The

© Springer Nature Switzerland AG 2020   97
L. Marques, *Capitalism and Environmental Collapse*,
https://doi.org/10.1007/978-3-030-47527-7_4

proliferation of this kind of waste gradually becomes the most salient and distinctive feature of the human being, especially since the mid-twentieth century. In other words, waste would be the human form of "marking" the whole Earth as our territory, the *mal propre* of our species, in Michel Serres' (2012) untranslatable play on words. As early as 1935, in *Tristes Tropiques*, Lévi-Strauss cast a retrospective look at the relationship between Western civilization and its growing waste generation:

> The great civilization of the West has given birth to many marvels; but at what a cost! As has happened in the case of the most famous of their creations, that atomic pile in which have been built structures of a complexity hitherto unknown, the order and harmony of the West depend upon the elimination of that prodigious quantity of maleficent by-products which now pollutes the earth. What travel has now to show us is the filth, our filth, that we have thrown in the face of humanity.

4.1 Preponderance of Waste

Waste is generated in all stages of the production/consumption cycle and is the preponderant form in each stage of this cycle. There is a regressive preponderance before final consumption: if we take any industrial product and compare it to all the waste generated in its production chain, beginning with the extraction or production of inputs, it will be easy to see that the final output is negligible in relation to all that was discarded. One way to calculate this disproportion was proposed by Joel Makower (2009): 94% of all the waste produced in the USA is industrial waste, including waste produced in the processing industry itself (76%) and waste produced in mining, fuel production, and metallurgy (18%). One can also speak of a chronological preponderance of waste: the useful life of an industrial product or of its components in the hands of a consumer is often tiny in comparison to the decades, centuries, or millennia of its existence as garbage.

4.2 Three Factors for the Increase in Waste

Twentieth-century capitalism has potentiated the predominance of waste in the production/consumption cycle. As pointed out by Serge Latouche (2007), three ingredients are necessary for consumer society to proceed with its diabolical rounds: advertising, credit, and the accelerated programmed obsolescence of products. Let us briefly examine these three factors.

(1) Programmed obsolescence, that is, the deliberate introduction (in the manufacturing process of certain products) of mechanisms or devices that shorten the life span of a product and, thus, accelerate its replacement rate. This question has come up since the first crises of industrial overproduction in the nineteenth century, but its birth as a joint corporate strategy goes back to the Phoebus cartel, celebrated in Geneva in 1924 among big lamp manufactures. They

aimed, among other things, to limit the lifespan of lamps to a thousand hours. Since the 1950s, programmed obsolescence has been the subject of detailed historical analyses (Galbraith 1958/1998; Packard 1962; Slades 2006; Latouche 2012). There are a number of planned obsolescence stratagems used by industries.

(2) Subjective obsolescence, or "neophilia," and even compulsive consumerism.

Starting with Edward Bernays (1891–1995), advertising techniques begin to manipulate people's desires and to program behavior; this has nothing to do with the dynamics of fashion or with changes in taste, which are common to all historical periods. There is a fine line between post-consumer waste and the countless objects already considered pre-consumer waste, the latter consisting of objects that are superfluous and almost always toxic; before polluting the garbage dumps, they pollute the material and mental world of the consumer. Pre-consumer waste functions as a ghost object: it is the pure stimulation of an ephemeral and pointless desire. Every time a new model of an electronic product is launched, the previous model suddenly "grows old," eroding the self-image of its owner who stops identifying himself with last year's object of desire. And this repeats itself continuously. The man of consumer society is comparable to the prisoners of Tartarus: Ixion, the Danaides, Tantalus, and Sisyphus. The discarded object will no longer be part of its owner's experience, except as the unconscious residue of a frustration. It turns out to be what it has always been: a fraction of the gigantic pre-debris that makes up the world of electronic waste, a toxin released gradually by its volatility or immediately by its incineration. The commodity fetishism analyzed by Marx does not cease to exist in contemporary capitalism. But today, it has gained an extra dimension by moving from the sphere of production to the sphere of consumption. While the capitalist production process produces a commodity fetishism, the capitalist consumption process produces its loss, after the act of acquisition. This is an equally "magical" process in which the object—once seemingly alive and endowed with a unique power of seduction and erotic transference—becomes rubbish, not by a loss in functionality, but because of post-purchase dysphoria which results in a divestment of meaning.

(3) The emergence of consumer credit and the change in the concept of credit.

The role of this type of credit in increasing and intensifying consumption (and hence waste) was well explained by Vance Packard (1960, p.139): "Spectacular gains were made in perfecting ways to help—and induce —Americans to buy more and more against future earnings." Regarding the change in the concept of credit, in 2013, Lord Adair Turner, former Director of the Financial Services Authority (the institution that regulates Britain's financial system), stated that only 15% of total UK financial flows are channeled into "investment projects." The remaining financial flows support assets that aim to facilitate the stabilization of the consumption life cycle (Tett 2013). According to Andy Haldane, Executive Director of Financial Stability of the Bank of England, the non-investment credit has exploded: "the size of private credit, relative to GDP, has doubled to 200% in the past 50 years" (as cited by Tett 2013).

4.3   *From* Mundus *to* WALL-E

As a result of these three factors which have grown in proportion throughout the second half of the twentieth century, we can imagine that if Marx were to rewrite *Capital*, he would dedicate a fourth volume to this at once obscure and omnipresent aspect of the "immense accumulation of goods" which is "the immense accumulation of waste." If commodities are, as Marx says, the starting point, the "elementary form" of wealth in capitalist society, waste reveals its degenerate form. Waste is nature degraded into a stable compound which capitalism then vomits back into nature, polluting it after having devastated it. In Latin, the word *mundus*, like the Greek *cosmos*, has meanings that refer to world, pure, and ornament. This triple semantic dimension of *mundus* and *cosmos* sustained the idea that the world was, *at once*, order and grace. The artist would imitate the world's order, for no other reason but its beauty. The industrial manipulation of molecules to create stable compounds coagulates the "everything flows" (panta rhei) of the Heraclitean world, destroys nature's cycle of death–transfiguration–rebirth, and interrupts the constant regeneration of the world, transforming, in short, the *mundus* into *immundus*. The cosmos of nature becomes the chaos of waste. Similarly, in English, *waste* also means desert and squander. Collapsed by the megacorporation *Buy n Large* (BnL), the Earth is reduced to this desert in the *WALL-E* (2008) animation by Andrew Stanton. The artist created what Syd Mead calls "reality ahead of schedule" (Cathcart 2008), an unretouched portrait of the capitalism of the twenty-first century.

4.4   The Global Increase in Waste

In a leaked internal memo of December 12, 1991, Lawrence Summers, Chief Economist of the World Bank (and later US Secretary of the Treasury), suggested that the World Bank finance the relocation of the most polluting companies to the so-called Third World (Stoett 1995):

> 'Dirty' Industries: Just between you and me, shouldn't the World Bank be encouraging MORE migration of the dirty industries to the LDCs [Least Developed Countries]? I can think of three reasons: The measurements of the costs of health impairing pollution depends on the foregone earnings from increased morbidity and mortality. From this point of view a given amount of health impairing pollution should be done in the country with the lowest cost, which will be the country with the lowest wages. I think the economic logic behind dumping a load of toxic waste in the lowest wage country is impeccable and we should face up to that.

Indeed, the rapid "peripheral" industrialization at the end of the twentieth century exacerbated the problem of pollution and waste in the so-called underdeveloped countries, creating perverse processes of symbiosis between land oligarchies, predatory capital, and military authoritarianism. This symbiosis stimulated the phenomena of mass migration, huge influx into cities, and the proliferation of slums in societies once stigmatized by colonization and slavery, as well as by a lack of social cohesion, education, resources, administrative efficiency, and infrastructure to process or recycle the new scale of urban and industrial waste. Less than three decades later, Lawrence Summers' "solution" shows its limits because the problem of waste and pollution in its various forms now reaches the fully industrialized countries, once prepared (thanks to massive investments in infrastructure and education since the nineteenth century) to keep the garbage problem under control. From the bottom of the Arctic Ocean to the heights of the venerable Mount Fuji, waste has become, in a word, ubiquitous and ever-increasing. The upward trend remains throughout the century, but the responsibilities of the rich and poor in generating garbage differ greatly (Silpa et al. 2018):

> The world is on a trajectory where waste generation will drastically outpace population growth by more than double by 2050. (…) Worldwide, waste generated per person per day averages 0.74 kilogram but ranges widely, from 0.11 to 4.54 kilograms. Though they only account for 16 percent of the world's population, high-income countries generate about 34%, or 683 million tonnes, of the world's waste.

4.5 Sewage and Municipal Solid Waste (MSW)

The figures are dire. The World Bank estimates that in 2000, 4.36 billion people did not have access to safe sanitation, defined as the use of an improved sanitation facility which is not shared with other households and where excreta are safely disposed in situ or transported and treated off-site. In 2015, this number increased to 4.47 billion people. According to the World Health Organization (WHO 2019), at least 10% of the world's population is thought to consume food irrigated by wastewater, and 1 in 3 people globally do not have access to safe drinking water. These deficiencies especially plague the so-called Third World countries. Lake Titicaca, between Bolivia and Peru, for example, with its more than 8500 square kilometers, is considered to be the lake most threatened by eutrophication in the world, largely due to untreated sewage. Eighteen percent of its pollution, in the vicinity of the city of Puno, comes from the feces and urine of the populations living around it. The rate of sewage treatment in Brazil is typical of this region of the world. In 2018, only 53% of the Brazilian population was served by sewage collection, and only 46% of sewage is treated.[1] The rest of the sewage goes into springs, streams and rivers, dams, beaches, and the ocean, without receiving any treatment. In 2013, this represented a daily discharge of eight billion liters of feces, urine, and other wastes.

In the twentieth century, the increase in world population did not quite reach a fourfold increment, while municipal solid waste (MSW) increased tenfold. By 2025 it will double again (Hoornweg and Bhada-Tata 2012). Daniel Hoornweg, Perinaz Bhada-Tata, and Chris Kennedy (2013) give a more detailed picture of this evolution:

---

[1] See "Sistema Nacional de Informações sobre Saneamento – SNIS 2018." http://www.tratabrasil. org.br/saneamento/principais-estatisticas/no-brasil/esgoto

As urbanization increases, global solid-waste generation is accelerating. In 1900, the world had 220 million urban residents (13% of the population). They produced fewer than 300,000 tonnes of rubbish (such as broken household items, ash, food waste and packaging) per day. By 2000, the 2.9 billion people living in cities (49% of the world's population) were creating more than 3 million tonnes of solid waste per day. By 2025 it will be twice that — enough to fill a line of rubbish trucks 5,000 kilometres long every day.

In 1992, the volume of MSW produced in industrialized countries grew at a rate of 3% per year (Douglas 1992). For the World Bank, in 2016, the world generated 2.01 billion tons (Gt) of MSW, with at least 33% of that—extremely conservatively—not managed in an environmentally safe manner. By 2030, it is expected to generate 2.59 Gt of MSW annually and by 2050, 3.40 Gt (Silpa et al. 2018). There is nothing to suggest that this trend will be reversed in this century. According to Daniel Hoornweg, Perinaz Bhada-Tata, and Chris Kennedy (2013), "by extending current socio-economic trends to 2100, we project that 'peak waste' will not occur this century." Waste generation will go from 3.4 billion tons in 2050 to 4.3 billion tons in 2100, a projection seconded by Joseph Stromberg (2013): "the global rate of trash production will keep rising past 2100 – a concern because waste can be a proxy for environmental stresses." Figure 4.1 shows three scenarios of waste generation along this century.

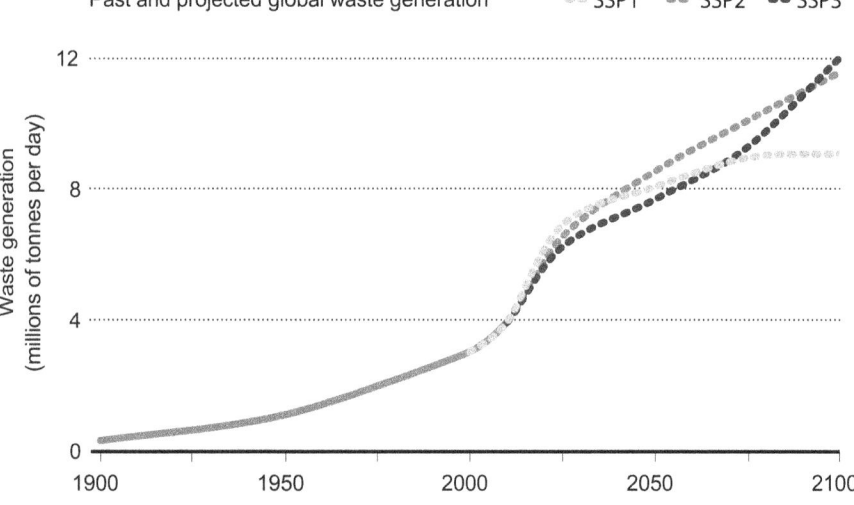

Fig. 4.1 Past (1900–2010) and projected (Up to 2100) Global Waste Generation. (Source: Hoornweg, Daniel; Bhada-Tata, Perinaz; & Kennedy, Chris. "Waste production must peak this century." *Nature*, 502, 31/X/2013, pp. 615–617)

In this graph, Daniel Hoornweg, Perinaz Bhada-Tata, and Chris Kennedy (2013) propose three projections (from bottom to up): "1. Shared Socioeconomic Pathway scenario (SSP1), the 7-billion population is 90% urbanized, development goals are achieved, fossil-fuel consumption is reduced and populations are more environmentally conscious." In this best-case scenario, waste generation should peak by 2075. 2. "SSP2 is the 'business-as-usual' scenario forecast, with an estimated population of 9.5 billion and 80% urbanization." In this scenario the authors predict that, by 2100, solid waste generation rates will exceed 11 million tons per day—more than three times today's rate. 3. "In SSP3, 70% of the world's 13.5 billion live in cities and there are pockets of extreme poverty and moderate wealth, and many countries with rapidly growing populations." In this scenario, solid waste generation should reach 12 million tons per day by 2100.

Obviously, industrialized countries generate much more waste per capita than other countries. The average person in the USA throws away their body weight in rubbish every month (Hoornweg et al. 2013). Between 1980 and 2005, the amount of MSW per capita increased by 29% in North America, 54% in the EU15, and 35% in OECD countries (Jöstrom and Östblom 2010). OECD countries generate more than 2 kg. per day per capita of MSW. At the other end, China generated, in the same year, 0.31 kg. per day per capita. For its part, the US EPA indicates that between 1960 and 2010, the daily production of MSW per capita in the USA almost doubled, from 1.2 kg. in 1960 to 2.3 kg. in 2010.

In Brazil, data from the Brazilian Institute of Geography and Statistics (IBGE) and from the Brazilian Association of Public Cleaning (Abrelpe 2011 and 2015) show that São Paulo is among the three largest garbage producers in the world after New York and Tokyo and that Brazil is among the ten largest. In the span of just 5 years (2011–2015), Brazil went from producing 62 million tons of MSW to 79.9 million tons, an increase of more than 30%. Of the total in 2015, 7.3 million tons were not even collected.

4.6 Plastic, the Throwaway Lifestyle

> *Humanity's plastic footprint is probably more dangerous than its carbon footprint.* (Charles Moore)

From the mid-nineteenth century onward, we see a transition from the first phase of the Industrial Revolution—when handmade manufacturing was replaced with industrial machine-based production, brutally reorganizing the labor force and boosting productivity—to the second phase, one that was characterized by a new ability to act and transform the molecular structures of matter. This transition stimulated and at the same time was made possible by a "Quiet Revolution," as Alan J. Rocke (1993) calls it, in organic chemistry:

> In 1860 there were about 3,000 well-characterized substances in the chemical literature; this number had grown steadily during the preceding several decades, with a consistent doubling period of about twenty years. But just about 1860 the trend dramatically accelerated, so that the doubling time thereafter was about nine years. It has remained so ever since.

As a consequence of this revolution, the chemical industry, under German leadership, would take the lead in the Industrial Revolution. Already in 1848, in the *Communist Manifesto*, Marx and Engels emphasize the "application of chemistry to industry and agriculture."[2] But it was only between the late nineteenth and early twentieth century that the petrochemical industry was born (from the rib of industrial chemistry). Its symbolic date goes back to 1907, when Leo Baekeland, a Belgian based in New York, invented Bakelite, which is synthesized from coal tar. Bakelite is the first in a series of plastics known as phenol resins, the series that inaugurates the Age of Plastic. The high technological density of plastic makes it inseparable from the history of the creation of large corporations, from Union Carbide and Carbon Chemicals Inc. (an industrial conglomerate formed in 1917 by several smaller industries that would, in 1939, come to absorb the Bakelite Corporation by Leo Baekeland) to Dow Chemical which, in turn, absorbed Union Carbide in 1999. Similar stories occur with Bayer, American Catalin Corporation, DuPont, etc.

The creation of the contemporary world's material universe is also a mental recreation. The launch of the magazine, *Plastics*, in New York in 1925 (Meikle 1997) was a milestone in twentieth-century history: its title consolidates the generic term used for these different polymers derived from petroleum. A semantic metamorphosis took place in this operation. *Plasma* and *plastics,* in Greek and Latin, refer to the modeled object and the art of modeling into clay, with their mythical resonances, from Prometheus to Yahweh. The word *plastic* would acquire undertones that are no less demiurgic. Just as the great sensation of the Paris Exposition Universelle in 1889 was the iron structure of the Eiffel Tower that stood at its entrance as a modern Arc de Triomphe, the great attraction at the 1939 New York World's Fair—whose motto, Dawn of a New Day, was the celebration of the future—was the nylon display by DuPont, announced as a substitute for silk and as man's "second skin." In the second post-war period, plastic begins to be presented as the solution to a life free from domestic work, one in which everything could be discarded after use. A photo of *Life* magazine from 1955 shows a couple euphorically discarding their home appliances under the title *Throwaway Living* and accompanied by the text: "Oh Joy, Oh Bliss! Disposable products are an innovative way to make life easier."

Throughout history, humans spun and wove animal and vegetable fibers to shelter themselves from the cold. Later, materials secreted by our own industry would be used to cover ourselves. In the same way, wood was the quintessential matter of human artifacts for millennia since the pre-Industrial Age. In Latin, the word *materia* has the meaning of both wood and matter. Hence, there was then a phenomenological continuity between "raw material" and manufactured objects. Man could recognize—in his house, his utensils, his clothes, and his art—the materials of the world which he inhabited and with which he was familiar: wood, stone, clay, iron, plant fibers, wool, and animal fur. From the mid-twentieth century on, the natural world that surrounds our human senses has started becoming strange; it begins to be hidden under a barrier of industrial products, derived from an artificial synthesis of matter.

---

[2] "Subjection of Nature's forces to man, machinery, application of chemistry to industry and agriculture, steam-navigation, railways, electric telegraphs, clearing of whole continents for cultivation, canalisation of rivers, whole populations conjured out of the ground – what earlier century had even a presentiment that such productive forces slumbered in the lap of social labour?" *The Communist Manifesto* (1848).

Remembering a comparison made by Christian Godin (2012), "today, our sensation of nature is more like the sensation that a prelingual deaf person has of music."

4.7   Plasticene: The World as a Continuum of Polymers

The Age of Plastics has manifested itself, in reality, as the Age of Waste (Andrady 2003; Weisman 2007, particularly Chap. 9 "Polymers Are Forever"). It has been a long time since plastic objects lost their glamour to become almost always synonymous with any cheap commodity—"made in China," inexpensive, ephemeral, and the umpteenth copy of a template that is generated infinitely. It is an object that is not even ugly (since ugliness belongs to the range of aesthetic values) and an object unworthy of patina, history, and memory, something that does not become waste because it is congenitally junk. In fact, it is the most ubiquitous waste on the planet. If nylon was hailed in 1939 as the second skin of man, today, as Jan Zalasiewicz asserts, "all the plastics that have ever been made are already enough to wrap the whole world in plastic film." (as cited by Reed 2015). Today we speak of plastisphere or plasticene, referring to the capacity of plastic to impact the planet's geology and oceans. Plastic is the third most manufactured product on the planet, after cement and steel. In 50 years (1964–2014), the production of plastic multiplied by almost 21. In 1964, 15 million tons of plastic were produced, and in 2014, it was up to 311 million tons, as shown in Fig. 4.2.

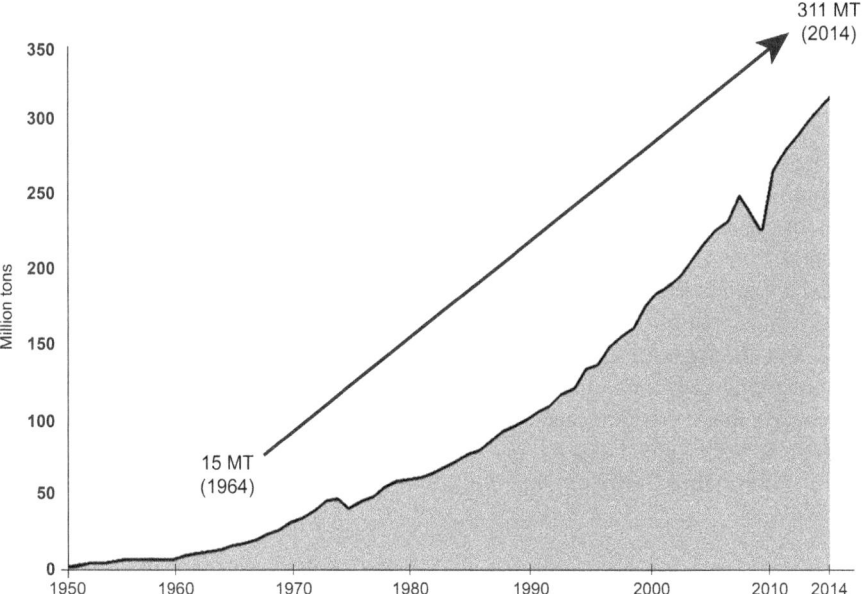

Fig. 4.2 Production of plastic in millions of tons (Mt). Based on Ellen MacArthur Foundation/McKinsey, *The New Plastics Economy—Rethinking the future of plastics*, 2016

According to Roland Geyer, Jenna Jambeck, and Kara Lavender Law (2017):

We estimate that 8300 million metric tons (Mt) as of virgin plastics have been produced to date. As of 2015, approximately 6,300 Mt of plastic waste had been generated, around 9% of which had been recycled, 12% was incinerated, and 79% was accumulated in landfills or the natural environment. If current production and waste management trends continue, roughly 12,000 Mt of plastic waste will be in landfills or in the natural environment by 2050.

In 2016, nearly 500 billion PET (polyethylene terephthalate) bottles were produced worldwide, with Coca-Cola accounting for a fifth of them. That means almost a million bottles produced per minute and an increase of almost 200 billion bottles compared to the production in 2004. And a report on packaging trends, made by Euromonitor International, estimates that 583.3 billion of these bottles will be produced in 2021 (Laville and Taylor 2017). The data firm IHS Markit has forecast that plastic production will grow on average 3.5–4% per year through at least 2035 (Bruggers 2020).

High-income OECD countries had been exporting much of their plastic waste (70% in 2016) to several countries in Asia, notably to China. Since 1992, China has imported 106 million tons of waste. With the Chinese government banning this trade in 2018, the disposal of plastic waste in exporting countries will tend to worsen (Brooks et al. 2018). Finally, the filters of the approximately six trillion cigarettes smoked globally each year often end up outside of trash bins (and all the more so, as cigarettes are banned from indoor environments). This is equivalent to about 750,000 tons of plastic per year poisoning the environment, not least with the carcinogenic residues present in cigarettes (Novotny 2014).

4.8   BPA and Phthalates

Created for the first time in 1891 by Alexander Dianin, bisphenol A (BPA) is a chemical compound synthesized by the condensation of acetone with two equivalents of phenol (an aromatic organic compound). In the second half of the 1950s, General Electric and Bayer industrialized and commercialized large-scale epoxy resins and polycarbonates from this compound; their versatility allows them to be used in a myriad of objects that surround the contemporary living world. Between 1997 and 2015, Frederick vom Saal and colleagues published more than 30 papers demonstrating the health damages caused by this compound. Animals in the uterus exposed to doses 20,000 times lower than those studied thus far presented malformations in their genital organs. In male dogs and humans, it is associated, among other malformations, with cryptorchidism or hypospadias,[3] whose occurrences have

---

[3] Cryptorchidism occurs when one or both testicles do not descend into the scrotum in the final stage of pregnancy. Hypospadias is a congenital defect characterized by the arrangement of the urethral meatus on the underside of the penis and not on the tip of the glans.

increased in percentage in recent decades. In 2007, in the Chapel Hill Bisphenol A Expert Panel Consensus Statement, 38 international researchers warned about several serious health disorders in humans caused by BPA (Saal et al. 2007):

> Examples include increases in abnormal penile/urethra development in males, early sexual maturation in females, an increase in neurobehavioral problems such as attention deficit hyperactivity disorder (ADHD) and autism, an increase in childhood and adult obesity and type 2 diabetes, a regional decrease in sperm count, and an increase in hormonally mediated cancers, such as prostate and breast cancers.

Scientists were also warning us, then, to the fact that "virtually everyone has measurable blood, tissue and urine levels of BPA that exceed the levels produced by doses used in the "low dose" animal experiments." In fact, a study conducted in the USA by the Centers for Disease Control and Prevention (CDC) in 2003 and 2004 detected BPA in 93% of the 2.517 urine samples of children 6 years and up.[4] A review aiming to summarize experimental studies describing the effects of BPA on the female reproductive health shows that exposure to this compound "has been correlated with alterations in hypothalamic-pituitary hormonal production, reduced oocyte quality due to perinatal and adulthood exposure, defective uterine receptivity and the pathogenesis of polycystic ovary syndrome" (Caserta et al. 2014). Other studies have linked it to obesity and cardiopathies, and a recent study shows its impact on animal hearing and perhaps on human hearing (Morin 2014). In 2009 and 2010, Canada and the European Union prohibited BPA from being present in baby bottles. In 2010, its use in cups and bottles, in addition to baby bottles, was banned by the US FDA, at the request of the corporations themselves who feared the implementation of more restrictive bills, which were then being debated in the US Congress. The ban has been in force since 2011 in several countries, including Brazil, but only for these three products. In France, since 2015, BPA can no longer be present in any food container (Foucart 2015). In the meantime, corporations have replaced bisphenol A (BPA) with bisphenol F (BPF) and bisphenol S (BPS), which recent studies suggest are equally harmful in relation to several of the malfunctions previously mentioned (Kuruto-Niwa et al. 2005; Viñas and Watson 2013). Cassandra Kinch et al. (2015) "show that bisphenol S, a replacement used in BPA-free products, equally affects neurodevelopment."[5]

Since the 1930s, the petrochemical industry has begun to develop a group of more than 25 chemical compounds called phthalates (esters of phthalic acid, themselves derived from naphthalene), used as additives to increase the malleability of plastics. Since there is no covalent bond between the phthalates and the plastics to which they are added, their release into the atmosphere increases as the plastic ages and decomposes. Some phthalates, like diethyl phthalate (DEP) and dimethyl phthalate (DMP), are volatile and are found in significant atmospheric concentra-

---

[4] See the National Health and Nutrition Examination Survey (NHANES III) in the site of the NIH.

[5] Regarding the harmful effects of bisphenol F and bisphenol S, see also the Inserm (Institut national de la santé et de la recherche médicale) bulletin 15/I/2015: "Bisphenol A. Alternative products that may also be harmful."

tions even in outdoor environments, with DEP being dangerously toxic. Even phthalates that have lower volatility, such as MBzP, BBzP metabolite, and DEHP, are found in important concentrations in domestic environments and act on organisms, causing endocrine dysfunctions similar to BPA, especially in pregnant women (Frederiksen et al. 2007). A study conducted at the University of Karlstad in Sweden (examining urine samples of 83 babies aged 2–6 months) shows that the presence of high concentrations of these phthalates added to domestic PVC floors is associated with asthma, allergies, and chronic diseases, as well as endocrine dysfunctions, in children (Carlstedt et al. 2013).

4.9 Plastic in the Oceans

The lethal impact of plastic waste on marine biology will be discussed in Chap. 11 (Items 11.1 "Overfishing and Aquaculture" and 11.5 "Plastic Kills and Travels Through the Food Web"). Let us analyze here only the quantitative aspects of the human plastic footprint which could be, as stated in the epigraph of this chapter, even more dangerous than our carbon footprint. Between 60% and 80% of all marine litter is plastic (Bejgarn et al. 2015). Pollution of the marine environment by plastic began to be better known and studied after 1988 when the US National Oceanic and Atmospheric Administration (NOAA) indicated the presence of large concentrations of plastic in the Pacific Ocean. In 1997, Charles Moore, a captain and oceanographer, Director of the Algalita Marine Research Foundation, discovered what came to be called the Great Pacific Garbage Patch (GPGP), a floating garbage area in the oceanic gyre of the North Pacific (between California and Hawaii), consisting mostly of plastic, of indefinite dimensions. In the last 30 years, the problem has been the subject of many studies, and there is evidence that it is getting worse. Laurent Lebreton et al. (2018) state that "at least 79 (45–129) thousand tonnes of ocean plastic are floating inside an area of 1.6 million km^2. (…) Ocean plastic pollution within the GPGP is increasing exponentially and at a faster rate than in surrounding waters." The problem extends, in fact, to all areas of the oceans. Jenna R. Jambeck et al. (2015) estimate that:

> 275 million metric tons (MT) of plastic waste was generated in 192 coastal countries in 2010, with 4.8 to 12.7 million MT entering the ocean. (…) Without waste management infrastructure improvements, the cumulative quantity of plastic waste available to enter the ocean from land is predicted to increase by an order of magnitude by 2025.

The authors estimate that, only in 2015, 9.1 million MT of plastic waste will have been thrown into the sea from land (not counting, therefore, garbage dumped by ships). Each year, 500 thousand tons of plastic are tossed into the Mediterranean Sea alone. Over the current decade (2015–2025), more than 80 million tons of plastic will accumulate in the sea.

   Since the plastic in the sea tends to fragment into tiny particles, much of this
material today is in the form of fragments between 5 mm and < 1 µm, generally
called microplastics (Erikson et al. 2014), which is even more threatening to marine
fauna. We are now discovering huge amounts of plastic stuck to ice and samples
containing up to 800,000 particles of plastic per cubic meter at the bottom of the
continental shelves of the Atlantic, Mediterranean, and Indian Oceans. Erik van
Sebille and colleagues (2015) conducted a global inventory of plastic particles floating in the ocean. According to their estimates:

> The accumulated number of microplastic particles in 2014 ranges from 15 to 51 trillion
> particles, weighing between 93 and 236 thousand metric tons, which is only approximately
> 1% of global plastic waste estimated to enter the ocean in the year 2010.

These fragments usually come from plastic bags and bottles, but also from synthetic
clothing fibers that are released in washing machines. According to Chris Bowler of
Tara Oceans, "the fact that we found these plastics is a sign that the reach of human
beings is truly planetary in scale." This omnipresence of plastic in the oceans,
including in Antarctic waters (Holman 2012), led Christina Reed (2015) to speak of
a dawn of the Plasticene.

4.10   Industrial Pesticides

Homer describes how Ulysses used smoke to control pests in his house and Pliny
recommended the use of arsenic as an insecticide. Such practices were, however,
limited, as pests were controlled by what Clive E. Edwards (1993) calls "cultural
methods" until the beginning of World War II. They cannot, therefore, be considered a historical precedent for industrial pesticides. The scope, scale, range, and
permanence, as well as the systemic noxiousness and lethality of the pesticides used
in the industrial era, are so different that they represent an absolutely new fact in the
history of human pollution of the environment.

4.11   Chemical Warfare and the War Already Lost

Organochlorine and organophosphorus insecticides, as well as herbicides based on
synthetic hormones, were born in the 1920s and 1940s as a result of research on
chemical warfare used during World War I by the two sets of belligerent powers. In
the interwar period, chemical weapons continued to be used by British aviation, for
example, in 1919 against the Bolsheviks and in 1925 against the city of Sulaymaniyah,
the capital of Iraqi Kurdistan. Italian aviation used them in 1935 and 1936 in their
attempt to exterminate the Ethiopian population, and, according to seemingly reliable documentation, the Bolshevik army used chemical weapons to decimate the

Tambov revolts, one of the 118 peasant revolts against the Red Army reported by Cheka, in February 1921 (Croddy et al. 2002).[6]

The example of the great German conglomerates created after World War I to give Germany its supremacy in the chemical industry is proverbial. As part of its scientific crew, the Degesch (Deutsche Gesellschaft für Schädlingsbekämpfung— German Corporation for Pest Control), created in 1919, had chemists like Fritz Haber (Nobel Prize) and Ferdinand Flury, who in 1920 developed the Zyklon A, a cyanide-based pesticide and immediate precursor of another insecticide, Zyklon B, which was patented in 1926 by Walter Heerdt and used successively in the gas chambers of the Auschwitz-Birkenau and Majdanek extermination camps. Another example is that of IG Farben: its dismemberment after 1945 resulted in Agfa, BASF, Hoechst, and Bayer. Working for this German industrial conglomerate, at one point the fourth largest corporation in the world, were chemists such as Gerhard Schrader (1903–1990), a Bayer employee responsible for the discovery and industrial feasibility of organophosphorus compounds that act on the central nervous system. These compounds are derived from pesticides such as Bladan and parathion (E 605) and chemical weapons such as tabun (1936), sarin (1938), soman (1944), and cyclosarin (1949), the first three of which were developed (yet not used) by the German army in World War II. After the war, Schrader was imprisoned for 2 years by the Allied powers and obliged to communicate the results of his research on organic phosphate esters.

4.12   A Growing Threat

For nearly 60 years, that is, at least since Rachel Carson's celebrated book, *Silent Spring* (1962), we have known that industrial pesticides have thrown the human species into an ecocide and, ultimately, suicidal war. This war is, therefore, already lost from the very beginning. As its name implies, an industrial pesticide is a chemical that seeks to exterminate a "pest," a term that in productivist paradigm refers to any species that competes with humans for the same foods or that poses a potential threat to productivity, to human health, or to species that serve as food for humans. Given the human inaptitude at exterminating target species, pesticides try to control populations of one or more of these target species or remove them from a given plantation or breeding area. The principle itself fully proves the insanity of industrial agriculture: our food is poisoned to prevent other species from eating it. The doses of poison, small in relation to the mass of the human body, do not kill us. But by shooting a species with a revolving machine gun, pesticides cause "collateral

---

[6] Eric Croddy quotes an order signed by Mikhail Tukhachevsky and Vladimir Antonov-Ovseyenko, the commanders in charge of suppressing the Tambov (*Antonovshchina*) rebellion, urging the use of chemical weapons: "forests where thieves are hidden must be unoccupied with the use of poisonous gas. This must be carefully calculated, so that the layers of gas penetrate the forests and kill all those who hide there" (p. 151).

damage": they kill or weaken non-target species, causing systemic imbalances which promote artificial selections. These artificial selections, in turn, might bolster the tolerance of the target species or facilitate an invasion by opportunistic species, which sometimes pose an equal or greater threat to plantations than the species targeted by the pesticide itself. Furthermore, in the medium and long term, pesticides intoxicate man himself, as demonstrated (at least since the 1990s) in a series of scientific studies (Dich et al. 1997, 2013), all the more so because, as the targeted species become tolerant to an active ingredient, humans are obliged to increase doses of pesticides and combine them with others into increasingly toxic cocktails.

The impacts of pesticides on various animal species will be discussed in some detail in Chaps. 10 and 11. Suffice it to say that there is a long-established consensus on the ineffectiveness of pesticides in terms of agricultural productivity. A 2002 FAO document, the *Report of the First External Review of the Systemwide Programme on Integrated Pest Management* (SP-IPM), had already documented the inability of industrial pesticides to reduce agricultural losses in relative terms, a phenomenon widely observed since the last third of the twentieth century:

> A comprehensive study by analyzing a huge volume of field trial data found that crop losses range from 25% to over 50% depending on the crop. While the productivity impacts of such high crop losses are significant, *it is disturbing that over the past three to four decades, crop losses in all major crops have increased in relative terms* (italics added).

Finally, in 2014, an international working group on systemic pesticides, the Task Force on Systemic Pesticides (TFSP), bringing together 29 researchers, states in its results that the systemic use of pesticides is, unequivocally and increasingly, a threat to both agriculture and ecosystems. Jean-Marc Bonmatin, of the National Centre for Scientific Research (CNRS) in France, one of the researchers who conducted the 4-year assessment, summarized the TFSP results (as cited by Carrington 2014):

> The evidence is very clear. We are witnessing a threat to the productivity of our natural and farmed environment equivalent to that posed by organophosphates or DDT. Far from protecting food production, the use of neonicotinoid insecticides is threatening the very infrastructure which enables it.

4.13 Increase in Consumption and Variety of Pesticides Since 2004

The pesticide market is growing. As the FAO document of 2002 shows, the average rate of increase in world consumption of pesticides during a 5-year period (1993–1998) was 5% per year in relation to the previous period (1983–1993). This growth reached a peak in 1998 (Tilman et al. 2001), and then, beginning in 2004, the global agrochemical market, dominated by six corporations, resumed its growth, as shown in Table 4.1.

Table 4.1 Sales of all pesticides, including non-crop products (in billions of dollars)

2003	2004	2008	2009	2013	2017	2018
29.3	32.6	46.7	52.1	60.6	61.2	65

Sources: For the years 2003 and 2004, see "Agrochemical and Biotech Corporations Spur Global Growth of Pesticides." For the years 2008 and 2009, see "Global Markets for Agrochemicals," January 2010. For the year 2013, see "Global Agrochemical Market Will Continue to Maintain Steady Growth." *AgroNews*, 28/X/2014. For the year 2017, see BCC Research, "Biopesticides: Global Markets to 2022." For the year 2018, see "Global Crop Protection Market Rose 6% in 2018." *Agrow Business*, 3/V/2019

According to BCC Research, "the global market for synthetic and biopesticides should reach US$ 79.3 billion by 2022 from US$ 61.2 billion in 2017 at a compound annual growth rate (CAGR) of 5.3% during 2017–2022."

4.14   The Examples of France, the USA, and South America

A report published on March 8, 2016, by the French Ministry of Agriculture shows an average increase of 5.8% in the use of pesticides by French farmers between 2011 and 2014 (this does not take into account the use of neonicotinoid products applied directly to seeds). In 2014 alone, France saw an average increase of 16% in the sales of pesticides, while sales of composts containing molecules suspected of being carcinogenic, mutagenic, or toxic for human reproduction increased by a rate of 13–22% in this same period (Valo 2016). The increase in the variety of active particles is equally startling. According to the EPA, in 2007 there were "more than 1,055 active ingredients registered as pesticides, formulated in thousands of pesticides available in the market."[7] In 1976, US President Gerald Ford signed the Toxic Substances Control Act (TSCA), a regulatory legal framework that sets levels for use of chemical agents considered safe for the human population. This law, which has been used as a reference for legislation outside the USA, lags behind in terms of what is known today about the toxic action of these substances on living organisms. Two years later, the US Congress authorized the EPA to use a conditional registration clause for pesticide licensing in cases of imminent public health threat, thereby shortening the regular procedures for pesticide approval which follow the Federal Insecticide, Fungicide, and Rodenticide Act (FIFRA), a law established in 1947 and amended in 1972. Corporations have been able to benefit from this margin, obtaining (in this emergency regime) approval from the EPA to use 65% of the 16,000 pesticides currently available in the North American market, as concluded by a 2013 Natural Resources Defense Council (NRDC) study. Moreover, the EPA admits to using the conditional registration clause in 98% of cases between 2004 and 2010, allowing for more than 11,000 pesticides to be sold in the USA in 2012 without being properly tested (Garric 2013).

---

[7] Cf. EPA, "Assessing Health Risks from Pesticides," 2007, last updated in 2012.

Pesticide use in the USA presents a clear upward trend. According to the EPA's *Pesticides Industry Sales and Usage 2008–2012 Market Estimates*, the most comprehensive report to date (Atwood & Paisley-Jones 2017), the pesticide market in the USA grew from 606 million pounds in 2005 to 762 in 2012, and "US expenditures at the user level for conventional pesticides[8] totaled nearly $14 billion in 2012 and nearly $13 billion in 2009." According to a 2019 Research Report provided by the Mordor Intelligence, these expenditures in North America have been estimated at $14.1 billion in 2017 and are projected to reach $19.1 billion by 2022, at a compound annual growth rate (CAGR) of 5.05% during the forecast period from 2017 to 2022. Sales of agrochemicals in Latin America, amounting to slightly more than US$ 4 billion in 2000, more than doubled in 2012, coming close to the levels in North America. In addition, corporations charge less for older products (some already banned in industrialized countries), in order to sell them in the poorest and most permissible markets. In Brazil, the situation is the worst in the world in terms of quantity of pesticides used and one of the worse in the world in terms of permissiveness. Brazil participates in only 4% of world agribusiness trade, but the pesticide market in Brazil is worth US$10 billion per year, or 20% of the global market. Figure 4.3 shows that between 2002 and 2014, the consumption of agrochemicals, measured according to the weight of the active ingredient, increased approximately 340%, from about 150 thousand tons to more than 500 thousand tons of active ingredients.

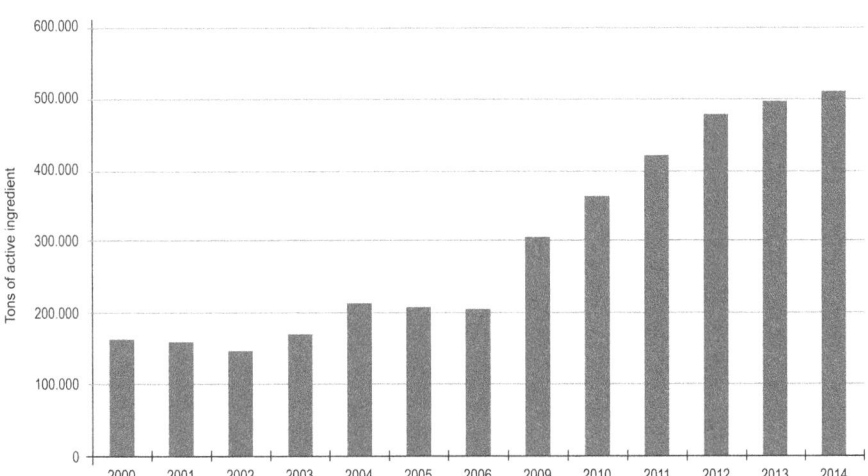

Fig. 4.3 Consumption of agrochemicals in Brazil in tons of active ingredient, 2000–2014. (Source: Ibama, cited in Larissa Mies Bombardi, *Geografia do Uso de Agrotóxicos no Brasil e Conexões com a União Europeia*. FFLCH – USP, São Paulo, November 2017, p. 33)

---

[8] Conventional pesticides are defined in this report as all active ingredients other than biological pesticides and antimicrobial pesticides.

Table 4.2 Five of the ten most commonly used agrochemicals in Brazil (in descending order)

Active ingredient	MRL Europe	MRL Brazil
1st Glyphosate	0.05 mg/kg	10 mg/kg
2nd 2.4-D	0.1 microgram	30 micrograms
3rd Acephate	0.1 microgram	No limits
5th Chlorpyrifos	0.1 microgram	30 micrograms
7th Atrazine	0,1 microgram	2 micrograms

Source: Larissa Mies Bombardi, *Geografia do Uso de Agrotóxicos no Brasil e Conexões com a União Europeia*. State University of São Paulo, November 2017

In 2017, agricultural monoculture in Brazil, which favors pest resistance, used 540,000 tons of active ingredients, more than triple the quantity used in 2002 and about 50% more than in 2010. Vicente Eduardo Soares de Almeida and colleagues (2017) showed that between 2000 and 2012, the accumulated increase in use of agrochemicals in Brazil was more than three times greater than the increase in agricultural productivity (t/ha) and more than 10 times higher than the population growth in that period. The use of agrochemicals (in kilos) by agricultural area (in hectares) more than doubled: in 2000, it was 6.09 kg/ha, and in 2012, it was 15.97 kg/ha. This explosive growth is boosted by federal and state tax exemptions. While for medications in Brazil, one pays 34% of taxes on the final value of the product, for pesticides, one pays only 22% (Rossi 2016). Larissa Mies Bombardi (2017) published a comparative study between Brazilian and European legislation on maximum residue limits (MRLs) allowed for each active ingredient in food and water samples. The overwhelming results of this study show the degree to which the agribusiness industry controls the Brazilian state. Table 4.2 lists five of the ten most commonly used agrochemicals in Brazil (in descending order), comparing the maximum residue limits (MRLs) allowed in Brazil and in Europe.

Exposure to these five pesticides has been linked to adverse neurodevelopmental effects, mostly in children who are at a critical period of vulnerability (see Chap. 15, Sect. 15.3 "Human Health Backfires"). Development of autism spectrum disorder, often with intellectual disability, was associated with prenatal exposure to glyphosate, chlorpyrifos, diazinon, and other pesticides (Ehrenstein et al. 2019). France announced that it will ban glyphosate by 2020. In 2015, the International Agency for Research on Cancer (IARC) claimed that it had "sufficient evidence" that glyphosate causes cancer in animals and "limited evidence" that it causes cancer in humans, classifying glyphosate as "probably carcinogenic to humans" (Group 2A) (Cressey 2015). Furthermore, its use has been associated with a reduction in progesterone in mammalian cells, abortion, and teratogenic changes via the placental route. These correlations were repeatedly observed by the scientific community and finally accepted in several lawsuits against Monsanto/Bayer in 2018 and 2019 in the USA and in Europe. Developed by Dow Chemical in the 1940s, 2.4-D is today an active ingredient in more than 1500 herbicides. It was banned in the province of

Ontario, Canada, in 2009, then in Australia in 2013, and in Vietnam in 2017.[9] And there have been repeated demands—unmet by authorities in the country—for 2.4-D to be banned in the USA. The reasons for such prohibitions are multiple. The IARC (International Agency for Research on Cancer) affirms[10]:

> The herbicide 2.4-D was classified as possibly carcinogenic to humans (Group 2B). (…) There is strong evidence that 2.4-D induces oxidative stress, a mechanism that can operate in humans, and moderate evidence that 2.4-D causes immunosuppression, based on in vivo and in vitro studies.

An endocrine disruptor, 2.4-D can prevent the normal action of estrogen, androgen, and, more conclusively, thyroid hormones. Dozens of epidemiological, animal, and laboratory studies have shown an association between 2.4-D and thyroid disorders (Sedbrook 2016). Recent findings suggest that adverse neurodevelopmental effects were also associated with early childhood exposure to chlorpyrifos (Jianqiu Guo et al. 2019). Belonging to a class of organophosphate pesticides, chlorpyrifos was first developed as a nerve gas in World War II. In 1965 it was added by Dow Chemical to the list of pesticides and is currently used in several crops on 100 countries. Long-term studies on large cohorts, notably in the USA, have shown that chlorpyrifos is an endocrine disrupter that acts on thyroid signaling, also interfering with acetylcholine receptors in the central nervous system. For this reason, chlorpyrifos has been banned in several countries in Europe. But it remains in imported fruits and vegetables, as well as in the urine of children and the umbilical cord of pregnant women, causing autism, early development of brain lesions, and an irreversible decrease in IQ (by an average of 2.5 points), notably in children whose mothers lived less than 2 kilometers from a spraying location. According to recent estimates, this type of exposure to this family of organophosphate pesticides is associated with 59,300 cases of intellectual disability per year in Europe (Horel 2019).

4.15   GMO = More Pesticides

It was only from 1996 onward that transgenic seeds began to be widely marketed. More than 20 years later, it is not possible to say, until proven otherwise, that genetically modified organisms (GMOs) cause negative impacts on humans, on other species, or on the environment *simply because they are genetically modified*. That said, GMOs designed to increase the tolerance of a specific plant to herbicides have proven to be harmful. Transgenic seeds, especially those of Monsanto/Bayer, are

---

[9] Cf. "APVMA [Australian Pesticides and Veterinary Medicines Authority]: Australia Bans Toxic Herbicide 2.4-D Products." *Sustainable Pulse*, 24/VIII/2013; "Govt bans 2.4-D, paraquat in Vietnam." *Vietnamnet*, 16/II/2017.

[10] Cf. IARC Monographs evaluate DDT, lindane, and 2.4-D. Press release n. 236, 23/VI/2015.

used today in almost all soy, cotton, and corn crops in the USA, Brazil, and other Latin American countries. The belief that genetically modified organisms have somehow contributed to a reduction in the use of herbicides and insecticides is known today to be false. It is well documented that transgenic seeds fostered the use of glyphosate in corn, soy, and cotton cultures between 1996 and 2008 (Benbrook 2009): "genetically engineered (GE) crops have been responsible for an increase of 383 million pounds of herbicide use in the U.S. over the first 13 years of commercial use of GE crops" (1996–2008).

As Natasha Gilbert (2013) shows, genetically modified seeds are victims of their own success. Because glyphosate is used exclusively and on its own (without other agrochemicals), it is more vulnerable to the vicious cycle that affects other pesticides: spontaneous weeds end up developing resistance to the aggressor more quickly, becoming "superweeds" that are hostile to it. According to Ian Heap, Director of the International Survey of Herbicide Resistant Weeds, since 1996, 24 "superweeds" resistant to *Roundup* have been identified in 18 countries, with the biggest agricultural impacts occurring in Brazil, Australia, Argentina, and Paraguay. The example of *Amaranthus palmeri*, an herb that competes with cotton, is instructive. Since 1990, North American farmers have adopted genetically modified cotton seeds to tolerate Monsanto's glyphosate *Roundup*, initially with great success. In 2004, *Amaranthus palmeri* was found in a county in the state of Georgia. By 2011, it had spread to 76 counties, causing losses of up to 50% in cotton plantations. This vicious cycle leads to the introduction of new seeds capable of tolerating even higher doses or even more aggressive cocktails of pesticides, such as the new corn seed genetically modified by Dow Chemical and capable of resisting the 2.4-D herbicide. A study by David Mortensen (2012) predicts that GMO seeds will increasingly demand herbicides. The study predicts that, as a result of genetically modified seeds, US herbicide use will increase from 1.5 kilograms per hectare in 2013 to over 3.5 kilograms per hectare by 2025. The same vicious cycle applies to insecticides, in particular the so-called Bt crops, plants that have been genetically modified since 1996 with sequences of the *Bacillus thurigiensis* genes to stimulate the proteins of this soil bacterium that have an insecticidal effect. A study on increasing resistance to Bt crops, published by Bruce E. Tabashnik, Thierry Brévault, and Yves Carrière (2013), shows that:.

> Although most pest populations remained susceptible, reduced efficacy of Bt crops caused by field-evolved resistance has been reported now for some populations of 5 of 13 major pest species examined, compared with resistant populations of only one pest species in 2005.

Particularly noteworthy is the new resistance of a beetle (*Diabrotica virgifera virgifera*) to an insecticidal toxin (Cry3Bb1) produced by a transgenic corn variety into which Monsanto introduced the *Bacillus thuringiensis* (Bt) gene. Faced with this new resistance, Monsanto announced the introduction of another gene into the next variety of transgenic corn, in a "holy war" against nature, one that is as lucrative as it is toxic and that, it must be repeated, is doomed to fail.

4.16 Persistent Organic Pollutants (POPs)

Pesticides, as well as solvents, dyes, preservatives, noncombustibles, and many other products, contain chemical compounds called persistent organic pollutants (POPs). These are industrial byproducts resistant to environmental degradation through chemical, biological, or photolytic processes. POPs are also characterized by low solubility in water and high solubility in lipids, which leads to their accumulation in the fatty tissues of individuals and their transmission along the food chain. These two characteristics—bioaccumulation and bioamplification—explain why POPs are also referred to by the abbreviation PBT (persistent, bioaccumulative, and toxic). Another feature of POPs is that they are semi-volatile, allowing them to travel long distances into the atmosphere—from where they receive the acronym LRAT (long-range atmospheric transport)—before forming deposits (Ritter et al. 1995). POPs are considered teratogenic, mutagenic, and carcinogenic. Furthermore, as Anne-Lise Bjorke Monsen (*apud* Sandelson 2013) states:

> POPs affect bone density. These types of contaminants that have been detected in farmed salmon have a negative effect on brain development and are associated with autism, ADHD (Attention Deficit Hyperactivity Disorder) and reduced IQ. We also know that they can affect other organ systems in the body such as the immune system and metabolism.

The fatty tissues of salmon store a cocktail of PCBs (polychlorinated biphenyls), dioxins, and other toxic substances. Moreover, salmon are fed flour made from small fish caught in the Baltic Sea, which is very polluted.

In May 1995, POPs entered into the UNEP Council agenda. Twelve POP products called the "dirty dozen" were initially listed in the Stockholm Convention of May 2001, which banned or regulated the production of these substances in some countries. The Convention entered into force in 2004 and was amended in 2009. Some countries have not ratified the Stockholm Convention, including the USA. Others were late signatories and have been unable to apply the legislation. Starting in 2015, production of hexabromocyclododecane (HBCD) was discontinued in Europe, but there were major derogations in some countries. Therefore, POPs continue, in part, to be manufactured, and, according to Roland Kallenborn (2006), "organochlorines have been found in virtually all environmental compartments on the globe." A major producer of DDT in the 1950s and 1960s, the then Soviet Union decided (following the trail opened up by Rachel Carson in 1962) to discontinue its production in the 1970s. Colossal stocks of that poison were buried in Tegouldet, in the Tomsk Oblast region (Western Siberia), contaminating the land and the Tom River. The case is not exceptional. Some 250,000 tons of pesticides are buried or left in the open in unrecorded deposits found in areas of the ex-Soviet Union. Piotr Tchernogrivov, leader of the Green Party in the Tomsk region, claims that nobody knows the map of these hotbeds of environmental poisoning. And according to the FAO, there are tens of thousands of unprotected storage sites for pesticides from Azerbaijan to Georgia, going through Ukraine and Tajikistan. The former Aral Sea, mentioned in Chap. 3, is now an open-air deposit of POPs.

4.17   Particulate Matter and Tropospheric Ozone

"Pollution is the largest environmental cause of disease and death in the world today, responsible for an estimated 9 million premature deaths." This is the diagnosis produced by the Lancet Commission on pollution and health, coordinated by Philip Landrigan (2017), and reporting on a series of studies on the impacts of air, water, and soil pollution on human health. In this context, air pollution is surely the most lethal because, as stated in the *Air Pollution, Climate and Health* report (WHO 2014):

> An estimated 7 million people per year die from air pollution-related diseases. These include stroke and heart disease, respiratory illness and cancers. The air pollutant linked most closely to excess death and disease is PM2.5 (particulate matter less than 2.5 micrometres in diameter), heavily emitted by both diesel vehicles and the combustion of biomass, coal and kerosene.

These numbers present a certain degree of uncertainty. Data from the Institute for Health Metric and Evaluation's (IHME) 2017 Global Burden of Disease Study provides a range from seven million to ten million total premature pollution-related deaths, attributing five million of those to air pollution. And for Richard Burnett et al. (2018), long-term exposure to outdoor fine particulate matter could kill nine million people yearly. Air pollution has many different sources. In addition to the "classical" soot particles resulting from the combustion of firewood and coal, there is pollution from the cement industry and industrial furnaces, as well as from incinerators, mining, pesticides, air conditioning equipment, and the wearing of polymers, brakes, tires, and other industrial materials. All these contribute to higher atmospheric concentrations of many harmful substances, including soot, lead, ammonia, arsenic, mercury, carbon monoxide, sulfates, sulfur dioxide, nitrate particles from motor vehicle exhaust, nitrogen oxides (NOx), POPs, volatile organic compounds (VOCs), and many other toxic substances. As defined by the EPA (Air Quality Criteria for Particulate Matter, October 2004), anthropogenic particulate matter (PM) is "a mixture of mixtures." They are microscopic pollutants in the air that can slip past our body's defenses, damaging the lungs, heart, and brain of human and other animals. In 2005, the World Health Organization updated (after the 1987 and 1997 guidelines) the safe levels of air quality for human organisms (1 micrometer or μm = 1 thousandth of a millimeter):

PM2,5   —   10 $\mu m/m^3$ average annual exposure
          —   25 $\mu m/m^3$ average annual exposure during 24 h

PM10   —   20 $\mu m/m^3$ average annual exposure
          —   50 $\mu m/m^3$ average annual exposure during 24 h

Today, only 12% of the world population living in urban areas resides in cities whose air quality meets these levels of safety. "About half of the urban population being monitored is exposed to air pollution that is at least 2.5 times higher than the levels WHO recommends" (WHO 2014). In most cities where there is sufficient data to compare the current situation to previous situations, atmospheric pollution is worsening. According to estimates from the WHO, 9 out of 10 people breathe air containing high levels of pollutants. The WHO believes that inhalation of particulate matter can cause cancer of the lung and bladder, contribute to cardiovascular diseases and asthma, and increase the risk of underweight births. Just breathing became as dangerous as smoking: "one third of deaths from stroke, lung cancer and heart disease are due to air pollution. This is having an equivalent effect to that of smoking tobacco" (WHO 2018).

The European Environment Agency estimates that between 2009 and 2011, 96% of the EU's urban population was exposed to concentrations of fine particulate matter exceeding the limits set by the WHO. Nine urban areas in the UK slightly exceeded concentrations of PM10 considered safe.[11] In March 2014, Paris reached a PM10 peak of 100 µg/m3, while 20 major Chinese cities live with concentrations of PM10 between 100 µg/m3 and 150 µg/m3. India's capital, New Delhi, smothers in air pollution from inhalable PM10 particles, ranging from 600 to 1300 micrograms per m^3, levels that conventional air quality monitoring scales do not even contemplate. According to Arvind Kumar, if such pollution levels were to hit Europe, cities would be evacuated. Scientists from the Chittaranjan National Cancer Institute (CNCI), in India, warned that about half of Delhi's 4.4 million schoolchildren (between 4 and 17 years age) had stunted lung development and the deficits in lung function will not be reversed as they complete the transition into adulthood (Ghosal and Chatterjee 2015).

Melinda Power et al. (2015) warn that "exposure to particulate matter (MP2.5) has been associated with elevated anxiety symptoms." Nanoparticles (having a diameter of even less than 0.1 µm), emitted by vehicles and industrial activities, are even more dangerous to health, as they penetrate deeply into the lungs and enter the bloodstream. There are 200 times more nanoparticles (between 0.2 and 1 µg) than particles between 1 and 10 µg. Measurements of these nanoparticles in the streets of Paris in 2013 and 2014 show peaks of six million of them per liter of air, a pollution equivalent to an enclosed 20 m^2 room occupied by eight smokers (Landrin, Van Eeckhout 2011).

4.18   Tropospheric Ozone

Nitrogen oxides—nitric oxide (NO), nitrogen dioxide (NO_2), and nitrous oxide (N_2O), carbon monoxide, and volatile organic compounds (VOC^{12}) released by explosion engines and by industries react with oxygen molecules in the air and,

---

[11] See "WHO names and shames UK cities breaching safe air pollution levels." *The Guardian*, 7/V/2014.

[12] A volatile organic compound (VOC) is defined by the US Environmental Protection Agency as "any compound of carbon, excluding carbon monoxide, carbon dioxide, carbonic acid, metallic carbides or carbonates, and ammonium carbonate, which participates in atmospheric photochemical reactions." Volatile organic compounds (VOCs) include hydrocarbons, such as the carcinogenic

through atmospheric photochemical reactions (UV), form ozone in the troposphere. In this lower layer of the atmosphere, ozone becomes a secondary pollutant, one that is formed by primary pollutants, such as N_2O and VOCs.

Three major consequences of this pollution are identified. The first is its impact on global warming, since ozone is a greenhouse gas. The second consequence relates to health, since ozone is a very harmful gas to animal and human health. In humans, ozone irritates the eyes and the respiratory system as a whole, from the mucous membranes of the nose to the alveoli of the lungs, causing nausea, cough, tears, headaches, and chest pain. It also has an adverse effect on the cardiovascular system. According to Robert Devlin et al. (2012), "ozone can cause an increase in vascular markers of inflammation and changes in markers of fibrinolysis and markers that affect autonomic control of heart rate and repolarization." This study grants greater biological plausibility to epidemiological research that associates mortality with exposure to higher concentrations of ozone in the troposphere. In an unpolluted environment, the human respiration system generally absorbs 10–15 ozone parts per billion parts of air (10–15 ppb). It is estimated that throughout the twentieth century, ozone concentrations in the troposphere of large cities have increased by 100–200%. And in the spring and summer, ozone levels reach 125 ppb in several cities, ten times more than in unpolluted air.

The third consequence (after its effects on the atmosphere and on the health of animal organisms) lies in the fact that ozone delays photosynthesis in some plants, making it difficult to fix carbon. According to the USDA, "ground-level ozone causes more damage to plants than all other air pollutants combined."[13] Amos Tai, Maria Val Martin, and Colette Heald (2014) estimated the individual and combined effects of 2000–2050 climate change and ozone trends on the production of four leading food crops (wheat, rice, maize, and soybean) that account for more than 50% of the calories humans consume worldwide. The authors find that "on average, 24%, 44%, 9.8%, and 46% of the observed sensitivities to killing degree days (KDD)[14] for wheat, rice, maize and soybean, respectively, arise from higher ozone associated with high KDD (albeit greater uncertainty for rice)." These impacts have been measured in certain crops in Asia. According to reports of a Japanese team, *Tropospheric Ozone. A Growing Threat* (2006), in 2000 the wheat harvest suffered losses of 25% in the Yangtze River basin in China, where there were average daily ozone concentrations of approximately 60 ppb between April and June. For rice, there was a harvest loss of 7% in the same year.

---

benzene (C_6H_6) released by combustion engines and dichloromethane (CH_2Cl_2), a chlorinated hydrocarbon used in various functions in the petrochemical industry.

[13] See "Effects of Ozone Air Pollution on Plants." USDA. Agricultural Research Service.

[14] Killing degree days (KDDs) are a commonly used measure for the cumulative warmth a crop has experienced over the growing season, defined for maize, for instance, as the sum of all daily average temperatures over the growing season in excess of 29 °C. KDDs decrease yield (e.g., by accelerating the plant through grain development or directly damaging plant tissue or enzymes) (see Butler and Huybers 2012).

4.19   Rare-Earth Elements (REEs)

Rare-earth elements (REEs) are playing a growing role in the planetary intoxication process. These chemical elements are widespread in the Earth's crust. Promethium is quite scarce, but cerium, for instance, is the 25th most abundant element of the 78 common elements in the Earth's crust. They are so called because it is usually rare to find them in high concentrations. In addition, it is difficult to separate the rocks in which they are embedded in and to refine them.[15] REEs are often called "critical materials" because they are essential for the manufacturing of screens and hard disks of laptops, televisions, smartphones, iPads, MP3, GPS, etc. They are also used in other things, including petroleum refining catalysts, catalytic converters of car engines, metal alloys for aircraft, scanners used in x-ray and magnetic resonance imaging apparatus, ultraviolet radiation filters, glass polishing and lenses for electronic items, cameras and telescopes, nickel metal hydride (Ni-MH) batteries, high-strength ceramics, hybrid car batteries, cerium-based plastic dyes, magnets for speakers and wind turbines, etc. Their use in the war industry is also important, with items such as night visors, "smart bombs," precision-guided weapons, white noise devices, etc.

Proven global rare-earth reserves amount to 120 million tons (Mt), of which 99 Mt are divided between China (44 Mt), Brazil (22 Mt), Vietnam (22 Mt), and Russia (12 Mt).[16] Mining of REEs increased from 132,000 t. in 2017 to 170,000 t. in 2018, with China accounting for 120,000 t. There are high concentrations of rare earths in indigenous reserves in the Brazilian Amazon, 40 million tons in Morro dos Seis Lagos, in the Balaio Indigenous Land, in the Amazon, and in Serra do Repartimento in the Yanomami Indigenous Land in Roraima, the latter already under strong pressure from gold miners (Farias 2013; Ives 2013).

4.20   Toxicity

The environmental cost of this mineral exploration is immense. For example, Molycorp's costs for repair of radioactive fluid leakage in the late 1990s in Mountain Pass, California, made it unfeasible to continue the project and led to the closure of the mine in 2002. Mining and refining these minerals, as well as the disposal of tailings from these processes, produce exposure to uranium and thorium. Contact with thorium provokes cancer of the pancreas, lungs, and blood. Mining, generally in the open air, is destructive to the environment. The refining process requires the use of toxic acids and produces tremendous amounts of chemical wastes that contaminate the soil, the atmosphere, and especially water. Tailings can prove to be devastating

---

[15] In the periodic table of the elements, 15 of the 17 elements of this group belong to the lanthanides (LNS, elements with an atomic number between $Z = 57$ and $Z = 71$), to which must be added scandium ($Z = 21$) and yttrium ($Z = 39$).

[16] Cf. USGS Rare Earths Statistics and Information (2019).

to organisms. "They are full of small, fine particles that can be absorbed into the water and ground surrounding a particular mine. Water can be contaminated in three ways: sedimentation, acid drainage, and metals deposition, and once contaminated is difficult to restore to its original quality."[17] Research on the toxicity of lanthanides showed that high concentrations of this group of elements in water inhibited the growth of monocellular algae by 50% (Tai et al. 2010). Other research indicates that children exposed to lanthanides present changes in their IQ level, vital capacity, blood pressure, and heart rate after exercise, suggesting that these elements have effects on the neuronal system.[18] According to Cindy Hurst (2010), who quotes an article published by the *Chinese Society of Rare Earths:*

> Every ton of rare earth produced, generates approximately 8.5 kilograms of fluorine and 13 kilograms of dust; and using concentrated sulfuric acid high temperature calcination techniques to produce approximately one ton of calcined rare earth ore generates 9,600 to 12,000 cubic meters of waste gas containing dust concentrate, hydrofluoric acid, sulfur dioxide, and sulfuric acid, approximately 75 cubic meters of acidic wastewater, and about one ton of radioactive waste residue (containing water).' Furthermore, according to statistics conducted within Baotou, where China's primary rare earth production occurs, 'all the rare earth enterprises in the Baotou region produce approximately ten million tons of all varieties of wastewater every year.

About half of the legal rare-earth production in China comes from the Bayan Obo mine, located north of Baotou, a city of 2.5 million people in Inner Mongolia, 650 km northwest of Beijing. The lands around Baotou were once used for the cultivation of wheat and maize. Today, the 10 km² reservoir created a few miles from the city and just over 10 miles north of the Yellow River watershed which supplies 150 million people with water has become a lethal cocktail of toxic substances. A Chinese engineer who preferred to remain anonymous told Keith Bradsher of the *New York Times* in 2010 that the sludge at the bottom of this reservoir has caused a slow expansion of radioactivity in groundwater, one that approaches the Yellow River at a rate of almost 300 m per year (Bradsher 2010, 2013).

4.21   Electronic Waste (E-Waste)

Twenty or thirty years ago, electronics were being made with 11 different elements. Today's computers and smartphones use something like 63 different elements. As pointed out by Thomas Graedel and colleagues (2013):

> A century ago, or even half a century ago, less than 12 materials were in wide use: wood, brick, iron, copper, gold, silver, and a few plastics. Today, however, substantial materials diversity in products of every kind is the rule rather than the exception. A modern computer chip, for example, employs more than 60 different elements.

---

[17] See MIT's website, *Mission 2016: The Future of Strategic Mineral Resources. Environmental Damage.*

[18] "The Neural Toxicity of Lanthanides: An Update and Interpretations." *RedOrbit*, 29/XI/2012.

These estimates give us an idea of the world of e-waste,[19] a type of waste that has the highest toxicity and that is the fastest growing, three times faster, for example, than MSW. E-waste may represent only 2% of solid waste, yet it can represent 70% of the hazardous waste that ends up in landfills in the USA (PACE report 2019).

The growth rate of e-waste is accelerating. In 2005, 40 million tons (Mt) of electronic waste were generated globally; in 2011, it was 41.5 Mt, an increase of 3.7% in 6 years. But over the next 5 years, between 2011 and 2016 (44.7 Mt), the increase was about 7.5%. And between 2016 and 2021 (52.2 Mt), there will be an estimated increase of about 17%. As shown in Fig. 4.4, e-waste increased much more than the proportional increase in world population between 2014 and 2021 (from 5.8 kg/inh in 2014 to 6.8 kg/inh in 2021).

Worldwide, 8.9 Mt of e-waste were recycled in 2016, which corresponds to 20% of all the e-waste generated in that year. According to the PACE report (2019), global e-waste production is on track to reach 120 million tons per year by 2050, if current trends continue. Although only 41 countries have e-waste statistics, the available data is not a surprise. In 2016, Asia was the largest generator of e-waste (18.2 Mt) and Africa and Oceania (2.2 Mt and 0.7 Mt, respectively) the smallest.

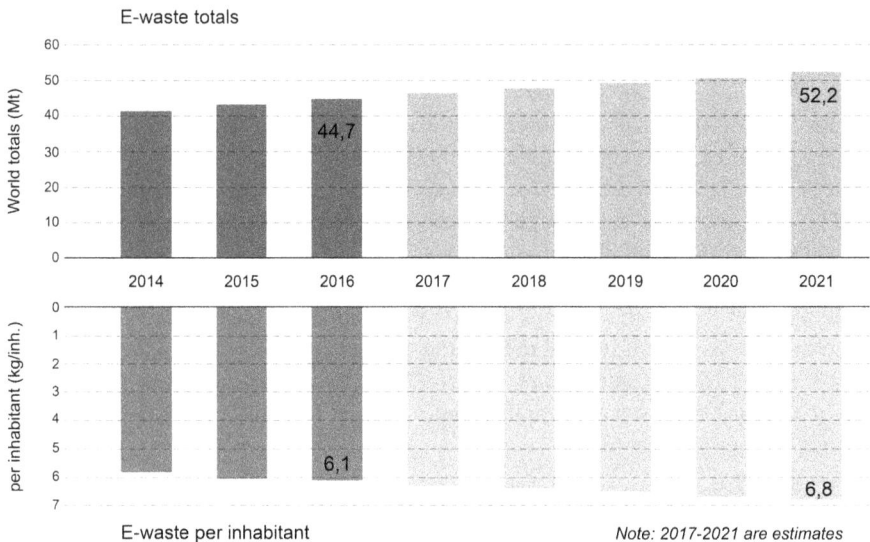

Fig. 4.4 Global electronic waste in million tons (Mt) and per capita (Kg/Inh), with projection until 2021. 2017–2021 are estimates. (Source: BALDÉ, C.P. et al., *The Global E-waste Monitor – 2017*, UNU, ITU & ISWA, Bonn/Geneva/Vienna, 2017)

---

[19] For the definition of e-waste, see Baldé et al. (2017): "Electronic waste, or e-waste, refers to all items of electrical and electronic equipment (EEE) and its parts that have been discarded by its owner as waste without the intent of re-use. (…) It includes a wide range of products – almost any household or business item with circuitry or electrical components with power or battery supply."

But in terms of production of e-waste per capita, the pyramid is reversed because in 2016 Oceania was the highest generator of e-waste per inhabitant (17.3 kg/inh), while Asia generated 4.2 kg/inh and Africa only 1.9 kg/inh.

About 400 million electronic products are discarded annually, with over 140 million cellphones disposed of every year in the USA alone (over 151 million according to the Environmental Protection Agency). Each smartphone contains 41 elements from the periodic table, and its incineration releases into the atmosphere particles of lithium (Li), yttrium (Y), lead (Pb), zinc (Zn), antimony (Sb), tantalum (Ta), cobalt (Co), beryllium (Be), nickel (Ni), and arsenic titanium (Ti) along with other harmful substances, including those contained in its plastic components. The same goes for computers. Their basic components contain polychlorinated biphenyls (PCBs), lead, mercury, cadmium, chromium, and dioxins. Incineration and acid treatments required to recycle these substances release toxic byproducts. In addition, plastic parts of a computer cannot be recycled since they contain flame retardants.

4.22   Export of E-Waste

Environmental catastrophes caused by corporations, such as in 1976 in Seveso, Lombardy, or in 1986 when 14 thousand tons of toxic ash were dumped by cargo ship *Khian Sea* in Haiti and into the ocean, led to the Basel Convention on the Control of Transboundary Movements of Hazardous Wastes and Their Disposal. As of October 2018, 186 states and the European Union are parties to the Convention. Haiti and the USA have signed the Convention, but not ratified it. Refusing to pay for the recycling costs provided in this treaty, corporations discard their e-waste in landfills or export about 80% of it to Asia and Africa. Available statistics are unable to track the precise amount of e-waste shipped to these continents. The data available, however, shows that Europe exports its electronic waste, usually illegally, to Eastern European countries and to Pakistan, India, China, Egypt, Senegal, the Ivory Coast, Benin, Nigeria, and Ghana. In 2015/2016, European countries were the origin of around 77% of e-waste imported into Nigeria. Jim Puckett, Chris Brandt, and Hayley Palmer (2018) affirm:

> The likely illegal exports of hazardous wastes from the EU flowed to the developing countries of Nigeria, Ghana, Tanzania, Ukraine, Pakistan, Thailand and Hong Kong. (…) Africa was by far the region of the world most targeted by EU e-waste exporters. (…) Extrapolation of the export rates to developing countries from our study from all of the 28 member states of the EU, gives a total of 352,474 metric tonnes exported per annum, which could fill 17,466 40-foot intermodal containers which would stretch back-to back on 18-wheel trucks for 401 km.

In Accra, the capital of Ghana, the Agbogbloshie market, home to more than 80,000 people who survive by extracting metals from e-waste, is a public health catastrophe.

Jindrich Petrlik et al. (2019) reported that the eggs of chickens that forage in Agbogbloshie contain extremely high levels of chemicals such as dioxins, brominated dioxins, polychlorinated biphenyl (PCBs), polybrominated diphenyl ethers (PBDE), and short-chain chlorinated paraffins (SCCPs): "An adult eating just one egg from a free-range chicken foraging in the Agbogbloshie scrap yard and slum would exceed the European Food Safety Authority (EFSA) tolerable daily intake (TDI) for chlorinated dioxins by 220-fold."

"If Europe is leaking, the U.S. is hemorrhaging" (Puckett et al. 2018): the USA produces more e-waste than any country in the world. In 2016, Americans generated about 14% of the world's electronic waste. The rate of recycling in the country is supposedly about 25%. But, according to a study carried out by the Basel Action Network (2018), the "recyclers" just ship abroad about a third of these materials, with almost a third of these exports (31%) probably being illegal under the legislation of the importing countries. Most of this exported equipment went to ten countries, including China, Taiwan, Pakistan, Mexico, Thailand, Cambodia, and Kenya. In 2018, Beijing imposed a ban on the import of plastic and e-waste, implementing China's "National Sword" policy. This ban led to a diversion of e-waste to Hong Kong's New Territories region near the Chinese border, as well as to some Southeast Asian countries, like Thailand, Vietnam, and Malaysia.

4.23   Conclusion

António Guterres, UN Secretary-General, and the scientific community as a whole incessantly warn that the relentless pace of climate change remains the most *systemic* threat to humankind along this century. But pollution caused by exposure to toxic air, water, soil, and industrial chemical compounds is still the largest *immediate* environmental threat to humans. It killed at least 8.3 million people, nearly one death in seven, and it was responsible for 275 million disability-adjusted life years in 2017. This is the central message of the Global Alliance on Health and Pollution study (GAHP, December 2019), which uses the most recent Global Burden of Disease data from the Institute of Health Metrics and Evaluation (IHME). It is hardly surprising that India and China, which are industrializing rapidly, are the worst affected countries, with over 2.3 million and 1.8 million premature pollution-related deaths in 2017, respectively. But other heavily populated countries, such as Nigeria, Indonesia, Pakistan, Bangladesh, the USA, the Russian Federation, Ethiopia, and Brazil, are also badly affected. Outdoor air pollution "is estimated to be responsible for five to ten per cent of the total annual premature mortality in the contiguous United States" (Dedoussi et al. 2020). As put by Rachael Kupka, Executive Director of GAHP, "pollution is a global crisis. It does not matter where you live. Pollution will find you."

References

ABRELPE, *Panorama dos Resíduos Sólidos no Brasil*, São Paulo, 2011 and 2015.

ALMEIDA, Vicente Eduardo Soares de et al., "Uso de sementes geneticamente modificadas e agrotóxicos no Brasil: cultivando perigos". *Ciência & Saúde Coletiva*, 22, 10, October 2017.

ANDRADY, Anthony L. (ed.). *Plastics and the Environment*. New Jersey, Wiley-Interscience, 2003

ATWOOD, Donald & PAISLEY-JONES, Claire, *Pesticides Industry Sales and Usage 2008–2012 Market Estimates*. EPA, 2017.

BALDÉ, C.P. et al., *The Global E-waste Monitor – 2017*, UNU, ITU & ISWA, Bonn/Geneva/Vienna, 2017.

BEJGARN, Sofia et al., "Toxicity of leachate from weathering plastics: An exploratory screening study with *Nitocra spinipes*". *Chemosphere*, 132, August 2015, pp. 114–119.

BENBROOK, Charles, *Impacts of Genetically Engineered Crops on Pesticide Use: The First Thirteen Years*". The Organic Center, 2009.

BOMBARDI, Larissa Mies, Geografia do Uso de Agrotóxicos no Brasil e Conexões com a União Europeia. FFLCH – USP, November 2017.

BRADSHER, Keith, "After China's Rare Earth Embargo, a New Calculus". *The New York Times*, 30/X/2010.

BRADSHER, Keith. "China Tries to Clean Up Toxic Legacy of Its Rare Earth Riches". *The New York Times*, 22/X/2013

BROOKS, Amy L., WANG, Shunli & JAMBECK Jenna R., "The Chinese import ban and its impact on global plastic waste trade". *Sciences Advances*, 4, 6, 20/VI/2018.

BRUGGERS, James, "Booming Plastics Industry Faces Backlash as Data About Environmental Harm Grows". *Inside Climate News*, 24/I/2020.

BURNETT, Richard et al., "Global estimates of mortality associated with long-term exposure to outdoor fine particulate matter". *PNAS*, 115, 2018, pp. 9592 – 9597.

BUTLER, Ethan E. & HUYBERS, Peter, "Adaptation of US maize to temperature variations" *Nature Climate Change*, 18/XI/2012.

CARLSTEDT, F.; JÖNSSON, B. A. & BORNEHAG, C. G. "PVC flooring is related to human uptake of phthalates in infants". *Indoor Air*, 23/II/2013.

CARRINGTON, Damian, "Insecticides put world food supplies at risk, say scientists". *The Guardian*, 24/VI/2014.

CASERTA, Donatella et al. "Bisphenol-A and the female reproductive tract: an overview of recent laboratory evidence and epidemiological studies". *Reproductive Biology and Endocrinology*, 12, 37, 2014.

CATHCART, Rebecca, "Borrowing an idea from Los Angeles, 2091". *The New York Times*, 22/V/2008.

CRESSEY, Daniel "Widely used herbicide linked to cancer". *Nature*, 24/III/2015.

CRODDY, Eric; PEREZ-ARMENDARIZ, Clarisa & HART, John. *Chemical and Biological Warfare. A comprehensive survey for the concerned citizen.* New York, Springer, 2002.

DEDOUSSI, Irene C. et al., "Premature mortality related to United States cross-state air pollution". *Nature*, 578, 7794, 13/II/2020, pp. 261–265.

DEVLIN, R. B. et al., "Controlled Exposure of Healthy Young Volunteers to Ozone Causes Cardiovascular Effects". *Circulation* (American Heart Association), 25/VI/2012

DICH, Jan et al. "Pesticides and Cancer". *Cancer, causes and control*, 8, 1997, pp. 420–443.

———. "Pesticide and prostate câncer. Again". *Ground Truth*, 23/I/2013.

DOUGLAS, T. "Patterns of land, water and air pollution by waste". *In*: NEWSON, M. (ed.). *Managing the Human Impact on the Natural Environment*. John Wiley & Sons, 1992, pp. 150–171.

EDWARDS, Clive E. "The impact of pesticides on the environment". *In*: PIMENTEL, David & LEHMAN, Hugh (eds.). *The Pesticide Question. Environment, Economics, Ethics.* New York, Routledge/Chapman, Hall, 1993.

EHRENSTEIN, Ondine S. von et al., "Prenatal and infant exposure to ambient pesticides and autism spectrum disorder in children: population based case-control study". *BMJ*, 364, 2019.

ERIKSON, Marcus et al. "Plastic Pollution in the World's Oceans: More than 5 Trillion Plastic Pieces Weighing over 250,000 Tons Afloat at Sea". *Plos One*, 10/XII/2014.

FARIAS, E., "Terras Indígenas da Amazônia alvos de pesquisas sobre terras raras". *Amazônia Real*, 21/X/2013.

FOUCART, Stéphane, "Bye bye BPA?". *Le Monde*, 6/I/2015.

FREDERIKSEN, H. et al. "Metabolism of phthalates in humans". *Molecular Nutrition and Food Research*, 51, jul. de 2007, pp. 899–911.

GAHP (Global Alliance on Health and Pollution) Global, Regional, and Country Analysis. December 2019.

GALBRAITH, John Kenneth. *The Affluent Society* (1958). New York, Houghton Mifflin, 1998.

GARRIC, Audrey, "Plus de 11.000 pesticides sont vendus aus États-Unis sans avoir été bien testés". *Le Monde*, 30/III/2013.

GEYER, Roland, JAMBECK, Jenna R., LAW, Kara Lavender, "Production, use, and the fate of all plastics ever made". *Science Advances*, 19/VII/2017.

GILBERT, Natasha, "A Hard Look at 3 Myths about Genetically Modified Crops". *Scientific American*, 1/V/2013.

GODIN, Christian. *La Haine de la Nature*. Seyssel, Éditions Champ Vallon, 2012.

GHOSAL, Aniruddha & CHATTERJEE, Pritha, "Landmark study lies buried. How Delhi's poisonous air is damaging its children for life". *The Indian Express*, 2/IV/2015.

GRAEDEL, Thomas et al., "On the materials basis of modern society". *PNAS*, 2/XII/2013.

GUO, Jianqiu et al., "Associations of prenatal and childhood chlorpyrifos exposure with Neurodevelopment of 3-year-old children". *Environmental Pollution*, 251, August 2019, pp. 538–554.

HOLMAN, Zoe, "Plastic debris reaches Southern Ocean, previously thought to be pristine". *The Guardian*, 27/IX/2012.

HOORNWEG, Daniel; BHADA-TATA, Perinaz, *What a Waste: A Global Review of Solid Waste Management* (World Bank, 2012).

HOORNWEG, Daniel; BHADA-TATA, Perinaz & KENNEDY, Chris. "Waste production must peak this century". *Nature*, 502, 31/X/2013, pp. 615–617.

HOREL, Stéphane, "Chlorpyrifos: les dangers ignorés d'un pesticide toxique". *Le Monde*, 17/VI/2019.

HURST, Cindy. *China's Rare Earth Elements Industry. What Can the West Learn?*. Institute for the Analysis of Global Security (IAGS), 2010.

IVES, Mike, "Boom in Mining Rare Earths Poses Mounting Toxic Risks". *Environment 360* 28/I/2013.

JAMBECK, Jenna R. et al. "Plastic waste inputs from land into the ocean". *Science*, 347, 6223, 13/II/2015, pp. 768–771.

JÖSTROM, M. & ÖSTBLOM, G. "Decoupling waste generation from economic growth – A CGE analysis of the Swedish case". *Ecological Economics*, 69, 7, 15/V/2010, pp. 1.545–1.552.

KALLENBORN, R. "Persistent organic pollutants (POPs) as environmental risk factors in remote high-altitude ecosystems". *Ecotoxicology and Environmental Safety*, 63, 1, 2006, pp. 100–107.

KINCH, Cassandra D. et al. "Low-dose exposure to bisphenol A and replacement bisphenol S induces precocious hypothalamic neurogenesis in embryonic zebrafish". *PNAS*, 112, 3, 20/I/2015.

KURUTO-NIWA, R. et al. "Estrogenic activity of alkylphenols, bisphenol S, and their chlorinated derivatives using a GFP expression system". *Environmental Toxicology and Pharmacology*, 19, 1, 2005, pp. 121–130.

LANDRIGAN, Philip J. et al., "The *Lancet* Commission on pollution and health". *The Lancet*, 391, 10119, 19/X/2017.

LANDRIN, S., Van EECKHOUT, L., « A Paris, la pollution équivaut à du tabagisme passif". *Le Monde*, 25/XI/2011.

LATOUCHE, Serge, *Petit traité de la décroissance sereine*. Paris, Fayard, 2007.
———. *Bon pour la casse*. Paris, Les liens qui libèrent, 2012.
LAVILLE, Sandra & TAYLOR, Matthew, "A million bottles a minute: world's plastic binge 'as dangerous as climate change'". *The Guardian*, 28/VI/2017.
LEBRETON, Laurent et al. "Evidence that the Great Pacific Garbage Patch is rapidly accumulating plastic". *Nature. Scientific Reports*, 8, 2018.
MAKOWER, Joel, *Strategies for the Green Economy: Opportunities and Challenges in the New World of Business,* McGraw Hill, 2009.
MEIKLE, Jeffrey L. *American Plastic. A Cultural History*. Rutgers University Press, 1997.
MORIN, Hervé, "Nouveaux soupçons sur le bisphénol A". *Le Monde,* 25/IV/2014.
MORTENSEN, David, "Navigating a Critical Juncture for Sustainable Weed Management". *BioScience*, 62, 1, January 2012.
NOVOTNY, Thomas, "Time to kick cigarette butts – they're toxic trash". *New Scientist*, 25/VI/2014.
PACE (Platform for Accelerating the Circular Economy), *A New Circular Vision for Electronics Time for a Global Reboot*, World Economic Forum, January 2019.
PACKARD, Vance. *The Waste Makers*. New York, 1960.
PACKARD, Vance. *The Waste Makers. L'art du gaspillage*. Paris, Calman-Lévy, 1962.
PETRLIK, Jindrich et al., *Weak Controls: European E-waste Poisons Africa's Food Chain*. IPEN and BAN, April 2019.
POWER, Melinda C. et al. "The relation between past exposure to fine particulate air pollution and prevalent anxiety: observational cohort study". *The British Journal of Medicine*, 24/III/2015.
PUCKETT, Jim, BRANDT, Chris & PALMER, Hayley, *Holes in the Circular Economy. Waste Electrical and Electronic Equipment (WEEE) Leakage from Europe*. Basel Action Network 2018.
REED, Christina, "Dawn of the Plasticene". *New Scientist*, 31/I/2015, pp. 28–32.
RITTER, L, SOLOMON, K.R. & FORGET, J., "Persistent Organic Pollutants". *The International Programme on Chemical Safety* (IPCS), December 1995.
ROCKE, Alan J. *The Quiet Revolution. Hermann Kolbe and the Science of Organic Chemistry.* University of California Press, 1993.
ROSSI, Marina, "Agrotóxicos: o veneno que o Brasil ainda te incentiva a consumir". *El País*, 11/IV/2016.
SAAL, Frederick S. vom et al. "Chapel Hill bisphenol A expert panel consensus statement: Integration of mechanisms, effects in animals and potential to impact human health at current levels of exposure". *Reproductive Toxicology*, 24, 2, August-September 2007, pp. 131–138.
SANDELSON, Michael, "Norway researchers' toxic salmon warning creates wave". *The Foreigner*, 25/XII/2013.
SEBILLE, Erik van et al. "A global inventory of small floating plastic debris". *Environmental Research Letters*, 10, 12, 8/XII/2015.
SERRES, Michel, *Retour au "Contrat naturel"*. Paris, Bibliothèque Nationale de France, 1998.
———. *Le Mal propre*. Paris, Le Pommier, 2012.
SILPA, Kaza, YAO, Lisa, BHADA-TATA, Perinaz & Van WOERDEN, Frank, *What a Waste 2.0: A Global Snapshot of Solid Waste Management to 2050. Urban Development*. World Bank, Washington, DC, 2018.
SLADES, Giles. *Made to Break. Technology and Obsolescence in America*. Cambridge University Press, 2006.
STOETT, Peter J., "Environmental Problems, Policies, and Prospects in Africa: A Continental Overview". In, O.P. DWIVEDI & Dhirendra K. VAJPEYI, *Environmental policies in the third world. A comparative analysis*. Greenwood Press, 1995.
STROMBERG, Joseph, "When Will We Hit Peak Garbage?" *Smithsonian Institution*, Washington, 30/X/2013.
TABASHNIK, Bruce E., BRÉVAULT, Thierry & CARRIÈRE, Yvez, "Insect resistance to Bt crops: lessons from the first billion acres". *Nature Biotechnology,* 31, 510–521, 10/VI/2013

TAI, Amos P. K.; MARTIN, Maria Val & HEALD, Colette L. "Threat to future global food security from climate change and ozone air pollution". *Nature Climate Change*, 4, 27/VII/2014, pp. 817–821.

TAI, P. et al. "Biological toxicity of lanthanide elements on algae". *Chemosphere*, 80, 9, August 2010, pp. 1031–1035.

TETT, Gillian, "West's debt explosion is real story behind Fed QE dance". *Financial Times*, 19/IX/2013.

TILMAN, David et al., "Forecasting Agriculturally Driven Global Environmental Change". *Science*, 292, 281, 2001.

VALO, Martine, "Des pesticides en doses toujours plus massives dans les campagnes". *Le Monde*, 9/III/2016.

VIÑAS, René & WATSON, Cheril S. "Bisphenol S disrupts estradiol-induced nongenomic signaling in a rat pituitary cell line: effects on cell functions". *Environmental Health Perspectives*, 17/I/2013.

WEISMAN, Alan. *The World Without Us*. London, Random House, 2007.

WHO (World Health Organization), Air Quality Guidelines for particulate matter, ozone, nitrogen dioxide and sulfur dioxide. Global update. Summary of Risk Assessment, 2005.

World Health Organization (WHO), "Air quality deteriorating in many of the world's cities". News Release, 7/V/2014.

World Health Organization (WHO), "How air pollution is destroying our health", November 2018.

World Health Organization (WHO), "Sanitation", 14/VI/2019.

Chapter 5
Fossil Fuels

In Chaps. 7 and 8, we will address the central issue of climate change, caused mainly by GHGs released into the atmosphere through the burning of fossil fuels and the destruction of forests which are large repositories of carbon. Expanding on the previous one, this chapter will discuss pollution caused by fossil fuels in the phases before consumption, especially pollution caused by petroleum, a compound mostly composed of nitrogen-containing hydrocarbons (95%), but that also contains heavy metals, sulfur, and nitrogen compounds. Due to its extreme harmfulness and its importance in the fossil fuel scenario, the next chapter will be dedicated to a discussion on coal.

5.1   Pollution in the Extraction and Transportation Processes

Russia is the world's largest source of pollution from oil spills. Estimates vary, but the Russian government itself acknowledges that about 1.5 million tons of oil (around 10 million barrels) are spilled each year (Thompson 2017). Other estimates, however, are much worse, as reported by Nataliya Vasilyeva (2014):

> Environmentalists estimate at least 1% of Russia's annual oil production, or 5 million tons, is spilled every year. That is equivalent to one Deepwater Horizon-scale leak [in the Gulf of Mexico] about every two months. Crumbling infrastructure and a harsh climate combine to spell disaster in the world's largest oil producer, responsible for 13% of global output. (…) Russian state-funded research shows that 10–15% of Russian oil leakage enters rivers; and a 2010 report commissioned by the Natural Resources Ministry shows that nearly 500,000 tons slips into northern Russian rivers every year and flow into the Arctic. The estimate is considered conservative: The Russian Economic Development Ministry in a report last year [2013] estimated spills at up to 20 million tons per year. That astonishing number, for which the ministry offered no elaboration, appears to be based partly on the fact most small leaks in Russia go unreported. Under Russian law, leaks of less than 8 tons are classified only as "incidents" and carry no penalties. Russian oil spills also elude detection because most happen in the vast swaths of unpopulated tundra and conifer forest in the north, caused either

© Springer Nature Switzerland AG 2020   131
L. Marques, *Capitalism and Environmental Collapse*,
https://doi.org/10.1007/978-3-030-47527-7_5

by ruptured pipes or leakage from decommissioned wells. (…) Even counting only the 500,000 tons officially reported to be leaking into northern rivers every year, Russia is by far the worst oil polluter in the world.

In North America, unconventional oil (shale oil or oil extracted from tar sands) has become abundant since 2005, so the USA and Canada no longer need to import large quantities of this fuel by sea. Its distribution by trains and pipelines, on the other hand, has led to an explosive increase in land-based accidents. According to Amanda Starbuck (2015):

Since 2010, over 3,300 incidents of crude oil and liquefied natural gas leaks or ruptures have occurred on U.S. pipelines. These incidents have killed 80 people, injured 389 more, and cost $2.8 billion in damages. They also released toxic, polluting chemicals in local soil, waterways, and air.

Between 2010 and 2014, therefore, there was an average of almost two incidents per day. Data from the Pipeline and Hazardous Materials Safety Administration (PHMSA) shows that the number of significant accidents in pipelines in the USA increased by 26.8% between 2006 and 2015 (Zukowski 2016). In railroad accidents, more crude oil was spilled in the USA in 2013 than in the sum of 38 years between 1975 and 2012: 4.3 million liters in 2013 versus 3 million liters in the previous four decades (Gerken 2014). Between 2010 and 2014, "train bomb" derailments and explosions occurred in Illinois, Maryland, Montana, Pennsylvania, Texas, Washington, Virginia, West Virginia, North Dakota, and Alabama, to name only the largest accidents in the country in these years.[1] In Quebec, Canada, the July 2013 catastrophe in the city of Lac-Mégantic and in the lake by the same name resulted in the spillage of 5.4 million liters of oil (34,000 barrels). Between 2011 and 2015, there were seven major spills in Alberta province, altogether leaking more than 13 million liters (82,000 barrels) of oil or of various blends of bitumen, water, and sand from pipelines into the environment.[2]

That said, the marine environment continues to be the biggest oil victim. In 2014, around 90% of all goods worldwide were transported by ship—a total of almost 10 billion tons (Gt) of goods per year (UNCTAD) up from almost 4 Gt in 1980. The lion's share is made up of crude oil (17%), according to the *World Ocean Review* (2017). The *World Ocean Review* (2010) discusses marine pollution caused by oil in its fourth chapter:

Oil pollution is one of the most conspicuous forms of damage to the marine environment. Oil enters the seas not only as a result of spectacular oil tanker or oil rig disasters, but also – and primarily – from diffuse sources, such as leaks during oil extraction, illegal tank-cleaning operations at sea, or discharges into the rivers which are then carried into the sea.

Table 5.1, published by the *World Ocean Review* (2010), shows that 80% of the marine pollution generated by oil is not a result of poor operating practices or of accidents which could hypothetically be avoided. This pollution is systematic: it is

---

[1] Fore more detailed information, see the site *The Right-to-Know-Network* (RTKNET).

[2] See "Alberta pipelines: 6 major oil spills in recent history". *CBC News Canada*, 17/VII/2015.

Table 5.1 Sources of marine pollution by oil

Natural sources	5%
Tanker traffic and other shipping operations (illegal discharges, tank cleaning, etc.)	35%
Volatile oil constituents which are emitted into the atmosphere during various types of burning processes and which then enter water Inputs from municipal and industrial effluents and from oil rigs	45%
Oil tanker disasters	~10%
Undefined sources	5%

Source: *World Ocean Review 2010*

a phenomenon that is inseparable from the thermo-fossil civilization which we call industrial capitalism, all the more so in its phase of extreme globalization.

According to data from the UN International Maritime Organization (IMO), oil tankers transport some 2900 million tons of crude oil and oil products every year around the world by sea. The IMO authorizes them to discharge oil at sea in concentrations below 15 parts per million (ppm), provided that it is done in non-sensitive areas. It is only when this concentration is exceeded that the discharge is considered illegal. The scale of this phenomenon is colossal. Something like 250,000 barrels of oil pollute the Persian Gulf each year, even without the occurrence of accidents.[3]

5.1.1   Spills from Ships (ITOPF) and Offshore Platforms

The International Tanker Owners Pollution Federation Limited admits that, excluding acts of war, "approximately 5.86 million tonnes of oil have been lost as a result of tanker incidents from 1970 to 2018" (Oil Tanker Spill Statistics, ITOPF 2018). ITOPF divides vessel spills into three types: (1) less than 7 tons, (2) from 7 to 700 tons, and (3) more than 700 tons of oil (respectively equivalent to less than 50 barrels of 159 l each, 50–5000 barrels, and more than 5000 barrels). Compared to the last decades of the twentieth century, the number of big spills has declined sharply in the twenty-first century. Even so, between 2000 and 2009, there were on average 3.2 big spills per year, and between 2010 and 2015, there were on average almost two big spills per year (1.8). According to ITOPF data, in the 3 years between 2015 and 2017, around 20,000 tons of petroleum polluted the oceans due to large- and medium-sized oil tanker accidents.

Of the 10,000 cases of accidental oil spills between 1970 and 2010 (two accidents every 3 days), 81% refer to losses of less than seven tons: "the vast majority of spills are small (i.e. less than 7 tonnes) and data on the number of incidents and quantity of oil spilt is incomplete due to inconsistent reporting of smaller incidents worldwide" (ITOPF). This "inconsistency" is suspicious because such data is provided by the oil companies themselves. Robert Howarth claims that they falsify or

---

[3] See GEO/SAT 2: "An estimated quarter of million barrels of oil pollute the Gulf each year."

omit facts in their reports to environmental protection agencies and ministries: "I was a specialist in an Alaskan tribe in the 1990s regarding the development of oil exploitation on offshore platforms. It was possible to demonstrate that the oil companies sent false information to the EPA" (as cited in *Le Monde* 3/XI/2012).

Even without counting the BP spill in the Gulf of Mexico in 2010, data provided by the Minerals Management Service (MMS) shows an increase in spills from offshore platforms in recent years in the USA. Analyzing this data, Alan Levin (2010) concludes: that:

> The number of spills from offshore oil rigs and pipelines in U.S. waters more than quadrupled this decade, according to government data. (…) From the early 1970s through the '90s, offshore rigs and pipelines averaged about four spills per year of at least 50 barrels. (…) The average annual total surged to more than 17 from 2000 through 2009. From 2005 through 2009, spills averaged 22 a year.

5.1.2   2010–2018: From the Gulf of Mexico to the China Sea

The Deepwater Horizon disaster in the Gulf of Mexico in 2010, which killed 11 workers, triggered the worst offshore oil spill in the US history. At its height, 88,522 square miles (229,270 km^2) of sea were closed to fishing because of the spill. It was originally estimated that 4.4 million barrels, with a margin of error of 20% (Crone and Tolstoy 2010), were spilled, admittedly over 87 days, in daily quantities of up to ten times those reported by British Petroleum. Its directors lied in all phases of the incident, and only on September 19 of that year did they announce that they had managed to close the well. But, according to a Bonny Schumaker film (*Drill, Spill, Repeat? Breaking Offshore Oil Drilling's Destructive Cycle*, 2014) and a number of depositions reported by the press (Kistner 2011), submarine pipelines damaged by the blast were still leaking in September 2011. In September 9, 2012, satellite images showed the presence of new leaks connected to the Macondo well or to its ducts; these were then proven by a chemical analysis of these residues. Combining satellite and in situ observations, and focusing on toxic-to-biota (i.e., marine organisms) oil concentration ranges, a new assessment published by Igal Berenshtein et al. (2020) demonstrates a much wider extent of the Deepwater Horizon spill beyond the satellite footprint, reaching the West Florida Shelf, the Texas Shores, the Loop Current system, and the Florida Keys. The authors warn that:

> Photoinduced toxicity studies demonstrate that the polycyclic aromatic hydrocarbons (PAHs) concentrations that cause 50% mortality (LC_{50}) or other deleterious effects (EC_{50}) to the test organisms are largely in the range of invisible and toxic oil. Oil at these concentrations for surface water (depth, 0 to 1 m) extended beyond the satellite footprint and the fishery closures, potentially exterminating a vast amount of planktonic marine organisms across the domain.

Between 2010 and 2013, a number of medium (7–700 tons) and larger (more than 700 tons) oil spills in the ocean or in rivers occurred due to accidents on offshore platforms, on oil tankers, or in coastal reservoirs. In 2013, Typhoon Haiyan caused a coastal spill of some 300 tons of oil in the Philippines, and in 2014, an oil tanker was shipwrecked in the Sundarbans delta in Bangladesh, causing an oil spill that spread over 10,000 km². The year 2017 was especially damaging to the sea and to marine life. In January, a collision of an oil tanker caused a spill of about 40 tons of oil off the Chennai coast in the extreme south of India. In May, the East River, which together with the Hudson River forms the island of Manhattan, was polluted with about 100 tons of an oil derivative as a result of an accident at Com Edison, which supplies power to New York. In September, an oil tanker, Agia Zoni II, was shipwrecked with 2500 tons of oil at a prominent point in the Aegean Sea between the port of Piraeus and the island of Salamis in Greece.[4] The Gulf of Mexico is a victim of repeated spills, and in October 2017, more than 16,000 barrels of oil spilled near the Louisiana coast from a submarine pipeline operated by LLOG Exploration (Caron 2017).

In January 2018, the collision between the Iranian tanker, *Sanchi*, and the Hong Kong freighter, CF *Crystal*, in the China Sea (300 km off of Shanghai), caused 32 deaths and an uncontrollable fire for 8 days. The tanker finally sank to a depth of 115 m with more than 100,000 tons of condensate oil, an ultra-light, particularly toxic, and volatile form of oil, as well as 2000 tons of fuel. One month after the accident, three ultra-light oil stains spread over 340 km² of the ocean's surface. The extent of damage to marine life has not yet been assessed, as there is no precedent for the spill of condensate oil in the sea and because the oil could be transported by the Kuroshio current. In any case, the damage is immense to the point of being compared to the disaster caused by the 1982 Exxon Valdez shipwreck, also because the Sanchi shipwreck area is a spawning ground for fish and an area of whale migration. The question is not whether it will occur, but when and to what extent will be the next disasters.

5.1.3   Acts of War

Before addressing the impacts of these oil spills, one must remember at least the two biggest cases of major oil spills as deliberate acts of war in the last 30 years. At the end of the Persian Gulf war, between January and February 1991, Saddam Hussein ordered several sabotage operations that resulted in fires in two mainland refineries, in an offshore loading terminal, in anchored tankers, and in over 730 oil wells, amounting to about 50% of all fires in oil wells throughout the industry's history. The fires consumed 10% of the world's daily oil consumption each day. By the end of 8 months, about one billion barrels of oil had burst into flames, an enormous amount, though equivalent to less than 10 days of global oil consumption in 2020. Furthermore, millions of barrels were spilled into the soil, and ten million barrels of oil were deliberately thrown into the Persian Gulf (twice as much as the 2010 leak

---

[4] See "Greek oil spill threatens popular Athens beaches". *BBC*, 14/IX/2017.

in the Gulf of Mexico). These numbers do not include the losses, fires, and oil spills caused by US bombings of Iraqi oil wells; these were captured by satellite images, but their actual loss and pollution balance has been kept confidential (Seacor 1994).[5]

Another example of pollution from acts of war was the Israeli bombing, on July 3, 2006, of the al-Jiyeh thermoelectric power plant off the coast of Lebanon, 30 km south of Beirut, causing the explosion of six oil tanks that fed the power plant and the largest ecological disaster in the Mediterranean. The leak and the fire lasted 10 days. According to Yacoub Sarraf, Lebanon's Environment Minister, all attempts to put out the fire or control the leak were prevented by the Israeli army. The result was the pollution of the sea with 100,000 (BBC and UNEP) to 200,000 (Bloomberg) barrels of oil: an oil smear covered 80 km of beaches in Lebanon and threatened those of Turkey and Cyprus, destroying much of the marine fauna, damaging the habitat of the green turtle (*Chelonia mydas*), and killing many bluefin tuna (*Thunnus thynnus*), two endangered species in the IUCN Red List of threatened species (Milstein 2006).

5.1.4   Acute and Long-Term Impacts

According to David Lusseau (*apud* Allix et al. 2011):

> In order to get an idea of the global balance [on an accident], two things need to be evaluated. The immediate and acute effects and the chronic effects. In the case of the Exxon-Valdez [the tanker that sank in 1989 on the Alaskan coast], these long-term effects on ecosystems could only be assessed over ten years.

In the case of the Gulf of Mexico, a team dove down more than 20 times to a depth of more than 1500 m and immediately found a *kill zone* in an area of 210 km² around the explosion. Just in the period between May and December 2010, 90 cetaceans were found dead on the beaches of the region. According to Davie Lusseau, the slaughter was probably much higher, since "in certain species only 1–3% of the dead are found." In fact, in 2015, 5 years after the explosion, various studies compiled by the Environmental Defense Fund (EDF), the National Audubon Society, the National Wildlife Federation, and Lake Pontchartrain Basin counted 201,000 pelicans (*Pelecanus occidentalis*) exposed to oil and the death of 1000 dolphins, 800,000 birds, and 20,000 to 60,000 kemp turtles (*Lepidochelys kempii*) (Schatzel 2015). In terms of impacts on human health, Wilma Subra (2011) observed the following symptoms on coastal city populations:

> Eye, nose and throat irritation, blurred vision and loss of vision, ear infections, sore throat, severe coughing for many months, croup like cough, hoarseness, difficulty breathing, shortness of breath, respiratory distress, pneumonia, lung irritation, asthma attacks, decreased lung function, chest pains and tightness, pulmonary edema, bleeding from nose, eyes and

---

[5] Werner Herzog's film, *Lektionen in Finsternis* (1992), offers a silent visual meditation on this devastated landscape.

ears, internal bleeding, blood in urine, rectal bleeding, blood in stool, diarrhea, nausea, vomiting, dizziness, weakness, loss of balance, headaches, kidney pain, lower back pain, abdominal pain, gastrointestinal disturbance, joint and muscle pain, weakness and fatigue, seizures, loss of weight, skin irritation, burning and lesions, nerves on edge, mental confusion, short term and long term memory loss, psychological damage, damage to liver and kidneys, immune system damage and suppression, central nervous system impacts, blood disorders, damage to red blood cells, aplastic anemia, leukemia, cardiovascular system stress, chest pain, heart palpitations, chronic obstructive pulmonary disease, hypertension, endocrine disruption, hormone level disruption, miscarriages, metal taste in mouth, depression.

5.2   The Devastation of Tropical Ecosystems

Everything pales, however, in comparison to the pollution caused by oil majors in tropical ecosystems. Since 1960 when it gained independence from the UK, Nigeria—the largest oil producer in Africa and the fifth in the Organization of the Petroleum Exporting Countries (OPEC)—has had its ecosystems destroyed by five oil corporations (Shell, Chevron, Mobil, Elf, and Agip). These operate in symbiosis with brutal dictatorships, one that a *Democracy Now* documentary (produced by Amy Goodman and Jeremy Scahill) has called *Drilling and Killing* (1998). As Chima Ubani claims in this documentary, oil corporations "are simply continuing what the Transatlantic slave trade and British colonialism did to us in the past." The socio-environmental devastation perpetuated by these companies has impaired the county to such a degree that the UN evaluates the time needed to restore its ecosystems at 30 years.

Shell has especially devastated the delta formed by the Niger and Benue rivers, one of the largest in the world and once also one of the richest in biodiversity. Explored since 1958, the oil basin of this delta now has 1183 oil exploitation fields, which extend over forests, mangroves, marshes, and offshore platforms. A UNEP report (*Environmental Assessment of Ogoniland*, 2011) documented 6817 oil spills in Nigeria between 1976 and 2001; these bombarded Nigerian ecosystems with 3 million oil barrels, of which 70% have never been recovered. A World Bank Document (*Defining an Environmental Strategy for the Niger Delta*, 1995) and other studies (Moffat and Lindén 1995) estimate that the actual amount of oil spilled into nature in Nigeria may be ten times greater than what is officially reported, reaching 100 million barrels. Ole Nielsen (2011) estimates that in 50 years (1960–2010), Shell leaked 13 million barrels (550 million gallons) of oil into this Nigerian delta, roughly equivalent to one Exxon Valdez per year. Nigeria's Department of Petroleum Resources estimates that in only 20 years (between 1976 and 1996), there were 4835 incidents in the country causing the spilling of 2.4 million barrels of oil, of which 1.8 million were in the Niger Delta.

Between 1972 and 1992, Texaco (incorporated into Chevron in 2002) polluted and devastated Ecuador's natural heritage by dumping 68 billion liters of toxic waste from oil and other chemicals into its forests and rivers. Chevron has caused vastly greater environmental and human disasters in the country than BP's spill in the Gulf of Mexico, leading to a dramatic increase in the incidence of cancer in the population. Chevron left in 1992, leaving more than 900 unlined waste pits that continue to contaminate groundwater and surface water. In 2017, Ecuador permitted oil exploration in 1030 ha of the Yasuni National Park. This park, consecrated by UNESCO as a biodiversity hotspot, is home to indigenous tribes, two of which are isolated, and is a refuge for more than 20 endangered mammal species. In all likelihood, therefore, this area will be affected by the state-owned Petroamazonas, with its Chinese funding.

Ironically, while Peru hosted the COP 20 in December 2014, the impacts of five oil spills in June 2014 from Petroperú's oil pipelines in the Peruvian Amazon forest were increasingly being felt. The Peruvian government is expanding its oil and gas extraction operations in the Amazon forest. According to the Carnegie Institution for Science, illegal logging and the deforestation wrought by these operations represent two-thirds of Peru's carbon emissions.[6]

5.2.1 Brazil (1975–2017)

The city of Cubatão near the coast of São Paulo was known as Death Valley in the 1980s because the toxic gases emitted from the region's petrochemical complex were responsible for countless deaths and teratogenic deformities. Numerous oil spills occurred in Brazil between 1975 and 2017. In 1984, 700,000 liters of gasoline spilled in the outskirts of Cubatão (Vila Socó) causing a fire that claimed more than 100 lives. In July 2000, 4 million liters of crude oil leaked from Petrobras' Repar refinery, polluting the Iguaçu river basin in Araucária, Paraná, in what was the largest of six accidents caused by Petrobras only in 2000. The Guanabara Bay, where the city of Rio de Janeiro is located, and the Campos Basin, where Petrobras's offshore platforms are located, have been the scenes of several oil spills. The biggest spills occurred in 2000, 2011, 2015, and 2017. A June 2017 technical report by the Brazilian Institute of the Environment and Renewable Natural Resources (Ibama, 43/2017) accuses Petrobras of dumping oil and grease into the sea—from its platforms in the Campos Basin—"well above the maximum limit allowed, reaching an amount 1925% greater than the reported result." The false information provided by Petrobras since 2008 has been consistent with the company's long history of environmental irresponsibility.

---

[6]"The Amazon oil spills overlooked by environmental leaders in Lima". *The Guardian*, 10/XII/2014.

5.3   Subsidies and Investments in the Oil Industry

According to the definition adopted here, post-tax consumer subsidies to fossil fuels occur when consumer prices are below supply costs plus environmental costs (or externalities) and general consumption taxes (Coady et al. 2019). Based on these factors, David Coady and colleagues found that subsidies amounted to US$ 4.9 trillion worldwide in 2013 and US$ 5.3 trillion in 2015 (6.5% of global GDP). Analyzing the first draft of this study, published in 2015 as an IMF working paper, Sir Nicholas Stern declared (as cited by Carrington 2015):

> This very important analysis shatters the myth that fossil fuels are cheap by showing just how huge their real costs are. There is no justification for these enormous subsidies for fossil fuels, which distort markets and damages economies, particularly in poorer countries.

In 2009, G20 leaders pledged to phase out fossil fuel subsidies. Over the last 10 years, these subsidies have grown and continue to increase, unmasking the real interests of state-corporations and their dependence on the fossil fuel industry. Even if more restrictive criteria for defining fuel subsidies are adopted, the estimates still remain scandalously high. For the IEA (World Energy Outlook 2010), "fossil-fuel consumption subsidies worldwide amounted to US$ 312 billion" in 2010. According to Elizabeth Bast et al. (2015), "G20 country governments are providing $452 billion a year in subsidies for the production of fossil fuels." A study conducted by Oil Change International (2013) showed a range of subsidies, "from at least $775 billion to perhaps $1 trillion or more in 2012."[7] In 2014, a document titled *Reforming Fossil Fuel Subsidies for an Inclusive Green Economy*, signed by the IMF, UNEP, Global Subsidies Initiative, and Deutsche Gesellschaft für Internationale Zusammenarbeit (GIZ), estimated that post-tax subsidies can be closer to US$ 2 trillion worldwide, which was then equivalent to about 2.9% of global GDP, or 8.5% of government revenues. In 2018, the Overseas Development Institute (ODI) showed that the G7 countries—the USA, Germany, France, the UK, Italia, Canada, and Japan—continued to heavily subsidize fossil fuels: "On average per year in 2015 and 2016 the G7 governments gave at least US$ 81 billion in fiscal support and US$ 20 billion in public finance, for both production and consumption of oil, gas and coal at home and overseas."[8]

The IPCC, *Special Report. Global Warming of 1.5 °C* (2018), warned that meeting a target of 1.5 °C requires annual average investments in energy (solar, wind, carbon capture and storage, and nuclear) of $2.38 trillion. This clearly implies not only eliminating subsidies to the fossil fuel industry but rapidly discontinuing their production. In reality, not only do the subsidies remain, but, above all, the investment efforts in low-carbon energies advocated by the IPCC are being directed toward bolstering the fossil fuel system. Indeed, according to a report titled *Banking*

---

[7] "No Time to Waste: The Urgent Need for Transparency in Fossil Fuel Subsidies". *Oil Change International*, 2013

[8] See "G7 fossil fuel subsidy scorecard: tracking the phase-out of fiscal support and public finance for oil, gas and coal", ODI, June 2018.

on Climate Change. Fossil Fuel Finance Report 2020, since the Paris Agreement, 35 global banks invested a combined $2.7 trillion in fossil fuel companies (Chap. 13 provides more information about this). Given that such investments assume an expectation of long-term profitability, this is further evidence that capitalism is not transitioning to less impactful forms of energy. Further proof of this is the fact that, since 2015, the six major French banks have favored fossil fuels. The fossil fuel industry benefitted from 43 billion euros of financing, or 71% of the total financing to energy companies (*Le Monde* 26/XI/2018). So, regardless of its price, oil consumption curve continued to rise at a rate of 1.3% in 2018 and 1.4% in 2019 (IEA, "Oil Market Report", November 2018).

5.4   Unconventional Oil and Gas: Maximized Devastation

The increasing demand for oil on a global scale in the context of conventional oil scarcity (projected or already underway) spurred a hunt for oil and gas through even more destructive procedures. They have exacerbated the destructiveness of global capitalism by provoking:

1. A regression to coal, a phenomenon examined in the next chapter.
2. An increase in the extraction of unconventional oil, defined by its low solubility and high viscosity (with API below 20°), found in geological formations of low permeability and absence of fluid movement. Its extraction is much more polluting and emits more GHG than conventional oil. The distinction between the encompassing and ambiguous terms "conventional oil" and "unconventional oil" is not always precise. Here we consider unconventional oil: tar sands, shale oil, tight oil, heavy oil, and deepwater oil.
3. An increase in the extraction of non-conventional gas (shale gas).
4. The exploration of oil in areas of greater environmental risk, notably in deep-sea areas under layers of salt (pre-salt oil), in areas close to coral reefs, and in the Arctic.
5. A bigger production of petroleum coke, known as petcoke.

5.4.1   Tar Sands and Petcoke

Petcoke is a refinery by-product of petroleum, technically defined as a "carbonization product of high-boiling hydrocarbon fractions obtained in petroleum processing (heavy residues)" (IUPAC Gold Book 1997). Petcoke has even higher carbon emissions than coal. According to Lorne Stockman, David Turnbull, and Stephen Kretzmann (2013):

> Petcoke has higher carbon content than coal. Different quality coals have different carbon content levels and different energy yields. The two most common types of coal are

bituminous and sub-bituminous. Using a median figure for these two coal types we find that petcoke emits 53.6% more CO_2 per ton than coal and 7.2% more CO_2 per unit of energy.

The proven tar sand reserves of Canada can yield roughly 5 billion tons of pet-coke—enough to fully fuel 111 US coal plants up to 2050. Furthermore, as stressed by Yu Dawei (2013), its burning emits more sulfur dioxide (SO_2) and nitrogen oxide (NO) than burning coal. Even when it is desulfurized, the sulfur remaining in pet-coke is still equivalent to the sulfur emitted in coal burning. International demand for petcoke, especially from China, India, Japan, Turkey, Italy, Spain, and Mexico, is increasing. Most assessments of the climate impact of tar sands or conventional oil production and consumption do not include petcoke emissions. Thus, the climate impact of oil production is being consistently underestimated.

Tar sand reserves in Alberta, Canada, amount to 166 billion barrels of oil, extracted from sand particles covered in a microscopic layer of water, which is then covered by thick oil. In Alberta, these particles are roughly made of 83% sand, 4% water, 3% clay, and 10% bitumen. Giant excavations remove 400 tons of sand at a time. The sand is mixed with hot water and then with solvents and other toxic sub-stances so as to separate and refine the bitumen. According to data from the Global Forest Watch, 775,000 hectares of land in the Alberta region have been deforested or degraded between the years 2000 and 2013, and more than 12.5 million hectares have experienced habitat disruption due to tar sands development. This extraction also consumes huge amounts of water: to produce 3.7 liters (one gallon) of oil from bituminous sands, 130 l of water are used. In total, the extraction, refining, and final transformation of bitumen into liquid petroleum rely on processes that destroy for-ests, lower the levels of rivers, produce lakes of toxic mud, and are intense emitters of greenhouse gases. Lorne Stockman, David Turnbull, and Stephen Kretzmann (2013) show that "emissions from tar sands extraction and upgrading are between 3.2 and 4.5 times higher than the equivalent emissions from conventional oil pro-duced in North America. According to James Hansen (2012):

> The tar sands contain enough carbon—240 gigatons—to add 120 parts per million (ppm) [from 393 ppm of CO_2 in 2012]. Tar shale, a close cousin of tar sands found mainly in the United States, contains at least an additional 300 gigatons of carbon. If we turn to these dirtiest of fuels, instead of finding ways to phase out our addiction to fossil fuels, there is no hope of keeping carbon concentrations below 500 ppm.—a level that would, as earth's his-tory shows, leave our children a climate system that is out of their control.

In addition to destroying forests, oil exploitation in Alberta's tar sands compromises Lake Cold, as well as the Peace River and Athabasca River basins, which together cover an area of 146,000 km². Erin N. Kelly et al. (2010) showed that "the oil sands industry releases the 13 elements considered priority pollutants (PPE) under the U.S. Environmental Protection Agency's Clean Water Act, via air and water, to the Athabasca River and its watershed." Open-pit toxic waste in decantation tanks along the Athabasca River extends for more than 60 km.[9] These open-air tanks release gases, leak into the river water, contaminate the soil, and reach the groundwater.

---

[9] See "Tar Sands Oil Extraction – The Dirty Truth" (YouTube).

Contact with this material is deadly for thousands of migratory birds, and the habitat loss impacts the plant and animal species that once made up the biological wealth of this territory. One of the most famous and most beautiful is the Canadian reindeer (*Rangifer tarandus caribou*), today included in Canada's Species at Risk Act (SARA). Alberta exports its oil mainly to the USA, but also to China, Japan, the European Union, Mexico, and other countries. Furthermore, in the growing European import of oil from the USA are important quantities of oil from Canadian tar sands, re-exported by the USA (Neslen 2015).

5.4.2   Hydraulic Fracturing or Fracking

Shale oil is a fine-grained sedimentary rock containing significant amounts of kerogen, a solid mixture of organic matter from which liquid oil and natural gas can be obtained through distillation. Shale oil mining occurs through the process of hydraulic fracturing, commonly known as fracking, a technology that has been used since the 1940s (Inman 2014; Allison and Mandler 2018), but that was not considered cost-effective until the increase of gas and oil prices in this century. Fracking consists of injecting enormous quantities of water mixed with chemical substances and sand or ceramic materials into the subsoil—under high pressure and at a depth between 3 and 5000 m. Such injections cause microcracks and fragmentation of the rocks. The sand injected allows for these cracks to stay open and steel pipes are then inserted through which gas or oil is released. Between 30% and 70% of the water injected returns to the surface (Dunlop 2014).

In the USA, the use of horizontal drilling with hydraulic fracturing to access previously uneconomic oil and gas deposits has been responsible for a radical change in the energy landscape. From 2006 to 2015, US oil production increased by 88% (Allison and Mandler 2018). In 2000, 23,000 hydraulic fracturing wells extracted about 102,000 barrels of oil per day (bpd), amounting to 2% of the country's oil production. By 2015, 300,000 of these wells brought to the surface more than 4.3 million bpd. In December 2018, US oil production reached 12.04 million bpd (EIA 2019). Outside the USA, however, hydraulic fracturing has been used on a commercial scale only in Canada, China, and Argentina, with apparently little chance of further expansion, either due to geological difficulties or to strong societal opposition, given the enormous environmental costs of the process (Nummi 2015).

5.4.3   The Six Major Harms of Hydraulic Fractioning

1. Earthquakes

Earthquakes can be caused primarily by the process of hydraulic fracturing in proximity to pre-existing faults and also by the disposal of fracking wastewater via underground injection. Increased earthquake activity has been observed in Western Canada, Lancashire (UK), Sichuan (China), Texas, Ohio, and Oklahoma. In November 2011, there was an M_w 5.7 earthquake in Oklahoma which was potentially linked to this injection (Keranen et al. 2013); in this state, hydraulic fractioning has caused earthquakes higher than a 3-degree magnitude, in addition to about 50 small earthquakes, originating from the hydraulic fractioning of the Eola field (Holland 2011). The states of Arkansas and Colorado also find themselves in a similar situation. And, according to Won-Young Kim (2013):

> Over 109 small earthquakes (M_w 0.4–3.9) were detected during January 2011 to February 2012 in the Youngstown, Ohio area, where there were no known earthquakes in the past. These shocks were close to a deep fluid injection well. The 14 month seismicity included six felt earthquakes and culminated with a M_w 3.9 shock on 31 December 2011.

2. Destruction of Habitats

Some 3.7 million wells have been drilled in the USA since 1859 (Allison and Mandler 2018). According to Jose et al. (2015), "at least 2 million of these wells have been hydraulically fracture-treated, and up to 95% of new wells drilled today are hydraulically fractured, accounting for more than 43% of total U.S. oil production and 67% of natural gas production."[10] It is easy to understand why the destruction of habitats caused by the exploration of oil and natural gas through fracking is much greater than the destruction caused by the exploration of conventional oil. Ian Dunlop (2014) provides a rough idea of the huge impact on the countryside caused by this activity:

> Once fracking has been carried out, production rises rapidly to a peak, but it then declines rapidly, too, often by 80 to 95 percent over the first three years, as the oil or gas around the fractured area is exhausted. As a result, the countryside has to be peppered with wells to maintain the production required to provide a return on investment, often several thousand wells in a single shale play.

Habitats are destroyed by the installation of cemented and/or asphalted surfaces for the construction of extraction equipment, tanks, and other storage containers that occupy, on average, 3.6 ha per perforation platform. To convey the oil produced from these platforms, roads and pipelines will rip apart the vegetation between the extraction area and the refinery zone. The Marcellus and Utica shales underlie 1.5 million acres (607,000 ha) of Pennsylvania state forest land; in 2017, over 130,000 of these acres (52,609 ha) were under lease for shale gas production (Allison and Mandler 2018). Major sources of land for fracking come from forested land and agricultural land, in this state. A study prepared by ECONorthwest (2019) estimates that about 24–38% of new wells affect forested land in Pennsylvania, and according to a study by Nels Johnson et al. (2010), the scenario for average growth of hydraulic fraction in Pennsylvania indicates that:

> By 2030, a range of between 34,000 to 82,000 acres [13,700 to 33,200 hectares] of forest cover could be cleared by new Marcellus gas development in the state. (…) Such clearings

---

[10] See "How is Shale Gas produced?" The US Department of Energy, 2013.

would create new forest edges where the risk of predation, changes in light and humidity levels, and expanded presence of invasive species could threaten forest interior species in 85,000 to 190,000 forest acres [340 to 768 km^2] adjacent to Marcellus development.

3. Increasing Amounts of Water Wasted

Analyzing water use in gas and oil wells that have been explored using different hydraulic fractioning methods, Tanya J. Gallegos and Brian A. Varela (2015) note that there is a trend toward "substantial increases in water volumes used to hydraulically fracture wells over time." Indeed, the amount of water used per well for hydraulic fracturing increased by up to 770% between 2011 and 2016 (Kondash et al. 2018):

> The steady increase of the water footprint of hydraulic fracturing with time implies that future unconventional oil and gas operations will require larger volumes of water for hydraulic fracturing, which will result in larger produced oil and gas wastewater volumes.

As early as 2010, the documentary *Gasland* by Josh Fox revealed that each hydraulic fractioning operation requires 3 million to 25 million liters of water, while other sources indicate that horizontal shale wells can require more than 30 million liters of water per well.[11] A same well can be hydraulically fractured up to 18 times, and each time, it will demand an identical amount of water. Transporting 3 million liters of water, new or used, requires an average of 200 truckloads.

4. Toxicity and Contamination of Groundwater

The water used in hydraulic fracturing is mixed with anti-corrosive, viscosifying, and gelling agents, such as diesel, bromides, chlorides, thiocyanomethyl thiobenzo-thiazole (TCMB), as well as Btex (benzene, toluene, and ethylbenzene). Benzene is known to be carcinogenic (Lombardi 2014). In addition, chemical reactions triggered by these substances in the rocks can release arsenic, barium, strontium, and uranium, elements found in waters that were used for hydraulic fractioning (Tollefson 2013). Studies show that 20–85% of these substances remain underground and may contaminate groundwater. Stephen Osborn et al. (2011) affirm that "In aquifers overlying the Marcellus and Utica shale formations of northeastern Pennsylvania and upstate New York, we document systematic evidence for methane contamination of drinking water associated with shale-gas extraction."

5. Atmospheric Pollution

Some of the fluids used in hydraulic fracturing return to the surface. The water used, together with its chemical substances—to which are added heavy metals and radioactive elements from the rocks that have been fractured—is usually stored in tanks which release toxins into the atmosphere. A study reports atmospheric concentrations of hydrocarbons considered carcinogenic in the vicinity of shale gas wells, due to the volatility of gases such as trimethylbenzene, xylene, and aliphatic hydrocarbons (McKenzie et al. 2012). Based on an analysis of the Texas Commission

---

[11] Cf. "Hydraulic Fracturing 101". *Earthworks*

on Environmental Quality (TCEQ), Paul Gallay (2012) warned that hydraulic fracturing in this state:

> Emit more smog-causing volatile organic compounds (VOCs) than all cars, trucks, buses and other mobile sources in the area combined. This wasn't true before the fracking boom: TCEQ's data shows that VOCs from oil and gas production have increased 60% since 2006. Ozone, a corrosive gas that can exacerbate asthma and other respiratory diseases, is created when VOCs from petroleum operations mix with heat and sunlight.

6. Methane Leaks into the Atmosphere: The Impact on Climate Change

More important than the five previous items, hydraulic fracturing results in the leakage of methane (CH_4), a primary component of natural gas, in all the operational phases of this industry: exploration, production, extracting, processing, transmission, and distribution. This releasing of methane into the atmosphere can occur in such high percentages that their impact on climate change becomes worse than that of other fossil fuels, including coal. As is known, methane remains in the atmosphere for only about 9–12 years, but it has a global warming potential (GWP) of up to 86 times higher than CO_2 over a 20-year horizon (GWP_{20}) and 34 times higher over a 100-year horizon (GWP_{100}), with inclusion of climate–carbon feedbacks (IPCC AR5 2013 Chap. 8 Table 8.7). Natural gas emits 50–60% less CO_2 when combusted in a power plant, compared with emissions from a coal plant. But shifting to natural gas from coal has immediate climatic benefits if, and only if, the cumulative leakage rate from natural gas production is below 3.2% (Alvarez et al. 2012; Tollefson et al. 2013; Pandey et al. 2019). Contrary to the numbers considered by the US EPA—which in 2009 identified a 2.4% rate of methane leakages into the atmosphere in the processes of shale gas extraction through fracking—several studies show higher percentages of leakages. Below, we report the results of eight of these studies, published from 2011 to 2020, in chronological order:

(a) Robert Howarth, René Santoro, and Anthony Ingraffea (2011): shale gas wells leak between 3.6% and 7.9% of methane over its exploitation period:

> Methane contributes substantially to the greenhouse gas footprint of shale gas on shorter time scales, dominating it on a 20-year time horizon. The footprint for shale gas is greater than that for conventional gas or oil when viewed on any time horizon, but particularly so over 20 years. Compared to coal, the footprint of shale gas is at least 20% greater and perhaps more than twice as great on the 20-year horizon and is comparable when compared over 100 years.

(b) Measurements at ground level and from airplanes in 2010 by NOAA and the University of Colorado in Boulder detected values of up to 4% and even up to 9% of methane leakage (Tollefson et al. 2013).

(c) Scot Miller et al. (2013):

> Regional methane emissions [in the U.S.] due to fossil fuel extraction and processing could be 4.9 ± 2.6 times larger than in EDGAR [Emission Database for Global Atmospheric Research], the most comprehensive global methane inventory. (…) Overall, we conclude that methane emissions associated with both the animal husbandry and fossil fuel industries have larger greenhouse gas impacts than indicated by existing inventories.

(d) Satellite measurements made in 2014 in the Four Corners region (southwestern USA), coordinated by Eric Kort et al. (2014), identify the presence of almost twice as much methane as measurements taken at ground level.

(e) Dana Caulton and colleagues (2014) revealed the existence of "a significant regional flux of methane over a large area of shale gas wells in southwestern Pennsylvania in the Marcellus formation." The methane release detected there was "2 to 3 orders of magnitude greater than US Environmental Protection Agency estimates for this operational phase" (drilling).

(f) Without identifying the sources (biogenic or fossil fuel industry), Alexander J. Turner et al. (2016) state that "U.S. methane emissions have increased by more than 30% over the 2002–2014 period. (…) This large increase in U.S. methane emissions could account for 30–60% of the global growth of atmospheric methane seen in the past decade."

(g) Ramón Alvarez et al. (2018) reassessed (see Alvarez et al. 2012) the magnitude of methane leaks from the US oil and natural gas supply chain. They found that in 2015, these industrial emissions were ~ 60% higher than the US EPA inventory estimate, "likely because existing inventory methods miss emissions released during abnormal operating conditions." The authors calculate that "methane emissions of this magnitude, per unit of natural gas consumed, produce radiative forcing over a 20-year time horizon comparable to the CO_2 from natural gas combustion."

(h) Assessing anthropogenic fossil CH_4 emissions during the 2003–2012 period, Benjamin Hmiel et al. (2020) warn the bottom-up inventories strongly underestimate CH_4 emissions from fossil fuel extraction, distribution, and use.

A study [Alvarez et al. 2018] using both ground-based facility-scale measurements and verification from aircraft sampling found that US oil and natural-gas CH_4 emissions (largely from the production and gathering industry segments) are ~60% higher than those reported by the US Environmental Protection Agency. (…) Our results imply that anthropogenic fossil CH_4 emissions now account for about 30% of the global CH_4 source and for nearly half of anthropogenic emissions, highlighting the critical role of emission reductions in mitigating climate change.

Methane also escapes from abandoned wells. The Pennsylvania Department of Environmental Protection estimates that there are 325,000 abandoned wells, but Denise Mauzerall estimates that there could be as many as 500,000 wells in that state (Davenport 2015).

5.5 Collapse by Detox or Overdose?

Large-scale exploration (notably in Canada and the USA) of the unconventional forms of oil extraction analyzed above allowed for a comfortable growth in global oil supply in the second decade of this century. This new abundance has eclipsed the warnings of several scholars on peak oil, that is, the imminence of an era of oil scarcity.

How long can such an abundance, however, last? To what extent does the present cornucopia of oil mask and only delay a new oil shortage already on the horizon in the third or fourth decade of the century? Given that oil, however abundant it may be, is a finite resource, and given that its consumption continues to grow, to what extent would this potential scarcity be structurally and technologically insurmountable? The data and projections on this are very variable and somewhat speculative. This is due, in part, to the fact that the discovery of new sources and new technologies for exploration is unpredictable and, in part, to the low reliability of data on the real number of reserves and oil resources that are still available and economically accessible, given the economic and geopolitical interests at stake. To my knowledge, it is not possible to foresee the future of oil supply. But it is possible, in any case, to imagine the consequences of the two possible scenarios. The first scenario is one of increasing oil scarcity and the second is one of continued abundance, say, over the next half century.

5.5.1 There Is No Peak in Oil Demand for the Foreseeable Future

In 2017, Fatih Birol, the IEA's Executive Director, declared: "We don't see a peak in oil demand any time soon. And unless investments globally rebound sharply, a new period of price volatility looms on the horizon" (IEA 2017). The conjectural fluctuations in the price of oil still follow the boom-bust-boom logic of supply due to technological innovations, rates of return on investment, financial speculation, instabilities or geopolitical motivations, and, to a lesser extent, environmentalist pressures. Obviously, demand also influences prices. But, the reverse is not true: oil prices have no power over the continued growth in demand for it, as shown in Fig. 5.1.

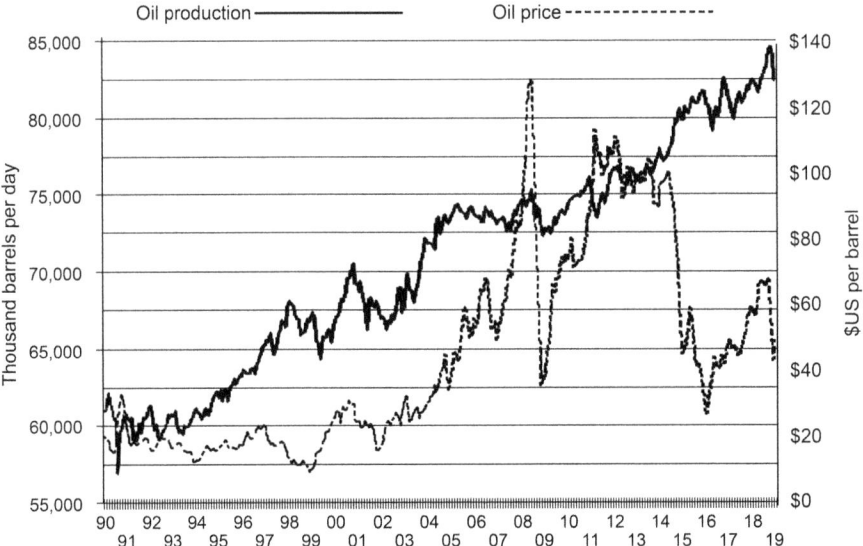

Fig. 5.1 World crude oil production (thousand barrels per day) and barrel prices (1990–2019). (Source: Energy Economist http://www.energyeconomist.com/a6257783p/archives/een050322. html)

Except for the roller coaster of oil prices, intimately linked to the 2008–2009 financial crisis, the growth in oil demand has little or nothing to do with price variations. Instead, it is mainly a function of the economic expansion inherent in the capitalist system and of the population growth. The growth in the demand for oil is, indeed, a constant of capitalism, both in the past and in a foreseeable future. In 1970, the world consumed 47 million barrels of oil per day (MMbb/d), and, in the third quarter of 2018, the milestone of 100 million barrels per day was passed. Also, in the long term, this demand should continue to grow—only if we take the premise that the global economy will not seriously fail until then, due to environmental or other reasons. Six institutions—EIA, BP, Exxon, MIT, IEA, and the Institute of Energy Economic, Japan (IEEJ)—foresee an increase of about 20–25% in global oil consumption until 2040, relative to 2015 (Andrews 2017). According to the OPEC World Oil Outlook 2018, a slightly more modest, but still very significant, increase in global oil demand will occur until 2040: from 97.2 Mb/d in 2017 to nearly 112 Mb/d, that is, an increase of 15%.

In this context, no hypothesis of a true energy transition is credible. However big may be the increasing share of solar and wind sources in the global energy supply, they only offer more energy (with obviously much lower, but not irrelevant, environmental costs) to the insatiable energy voracity of capitalism. In the absence of much stronger environmentalist pressures, the increased supply of alternative energy is unlikely to reduce the use of fossil fuels in a meaningful way.

Are the current investments in oil production sufficient to keep pace with the projected growth in demand? The IEA's *Oil 2019* report projects a steady growth in shale oil production in the USA, from 6.3 million barrels/day (mb/d) in 2018 to 9.59 mb/d in 2024. In Brazil, oil production surpassed the 1 billion barrel mark in 2019, and analysts expect a 25% growth by 2022 from nearly 2.6 mb/d in 2018 (Wilczynski et al. 2019). But according to OPEC, to increase the supply of oil in order to meet the projected demand until 2040 will require investments of about US$ 11 trillion. As seen above, Fatih Birol warns that in the absence of an immediate and vigorous wave of new investments, supply will not meet the expected increase in demand after 2020. This same prognosis of ephemeral abundance of both gas and shale oil is reiterated by Ian T. Dunlop and Richard Heinberg, among many other experts.

Other two variables that hinder any reasonable prediction on the future of the oil supply are (1) the evolution of the rate of energy return on energy invested (EROEI, sometimes EROI), or net energy (Hall 2008, 2017; Gupta and Hall 2011; Hall et al. 2014), and (2) the speed in the decline of conventional oil, the object of several projections, especially from the Association for the Study of Peak Oil and Gas—ASPO (Campbell 2004; Heinberg 2005, 2007, 2009, 2013, 2014; Heinberg and Campbell 2006). For these scholars, peak oil for liquid or conventional oil generally falls between 2005 and 2020 (Deffeyes 2001, 2006, 2010). In his book *Extracted. How the Quest for Mineral Wealth is Plundering the Planet. A Report to the Club of Rome* (2013), Ugo Bardi (2014) prefers to maintain an indefinite position on the peak in global conventional oil production: "We still don't know for sure if this global peak has occurred, because it is masked by production oscillations generated by market factors. But we may be very close to it." His book, however, was published in 2014. Today, in 2019, it seems more likely than ever that we would already be experiencing an era of irreversible energy scarcity in the absence of unconventional forms of oil extraction. There are, indeed, still predictions that delay this global peak beyond 2020, as proposed by the Cambridge Energy Research Associates and even to 2037 (USGS), but they are now the minority. If we confirm Fatih Birol's predictions, as from the third decade of the century, an imbalance between supply and demand of oil *in general* (conventional and nonconventional) may, however, return forcefully to the international scene, as the eventual effects of the peak in unconventional oil begin to be felt.

In addition, we must consider the admirable environmental campaigns in favor of divestment in the fossil fuel industry, such as *Keep it in the ground*, promoted by the Guardian Group, Greenpeace, and 350.org. A significant increase in political and social pressure for divestment in the coming years could, in the end, imply a depreciation in value of a good part of these fuel reserves, as they become converted into "stranded assets." But how successful will these messages be in the face of the business plans of the state-corporations holding the world's largest oil reserves? Will state-corporations be able to maintain their noticeable progress in the current scenario of price recovery and, hence, profitability of investments (maintaining subsidies)? The valuation of global proven oil reserves, according to Jude Clemente (2016), at US\$ 107 trillion (at a value of US\$ 63 per barrel) continues, nonetheless, to be a major investment attraction. We should not forget that between 2016 and 2018 alone, 33 global banks invested a combined \$1.9 trillion in fossil fuel companies (see, above, item 5.3, Subsidies and Investments in the Oil Industry).

All these combined variables and unknowns seem to reciprocally counterbalance each other and allow one to easily conclude that even in the short term—that is, throughout the second quarter of the century—the energy destiny of contemporary societies has never been so uncertain. As pointed out by Ugo Bardi (2017), "the question remains very complicated and difficult to assess; the only certainty is that we risk multiple collapses unless we manage to get rid of fossil fuels." That said, states and corporations, but also societies, addicted to the core to oil, have not hitherto been willing to abandon it, despite the scientific certainty of the environmental collapse that is coming from its exploration.

5.5.2 The Hypothesis of Scarcity

If this addiction is maintained (and that is of course a big if), four uncertainties must be considered: (1) How much is there, in fact, of remaining proven oil reserves that are technically and economically recoverable?[12] (2) Is it true that we are close to, or

---

[12] By reserves, we mean a subset of the remaining ultimately recoverable resources (RURR or URR) at all times with present and future technology, regardless of economic conditions. (McGlade & Ekins 2015).

have even passed, peak oil (for conventional liquid oil reserves), as various experts estimate? (3) In this case, would there be unconventional fossil fuel reserves and the necessary technology to offset the decline in conventional liquid oil, so as to avoid a bottleneck in the energy flow in this period? (4) Will societies be ideologically and politically able to abandon oil before it abandons or destroys them?

As far as I can tell, no one has solid answers to these four questions, and especially not to the fourth one, the only one that is truly crucial in our time. As for the first question, according to the 2018 OPEC Annual Statistical Bulletin, world crude oil reserves in 2017 amounted to 1492 billion barrels of oil. This estimate is slightly higher than what was proposed by Christophe McGlade and Paul Ekins (2015): 1294 billion barrels of oil (plus 192 trillion m^3 of gas, 728 Gt of coal, and 276 Gt of lignite). Let us assume (first uncertainty) that the reality is halfway between the OPEC and McGlade–Ekins estimates and so our current reserves would amount to about 1.4 billion barrels. In this case, if global oil consumption fluctuates between 105 and 110 million barrels per day (given that global consumption increased around 1.2 million b/d per year between 2010 and 2014), global oil reserves would be exhausted by 2050. Obviously, a global oil-based economy would become unfeasible long before the last barrel is extracted.

Let us also assume (second uncertainty) that a peak in conventional liquid oil is imminent or has already occurred in the second decade of the century, as proposed by most of the aforementioned scholars. Let us further assume (third uncertainty) that the projections of the IEA and other experts are confirmed and that unconventional oil from the USA peaks in the third decade or that, even if these reserves do not decline so soon, there is a physical, technological, or financial bottleneck in the transition from conventional to unconventional oil. Once these conditions are met, they will mean the end of the oil era in the first half of this century, due to one or more of the following reasons: physical scarcity, technological bottlenecks, or financial unfeasibility of its production by a 1:1 EROEI rate.

Cornucopian economists will argue, as usual, that the price system and markets have always circumvented bottlenecks in the scarcity of a natural resource by replacing it with another resource. But oil is not a natural resource like any other. And however fast may be the technological advance toward low-carbon renewable energy, it is hard to imagine that a global thermo-fossil economy, whose primary energy supply today comes from fossil fuels (80%), will be capable of emancipating itself from its reliance on fossil fuels over the next decade. Moreover, oil companies and the entire industrial, agricultural, and financial system have done and will continue to do their utmost to delay the end of the fossil fuel era so as to exploit as much as possible the global oil reserves.

It is not difficult to realize that a structural scarcity of oil implies the unfeasibility of consumer society and, in general, of contemporary global capitalism which is crucially dependent on oil, starting with the transport of seaborne containers. According to data provided by Statista, the quantity of goods carried by these containers has risen from around 102 million tons in 1980 to about 1.83 billion tons in 2017. The trade in soft commodities itself is inconceivable without intercontinental mobility and without massive doses of petroleum-based industrial fertilizers. In

Chap. 11 (item 11.2, Hypoxia and Anoxia), we will see that between 1950 and 1998, global use of petrochemical fertilizers increased more than tenfold and more than four times per capita. The prognosis outlined by Colin Campbell (2012) is that by 2050 the world oil supply will be enough to support no more than half of the planet's current population in its present way of life. And afterward? The growing scarcity of oil opens up a range of more or less unfathomable scenarios. It seems excessive to admit, in the end, Richard Duncan's (1996) Olduvai theory, which predicts a forthcoming decline to a "Post-Industrial Stone Age." But there is no doubt that the fate of our societies is "largely underwritten by energy subsidies and will ultimately be constrained by larger forces of nature" (Hall 2017). In any case, if the current rates of GHG emissions are maintained, these forces of nature are already leading us to an environmental collapse well before the end of oil, whatever the date of this end may be.

That being said, the hypothesis of a structural scarcity of oil in the next few decades is optimistic, first of all because this scarcity would, in extremis, make climate change less catastrophic and pollution levels less lethal. Without underestimating the risk of wars and the immense suffering that such scarcity will likely cause to global society, it is undeniable that, after the most acute phase of withdrawal, there will be everything to gain if such scarcity produces a "therapeutic" shock, a detox through forced abstinence. "Weaning" ourselves from fossil fuels would take us back to local food agriculture; it would force us to abandon the automatisms of contemporary consumerism and to redefine our real energy needs, conditioning them to the security and to the possibilities of ecosystems and of the Earth system as a whole.

5.5.3   The Hypothesis of an Overdose

Let us now examine the opposite hypothesis: our societies will remain addicted to oil and it will remain relatively available for the next 30 years. The collapse of the global economic system from structural scarcity of oil appears to be far less brutal than the global environmental collapse from overdose. As already mentioned, estimates of the remaining global fossil fuel reserves, including conventional and non-conventional oil, vary widely. The BP *Statistical Review of World Energy* estimates for 2017 are even higher than those proposed by OPEC, reported above (1492 billion barrels):

> Global proved oil reserves in 2016 rose by 15 billion barrels (0.9%) to 1,707 billion barrels, which would be sufficient to meet 50.6 years of global production at 2016 levels. The increase came largely from Iraq (10 billion barrels) and Russia (7 billion barrels). (…) OPEC countries currently hold 71.5% of global proved reserves

If this is so, with the probable discovery of other reserves, exploration of oil in the Arctic, and further technological improvements that enable exploitation of larger fractions of oil in the reservoirs, oil demand could be met for at least another

50 years. In this case, climate change due to increasing atmospheric concentrations of GHGs—with the growing weight of climate feedback loops (see Chap. 8)—will continue to plunge us into an abysmally worse future. As Christophe McGlade and Paul Ekins (2015) affirm:

> It has been estimated that to have at least a 50% chance of keeping warming below 2°C throughout the twenty-first century, the cumulative carbon emissions between 2011 and 2050 need to be limited to around 1,100 gigatonnes of carbon dioxide (Gt CO_2). However, the greenhouse gas emissions contained in present estimates of global fossil fuel reserves are around three times higher than this, and so the unabated use of all current fossil fuel reserves is incompatible with a warming limit of 2°C.

Therefore, unless our societies, anesthetized by what fossil fuels have provided, find in themselves the courage and political lucidity to enforce the decision of abandoning fossil fuels (before being destroyed by them), unless they redefine how much energy is necessary for a civilized life—which would presuppose an anthropological redefinition of the very notion of civilization—they will end up prey to three perverse mechanisms. Because, the longer the survival of fossil fuels:

1. The more difficult and costly will be the energy transition, since a matrix transition requires an abundant and cheap energy stock. Eleven years ago, in 2009, the IEA warned:

   > We calculate that each year of delay before moving onto the emissions path consistent with a 2°C increase would add approximately US$ 500 billion to the global incremental investment cost of US$ 10.5 trillion for the period 2010-2030. A delay of just a few years would probably render that goal completely out of reach. If this were the case, the additional adaptation costs would be many times this figure.

   Eleven years have passed since this warning, and the global economic trajectory continues to be one of the increases in fossil fuel consumption and in its corresponding GHG atmospheric concentrations. This implies that, if IEA calculations are correct, the cost of energy transition has increased by at least US$ 5.5 trillion, in theory going from US$ 10.5 to US$ 16 trillion.

2. The greater will be the socioeconomic disorganization that the world population (perhaps about 50% above the 2020 population) will be plunged into when fossil fuel depletion finally occurs.
3. The more the biosphere will suffer from the consequences of devastating processes and of climate changes to the point where the imbalances will not only prevent an energy transition, but probably the survival of any society worthy of that name. As we shall see in the next chapters, given the current trajectory, "eventually large swathes of Africa, Australia, China, Brazil, India and the U.S. will become uninhabitable for at least part of the year." These words by Steven Sherwood of the University of New South Wales, reported by Michael Le Page (2012), have now become commonplace in the scientific community. We know today, after the 2018 and 2019 heat waves in Europe, India, Japan, and Australia, and in several other regions, that this prognosis may come true already in the next decade.

References

ALLISON, Edith & MANDLER, Ben, "Abandoned Wells. What happens to oil and gas wells when they are no longer productive?". *Petroleum and Environment*, AGI (American Geosciences Institute), 2018.

ALLIX G., FOUCART, S. & IMBERT, C., "Dossier Marée Noire record de 2010 aux États Unis", *Le Monde*, 19/IV/2011.

ALVAREZ, Ramón A. *et al.* "Greater focus needed on methane leakage from natural gas infrastructure". *PNAS*, 109, 17, 2012, pp. 6.435-6.440

ALVAREZ, Ramón A. *et al.*, "Assessment of methane emissions from the U.S. oil and gas supply chain". *Science* 361, 2018, pp. 186–188.

ANDREWS, Roger, "The gulf between the Paris Climate Agreement and energy projections". *Energy Matters*, 18/I/2017.

BARDI, Ugo, *Extracted. How the Quest for Mineral Wealth is Plundering the Planet. A Report to the Club of Rome* (2013). Vermont, Chelsea Green Publisher, 2014.

———. *The Seneca Effect. Why Growth is Slow but Collapse is Rapid.* Springer, 2017.

BAST, Elizabeth *et al.*, *Empty promises. G20 subsidies to oil, gas and coal production.* Oil Change International. November 2015.

BERENSHTEIN, Igal *et al.*, "Invisible oil beyond the *Deepwater Horizon* satellite footprint". *Science Advances*, 6, 7, 12/II/2020.

CAMPBELL, Colin J. *The Coming Oil Crisis.* Multi Science Publishing, 2004.

———. "Changes in World's Energy Supply". *New Energy Era Forum 2012.* Skibbereen, Ireland <https://www.youtube.com/watch?v=wcZwpVVDP2s>.

CARON, Christina, "How a 672,000-Gallon Oil Spill Was Nearly Invisible". *The New York Times*, 29/X/2017.

CARRINGTON, Damian, "Fossil fuels subsidised by $10m a minute". *The Guardian,* 18/V/2015.

CAULTON, Dana R. *et al.* "Toward a better understanding and quantification of methane emissions from shale gas development". *PNAS*, 14/IV/2014.

CLEMENTE, Jude, "How Much Oil Does the World Have Left?" *Forbes*, 25/VI/2016.

COADY, David *et al.*, "Global Fossil Fuel Subsidies Remain Large: An Update Based on Country-Level Estimates". IMF Working Paper, May 2019.

CRONE, T.J. & TOLSTOY, M., "Magnitude of the 2010 Gulf of Mexico Oil Leak", *Science*, 330, October 2010, p. 634.

DAVENPORT, Matt, "Accounting for Methane emissions from oil and gas wells". *Chemicals and Engineering News*, 5/I/2015.

DAWEI, Yu, "As US Refuses a Dirty Fuel, China Only Too Ready to Increase Imports". *The New York Times*, 12/III/2013.

DEFFEYES, Kenneth S. *Hubbert's Peak: The Impending World Oil Shortage.* Princeton, 2001.

———. *Beyond Oil: The View from Hubbert's Peak.* Princeton, 2006.

———. *When Oil Peaked.* New York, Hill and Wang, 2010.

DUNCAN, Richard C. *The Olduvai Theory. Sliding Towards a Post-Industrial Stone Age.* Institute on Energy and Man, 1996.

DUNLOP, Ian T., "Fracking: The Boom and its Consequences". In Ugo Bardi, *Extracted. How the Quest for Mineral Wealth is Plundering the Planet. A Report to the Club of Rome* (2013). Vermont, Chelsea Green Publisher, 2014.

ECONorthwest, *The Economic Costs of Fracking in Pennsylvania*, Oregon, 14/V/2019.

EIA (Energy Information Administration), U.S. Oil and Natural Gas Wells by Production Rate, 20/XII/2019.

GALLAY, Paul, "Gas Industry Spin Can't Cover Up Problems Caused by Fracking". *EcoWatch*, 3/IV/2012.

GALLEGOS, Tanya J. & VARELA, Brian A., "Trends in Hydraulic Fracturing Distributions and Treatment Fluids, Additives, Proppants, and Water Volumes Applied to Wells Drilled in the U. S. from 1947 through 2010. Data Analysis and Comparison to the Literature". USGS Survey Scientific Investigations Report 2015.

GERKEN, James, "Oil Trains Spilled More Crude Last Year than in the Previous 38 Years Combined". *The Huffington Post*, 22/I/2014.

GUPTA, Ajay & HALL, Charles. "A Review of the Past and Current State of EROI Data". *Sustainability*, 3, 2011, pp. 1796-1809.

HALL, Charles A. S. "Provisional Results from EROI Assessments". *The Oil Drum*, 2008.

———. "Will EROI be the Primary Determinant of Our Economic Future? The View of the Natural Scientist versus the Economist". *Joule*, 1, 4, 20/XII/2017.

HALL, Charles A.S., LAMBERT, Jessica G. & BALOGH, Stephen B., "EROI of different fuels and the implications for society". *Energy Policy*, 64, I/2014, pp. 141-152.

HANSEN, James, "Game over for the climate". *The New York Times*, 9/V/2012.

HEINBERG, Richard. *The Party's over. Oil, War and the Fate of Industrial Societies* (2003). Gabriola Island, New Society Publishers, 2005.

———. *Peak Everything: Waking Up to the Century of Declines*. Gabriola Island, New Society Publishers, 2007.

———. *Searching for a Miracle. 'Net Energy' Limits and the Fate of Industrial Society*. Post Carbon Institute, 2009.

———. *Snake Oil: How Fracking's False Promise of Plenty Imperils Our Future*. Post Carbon Institute, 2013.

———. "Paul Krugman's Errors and Omissions". *Post Carbon Institute*, 21/IX/2014.

HEINBERG, Richard & CAMPBELL, Colin J. *The Oil Depletion Protocol. A Plan to Avert Oil Wars, Terrorism and Economic Collapse*. Gabriola Island, New Society Publishers, 2006.

HMIEL, Benjamin *et al.*, "Preindustrial $^{14}CH_4$ indicates greater anthropogenic fossil CH_4 emissions". *Nature*, 578, 19/II/2020, pp. 409-412.

HOLLAND, Austin. *Examination of Possibly Induced Seismicity from Hydraulic Fracturing in the Eola Field, Garvin County, Oklahoma*. Oklahoma Geological Survey, August 2011.

HOWARTH, Robert W.; SANTORO, René & INGRAFFEA, Anthony. "Methane and the greenhouse gas footprint of natural gas from shale formation". *Climatic Change*, 106, May 2011, pp. 679-690.

IEA, World Energy Outlook, November 2010.

IEA, "Global supply to lag demand after 2020 unless new investments are approved soon". 6/III/2017.

INMAN, Mason. "Natural gas. The fracking fallacy". *Nature*, 516, 7.529, 4/XII/2014

IUPAC (International Union of Pure and Applied Chemistry), Gold Book, London, Royal Society of Chemistry 1997.

JOHNSON, Nels, *et al.*, *Pennsylvania Energy Impacts Assessment Report 1: Marcellus Shale Natural Gas and Wind*. Nature Conservancy, 15/XI/2010.

JOSE, Coleen, WALL, Kim, HINZEL, Jan Hendrik, "This dome in the Pacific houses tons of radioactive waste – and it's leaking". *The Guardian*, 3/VII/2015.

KELLY, Erin N. *et al.*, "Oil sands development contributes elements toxic at low concentrations to the Athabasca River and its tributaries". *PNAS*, 30/VIII/2010.

KERANEN, Katie *et al.* "Potentially induced earthquakes in Oklahoma, USA: Links between wastewater injection and the 2011 5.7 earthquake sequence". *Geology*, 26/III/2013.

KIM, Won-Young, "Induced seismicity associated with fluid injection into a deep well in Youngstown, Ohio". *Journal of Geophysical Research. Solid Earth*, 19/VII/2013

KISTNER, Rocky, "The Macondo Monkey on BP's Back" *Huffington Post*, 30/IX/2011.

KONDASH, Andrew J., LAUER, Nancy E. & VENGOSH, Avner, "The intensification of the water footprint of hydraulic fracturing". *Science Advances*, 4, 8, 15/VIII/2018.

KORT, Eric *et al.* "Four corners: The largest US methane anomaly viewed from space". *Geophysical Research Letters*, 9/X/2014.

LE PAGE, Michael, "Global Warning". *New Scientist*, 216, 17/XI/2012.

LEVIN, Alan, "Oil spills escalated in this decade", *USA Today*, 08/VI/2010.

LOMBARDI, Kristen, "Benzene and worker cancers. An American tragedy". The Center for Public Integrity. Columbia University, 4/XII/2014.

McGLADE, Christophe & EKINS, Paul. "The geographical distribution of fossil fuels unused when limiting global warming to 2°C". *Nature*, 517, 8/I/2015, pp. 187-190.

McKENZIE, Lisa M.; WITTER, Roxana Z.; NEWMAN, Lee S. & ADGATE, John L. "Human health risk assessment of air emissions from development of unconventional natural gas resources". *Science of the Total Environment*, 10/II/2012.

MILLER, Scot *et al.* "Anthropogenic emissions of methane in the United States". *PNAS*, 10/XII/2013.

MILSTEIN, Mati, "Lebanon Oil Spill Makes Animals Casualties of War" *National Geographic*, 31/VII/2006.

MOFFAT, David & LINDÉN, Olof, "Perception and reality: Assessing priorities for sustainable development in the Niger River Delta". *Ambio. A Journal of the Human Environment*, 24, 7-8, December, 1995, pp. 527-538.

NESLEN, Arthur, "Tar sands alarm as US crude exports to Europe rise". *The Guardian*, 8/XII/2015.

NIELSEN, Ole, "Nigerian oil spills again". *Olelog. What on Earth*, 27/XII/2011.

NUMMI, Esa, "Can the Fracking Boom Spread Beyond the United States?". *Advancing Mining*, 8/XII/2015.

Oil Change International, "No Time to Waste: The Urgent Need for Transparency in Fossil Fuel Subsidies". 2013.

OSBORN, Stephen G. *et al.* "Methane contamination of drinking water accompanying gas-well drilling and hydraulic fracturing". *PNAS*, 108, 20, 2011, pp. 8.172-8.176.

PANDEY, Sudhanshu *et al.*, "Satellite observations reveal extreme methane leakage from a natural gas well blowout". *PNAS*, 116, 52, 26/XII/2019, pp. 26376-26381.

SCHATZEL, Emily Guidry, "Five Years after BP Oil Spill: Focus Should Be on Continued Need for Restoration". *National Wildlife Federation*, 16/IV/2015.

SEACOR, Jessica E. "Environmental Terrorism: Lessons from the Oil Fires of Kuwait". *American University International Law Review*, 10, 1, 1994.

STARBUCK, Amanda, "Map Displays Five Years of Oil Pipeline Spills". *Center for Effective Government*, 22/VI/2015.

STOCKMAN, Lorne, TURNBULL, David & KRETZMANN, Stephen, *Petroleum coke: the coal hiding in tar sands*. Oil Change International, January 2013.

SUBRA, Wilma, Summary of Human Health Impacts of the BP Deepwater Horizon Disaster – Given at the Gulf Coast Leadership Forum. *Lean*, 19/IV/2011.

THOMPSON, John, "Russia's environmental aspirations marred by Arctic oil spills". *Arctic Deeply*, 19/IV/2017.

TOLLEFSON, Jeff, "Secrets of fracking fluids pave way for cleaner recipes". *Nature*, 501, 12/IX/2013.

TOLLEFSON, Jeff *et al.* "Methane leaks erode green credentials of natural gas". *Nature*, 493, 7.430, 2/I/2013.

TURNER, Alexander J. *et al.* "A large increase in U.S. methane emissions over the past decade inferred from satellite data and surface observations". *Geophysical Research Letters*, 2/III/2016.

VASILYEVA, Nataliya, "Constant oil spills devastate Russia". *The Seattle Times,* 24/XII/2014.

WILCZYNSKI, Pawel, FIASCA, Ricardo, PAGKALOU, Evelina, "Brazil has the potential for a 70% oil production increase by 2035—but is it likely?". McKinsey and Company, 11/IV/2019.

ZUKOWSKI, Dan, "220 'Significant' Pipeline Spills Already This Year Exposes Troubling Safety Record". *EcoWatch*, 25/X/2016.

Chapter 6
The Regression to Coal

In its old age, global capitalism is experiencing a regression to coal, the most pollut-ing of the fossil fuels and the most important factor responsible for increasing atmo-spheric concentrations of CO_2. "Like it or not, coal is here to stay for a long time to come." Whoever follows some recent declarations of BlackRock, joining the long list of investors taking a harder stance on coal investments, will perhaps say that this 2013 declaration by Maria van der Hoeven, IEA's former executive director, as well as the title of this chapter, proposed in 2015, no longer reflect the planet's energy scenario. Unfortunately, this title remains true not only today, in 2020, but even in the foreseeable future. Firstly, because the coal rush that has occurred since the oil shocks of the 1970s is not transitory. It is a phenomenon of civilization; a central chapter of the history of the "Great Acceleration" begun in the second post-war period (Steffen et al. 2015) and also, particularly, of the energy voracity of capital-ism in the last half century.

Secondly, because of the worldwide effects of coal on the climate, the atmo-sphere, the water, and the health of organisms will be felt for a long time, even if we were to stop burning coal tomorrow. Thirdly, because the recent development of low-carbon renewable energies did not lead to a *relevant* reduction in overall coal consumption. The adjective "relevant" here refers to a decrease in speed and scale that is compatible with a scenario of average global warming equal to or lower than 3 °C above the pre-industrial period. Already in 2016, Hans-Joachim Schellnhuber affirmed that "by 2025 we will have to have closed down all coal-fired power sta-tions across the planet. And by 2030 you will have to get rid of the combustion engine entirely. That decarbonization will not guarantee a rise of no more than 1.5C but it will give us a chance" (as cited by McKie 2016). Two years later, Ottmar Edenhofer et al. (2018) showed that the activity of coal-fired power plants currently in existence entails GHG emissions above the targets set by the signatories of the Paris Agreement:

© Springer Nature Switzerland AG 2020
L. Marques, *Capitalism and Environmental Collapse*,
https://doi.org/10.1007/978-3-030-47527-7_6

Unless these power plants are retired well before their expected life-time, which would increase mitigation costs and constitute a formidable political challenge, their associated emissions jeopardize the achievement of the (I)NDC targets as well as effective long-term climate change mitigation.

6.1 The Coal Saga

In 2004, 26% of the primary energy consumed by mankind came from coal. The *BP Statistical Review of World Energy* 2014 reported that in 2013 "coal's share of global primary energy consumption reached 30.1%, the highest since 1970." This increasing share of coal in the overall rise in energy consumption is reflected in the correlative escalation of world coal production in the twenty-first century, as shown in Table 6.1.

Table 6.1 World production of coal between 2003 and 2013 (Gt)

2003	2004	2005	2006	2007	2008	2009	2010	2011	2012	2013
5.3	5.71	6.03	6.34	6.57	6.79	6.88	7.22	7.69	7.79	7.83

Sources: *BP Statistical Review of World Energy*. June 2012 and *World Coal Association* http://www.worldcoal.org/resources/coal-statistics/

Steven Davis and Robert Socolow (2014) call our attention to the fact that more coal-fired power plants were constructed globally between 2003 and 2013 than in any previous decade. This acceleration, in the last years of this decade (2003–2013), is shocking: "Worldwide, an average of 89 gigawatts per year (GW year^{-1}) of new coal generating capacity was added between 2010 and 2012, 23 GW year^{-1} more than in the 2000–2009 time period and 56 GW year^{-1}more than in the 1990–1999 time period." There was also a significant per capita increase in this production, as world population increased during the same period from 6.35 billion to 7.1 billion (around a 10% increase). In 2003, 840 kg of coal were produced per capita. And in 2013, more than one ton of coal was produced per capita.

As is well known, much of the growth in global coal production and consumption in the twenty-first century is due to China, a country that is responsible for about 50% of global demand for coal. Coal consumption during the 12th 5-Year Plan (2011–2015) was 5.6 times greater than during the 6th 5-Year Plan (1981–1985). Between 2000 and 2013 alone, Chinese consumption rose from 1.36 gigatons (Gt) to 4.42 Gt, a frantic increase at an average annual rate of 12%. Stephen Ansolabehere and colleagues asserted in 2007: "China is currently constructing the equivalent of two, 500 megawatt, coal-fired power plants per week and a capacity comparable to the entire UK power grid each year. One 500 megawatt coal-fired power plant produces approximately 3 million tons/year of CO_2." In 2012, China continued to build one power plant of comparable proportions per week (Pearce 2013).

6.1.1   2014–2016

Then, suddenly, against the expectations of the IEA, global coal consumption in 2014 and 2015 declined slightly. According to the World Energy Council (2016):

> Global coal consumption increased by 64% from 2000 to 2014. That classified coal as the fastest growing fuel in absolute numbers within the indicated period. 2014 and 2015 witnessed the first annual decrease in global thermal coal production of 0.7% and 2.8% respectively, since 1999.

In 2014, Chinese consumption fell slightly to 4.12 Gt, against 4.24 Gt in 2013, a drop of 2.9%. And in 2015 there was a decline of 3.6% from 2014. It was enough for Ye Qi, Nicholas Stern, and colleagues (2016) to announce the decoupling scenario that economists had always dreamed of: "China's post-coal growth."

> China's coal consumption has indeed reached an inflection point much sooner than expected, and will decline henceforth — even though coal will remain the primary source of energy for the coming decades. We suggest that China has entered the era of post-coal growth. (…) The end of coal-fired growth in China does not mean that coal will cease to be a major energy source; it means that it is entering a phase of development when China's economic growth — and the improving living standards of its population — will not depend on rising coal consumption.

Thanks to reduced coal consumption in China, but also in the United States and some European countries, the 2014–2016 triennium actually saw a reduction in world coal consumption after more than a decade of spectacular growth. The 2017 edition of *BP Statistical Review of World Energy* reports that global coal production fell by 6.2% and consumption by 1.7% in 2016 (from 2015), the second consecutive year of decline. Moreover, the share of coal in global primary energy production fell to 28.1% in 2016. As mentioned before, it had reached 30.1% in 2013, the highest point in the curve.

6.2   A Still Distant Peak Coal

This declining trend, however, was short-lived, refuting not only the prognoses of Ye Qi and Nicholas Stern but also a *Greenpeace* paper of 2015, titled *Coal's terminal decline*. Again, the facts belie those who believe that industrial capitalism can reinvent itself and overcome its first love in its decrepit old age. As Spencer Dale, BP's chief economist, informs us: in 2017, world production of coal increased 3.2%, driven by India (4.8%), China (3.6%), and the United States (6.9%) output. India recorded the fastest growth, as demand both inside and outside of the power sector increased. As mentioned in the Introduction, in 2017 alone, the five largest US banks loaned US$ 1.5 billion to the major coal corporations, Peabody Energy, Arch Coal, and Alpha Natural Resources (Flitter 2018).

In 2018 global coal demand increased by 1.1%, continuing the rebound that began in 2017 after 3 years of decline. "The main driver was coal power generation, which rose almost 2% in 2018 to reach an all-time high" (IEA 2019). In 2018, the Chinese government abruptly reduced subsidies for photovoltaic expansion projects and suspended, for 2 years, the ban on the construction of new coal-fired thermo-electric power plants (see Climate Action Tracker, China 2018). In 2018, coal remained the largest source of electricity and the second-largest source of primary energy worldwide. The 2019 IEA coal report affirms that electricity generation from coal will rise over that period (although at less than 1% per year) and concludes that, even if global coal demand is expected to decline in 2019, global coal consumption "will remain broadly stable over the next five years, supported by robust growth in major Asian markets" (IEA 2019).

The permanence of coal in the international energy system is guaranteed by the banking system. Indeed, according to a new report, titled *Investments in Coal Power Expansion* (2019), a joint research signed by BankTrack, Urgewald, and three other NGOs,[1] a further 670,000 MW from coal-fired power plants, forecast in the corporate and government plans of 59 countries, is expected to be added in 2019. Financial institutions have channeled US$ 745 billion to 258 coal plant developers between January 2017 and September 2019 in the form of loans, investments, and underwriting. Since January 2017, 307 commercial banks have provided US$ 159 billion in direct loans to coal plant developers. The top three lenders are Japanese banks as Japan plans to build as many as 22 new coal-burning power plants at different sites in the next 5 years. "Together the 22 power plants would emit almost as much CO_2 annually as all the passenger cars sold each year in the United States" (Tabuchi 2020). Despite declarations of the contrary, the simplest arithmetic shows that we continue in full swing in the age of regression to coal, with 3 years of a slight decline (2014–2016) in a series of 15 years (2003–2018) of constant growth in the production and consumption of this fuel.

No indicator can posit, for the next 15 years (2021–2035), a decline in coal consumption symmetrical to that of its growth in the years 2003–2018. In its *World Energy Outlook 2017*, for example, the IEA projects no decline in global coal until 2040. In its "New Policy" scenario, based on the assumption of a 30% increase in world energy consumption by 2040, the IEA predicts a 4% increase in coal consumption in the same period. It predicts, in short, only a slowdown in the rate of this growth: +900 GW between 2000 and 2015 against +400 GW between 2016 and 2040. The *International Energy Outlook 2019, With Projections to 2050*, published in September 2019 by the US Energy Information Administration (EIA), reinforces this assessment:

> Worldwide coal production holds steady at about 8 billion short tons, or 160 quadrillion British thermal units (Btu), per year through 2040. Increased coal use in India and other Asian countries that are not part of the OECD helps drive consumption to more than 9 billion short tons (175 quadrillion Btu) by 2050.

One of the greatest concerns is the global balance between abandoning old coal-fired thermoelectric plants in certain countries and the construction of new ones in

---

[1] Les Amis de la Terre France, Rainforest Network & Re:Common, 2019.

other countries in the next few years. The average age of coal-fired power plants in Asia in 2019 is just 12 years (Harvey 2019). Considering that the lifetime of a coal-fired power plant is 30–50 years, the only way to make "coal's terminal decline" (desired by Greenpeace in 2015) a reality is through a political decision of societies. Otherwise, the already active coal plants, added to the ones in construction, condemn us to yet another 30 years of coal, implying what is designated by the term "committed GHG emissions," referring to emissions that have not yet actually been released into the atmosphere but that will inevitably be so. According to Steven Davis and Robert Socolow (2014), mentioned above, "despite international efforts to reduce CO_2 emissions, total remaining commitments in the global power sector have not declined in a single year since 1950 and are in fact growing rapidly."

Pieter van Breevoort et al. (2015) showed that in 2005 there were projects underway for the construction of 2440 new coal-fired generating units that would provide an additional 1428 GW of energy. Many of these projects were frozen or permanently abandoned throughout 2016. The most reliable data on the current state of these new coal-fired power plants—planned or under construction—comes from Coal Swarm's document, *Global Coal Plant Tracker*. In terms of additional capacity for electricity generation through coal combustion, in January 2016, plants whose construction was already authorized would provide an increase of 1,089,700 GW, against an increase of 569,600 GW in January 2017, as shown in Table 6.2.

This information was updated in October 2017. There were then 267 thermoelectric coal plants being constructed in the world: 154 were completely new and 113 were expansions of existing ones. Eighteen countries are responsible for 93% of this increase in coal plants (see Table 6.3):

In general, between 2010 and 2017, the electricity-generating capacity of coal was bolstered more than it was deactivated, as shown in Fig. 6.1:

Table 6.2   Change in global coal plant pipeline, January 2016 to January 2017 (MW)

	January 2016	January 2017	Change
(1) Announced	487,261	247,909	−49%
(2) Pre-permit	434,180	222,050	−49%
(3) Permit	168,230	99,637	−41%
(4) (1) + (2) + (3)	**1,089,671**	**569,601**	−48%
(5) Started construction (past 12 months)	169,704	65,041	−62%
(6) In construction	338,458	272,940	−19%
(7) On hold	230,125	607,367	164%
(8) Completed (past 12 months)	108,029	76,922	−29%
(9) Retired (past 12 months)	36,667	27,041	−26%
(10) Operating	1,914,579	1,964,460	3%

Source: Christine Shearer, Nicole Ghio, Lauri Myllyvirta, Aiqun Yu, and Ted Nace, *Boom and Bust 2017 Tracking the Global Coal Plant Pipeline*, March 2017 https://endcoal.org/wp-content/uploads/2017/03/BoomBust2017-English-Final.pdf
Note: These include coal-generating plants of at least 30 MW. According to the Platts WEPP databank, there are around 27,060 MW coming from plants that generate less than 30 MW of energy

Table 6.3 Thermoelectric coal plants being constructed in the world

	New plants	Expansions	Total
China	74	46	120
India	19	26	45
Japan	7	7	14
Indonesia	9	4	13
Philippines	5	6	11
Vietnam	6	4	10
South Korea	1	6	7
Poland	4	1	5
Pakistan	4	0	4
Turkey	3	1	4
Malaysia	2	1	3
Bangladesh	1	1	2
Botswana	1	1	2
Thailand	1	1	2
Iran	1	0	1
Brazil	1	0	1
Morocco	1	0	1
Germany	1	0	1
Total	**141**	**105**	**246**

Source: Adam Morton, "The world is going slow on coal, but misinformation is distorting the facts", *The Guardian*, 16/X/2017, based on data from CoalSwarm https://www.theguardian.com/environment/2017/oct/16/world-going-slow-coal-misinformation-distorting-facts

In 2018, according to "The world's coal power plants" published by Carbon Brief (2019), coal-fired power plants were producing 2,024,100 MW of energy, maintaining a 3% increase in relation to January 2017. Further capacity was being built for 236,000 MW and planned for 336,000 MW, totaling 572,000 MW. As can be seen, there is a clear deceleration in all phases of the implementation of new coal thermoelectric plants and in the expansion of existing ones. Nevertheless, the least one can say is that it is premature to affirm that we are in the process of overcoming the historical phase of regression to coal, a phase that was initiated with the two oil "shocks" in the 1970s.

Furthermore, China's efforts to reduce its domestic coal consumption were offset by its export of coal-fired power plants. Since 2015, Chinese banks and corporations have been involved in at least 79 coal-fired power plant projects outside of China, having a total capacity of generating more than 52 GW. This exceeds the 46 GW from the plants that are set to be shut down in the United States by 2020.

In Germany, the much vaunted *Energiewende*, an energy transition program toward renewable and low-carbon energies, proved to be a real failure. The country has abandoned its GHG reduction targets for 2020 and, since 2009, its CO_2-eq emissions (more than 90% of which come from energy production) have been parked at 900 $MtCO_2$-eq, as shown in Fig. 6.2.

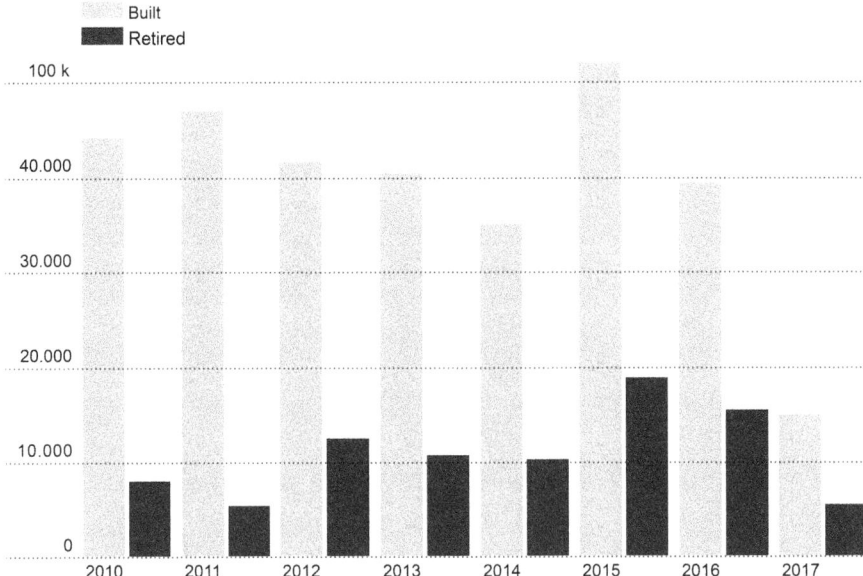

Fig. 6.1 Global coal power capacity (MW), 2010–2017: built (left) and retired (right). Source: Adam Morton, "The world is going slow on coal, but misinformation is distorting the facts", *The Guardian*, 16/X/2017, based on Global Coal Plant Tracker, July 2017

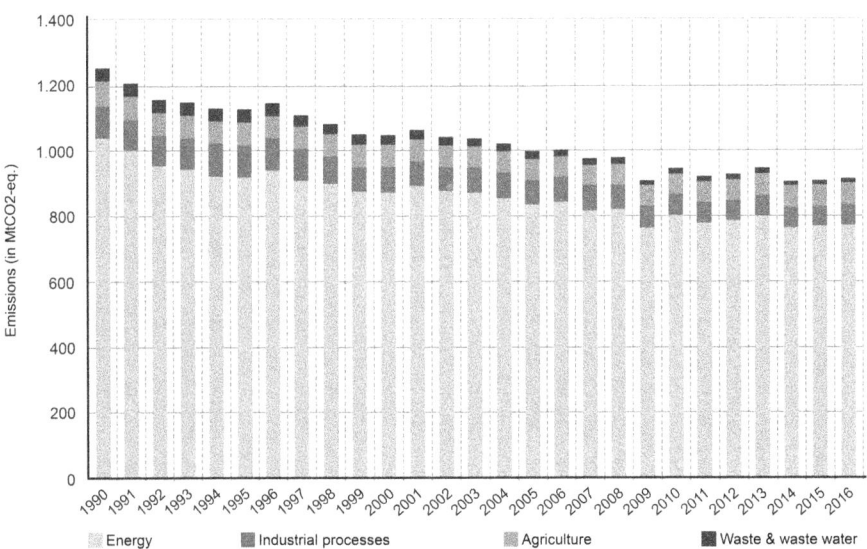

Fig. 6.2 Emissions of greenhouse gases in Germany (1990–2016) in MtCO$_2$-eq. (Source: Umwelt Bundesamt (UBA) Nationales Treibhausgasinventar 2018 https://www.umweltbundesamt.de/presse/pressemitteilungen/klimagasemissionen-stiegen-im-jahr-2016-erneut-an)

In 2017, fossil fuels reigned supreme in Germany, accounting for 80.5% of the country's primary energy supply. Coal accounts for 22.1% of this consumption. In electricity production, coal still occupies 37%. According to Craig Morris (2018), a specialist in energy transition, "the share of renewables in 2020 may not look so different from 2017." Germany remains the largest producer and consumer of coal in Europe and continues to subsidize its coal industry and expand the Hambach and Garzweiler lignite mines, destroying villages and ancient forests, such as the Hambach forest, between Düsseldorf, Cologne, and Aachen (Hicks 2019). Abandoning accelerated coal would thwart the immediate earning prospects RWE in Germany and other private- and state-owned mega-corporations around the world that own or operate these assets and that naturally are opposed to and will continue to oppose the end of coal.

6.2.1   The Second Childhood of the Industrial Revolution

Air pollution is often associated with coal burning in the early stages of the Industrial Revolution. It brings to mind Charles Dickens's descriptions of his city, as in *Bleak House* (1853): "Smoke lowering down from chimney-pots, making a soft black drizzle, with flakes of soot in it as big as full-grown snowflakes — gone into mourning, one might imagine, for the death of the sun." In 1865, William Stanley Jevons wrote in his famous *The Coal Question*: "Coal in truth stands not beside but entirely above all other commodities." Fast forward to our days where coal-fired power plants are producing 2,024,100 MW of energy, as seen above, we realize that global capitalism has transformed several metropolises around the world into "bleak houses." It is almost unbelievable that between 1973 and today we have regressed to a new golden age of coal and that, if the IEA and EIA projections for 2040 and 2050 are confirmed, this age still has a long life ahead of it. Coal is, in fact, the mockery of the sagas composed to the greater glory of the permanent technological revolution of capitalism. In the world imagined by these sagas, a new age of mankind, driven by cleaner, renewable, and efficient energies—solar, hydrogen, nuclear fusion, geothermal, wind, sea, etc.—would already be revealing itself. In the real world, technological advances are combined with what appears to be today a second childhood of the Industrial Revolution. As Luis Fernando Veríssimo writes in one of his chronicles about the unfulfilled promises of technology: "what we miss the most about the past is its future."

6.3   The Most Polluting Fossil Fuels

If coal is disadvantageous when compared to oil and gas in terms of energy (one ton of oil is equivalent to 1.5 tons of anthracite or hard coal and to 3 tons of lignite), it is all the more so environmentally. Coal pollutes the air, water, and soils at all stages of its industrial cycle, from extraction to transportation, to washing, burning, and the discharge of tailings.

6.3.1   Extraction

In the three types of deposits and methods of exploration—open pit-mining, underground mining, and mountaintop removal (MTR)—the pollution produced by coal mining is immense, either due to the impact in the area, to its heavy use of water, or to the production of waste residue. The MIT Union of Concerned Scientists describes "How Coal Works":

> While coal mining has long caused environmental damage, the most destructive mining method by far is a relatively new type of surface mining called mountaintop removal, or MTR. (…) MTR requires stripping all trees from a mountaintop and blasting away the top several hundred feet with explosives. The resulting debris is then dumped into adjacent valleys, burying streams and severely—and irrevocably—impacting local environments.

The extracted coal is dipped into an intermediate density liquid to separate it from the soil and rocks (float and sink testing) and for sorting. It is then rinsed with water and toxic substances to remove this liquid. The liquid resulting from this rinsing (coal slurry), composed of used water (black water) and other wastes from this process, cannot be recycled or utilized and is disposed of in wastewater ponds that are not impervious and that pollute the soil, water, and atmosphere.

Coal is further fractioned in situ in a process that emits particles of sulfur dioxide, nitrogen oxides, and carbon monoxide, among other chemicals. Finally, it is transported by trucks, trains, or conveyers to processing plants, coke plants, thermoelectric plants, etc. In the Santa Catarina mines, in Brazil, each ton of run-of-mine (ROM) coal "generates about 60% of solid wastes (thick and fine tailings) and approximately 1.5 m^3 of acidic effluents" (Nascimento et al. 2002).

6.3.2   Water and "Airpocalypse"

Freshwater supplies are heavily polluted by coal mines and power plants, sometimes with long-lasting effects. Groundwater pollution from the coal industry occurs through a process known as acid mine drainage, that is, by rainwater infiltration into tailings generated from coal extraction and rinsing. Runoff of oxidized iron sulfide (FeS_2) "can change the pH of nearby streams to the same level as vinegar."[2] In rivers and lakes, this contamination reduces the fecundity of fish and other animals and obviously compromises the potability of water. According to a 2013 EPA report, in the United States, "power plants contributed to the degradation of 399 bodies of

---

[2] See Union of Concerned Scientists, "Coal and Water Pollution."

water that are drinking water sources." Abandoned coal mines, even those abandoned a century ago, continue to contaminate waters. Coal tailings alone are responsible for 50% to 60% of all pollution entering US river waters.

The burning of coal emits a series of pollutants that are extremely damaging to human health and to the biosphere. These pollutants are contained in coal ash, either fly ash or bottom ash (ash that falls to the bottom of the boiler). Coal ash contains variable amounts of silica (SiO_2), lead, arsenic, copper, chromium, mercury, selenium, cobalt, thallium, etc. Exposure to coal ash can lead to bronchitis, rheumatoid arthritis, heart damage, lung cancer, and hormonal and neurological disorders, among many other serious health conditions. Since the purest coal has already been mined and corporations favor mining the cheapest coal they can find, the coal burned in thermoelectric plants is becoming dirtier. Even before its burning, when it is washed, stored, and transported, the coal releases important doses of mercury and arsenic into the air and water.

Although soot particles are a byproduct of almost all combustion processes, coal is one of its largest producers. In their structure, these particles contain the so-called polycyclic aromatic hydrocarbons (PAHs), including benza(a)anthracene ($C_{18}H_{12}$), a mutagenic compound that can cause changes in tissue proliferation, such as bone marrow, lymph organs, gonads, and intestinal epithelium. This carcinogenic compound can also cause tumors in animals and mutation in bacteria, in addition to being, by its lipophilic property, a typical substance characterized by bioaccumulation and bioconcentration.

Ash fly results from the melting and calcination of the noncombustible mineral impurities of coal. We have already covered the effects of particulate pollution in Chap. 4. In the case of coal particles, ranging from 0.5 to 100 micrometers and suspended in the gaseous stream, they absorb toxic substances such as arsenic, barium, beryllium, boron, cadmium, chromium, thallium, selenium, molybdenum, mercury, and sulfur dioxide before entering into the lungs. Sulfur dioxide has an irreversible effect on respiratory capacity, causing asthma, recurrent cough, and other problems.

The combination of soot and ash fly particles constitutes smog, famous since the lethal crisis in Donora, Pennsylvania, in October 1948 which killed 20 people and sickened half the city, as well as the Great Smog of December 1952 in London which caused 12,000 deaths and led to the Clean Air Act enacted by the English Parliament in 1956 (Bell et al. 2004). The situation today in China is, of course, much worse. In the winter of 2013–2014, as a result of the intense activity of coal-fired thermoelectric plants, smog affected 15% of the country's territory, including the large urban areas of Beijing, Shanghai, Harbin, Chengdu, and even Lhasa.[3] In January 2014, in Beijing, particle concentrations of 2.5 micrometers in diameter (PM 2.5) reached levels 26 times above the ceiling recommended by the WHO for exposure within 24 h. In November and December 2015, the Chinese government issued two red alerts (the worst of four warning types) in ten cities in northeastern

---

[3] "Pollution en Chine: pour la première fois, un citoyen poursuit le gouvernement". *Le Monde*, 25/II/2014.

China, including Beijing, whose atmosphere had concentrations of particulate matter that were 30 times higher than those recommended by the WHO for 24-h exposure.

References

ANSOLABEHERE, Stephen et al., "The Future of Coal". *An Interdisciplinary MIT Study.* MIT, Cambridge, (Mass.) 2007.

BELL, Michelle L, DAVIS, Devra L, & FLETCHER, Tony, "A retrospective assessment of mortality from the London smog episode of 1952: the role of influenza and pollution". *Environmental Health Perspectives*, 112, 1, 2004, pp. 6–8.

BREEVOORT, Pieter van et al., "The Coal Gap: planned coal-fired power plants inconsistent with 2°C and threaten achievement of INDCs". Climate Action Tracker, 1/XII/2015.

DAVIS, Steven J, SOCOLOW, Robert H, "Commitment accounting of CO_2 emissions". *Environmental Research Letters*, 26/VIII/2014.

EDENHOFER, Ottmar et al., "Reports of coal's terminal decline may be exaggerated". *Environmental Research Letters*, 7/II/2018.

FLITTER, Emily, 'Think the Big Banks have abandoned coal? Think again". *The New York Times*, 28/V/2018.

HARVEY, Fiona, "Global coal use up as greenhouse gas emissions rise". *The Guardian*, 26/III/2019.

HICKS, Celeste, "Why we want to save the Hambach Forest". Friends of the Earth, 20/V/2019.

IEA 2019, "Coal 2019. Analysis and Forecasts to 2024". International Energy Agency (IEA), Paris, December 2019.

McKIE, Robin, "Scientists warn world will miss key climate target". *The Guardian,* 6/VIII/2016.

MORRIS, Craig, "German power sector: coal and nuclear down, renewables up in 2017". *REneweconomy*, 16/I/2018.

NASCIMENTO, Flávia M. F. et al. "Impactos ambientais nos recursos hídricos da exploração de carvão em Santa Catarina". RLGeo, 2002.

PEARCE, Fred, "A new course for global emissions?" *New Scientist*, 9/IX/2013.

QI, Ye et al., "China's post-coal growth". *Nature Geosciences*, 9, 25/VI/2016, pp. 564–566.

STEFFEN, Will et al., "The trajectory of the Anthropocene: The Great Acceleration". *The Anthropocene Review*, 2015, 2(1), pp. 81-98.

TABUCHI, Hiroko, "Japan Races to Build New Coal-Burning Power Plants, Despite the Climate Risks". *The New York Times*, 3/II/2020.

VAN DER HOEVEN, Maria, "Global coal demand growth slows slightly, IEA says in latest 5-year outlook". IEA, Paris, 16/XII/2013.

Chapter 7
Climate Emergency

In the first six chapters of this book, we address the three major causes of the ongoing socio-environmental collapse: (1) the decline in forest cover, in water resources, and in soils, driven by the globalized corporate food system; (2) widespread pollution of the environment and intoxication of organisms, and (3) the increasing consumption of fossil fuels, with their equally devastating impacts on ecosystems. In this chapter and the next one, we will address the crux of the planetary environmental crisis: the increasing Earth's Energy Imbalance (EEI), or the climate emergency, which is deemed the most systemic and impending threat to humankind and to the biosphere in general. This chapter will address the problems on which there is scientific consensus, notably expressed in the IPCC's Assessment Reports, while the next one will deal more broadly with the uncertainties about the possible outcomes of these ongoing processes. Chapters 10, 11, and 12, which conclude the first part of this book, will discuss the immediate consequences of climate change to humanity and to biodiversity. This chapter and the next one occupy, therefore, a central position in the first part of this book and the preceding chapters can be understood as a long introduction to them.

7.1 The Peak in Global Greenhouse Gas Emissions Is Not Yet in Sight

The physiochemical processes involving carbon release and removal from the atmosphere and oceans were in balance before the Industrial Revolution. In less than two centuries, and especially after 1950 (Steffen et al. 2015), deforestation and the burning of fossil fuels caused an increasing share of carbon—accumulated and stored in forests and underground during historical periods and preceding geological times— to be released into the Earth system. We saw in chapter two that between 1800 and 2010, the growing globalization of capitalism was responsible for the destruction of

© Springer Nature Switzerland AG 2020
L. Marques, *Capitalism and Environmental Collapse*,
https://doi.org/10.1007/978-3-030-47527-7_7

ten million km² of forests on the planet. We also saw that between 2001 and 2018 alone, there was a global decrease in tree cover of 3.61 million km² (Fig. 2.3). Due to deforestation, heavily concentrated in the last 50 years, only 450 Gt of carbon was stored by the terrestrial biomass in 2018, a reduction of more than 50% compared to 916 Gt of carbon present in the hypothetical absence of land use change, under current climate conditions (Erb et al. 2018).

Regarding CO_2 emissions from 1750 to 2018, more than 1600 $GtCO_2$ have been emitted into the atmosphere, especially by industrialized societies. The curve of increase in these emissions is so steep that already in 2014, when a combined 1480 $GtCO_2$ had been released into the atmosphere, more than half of this release had occurred since 1988 (ironically, the year when, thanks to James Hansen, the evidence of human-caused warming first became widely known), as shown in Fig. 7.1.

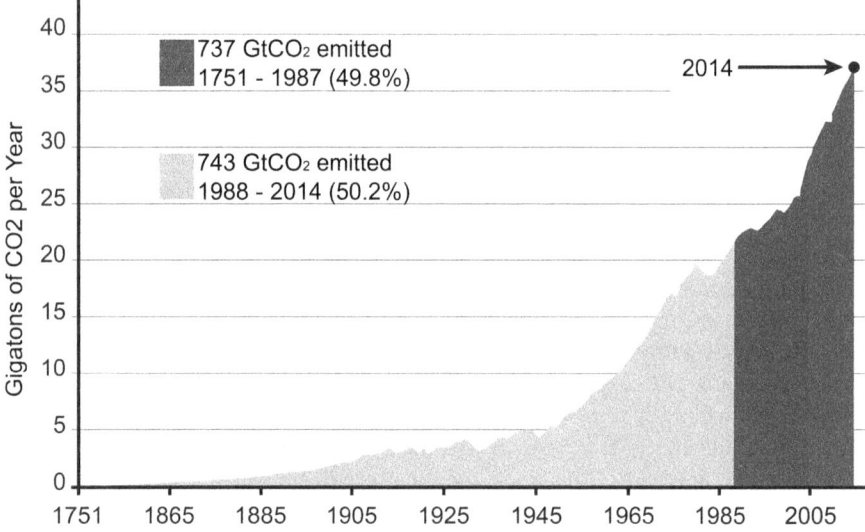

Fig. 7.1 Industrial CO_2 emissions between 1751 and 2014. From 1751 to 1987, 737 $GtCO_2$ (49.8%) were emitted. Between 1988 and 2014, 743 Gt (50.2%) were emitted. Total = 1480 $GtCO_2$. (Source: Based on Frumhoff et al. (2015); Boden et al. (2013); Le Quéré et al. (2015)

More than half of the anthropogenic CO_2 emissions since the Industrial Revolution has occurred in the last three decades, dominated by a runaway globalization of corporate capitalism. The unprecedented concentration of income generated by this process is expressed in the extreme carbon inequality: the richest 10% of people around the world are now responsible for almost 50% of global CO_2 emissions, while the poorest 50% are responsible for only 10% of them (Gore 2015). We will discuss this in detail in chapter nine (Demography and Democracy), but it is worth remembering at the outset that, according to Timothy Gore, from Oxfam, the richest 10% have average carbon footprints 11 times higher than the poorest half of the population and 60 times higher than the poorest 10%. An average Australian or American generates 3.5 times the global average, almost 17 tons of CO_2 per capita every year, or more than twice the amount of someone in Europe and China (Jackson and Canadell 2019). If regulations were to force the richest 10% to cut their CO_2 footprint to the level of the European average citizen, global emissions would be reduced by 30% (Anderson 2019). The NGO Atmosfair calculates that a simple return flight from London to New York generates almost one ton of CO_2 per passenger. This estimate does not include emissions generated by building the airplane, nor any other emissions indirectly generated by this trip. It includes only the CO_2 emitted by burning jet fuel. The average citizen of 56 countries in Africa and South America emits less CO_2 over the whole year than each passenger during that flight alone (Kommenda 2019). The unbridled consumption of the richest 10% explains why there is no deceleration in the increase in global CO_2 emissions. Figure 7.2 shows the evolution of these emissions from the burning of fossil fuels between 1965 and 2018.

Note that in 2017 and 2018, the rate of increase in these emissions was the highest in the second decade and occurred at rates similar to those of the first decade. CO_2 emissions went up in 2018 by more than 2%, the fastest rate of increase in the last 7 years. The USA, China, and India increased their emissions by 2.5%, 4.7%, and 6.3%, respectively. Using aircraft measurements over Canadian oil sands, John Liggio and colleagues (2019) indicate that:

> CO_2 emission intensities for oil sand facilities [in Canada] are 13–123% larger than those estimated using publically available data. This leads to 64% higher annual GHG emissions from surface mining operations, and 30% higher overall oil sand GHG emissions (17 Mt) compared to that reported by industry.

Between 1990 and 2017, global CO_2 emissions increased by 63% (GCP, Global Carbon Budget 2018). Anthropogenic emissions of all GHG, expressed in terms of global warming potential of CO_2 (CO_2-eq), increased by almost the same proportion. In 2017, they reached 55.1 $GtCO_2$-eq, having increased 55% in relation to 1990 and 40% in relation to 2000, as shown in Fig. 7.3.

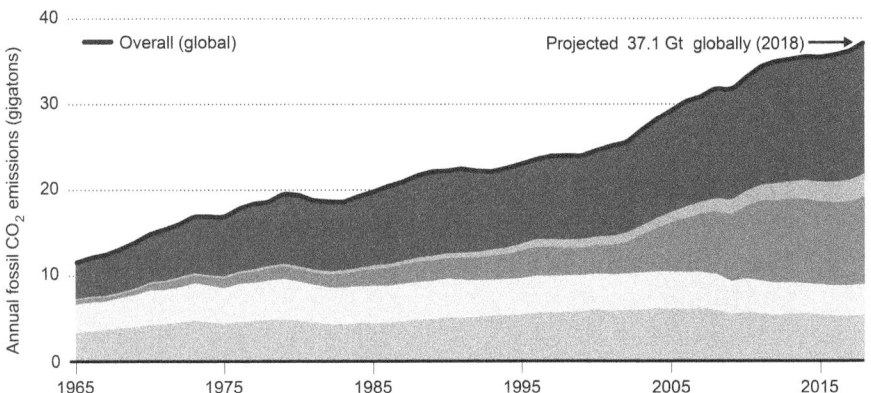

Fig. 7.2 CO_2 global emissions in Gt (1965–2018) from fossil fuels. From bottom to top: USA, Europe, China, India, all others. (Source: Figueres et al. 2018, based on Global Carbon Project)

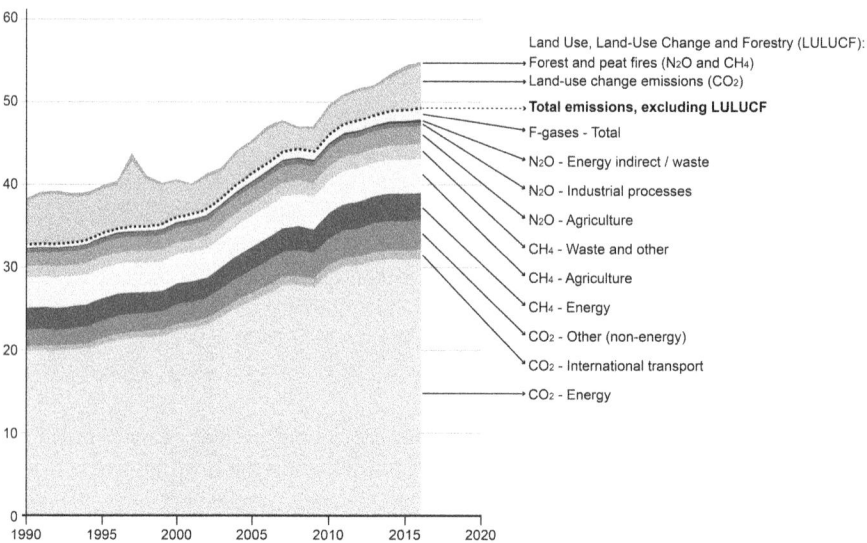

Fig. 7.3 Global greenhouse gas emissions, per type of gas and source, between 1990 and 2017, in gigatons of CO_2-eq (CO_2 equivalent). Source: Olivier J.G.J. & Peters J.A.H.W., Trends in global CO2 and total GHG emissions: 2018 report. PBL Netherlands Environmental Assessment Agency, The Hague, 2018, with data from Emission Database for Global Atmospheric Research (EDGAR v4.3.2 database). Observation: LULUCF = Land use, land use change and forestry

According to the IPCC SR1.5C (2018), global emissions of CO_2 must peak by 2020 to ensure a reasonable chance that global warming will not exceed 1.5 °C above the pre-industrial period until 2100. But, as seen in the previous two chapters, low-carbon renewables are not replacing fossil fuels. They are just diversifying and adding more alternatives to society's energy greed. There are no convincing signs that the fossil fuel consumption curve and, hence, global GHG emissions will begin its decline phase in a foreseeable future. Global CO_2 emissions hit an all-time high in 2019 (+0.6% from 2018 or about 43.1 $GtCO_2$, according to the Global Carbon Project), and all projections indicate an increase in fossil fuel consumption over the next decade (Andrews 2017).

7.2   The Current CO_2 Atmospheric Concentration Is Unprecedented Over the Past 3 Million Years (and Its Rate of Increase Is Unprecedented Over at Least the Past 55 Million Years)

Thanks to the measurement method created by Charles David Keeling (1928–2005) at the Mauna Loa observatory in Hawaii, the Keeling curve has been used since 1958. It measures the concentration of CO_2 in the atmosphere in parts per million (ppm) at an altitude of 3400 m in the northern subtropics. Other centers for measuring these concentrations exist in other countries and latitudes, with very close results. In 1750, these concentrations were at about 227 parts per million (ppm). In 2019, they had momentarily reached 415 ppm. Since 2000, the increase in atmospheric CO_2 concentrations has been around 20 ppm per decade, which is up to 10 times faster than any sustained rise in CO_2 during the past 800,000 years (Steffen et al. 2018). In 2013, when these concentrations hit 400 ppm, Ron Prinn (2013) warned that there was more CO_2 in the Earth's atmosphere than ever before in the past three million years (see also Rahmstorf 2017). Matteo Willeit and colleagues (2019) confirmed this estimate: "the current CO_2 concentration is unprecedented over the past three million years," that is, it is equivalent only to the Pliocene (the epoch in the geologic timescale that extends from 5.33 million ton 2.58 million years BP).

More worrying than the magnitude of these CO_2 concentrations is its rate of increase, as shown in Table 7.1.

Between 1959 and 1997, there was no annual increase above 2.2 ppm. Between 1998 and 2016, there were six annual increases above 2.2 ppm. In this most recent period, there were three real jumps of 2.93, 3.03, and 2.77 ppm in 1998, 2015, and 2016, respectively, and the estimated margin of uncertainty at these growth rates is 0.11 ppm/ year. Finally, between May 2018 and May 2019, there is a jump of about 3.5 ppm. Three of the four highest annual increases in CO_2 atmospheric concentrations have occurred in the past 4 years (2015–2018). "The rate of CO_2 growth over the last decade is 100–200 times faster than what the Earth experienced during the transition from the last Ice Age. This is a real shock to the atmosphere" (Tans 2017). According to a review published in 2013 by the US National Research Council of the National Academies (Somero et al. 2013):

> Past increases in CO_2 occurred over periods of hundreds of thousands to millions of years, and thus differ considerably from the very rapid present day increase related to human activities. The current rate of increase in the level of atmospheric carbon dioxide is unprecedented over at least the past 55 million years. The rate is far greater than occurred in even the most rapid events known from Earth history, and each of these past events were accompanied by important changes in ocean chemistry and mass extinctions of ocean or terrestrial life or both.

Table 7.1 Atmospheric CO_2 average growth rates (1959–2014)

Decade	Atmospheric CO_2 average growth rates (Parts per million / year)
2005–2014	2.11
1995–2004	1.87
1985–1994	1.42
1975–1984	1.44
1965–1974	1.04
1959–1964 (6 years only)	0.73

Source: CO_2 Acceleration https://www.co2.earth/co2-acceleration (Concentrations apply to the lower 75–80% of the atmosphere, known as the troposphere)

Ken Caldeira (2012) presents this same worry: "In geologic history, transitions from low to high-CO_2 atmospheres typically happened at rates of less than 0.00001 degree a year. We are re-creating the world of the dinosaurs 5,000 times faster."

496 ppm CO_2-eq (2018)

If we consider the CO_2 equivalent (ppm CO_2-eq), that is, the atmospheric concentrations of all GHGs expressed in terms of the atmospheric warming potential of CO_2, the current atmospheric concentration (2019) is not around 410 ppm, but 496 ppm CO_2-eq. (2018), with a 30% increase in these concentrations in just 40 years: 1979 = 382 ppm; 1990 = 417 ppm; 2010 = 468 ppm; 2018 = 496 ppm (NOAA/AGGI, Spring 2019).

Between 2011 and 2018, the average rate of increase of these concentrations was 1.6% per year. If this rate is maintained (without further acceleration), the doubling of these concentrations from 496 ppm to 992 ppm will occur in 45 years, that is, by the mid-2060s, implying a likely average global temperature increase of 3–4 °C above the pre-industrial period. This level of warming is deemed incompatible with any civilization along the lines of what we know today.

7.3   Between 1.2 °C and 1.5 °C and Accelerating

As the IPCC Special Report 1.5, 2018 asserted for the umpteenth time, "the total amount of CO_2 emitted until we reach the threshold of zero global emissions largely determines the amount of warming to which we are committed. We know, therefore, that much more warming is coming. But any reflection on the future scenarios of the climate system and, thus, on the habitability of our planet, must begin with a brief description of the current state of the global climate system. "The average temperature of the last 5 years [2014–2018] was 1.1°C higher than the pre-industrial average (as defined by the IPCC)" (Copernicus Climate Change Service 2019). The 2019 global temperature was 1.2 °C warmer than in the 1880–1920 base period, according to the Goddard Institute for Space Studies global temperature analysis (Hansen et al. 2020). Global average warming may exceed 1.2 °C by 2020 as "monthly temperatures over the past 12 months [October 2018 – September 2019] have averaged close to 1.2°C above the pre-industrial level." (Copernicus Climate Change Service, "Surface air temperature for September 2019"). Figure 7.4 shows this escalation in global warming in 2019 of 1.2 °C above the average of the 1880–1920 base period.

As can be seen, despite the interannual temperature variations caused by El Niño and other geophysical phenomena, global warming has become, more than ever before, a certainty that is empirically measurable. This can also be inferred from the following evidence, based on data from GISS-NASA, NOAA, MET Office, and WMO:

1. Global temperature hasn't been cooler than the twentieth-century average since 1976.

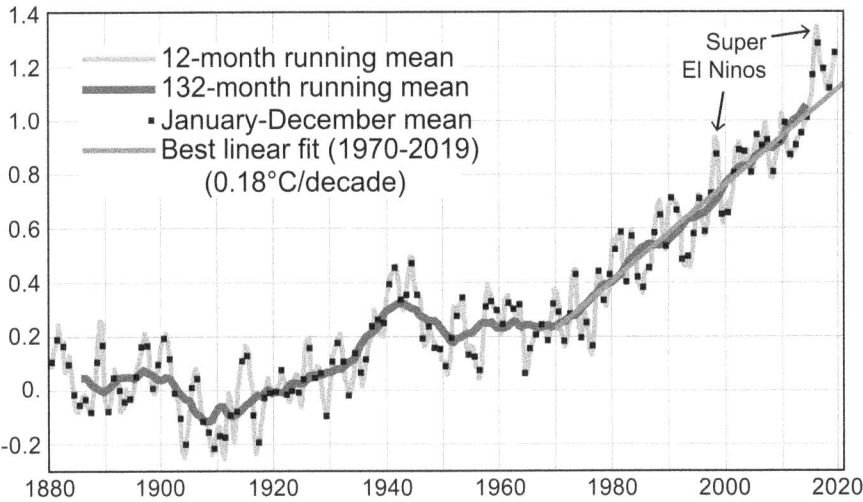

Fig. 7.4 Global surface temperatures 1880–2019 relative to 1880–1920. Source: James Hansen, Makiko Sato, Reto Ruedy, Gavin A. Schmidt, Ken Lo & Michael Hendrickson, "Global Temperature in 2019". 15/I/2020. http://www.columbia.edu/~jeh1/mailings/2020/20200115_Temperature2019.pdf

2. "Each decade from the 1980s has been successively warmer than all the decades that came before. 2019 concludes the warmest 'cardinal' decade (those spanning years ending 0–9) in records that stretch back to the mid-19th century" (MET Office, Press Office, 15/I/2020).
3. Of the 22 warmest years on record, 20 have occurred since 1998.
4. The last 6 years (2014–2019) rank as the warmest on record.
5. 2017 and 2019 were the warmest non-El-Niño years on record.
6. Already in 1999, Michael Mann, Raymond Bradley, and Malcom Hughes had shown, with the famous graphic dubbed hockey stick (Mann 2012), that in the late twentieth century, temperatures were warmer than during the past millennium. More recent research (Marcott et al. 2013), not only confirms the iconic hockey stick but also shows that the global temperature variation observed in the Anthropocene (the proposed current geological epoch), especially since the 1980s, is greater than any other variation throughout the Holocene (11.700 years ago – to 1950). During these almost 12 millennia, natural climate variability has not reached 0.5 °C above or below the 1961–1990 average. Moreover, there is no evidence that the temperature variations observed during the Holocene—the so-called Medieval Climate Anomaly and the "Little Ice Age"—were global phenomena. Based on PAGES 2 k, a compilation including nearly 700 records from trees, ice, sediment, corals, cave deposits, and other sources of information, Raphael Neukom and colleagues (2019) can affirm that the warming observed in the late twentieth century alone occurred simultaneously in more than 98% of the globe. "This provides strong evidence that anthropogenic global warming is not only unparalleled in terms of absolute temperatures, but also unprecedented

in spatial consistency within the context of the past 2,000 years." Actually, the Earth's current climate has already surpassed the warmest period of the Holocene (Marcott et al. 2013) and may be as warm as it was during the Eemian, the prior interglacial period (Hansen et al. 2016).

7. Current warming exceeds the Holocene's natural variability not only in magnitude, but also in speed. It is much greater than the warming that took place in geological times prior to the Holocene (Aengenheyster et al. 2018).

8. The acceleration of warming in the twenty-first century is absolutely exceptional, even in relation to the late twentieth century. The pace of warming multiplied by a factor of 2.5 in the 2008–2017 period, reaching a rate of 0.43 °C per decade compared to the 1970–2014 period when it still rose at a rate of 0.17 °C per decade in relation to the 1951–1980 period, as shown in Fig. 7.5.

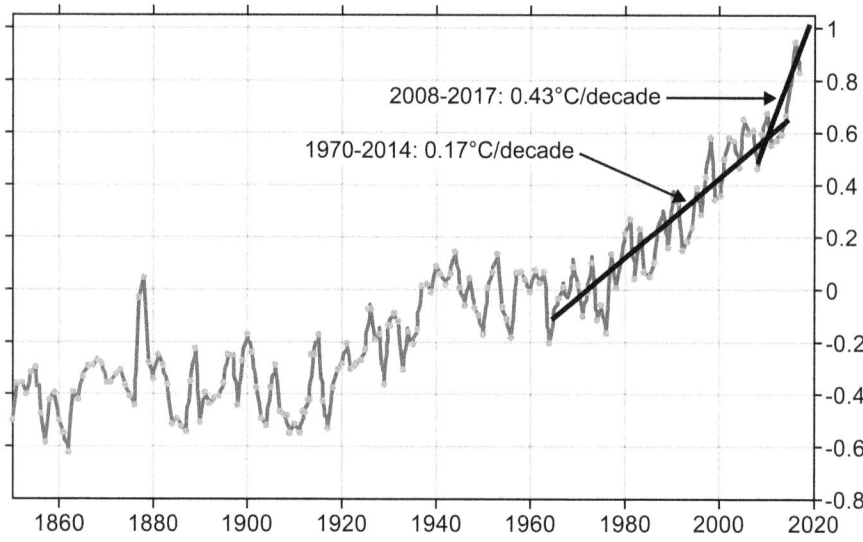

Fig. 7.5 Acceleration of the 10-year average global warming rate in 2008–2017 compared to 1970–2014, relative to the 1951–1980 average. Land data prepared by Berkeley Earth and combined with data adapted from the UK Hadley Centre. Global temperature anomalies relative to 1951–1980 average. (Source: Based on Climate Change Data Center from the Chiangmay University http://ccdatacenter.org/PageFact.aspx?FactPageID=8&Categories=YES)

7.3.1 Spatial Distribution: Global vs. Regional Warming

There is, therefore, a consistent and growing body of evidence showing that the rise in the Earth's average surface temperatures is not only occurring, but also accelerating. However, as is well-known, the Earth has not warmed evenly over the past century. Two major differentials should be noted in the spatial distribution of warming: (a) warming over continents is faster than over oceans (and, therefore, it tends to be

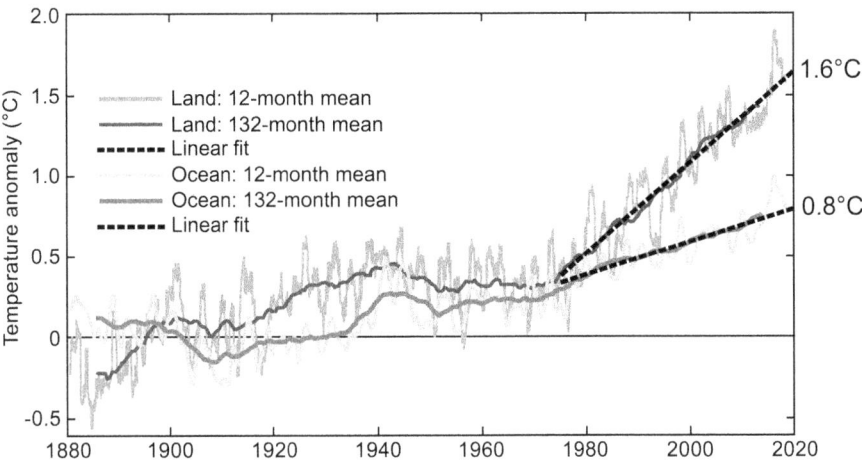

Fig. 7.6 Temperature anomalies relative to 1880–1920 for global land and ocean areas. Source: James Hansen, Makiko Sato, Reto Ruedy, Gavin A. Schmidt & Ken Lo, "Global Temperature in 2018 and Beyond". GISS/NASA, 6/II/2019. http://csas.ei.columbia.edu/2019/02/06/global-temperature-in-2018-and-beyond/. Observation: Ocean temperatures refer to open oceans, ice-free all year. The dotted line indicates the annual averages and the continuous line represents the 5-year averages

faster in the Northern Hemisphere than in the Southern Hemisphere); and (b) warming at high latitudes is also faster than global average warming. Figure 7.6 discriminates between terrestrial and ocean warming between 1880 and 2018 relative to the 1880–1920 base period.

Compared to the 1880–1920 period, global land warming reached 1.8 °C in 2016, dropping to 1.6 °C in 2018, while global ocean warming, naturally much slower, reached 1 °C in 2016 and 0.8 °C in 2018. From the late 1970s on, land and sea warming rates have become more clearly divergent, with land warming increasing at a much faster rate. Consequently, in various terrestrial regions of the planet, especially at high latitudes, the warming that occurred until 2019 has already exceeded the 2 °C threshold, when compared to the pre-industrial period. According to data from Berkeley Earth, roughly one-tenth of the planet surface has already warmed by more than 2 °C when the last 5 years (2014–2018) are compared with the mid- to late 1800. Regarding the higher warming velocity at high latitudes when compared to the global average, measurements at the Byrd station showed that average temperatures in Central West Antarctica had already risen by 2.4 °C (+/−1.2 °C) between 1958 and 2010 alone (Bromwich et al. 2013).[1]

Over the past 50 years, the Arctic region has warmed more than any other region on Earth, a phenomenon known as Arctic amplification. According to NOAA, in the first half of 2010, air temperatures in the Arctic were 4 °C warmer than the

---

[1] See also "West Antarctica Warming Three Times Faster Than Global Average, Threatening To Destabilize This Unstable Ice Sheet." *Climate Progress*, 27/XII/2012

1968–1996 reference period. Arctic atmospheric changes from 1971 to 2017 indicate warming: by 2.7 °C at the annual scale, by 3.1 °C in the cold season (October–May), and by 1.8 °C in the warm season (June–September) (Box et al. 2019). Due to these major climate anomalies, we find unprecedented weather patterns in the Arctic. In 4 out of the past 5 years (2014–2018), the North Pole has experienced the most intense heat waves ever recorded (historical records began in 1958), with temperatures of 17–19 °C above normal (Samenow 2018). In February 2016, parts of the Arctic reached temperatures above freezing that were more than 16 °C above the average for this month. In February 2018, although it was still practically without sunlight, Siberia's temperature reached 35 °C above the historical average for that month. In the middle of a February 2018 winter night, the temperature in Greenland stayed above zero, three times the average of any previous year, for 61 successive hours (Watts 2018). In Finland and Scandinavia, temperatures have exceeded several historical records, with temperatures of around 33 °C sometimes at latitudes above 70°N. On the night of July 18 to July 19, 2018, the temperature did not drop below 25.2 °C at 70.7°N, the hottest night ever seen in northern Scandinavia, on the shore of the Barents Sea in the Arctic Ocean.

Outside the Arctic, temperatures in the Northern Hemisphere are also warming at an average rate faster than the Southern Hemisphere, because, as already mentioned, it contains more land. According to the Japan Meteorological Agency, the rise in temperatures in the Northern Hemisphere has been consistently double that of the Southern Hemisphere; in 2018, this rise corresponded, respectively, to 0.41 °C and 0.20 °C above the 1981–2010 average. During the last decade (2008–2017), the average annual temperature in European land areas was between 1.6 °C and 1.7 °C above the pre-industrial level, but strong temperature anomalies are also found in the Southern Hemisphere. In the winter of 2015, temperatures reached 3–4 °C above normal in Brazil (Nobre et al. 2019). In 2014, the Southeastern region of Brazil had already registered temperatures between 1 °C and 2 °C higher than the 1961–1990 average (Marengo 2014). And in the Northeastern region of the country, the average temperature increased by 2.5 °C in the last decades compared to this same period.

As shown also in Fig. 7.6, from 2014 onward, the rise in the temperature curve of ocean, as well as terrestrial areas, becomes steeper. This acceleration in ocean heating has been observed at increasing depths. In the first decade of the twenty-first century, about 30% of the heating was already occurring beyond a depth of 700 meters (Balmaseda et al. 2013). There was a surge in ocean warming in 2017, by far the hottest year on record for the oceans, when the top two thousand meters of the oceans were 1.51×10^{22} Joules hotter than in 2015, the second hottest year of these records. As a reference, the increase in oceanic heat in 2017 alone was 699 times greater than all the electricity generated in China in 2016 (Cheng and Zhu 2018). "The ocean has been heating at a rate of around 0.5–1 watt of energy per m^2 over the past decade, amassing more than 2×10^{23} joules of energy — the equivalent of roughly five Hiroshima bombs exploding every second — since 1990" (Katz 2015). The linear trend of ocean heat content (OHC) at a depth of 0–700 m during 1992–2015 shows a warming trend four times stronger than the 1960–1991 period (Cheng et al. 2017).

7.4 Suffering and Greater Lethality due to the Current Warming

An average global warming of 0.74 °C already in 2005 (reported by the IPCC in 2007[2]) was capable of triggering unbelievable socio-environmental crises in the history of civilization. The suffering inflicted on humans and on so many other life-forms by the current warming is tremendously greater. It cannot, however, be grasped by abstract trends, such as a global average warming of 1–1.5 °C. In the world of concrete experience, these references are translated in many tragic ways. As stressed by the World Health Organization (WHO), vectors, pathogens, and hosts survive and reproduce within a range of optimal climatic conditions. Changes in infectious disease transmission patterns are a likely major consequence of climate change. The increasing burden of insect-borne or tick-borne diseases caused by bacteria, viruses, and protozoa is now being exacerbated by higher temperatures. In July 2016, an outbreak of anthrax in Siberia killed more than 2000 reindeers. Populations of insects such as *Aedes aegypti* and *Aedes albopictus* or arachnids, like ticks, are clearly increasing at low latitudes and spreading toward higher latitudes, thanks to milder winters and longer and warmer summers. These vectors transmit a number of serious, and sometimes lethal, diseases that tend to acquire epidemic proportions. Global warming is also fueling a rise in dangerous fungal infections. As pointed out by Arturo Casadevall, "as the climate has gotten warmer, some of these organisms, including *Candida auris*, have adapted to the higher temperature, and as they adapt, they break through human's protective temperatures" (Andrei 2019).

Global warming also increases ozone concentrations and atmospheric pollution and already decreases yields of major crops in some regions. Global maize and wheat production, for instance, declined by 3.8 and 5.5%, respectively, between 1980 and 2008, relative to a counterfactual without climate trends (Lobell et al. 2011). Furthermore, according to Kristie Ebi and Irakli Loladze (2019):

> Increased concentrations of CO_2—by directly affecting plants—worsen the nutritional quality of food by decreasing protein and mineral concentrations by 5–15%, and B vitamins by up to 30%. Higher CO_2 concentrations increase photosynthesis in C_3 plants (e.g., wheat, rice, potatoes, barley), which can increase crop yields. But those increases come at the cost of lower nutritional quality as plants accumulate more carbohydrates and less minerals (e.g., iron and zinc), which can negatively affect human nutrition.

Current warming is the indirect cause of an increasing number of deaths from droughts, floods, landslides, extreme weather events, forest fires, agricultural crises, malnutrition, forced displacement, etc. Moreover, it is the direct cause of heat waves, defined as a combined measure of temperature and relative humidity (heat index). An extreme heat wave occurs when summertime temperatures are much

---

[2] Cf. IPCC, Fourth Assessment Report, 2007, p. 2: "The 100-year linear trend (1906–2005) of 0.74 [0.56 to 0.92]°C is larger than the corresponding trend of 0.6 [0.4 to 0.8]°C (1901–2000) given in the Third Assessment Report."

hotter and/or humid than is typical for a location. Increasingly hot summers are now taking place with more frequent and extreme heat waves, such as the heat wave that swept Europe in 2003, Australia in 2012–2013 and 2019–2020, the "Lucifer" heat wave in Europe in 2017, and the heat waves that hit the Northern Hemisphere in 2018, 2019 and 2020. Warmer winters also follow, sometimes (since the 1990s) with exceptional cold spikes in North America and Eurasia (Overland et al. 2016; Kretschmer 2018). But, the disproportion between hot and cold record temperatures is stark. In 2018, a year without El Niño, weather stations recorded 40 cold records and 430 absolute heat records (Wong 2019).

Today, exposure, for just a few hours, to an extreme heat wave can cause damage to the brain and other vital organs or even death by hyperthermia, heat exhaustion, or heat stroke. This is especially the case after 2003, when the first of these well-documented phenomena in Europe caused about 70,000 premature deaths. From June to October 2010, a second heat wave with several record-breaking temperatures devastated the Sahel, the Eastern United States, the Middle East, Eastern Europe, Russia, and China, causing deadly forest fires. In several regions of Russia, one of the countries most affected by these fires, temperatures in June and July 2010 ranged from 36 °C to 42.3 °C (Belogorsk), with 44 °C being the highest temperature ever recorded in Russia on July 11, 2010, in Yashkul, near the Black Sea. Since 2015, the World Meteorological Organization (WMO) has repeatedly acknowledged unbearable temperatures for the human body, between 45 °C and 54 °C, in China, India, Pakistan (53.7 °C), Afghanistan, Iran, Iraq, Jordan, Oman, Turkey, France (45.9 °C), Spain (47.7 °C), Australia (49.9 °C), and Kuwait (Mitribah, 2016, 54 °C), the first country to make it illegal to work in the sun from 11 a.m. to 5 p.m. between July 1 and August 31. According to the WMO Archive of Weather and Climate Extremes, the Mitribah, Kuwait temperature is now accepted as the highest temperature ever recorded for the continental region of Asia. In June 2019, temperatures reached 48 °C in New Delhi, the highest ever recorded in the capital in this month, and hundreds of people were killed in the eastern state of Bihar, where a curfew was imposed, preventing people from going outside. The *Nature* editorial of June 22, 2017, reflects the concern that heat waves are increasingly exceeding human thermoregulatory capacity:

> From extreme rainfall to rising sea levels, global warming is expected to wreak havoc on human lives. Sometimes, the most straightforward impact – the warming itself – is overlooked. Yet heat kills. The body, after all, has evolved to work in a fairly narrow temperature range. Our sweat-based cooling mechanism is crude; beyond a certain combination of high temperature and humidity, it fails. To be outside and exposed to such an environment for any length of time soon becomes a death sentence.

Camilo Mora and colleagues (2017) identified a global threshold beyond which daily mean surface air temperature and relative humidity become deadly and showed that around 30% of the world's population (13.6% of land area) is currently exposed to climatic conditions exceeding this deadly threshold for at least 20 days a year. Also, the "Lancet Countdown," a collective effort published under the coordination of Nick Watts (2017), affirms that:

> The evidence is clear that exposure to more frequent and intense heatwaves are increasing, with an estimated 125 million additional vulnerable adults exposed to heatwaves from 2000 to 2016. Higher ambient temperatures have resulted in estimated reduction of 5.3% in labour productivity, globally, from 2000 to 2016.

The current trend toward more intense, expansive, longer-lasting, and more frequent heat waves is clear. James Hansen, Makiko Sato, and Reto Ruedy (2012) demonstrated a higher probability of extremely hot summers compared to the years 1951–1980 and "the emergence of a category of 'extremely hot' summers, more than 3σ [standard deviations] warmer than the base period mean" (1951–1980). Moreover, while these extreme summer heat waves covered much less than 1% of Earth's surface during the base period, they typically covered about 10% of the land area in 2012. Concerning the number and the recurrence of extreme heat waves, "worldwide, the number of local record-breaking monthly temperature extremes is now on average five times larger than expected in a climate with no long-term warming" (Coumou et al. 2013). Events that would occur twice a century in the early 2000s are now expected to occur twice a decade (Christidis et al. 2014). Referring to the European heat waves in the summer of 2018, Peter Stott declared at the COP24 (December 2018) that "the intensity of this summer's heatwave is around 30 times more likely than would have been the case without climate change."[3]

The future will bring more severe and more lethal heat waves. Mora and colleagues (2017) predict that a global warming of 2 °C could expose 48% of the population to deadly heat waves. They project that in 2100, this percentage will be about 74% of the global population, maintaining the current scenario of rising emissions with global warming at 4 °C above the pre-industrial period. The heat lethality threshold will then be reached for 365 days a year in the equatorial region, making it, therefore, uninhabitable and will condemn what remains of the Amazon rainforest and of rainforests in general to final disappearance. The authors conclude: "An increasing threat to human life from excess heat now seems almost inevitable, but will be greatly aggravated if greenhouse gases are not considerably reduced." Commenting on his own work, Camilo Mora concludes: "for heat waves, our options are now between bad or terrible. Many people around the world are already paying the ultimate price of heatwaves" (as quoted by Chow 2019). According to the IPCC's Special Report 1.5 °C (2018), the difference between a warming of 1.5 °C and a warming of 2 °C in the tropics is equivalent to a difference in heat waves lasting 2 or 3 months. In Europe, a heating of 1.5 °C or 2 °C is expected to cause the July 2018 heat wave to repeat itself every 2 years or every year, respectively. In the USA, by the mid-twenty-first century (2036–2065), the annual numbers of days with heat indices exceeding 37.8 °C and 40.6 °C are projected to double or triple under RCP4.5 and RCP8.5 emission scenarios, respectively, compared to a 1971–2000 baseline (Dahl et al. 2019).

---

[3] Cf. MET Office, "2018 UK summer heatwave made thirty times more likely due to climate change," 6/XII/2018

7.5   2 °C: A Social–Physical Impossibility

In a historic Ted Talk, titled "Why I must speak about climate change" (2012), James Hansen, former director of NASA's Goddard Institute for Space Studies, made the most didactic and decisive argument on how the Earth's system is already locked into successive temperature increases:

> There is a temporary energy imbalance. More energy is coming in than going out until Earth warms up enough to again radiate to space as much energy as it absorbs from the Sun. More warming is in the pipeline. It will occur without adding any more GHG.

More recently, James Hansen and colleagues (2016) reiterated the inevitability of this future warning: "Earth is out of energy balance with present atmospheric composition, implying that more warming is in the pipeline." Let us repeat once more: as long as this Earth Energy Imbalance (EEI) persists, the planet's temperature will inevitably increase, even "without adding any more GHG," as James Hansen states. This imbalance is already huge. According to Kevin Trenberth and colleagues (2016, 2018), the Earth's current Energy Imbalance (EEI) amounts to about 1 Watt/m^2 (0.9 +/− 0.3 W/m^2) or about 500 terawatts globally for the period of 2005–2014. How much is a terawatt (TW) or a million megawatts? In 2008, for example, humans used energy (this includes all types of energy, not just electricity) at an average rate of about 16.5 TW of power.[4] "Even during the solar minimum," James Hansen (2018) says, "it was equivalent to the energy of 400,000 Hiroshima atomic bombs per day every day of the year." Moreover, simulations done by Catherine Ricke and Ken Caldeira (2014)—on the time series of marginal warming for the first 100 years after CO_2 emission—show that "maximum warming occurs a median of 10.1 years after the CO_2 emission event." In other words, the more than 41Gt of CO_2 emitted by mankind in 2019 will have its full impact on global warming in the late 2020s. In short, we are warming the planet on credit.

7.5.1   The Imminence of 1.5 °C

The IPCC and almost all statements on global warming take the "pre-industrial" period as their reference period and place it in the years 1850–1900. Michael Mann (2015) correctly disputes this conventional definition:

> [T]he base year implicitly used to define "pre-industrial" conditions is 1875, the mid-point of that interval. Yet the industrial revolution, and the rise in atmospheric CO_2 concentrations associated with it, began more than a century earlier. (…) It is evident that roughly 0.3 °C greenhouse warming had already taken place by 1900, and roughly 0.2 °C warming by 1870. While that might seem like a minor amount of warming, it has significant implications for the challenge we face in stabilizing warming below 2 °C, let alone 1.5 °C.

---

[4] See Climate Central, "Helpful Energy Comparisons, Anyone? A Guide to Measuring Energy," 14/VII/2011.

If Mann's adjustment is adopted, the current global warming average of 1.2 °C (2019) should be about 1.5 °C above the "true" pre-industrial period (1750–1850). But even if we adopt the conventional base period (the second half of the nineteenth century), the strong 2016 El Niño effect caused the planet's average temperature to approach or even momentarily surpass the Paris Agreement target: in February 2016, for the first time, monthly average global warming exceeded 1.5 °C above the 1880–1899 average, and the year 2016, as a whole, was hotter than this same period by an average of 1.24 °C.

In any case, a global warming average above 1.5 °C relative to the 1850–1900 average is considered imminent. The IPCC (SR1.5 2018) projects that "global warming is likely to reach 1.5°C between 2030 and 2052 if it continues to increase at the current rate." This prediction was considered conservative by many scientists. As Michael Mann warned: "We are closer to the 1.5°C and 2°C thresholds than they [IPCC, SR1.5 2018] indicate and our available carbon budget for avoiding those critical thresholds is considerably smaller than they imply. In other words, they paint an overly rosy scenario by ignoring some relevant literature" (Waldman 2018). Also for the European Environment Agency, the 1.5 °C mark can be reached much earlier than the IPCC suggests: "If the concentrations of the different greenhouse gases continue to increase at current rates, peak concentration levels required to stay below a temperature increase of 1.5°C above pre-industrial levels could be reached within the next 3–13 years."[5] The IPCC's prediction can be considered too cautious also because, as seen above (Fig. 7.5), global warming is not increasing at the same rate. It is currently (2008–2017) accelerating and unfolding at a rate of 0.43 °C per decade. According to the MET Office's Decadal Forecast (January 2020), during the 5-year period 2020–2024, global average temperature is expected to remain between 1.06 °C and 1.62 °C (90% confidence range) above the pre-industrial period (1850–1900) with a small (~10%) chance of 1 year temporarily exceeding 1.5 °C. This first possible breach of the 1.5 °C threshold in the global annual average by 2024 would still be momentary, that is, it is subject to the usual interannual temperature variations. But it may become irreversible as early as the second half of the third decade (Henley and King 2017; Stone 2017; Xu et al. 2018).[6] Kevin Trenberth expresses the same view. He declared in 2016: "I don't see at all how we're going to not go through the 1.5 degree-number in the next decade or so" (Schlossberg 2016).

More important than predicting the date—after 2025 or after 2030 (IPCC)—when we will irreversibly go beyond 1.5 °C, is understanding that avoiding this threshold now seems to be a nearly impossible task. For Chris Field, co-director of the IPCC, "the 1.5°C goal now looks impossible or at the very least, a very, very difficult task. We should be under no illusions about the task we face" (McKie 2016). Drew Shindell, co-author of the IPCC SR1.5 (2018), affirmed in September

---

[5] See EEA's Indicator Assessment, published on 5/XII/2019, last modified 25/II/2020. https://www.eea.europa.eu/data-and-maps/indicators/atmospheric-greenhouse-gas-concentrations-6/assessment-1

[6] See also Climate Central Research Report, "Flirting with the 1.5 °C Threshold." 20/IV/2016

2018: "It's extraordinarily challenging to get to the 1.5°C target and we are nowhere near on track to doing that. While it's technically possible, it's extremely improbable, absent a real sea change in the way we evaluate risk. We are nowhere near that" (Milman 2019). With the US decision to leave the Paris Agreement, the further increases in coal consumption in 2017 and 2018 (see Chap. 6), and the brutal regression in climate mitigation policies in Brazil, Australia, Japan, and some Eastern European countries, the current chances of slowing down global warming are virtually nil.

That said, according to the laws of physics, what is the least amount of warming possible for this century? In November 2017, Richard Millar and colleagues maintained that keeping global warming below 2 °C, this century was not yet a geophysical impossibility:

> Limiting cumulative post-2015 CO_2 emissions to about 200 GtC would limit post-2015 warming to less than 0.6 °C in 66% of Earth system model members of the CMIP5 [Coupled Model Intercomparison Project Phase 5] ensemble with no mitigation of other climate drivers, increasing to 240 GtC with ambitious non-CO_2 mitigation. (…) Assuming emissions peak and decline to below current levels by 2030, and continue thereafter on a much steeper decline, which would be historically unprecedented but consistent with a standard ambitious mitigation scenario (RCP2.6), results in a likely range of peak warming of 1.2–2.0 °C above the mid-nineteenth century.

This projection has been the subject of much criticism. Andrew Schurer and colleagues (2018), for example, criticized the author's starting point, that is, the use of an observational dataset (HadCRUT4) that does not include data from the Arctic, which has been found to be warming at a rate much faster than the global mean. In fact, in 2017 global warming was already about 1.1 °C above the pre-industrial period, not "about 0.9°C from the mid-nineteenth century," as the authors propose. Furthermore, Millar and colleagues do not seem to take due account of the recent increase in atmospheric concentrations of methane and their potential positive feedbacks on the global climate system in a post-1.5 °C world (Wadhams 2015, 2016; Dean et al. 2018; Beckwith 2019; see Chap. 8, particularly Sect. 8.2 The Arctic Methane Conundrum). The increased vulnerability of all forests to fires and their diminishing ability to function as a carbon sink are other positive feedbacks that are not mentioned in the study in question. In fact, in terrestrial ecosystems, the vegetation canopy, from tropical and boreal forests to semi-arid vegetation (Poulter et al. 2014), is largely responsible for land sink, that is, the removal of carbon from the atmosphere by plants through photosynthesis. According to Simon L. Lewis (2009), in the years of the first decade of the century, tropical forests were absorbing 18% of the CO_2 added to the atmosphere each year from the burning of fossil fuels. In the tropics, however, biomass mortality increased, leading to a shortening of carbon residence times. Thus, the benefit once provided by the net increase in biomass has been reversed and is declining in recent years, as shown by Roel Brienen and colleagues (2015):

> Atmospheric carbon dioxide records indicate that the land surface has acted as a strong global carbon sink over recent decades, with a substantial fraction of this sink probably located in the tropics, particularly in the Amazon. (…) We find a long-term decreasing trend

of carbon accumulation [in the Amazon forest]. Rates of net increase in above-ground biomass declined by one-third during the past decade compared to the 1990s.

7.5.2 *"The Great Deceleration"?*

Suppose for a moment, however, that Millar and colleagues are correct in their affirmation that keeping global warming below 2 °C is not yet a "geophysical impossibility." That would be great news. The bad news is that, given society's weak political resistance and the current strong power of corporations to impose their paradigms, worldviews, and business plans on humanity, a global average warming below 2 °C is now considered a social–physical impossibility. The assumption that the Great Acceleration, which began around 1950 (Steffen et al. 2015), might be converted into a "Great Deceleration" while maintaining the political *status quo* is illusory. The ten editions of the *Emissions Gap Report* produced by the United Nations Environment Programme (UNEP) measure the gap between what countries have pledged to do to reduce their GHG emissions and what they need to do to meet the UNFCCC goals. Niklas Höhne and colleagues (2020) show that "the gap has widened by as much as four times since 2010." The hypothesis envisioned by Millar and colleagues that we will not exceed a 200 GtC (about 730 $GtCO_2$) carbon budget, based on 2015 emissions, is clearly disconnected from reality. Global annual GHG emissions increased by 14% between 2008 and 2018. In the present scenario, nothing in the logic of the capital accumulation driving our thermo-fossil civilization toward its end allows for such a conjecture. From 2016 to 2018 alone, about 120 $GtCO_2$ and about 160 $GtCO_2$-eq were emitted globally. These CO_2 emission levels are projected to grow 0.6% in 2019. Let us not fool ourselves. After an important, but probably ephemeral, reduction in 2020 caused by the COVID-19 pandemic, global CO_2 emissions will remain substantially unchanged or even increase further in the next years. It is practically certain, therefore, that between 2016 and 2024 at least an additional 320 $GtCO_2$ will have been emitted globally; this is about 44% of the carbon budget allowed by the calculations of Richard Millar and colleagues in order to not exceed a warming of 2 °C above the pre-industrial period. Figure 7.7 shows how radical the reduction in anthropogenic emissions must be or should have been since 2016, in order to keep global average warming between 1.5 °C and 2 °C until 2100.

Depending on the chosen probability, a warming below 2 °C would imply that future emissions be limited to a range between 150 and 1050 $GtCO_2$. Christiana Figueres and colleagues (2017) work with the arithmetic mean of these two values (600 $GtCO_2$). To reach this limit, net emissions should be zeroed by 2040 and remain zero throughout the century. In this way, if we had started the reduction curve in 2016, we would have until 2045 to permanently clear emissions. Nothing indicates that we will begin a steep decline of CO_2 emissions between 2020 and 2024. If we finally start the decline in net CO_2 emissions by 2025, we will have to zero them by 2035 and keep them at zero after that. As Christiana Figueres and

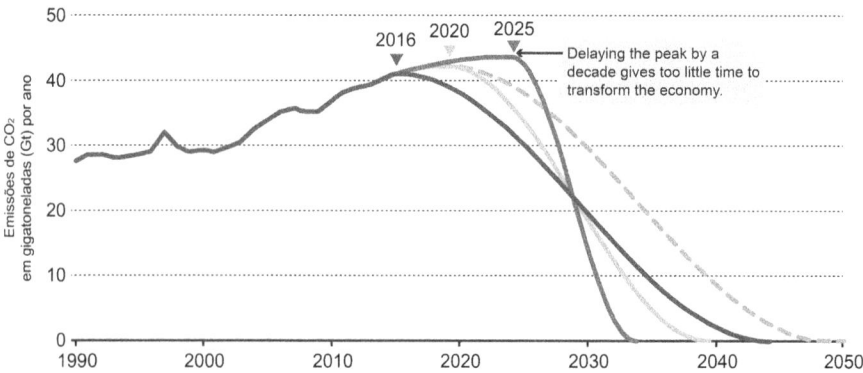

Fig. 7.7 Initial and final dates of CO_2 emission reduction to net zero. Source: Based on Christiana Figueres, Hans Joachim Schellnhuber, Gail Whiteman, Johan Rockström, Anthony Hobley & Stefan Rahmstorf, "Three years to safeguard our climate", *Nature*, 28/VI/2017, based on data from The Global Carbon Project

colleagues affirmed in 2017, "delaying the peak by a decade gives too little time to transform the economy."

As is well-known, while the diplomats' applause still resounded, the Paris Agreement goal to limit global warming "well below" 2 °C above the pre-industrial period was immediately denounced by James Hansen and others as a fictional play (Milman 2015):

> It's a fraud really, a fake. It's just bullshit for them to say: 'We'll have a 2C warming target and then try to do a little better every five years.' It's just worthless words. There is no action, just promises. As long as fossil fuels appear to be the cheapest fuels out there, they will be continued to be burned.

Immediately thereafter, in January 2016, 11 eminent scientists affirmed that the Paris Agreement raised a false hope and that the 2 °C warming threshold had become unattainable without a massive use of geo-engineering (Bawden 2016)[7]:

> The time for the wishful thinking and blind optimism that has characterized the debate on climate change is over. The time for hard facts and decisions is now. Our backs are against the wall and we must now start the process of preparing for geo-engineering. We must do this in the knowledge that its chances of success are small and the risks of implementation are great.

Successive facts confirm these verdicts. Continuing to invoke the pledges made by its signatories as an apotropaic formula cannot and should not conceal the emptiness of this agreement. Such lack of credibility is underlined by David Victor and colleagues (2017):

---

[7] Signed by Paul Beckwith (University of Ottawa), Stephen Salter (University of Edinburgh), James Kennett (University of California), Hugh Hunt (Cambridge), Alan Gadian (University of Leeds), Mayer Hillman (Institute of the Policy Studies), John Latham (University of Manchester), Aubrey Meyer (Global Commons Institute), John Nissen (Arctic Methane Emergency Group), Kevin Lister, and Peter Wadhams (University of Cambridge)

No major advanced industrialized country is on track to meet its pledges to control the greenhouse-gas emissions that cause climate change. Wishful thinking and bravado are eclipsing reality (…) National governments are making promises that they are unable to honour.

Peter Wadhams endorses this statement, reaffirming his disbelief in the ability of governments—partners in corporations' expansive strategies and also dependent on them—to fulfill their own commitments: "The [UK] government can make a commitment saying in 30 years time it will reduce our CO_2 emissions by 80%. They can quote any number they like because they have no intention of keeping to it."[8] On May 2019, the UK Parliament declared a "climate change emergency." Hundreds of cities, towns, and countries have also declared "climate emergencies" during 2019. Oxford Dictionaries has declared "climate emergency" the word of the year for 2019 and defined it as "a situation in which urgent action is required to reduce or halt climate change and avoid potentially irreversible environmental damage resulting from it." But, again, these declarations are not legally binding, and there is no clear definition of what they imply in economic and political terms (see *Nature* 9/V/2019, p. 165).

Australia, Canada, Japan, South Korea, and South Africa are not even on track to achieve their now plainly inadequate 2015 pledges (Höhne et al. 2020). Furthermore, we must keep in mind that the Paris Agreement is far from being universal. In 2019 it had not yet been ratified by 13 countries which together produce almost a quarter of world's oil, including Angola, Libya, Iraq, Iran, and Kuwait. With the US decision to abandon the agreement, more than a third of the world's oil production comes from countries that cannot be accused of not complying with the Paris Agreement because they do not even officially recognize it. Finally, the 2018 tragic election of a far-right government in Brazil, the seventh largest CO_2 emitter on the planet, is another major blow to the effectiveness of global governance.

The dynamics of increased warming, driven by growing GHG emissions and by climate positive feedbacks (see Chap. 8), combined with the failure of the Paris Agreement, explain the scientific community's skepticism as to the possibility of not exceeding the 2 °C target. Three years ago, Adrian Raftery and colleagues (2017) estimated that there was only a 5% chance that average global warming would be below 2 °C by 2100, compared to the pre-industrial period. Also, in 2017, Andrew Simms surveyed the opinion of a number of experienced climatologists on what chances remain of us not exceeding a global warming of 2 °C above the pre-industrial period:

In short, not a single one of the scientists polled thought the 2C target likely to be met. Bill McGuire, professor emeritus of geophysical and climate hazards at University College London, is most emphatic. "My personal view is that there is not a cat in hell's chance."

The IPCC Special Report 1.5 (2018) shows that a warming of 2 °C causes far more deleterious effects on human societies and biodiversity than a warming of 1.5 °C,

---

[8] Cf. Peter Wadhams, Scientists Warnings.org, 11/XII/2018 https://www.youtube.com/watch?v=AEM2NhPw%2D%2DU

which is already considered very dangerous. As might be expected from an IPCC report, its findings reflect the most representative positions of the scientific community. In fact, as early as 2017, Sir Brian Hoskins had stated to Andrew Simms of *The Guardian*: "We have no evidence that a 1.9°C rise is something we can easily cope with, and 2.1°C is a disaster." It is certainly a disaster for its reasonably known effects, based on the climate models, but perhaps and most importantly, it is to be feared for its still uncertain or unknown effects (discussed in the next chapter). Also, for the scientists at RealClimate, a "global warming of 2°C would leave the Earth warmer than it has been in millions of years."[9] Matteo Willeit and colleagues (2019) give chronological precision to this statement: "global temperature never exceeded the preindustrial value by more than 2°C during the Quaternary" (the last 2.58 million years). Hence, it now seems inevitable that in the near future, the Earth will experience an average temperature never faced by our species and, a fortiori, by agriculture and by the human civilizations that flourished in the 11.7 millennia of relative climatic stability of the Holocene.

7.5.3   A 2 °C Warming Will Be Reached in the Second Fourth of the Century

If keeping the Earth's temperature below 2 °C now seems like a social–physical impossibility, one question arises: when will average planetary temperatures exceed the columns of Hercules of climate, this 2 °C limit of global warming that we could never allow ourselves to exceed? Without any intention of weighing the uncertainties that hinder a more precise answer to this question, projections on a warming of 2 °C above the pre-industrial period converge to the second quarter of this century and, more precisely, to the period between 2035 and 2045. According to the Potsdam Institute for Climate Impact Research (2012), "we could see a 2°C world in the space of one generation." Michael Mann (2014) projects that, maintaining the baseline scenario, we may exceed the 2 °C limit from 2036 onward. According to Mann, to avoid this, it would have been necessary to keep atmospheric CO_2 concentrations below 405 ppm, while we are already well above 412 ppm. The MET Office forecasts the 2020 annual average CO_2 concentration at Mauna Loa Observatory (Hawaii) to be 414.2 ± 0.6 ppm. For the first time in millions of years, we have hit 416.08 on February 10, 2020, up from 411.97 a year ago and 314 ppm in 1958 when the observations began at Mauna Loa. If the current rate of increase in atmospheric CO_2 concentrations is maintained, by 2030–2035, we will have gone beyond the fateful 450 ppm limit, largely correlated with a global warming of 2 °C above the pre-industrial period. This correlation is restated, for example, in a document titled *Millennium Ecosystem Assessment Findings* (2005):

---

[9]Cf. *RealClimate,* "Hit the brakes hard", 29/IV/2009

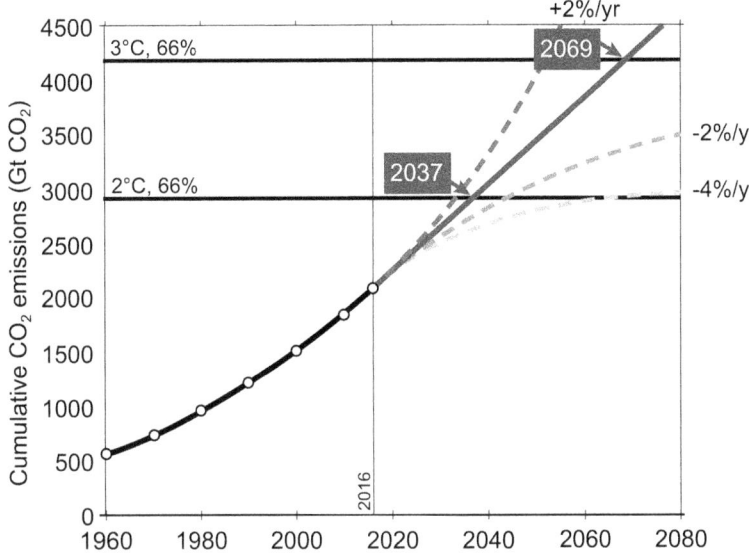

Fig. 7.8 Cumulative global CO_2 emissions and temperature (1960–2080). Source: Global Carbon Project. Observation: the boxes show the year that the exceedance budgets are surpassed assuming 2016 emission levels (continuous line). The dotted lines represent future pathways in scenarios with an increase in emissions after 2016 at a rate of 2% a year or a decrease in emissions at a rate of 2% or 4% a year, always in relation to 2016

> The balance of scientific evidence suggests that there will be a significant net harmful impact on ecosystem services worldwide if global mean surface temperature increases more than 2 °C above pre-industrial levels (*medium certainty*). This would require CO_2 stabilization at less than 450 ppm (https://slideplayer.com/slide/711361/).

The IPCC Fourth Assessment Report (AR4 2007) reiterates this prognosis, even if less explicitly: "any CO_2 stabilisation target above 450 ppm is associated with a significant probability of triggering a large-scale climatic event."[10] Figure 7.8, proposed by the Global Carbon Project (GCP), reaffirms all these estimates.

Sonia Seneviratne and colleagues (2016) point out that a 2 °C target for global warming implies increases much higher than 2 °C over most land regions. Thus, although the authors consider that a 2 °C global warming above the industrial period will occur only by mid-2040, this warming threshold was already crossed in the Arctic around the year 2000 and should be crossed regionally by 2030 "in the Mediterranean, Brazil and the contiguous U.S., under the business-as-usual (RCP8.5) emissions scenario."

---

[10] Cf. IPCC AR4 (2007), Working Group II: Impacts, Adaptation and Vulnerability

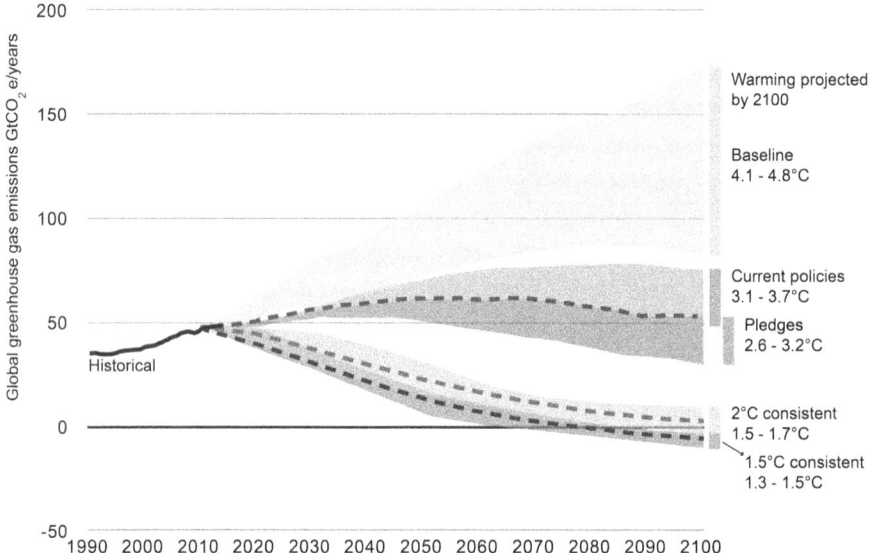

Fig. 7.9 Historical and projected global greenhouse gas emissions ($GtCO_2$-eq) and global warming by 2100. (Source: Climate Action Tracker http://climateactiontracker.org/global.html)

7.6   The Current Trajectory Leads to a Warming Beyond 3 °C

If we maintain the current trajectory of increasing GHG emissions, deforestation, biodiversity decline, and pollution, the destruction of the environmental foundations that enable the most basic functions of contemporary societies is unavoidable. An environmental collapse of unfathomable dimensions is indeed inevitable with a global warming of 2.6–4.8 °C, which should be reached by the end of this century, as shown in Fig. 7.9.

Figure 7.9 shows the various projections of increase in average surface temperatures until 2100 in relation to the pre-industrial period and in correlation with GHG emission levels. They can be summarized in three scenarios:

1. Scenario one, consistent with the Paris Agreement targets—a global warming "well below" 2 °C (between 1.3 °C and 1.7 °C) by 2100 above the pre-industrial period—assumes that the four conditions set by the IPCC Special Report 1.5 (2018) are fulfilled: (a) drastic and immediate reductions in CO_2 emissions are at a pace of more than 7% per year; (b) by 2030, global CO_2 emissions should be reduced to 1977 levels (18 $GtCO_2$), which means returning per capita emissions to those recorded for 1955 (Marland et al. 2019; see also Introduction); (c) net global emissions should be zeroed by 2050; and (d) after 2050 it would be necessary to remove carbon from the atmosphere at a rate of 12 $GtCO_2$ per year (400–800 Gt by 2100) (Lenzi et al. 2018). For reference, China CO_2 emissions in 2018

reached 10.3 Gt. It would be necessary, therefore, after 2050, to remove every year from the atmosphere almost 20% more than China's carbon emissions in 2018. Despite a spate of proposals of unknown risk, there is no technology available for the required scale of this CO_2 sequestration. Without a true change in the civilizational paradigm, including democratic global governance and a complete reorientation of strategic investments in the energy, food, and transportation systems, there is absolutely no chance of achieving the goals agreed upon in Paris.

2. Scenario two corresponds to the fulfillment of voluntary pledges assumed by the signatory countries of the Paris Agreement. It would lead to a global warming between 2.6 °C and 3.2 °C, considered catastrophic. The USA withdrew from the Agreement in 2019; Brazil is intensifying its oil production and the destruction of its forests, and peak oil demand will not be reached in the short term. Indeed, according to EIA's Short-Term Energy Outlook (11/III/2020), global oil and liquid fuels demand will rise by less than 0.4 million b/d in 2020 and by 1.7 million b/d in 2021. Therefore, this level of warming cannot be considered a likely scenario.

3. Scenarios 3 and 4, thus, remain the most likely under current conditions. They lead to an average global warming between 3.1 °C and 4.8 °C by 2100. It should be kept in mind that these are global *average* temperatures. At high latitudes and over land, mostly across the Northern Hemisphere, the warming will likely be double that. According to most models, 4 °C global average surface temperature implies 5–6 °C global land mean. On the hottest days, there will be an increase of 6–8 °C in China, 8–10 °C in Central Europe, and 10–12 °C in New York.

In fact, according to the IPCC AR5 Working Group III (2014), in the absence of strong climate mitigation policies capable of altering this trajectory (RCP 8.5 W/m^2), an average global warming of 4.1–4.8 °C above the pre-industrial period will probably be reached during this century. These IPCC projections reiterate the UNEP's updated *Emissions Gap Report* (2012), according to which current GHG emission trends are consistent with emission pathways that reach warming in the range of 3.5–5 °C by 2100. The same prognosis is proposed by the World Energy Outlook (IEA 2012) and by *Turn Down the Heat: Why a 4 °C Warmer World Must be Avoided* (2013), a report by the Potsdam Institute for Climate Impact Research and Climate Analytics. According to this report:

> New assessments of business-as-usual emissions (…) as well as recent reevaluations of the likely emission consequences of pledges and targets adopted by countries, point to a considerable likelihood of warming reaching 4 °C above pre-industrial levels within this century. (…) The most recent generation of energy-economic models estimates emissions in the absence of further substantial policy action (business as usual), with the median projections reaching a warming of 4.7 °C above pre-industrial levels by 2100, with a 40% chance of exceeding 5 °C.

This report also warns that in this scenario (RCP8.5 W/m^2), "warming continues to increase until the end of the century with global-mean land surface temperature for the Northern Hemisphere summer reaching nearly 6.5°C above the 1951–80 baseline by 2100." There is now a growing consensus, also reinforced by the IPCC, that a warming of 3 °C is considered "beyond adaptation." This is also what Jean Jouzel, former Vice President of IPCC (2002–2015), affirms in an interview with *Le Monde* in 2019: "I believe that we will not be able to adapt to a warming of 3°C and that we will experience major conflicts" (Herzberg 2019).

This emerging consensus fosters a new interpretation of the terms of the United Nations Framework Convention on Climate Change (UNFCCC 1992). As is well-known, the ultimate objective of this Convention is to "prevent Dangerous Anthropogenic Interference (DAI) with the climate system."[11] In line with this goal, the IPCC Third Assessment Report (2001) produced the famous five Reasons for Concern (RFCs), further updated by the Fourth Assessment Report (2007) and again by Joel Smith and colleagues (2009).[12] Returning to these two central notions of climate science—DAI and RFCs—Yangyang Xu and Veerabhadran Ramanathan (2017) interpret the Paris Agreement in terms of three climate risk categories:

> We are proposing the following extension to the DAI risk categorization: warming greater than 1.5 °C as "dangerous"; warming greater than 3 °C as "catastrophic?"; and warming in excess of 5 °C as "unknown??," with the understanding that changes of this magnitude, not experienced in the last 20+ million years, pose existential threats to a majority of the population. The question mark denotes the subjective nature of our deduction and the fact that catastrophe can strike at even lower warming levels.

Regarding a warming beyond 3 °C, the first major concern, state the authors, "is the issue of tipping points" (see also Chap. 8):

> Several studies have concluded that 3 to 5 °C global warming is likely to be the threshold for tipping points such as the collapse of the western Antarctic ice sheet, shutdown of deep water circulation in the North Atlantic, dieback of Amazon rainforests as well as boreal forests, and collapse of the West African monsoon, among others.

7.6.1 Deadlines and Impacts

There is a collective effort to specify the most likely deadlines by which the planet's average temperatures could exceed 3 °C or 4 °C, maintaining the current trajectory of emissions and deforestation. Projections converge on the third quarter of the century, as shown below:

---

[11] See UNFCCC, article 2 – Objective: "The ultimate objective of this Convention (...) is to achieve (…) stabilization of greenhouse gas concentrations in the atmosphere at a level that would prevent dangerous anthropogenic interference with the climate system." See also Mann (2009).

[12] The five Reasons for Concern (RFC) are (1) unique and threatened systems; (2) frequency and severity of extreme climate events; (3) global distribution and balance of impacts; (4) total economic and ecological impacts, and (5) risk of irreversible large-scale and abrupt transitions. Cf. IPCC. Climate Change 2007: Working Group II: Impacts, Adaptation and Vulnerability (19.3.7 Update on "Reasons for Concern").

1. As seen above (Fig. 7.8), the Global Carbon Project (GCP) estimates that, maintaining the 2016 level of CO_2 emissions, there is a 66% chance of reaching a global warming of 3 °C by 2069.
2. Richard A. Betts and colleagues (2011) project that our currently fossil intensive scenario "would lead to a warming of 4°C relative to pre-industrial during the 2070s. If carbon-cycle feedbacks are stronger, which appears less likely but still credible, then 4°C warming could be reached by the early 2060s in projections that are consistent with the IPCC's 'likely range.'"
3. Xu and Ranamathan (2017) affirm that, with unchecked emissions, "there is a low probability (5%) of warming becoming catastrophic [>3°C] by 2050."

Many scientists, including Kevin Anderson (2011) and the teams from the Potsdam Institute for Climate Impact Research (2012), Carbon Brief (McSweeney 2018), and the project High-End cLimate Impact and eXtremes (HELIX), supported by the European Commission (Richardson & Bradshaw 2017), attempt to describe the world created by a global average warming between 3.1 °C and 4.8 °C above the pre-industrial period (scenarios 3 and 4). It is very difficult to imagine such a world, almost unrecognizable to the world in the twentieth century, but it is possible to highlight some of its most likely features:

1. A dramatic increase in the intensity and frequency of high-temperature extremes in Tropical South America, North and Central Africa, Mediterranean, and the Middle East (4 °C) (Potsdam 2012).
2. "The warmest July in the Mediterranean region could be 9°C warmer than today's [2012] warmest July" (4 °C) (Potsdam 2012).
3. An increase of about 150% in acidity of the ocean. The death of entire coral reef ecosystems could occur well before 4 °C is reached (Potsdam 2012).
4. Global marine heat wave days per year multiply by 41 (3.5 °C) (McSweeney 2018).
5. Probability of an ice-free Arctic summer at least once before hitting temperature limit (3 °C) = 100% (McSweeney 2018).
6. Probability of an ice-free Arctic summer in any 1 year (3 °C) = 63% (McSweeney 2018).
7. Average drought length (3 °C) = 10 months; Eastern Europe = 8 months; Southern Europe (3 °C) = 12 months (McSweeney 2018).
8. Proportion of living creatures projected to lose over 50% of their climatic range (4.5 °C) (McSweeney 2018): invertebrates = 68%; vertebrates = 44%; plants = 67%.
9. An estimated 1.8 billion people will potentially reach unprecedented levels of vulnerability to food insecurity (4 °C) (Richardson and Bradshaw 2017).
10. The most serious of all these catastrophic impacts anticipates and introduces the next chapter. It is the widely shared opinion in the scientific community that a 4 °C warmer world "would be an interim temperature on the way to a much higher equilibrium level" (Anderson 2011). In this case, the planet's climate system would be doomed to evolve, driven by positive feedback loops, toward a global warming above 5 °C. As mentioned above, this warming has been cat-

egorized "as 'unknown??', with the understanding that changes of this magnitude, not experienced in the last 20+ million years, pose existential threats to a majority of the population" (Xu and Ramanathan 2017). And it should be remembered that the authors also add: "The question mark denotes the subjective nature of our deduction and the fact that catastrophe can strike at even lower warming levels."

References

AENGENHEYSTER, Matthias *et al.*, "The point of no return for climate action: effects of climate uncertainty and risk tolerance, *Earth System Dynamics*, 9, 2018, pp. 1085–1095.

ANDERSON, Kevin, *Going beyond Dangerous Climate Change: Exploring the Void Between Rhetoric and Reality in Reducing Carbon Emissions.* Public Lecture. London School of Economics, 21/X/2011. https://pt.slideshare.net/DFID/professor-kevin-anderson-climate-change-going-beyond-dangerous

———. "Climate's Holy Trinity". Public Lecture. Oxford Climate Society, 24/I/2019. https://www.youtube.com/watch?v=7BZFvc-ZOa8.

ANDREI, Mihai, "Global warming fuels rise of dangerous fungal infections". *ZME Science*, 23/VII/2019.

ANDREWS, Roger, "The gulf between the Paris Climate Agreement and energy projections". *Energy Matters*, 18/I/2017.

BALMASEDA, Magdalena A.; TRENBERTH, Kevin; KÄLLEN, Erland *et al.* "Distinctive climate signals in reanalysis of global ocean heat content". *Geophysical Research Letters*, 40, 9, 16/V/2013, pp. 1754–1759.

BAWDEN, Tom, "COP21: Paris deal far too weak to prevent devastating climate change, academics warn". *The Independent*, 8/I/2016.

BECKWITH, Paul, "Rise of the methane over time and latitude" (two videos), 2019. https://paul-beckwith.net/2019/03/01/rise-of-the-methane-over-time-and-latitude-two-videos/

BETTS, Richard A. *et al.*, "When could Global Warming Reach 4°C?" *Philosophical Transactions of the Royal Society*, 369, 2011, pp. 67–84.

BODEN T.A., MARLAND, G., ANDRES R.J., "Global, regional, and national fossil-fuel CO2 emissions. Carbon Dioxide Information Analysis Center", Oak Ridge National Laboratory, U.S. Department of Energy, Oak Ridge, 2013.

BOX, Jason *et al.*, "Key indicators of Arctic climate change: 1971–2017". *Environmental Research Letters*, 14, 2019.

BRIENEN, R. J. W *et al.* "Long-term decline of the Amazon carbon sink". *Nature*, 519, 19/III/2015, pp. 344–348.

BROMWICH, David H. *et al.*, "Central West Antarctica among the most rapidly warming regions on Earth". *Nature Geoscience*, 6, 2013, pp. 139–145.

CALDEIRA, Ken, "The Great Climate Experiment. How far can we push the planet?" *Scientific American*, September 2012.

CHENG, Lijing *et al.*, "Improved estimates of ocean heat content from 1960 to 2015". *Science Advances*, 10/III/2017.

CHENG, Lijing & ZHU, Jiang, "2017 was the warmest year on record for the global ocean". *Advances in Atmospheric Sciences*, 35, 3, March 2018, pp. 261–263.

CHOW, Lorraine, "10 Worst-Case Climate Predictions if We Don't Keep Global Temperatures Under 1.5 Degrees Celsius". *EcoWatch*, 2/I/2019.

CHRISTIDIS, Nikolaos; JONES, Gareth S. & STOTT, Peter A. "Dramatically increasing chance of extremely hot summers since the 2003 European heatwave". *Nature Climate Change*, 8/XII/2014.

Copernicus Climate Change Service, "Last four years have been the warmest on record – and CO_2 continues to rise". 7/I/2019.

COUMOU, Dim & ROBINSON, Alexander. "Historic and future increase in the global land area affected by monthly heat extremes". *Environmental Research Letters*, 8, 034018, 14/VIII/2013.

DAHL, Kristina *et al.*, "Increased frequency of and population exposure to extreme heat index days in the United States during the 21st century". *Environmental Research Communications*, 1, 2019.

DEAN, Joshua F. *et al.*, "Methane Feedbacks to the Global Climate System in a Warmer World". *Reviews of Geophysics*, 56, 15/II/2018.

EBI, Kristie L. & LOLADZE, Irakli, "Elevated atmospheric CO2 concentrations and climate change will affect our food's quality and quantity" *The Lancet*, July 2019.

ERB, Karl-Heinz *et al.*, "Unexpectedly large impact of forest management and grazing on global vegetation biomass". *Nature*, 553, 4/I/2018.

FIGUERES Christiana *et al.*, "Three years to safeguard our climate", *Nature*, 28/VI/2017.

FIGUERES, Christiana *et al.*, "Emissions are still rising: ramp up the cuts". *Nature*, 5/XII/2018.

FRUMHOFF, Peter C., HEEDE, Richard, ORESKES, NAOMI, "The climate responsibilities of industrial carbon producers". *Climatic Change*, 132, 2, September 2015, pp. 157–171.

GORE, Timothy, "Extreme Carbon Inequality". Oxfam, 2/XII/2015.

Global Carbon Budget (2018), see Le Quéré, Corinne et al. Global Carbon Budget, Global Carbon Project, 2018.

HANSEN, James; SATO, Makiko & RUEDY, Reto. "Perception of climate change". *PNAS*, 29/III/2012.

HANSEN, James *et al.* "Ice melt, sea level rise and superstorms: evidence from paleoclimate data, climate modeling, and modern observations that 2°C global warming could be dangerous". *Atmospheric Chemistry and Physics. An interactive open-access journal of the European Geosciences Union*, 16, 22/III/2016.

HANSEN, James, "Climate Change in a Nutshell: The Gathering Storm", 18/XII/2018. <http://www.columbia.edu/~jeh1/mailings/2018/20181206_Nutshell.pdf>

HANSEN, James *et al.*, "Global Temperature in 2018 and Beyond". Columbia University, 6/II/2019. http://www.columbia.edu/~jeh1/mailings/2019/20190206_Temperature2018.pdf

HANSEN, James et al., "Global Temperature in 2019". Goddard Institute for Space Studies, 15/I/2020.

HENLEY, Benjamin J. & KING, Andrew D., "Trajectories toward the 1.5°C Paris target: Modulation by the Interdecadal Pacific Oscillation". *Geophys. Research Letters* 8/V/2017.

HERZBERG, Nathaniel, "Jean Jouzel: 'L'effondrement n'est pas imminent. Je nous vois griller à petit feu". *Le Monde*, 2/VI/2019.

HÖHNE, Niklas *et al.*, "Emissions: world has four times the work or one-third of the time". *Nature,* 579, 7797, 5/III/2020, pp. 25–28.

IPCC, *Global Warming of 1.5. Special Report*, October 2018.

JACKSON, Rob & CANADELL, Pep, "What's 'Fair' when it comes to Carbon Emissions?". *The New York Times*, 4/XII/2019.

KATZ, Cheryl, "How long can oceans continue to absorb Earth's excess heat?". *Yale360*, 18/III/2015.

KOMMENDA, Niko, "Carbon calculator: how taking one flight emits as much as many people do in a year". *The Guardian*, 19/VII/2019.

KRETSCHMER, Marlene, "More-Persistent Weak Stratospheric Polar Vortex States Linked to Cold Extremes". BAMS, January 2018.

LENZI, Dominic *et al.*, "Don't deploy negative emissions technologies without ethical analysis". *Nature*, 19/IX/2018.

LE QUÉRÉ, Corinne et al., *Global carbon budget 2014. Earth System Science Data*, 7, 2015, pp. 47–85.

LEWIS, Simon L. "Increasing carbon storage in intact African tropical forests". *Nature*, 457, 19/II/2009, pp. 1003–1006.

LIGGIO, John *et al.*, "Measured Canadian oil sands CO_2 emissions are higher than estimates made using internationally recommended methods". *Nature Communications*, 23/IV/2019.

LOBELL, David B., SCHLENKER, Wolfram, COSTA-ROBERTS, Justin, "Climate Trends and Global Crop Production Since 1980". *Science*, 29/VII/2011.

MANN, Michael E, "Defining dangerous anthropogenic interference". *PNAS*, 17/III/2009, pp. 4065–4066.

———. *The Hockey Stick and the Climate Wars*: *Dispatches from the Front Lines*. Columbia Univ. Press, 2012.

———. "Earth Will Cross the Climate Danger Threshold by 2036". *Scientific American*, 1/IV/2014.

———. "How Close Are We to 'Dangerous' Planetary Warming?". *The Huffington Post*, 23/XII/2015.

MARCOTT, Shaun A. *et al.* "A Reconstruction of Regional and Global Temperature for the Past 11,300 Years". *Science*, 339, 6124, 8/III/2013, pp. 1198–1201.

MARENGO, José A., "O futuro clima do Brasil", *Revista USP*, 103, 2014, pp. 25–32.

MARLAND, Gregg; ODA, Tom & BODEN, Thomas A., "Cut emissions per capita to 1955 levels". *Nature*, 565, 31/I/2019, p. 567.

McKIE, Robin, "Scientists warn world will miss key climate target". *The Guardian,* 6/VIII/2016.

McSWEENEY, Robert, "The Impacts of climate change at 1.5C, 2C and beyond". Carbon Brief, 4/X/2018.

MILLAR, Richard J. *et al.*, "Emission budgets and pathways consistent with limiting warming to 1.5 °C". *Nature Geoscience*, 18/IX/2017.

MILMAN, Oliver, "James Hansen, father of climate change awareness, calls Paris talks 'a fraud'". *The Guardian*, 12/XII/2015.

———. "World 'nowhere near on track' to avoid warming beyond 1.5C target". *The Guardian*, 25/IX/2019.

MORA, Camilo *et al.*, "Global risk of deadly heat". *Nature Climate Change*, 19/VI/2017.

NEUKOM, Raphael *et al.*, "No evidence for globally coherent warm and cold periods over the preindustrial Common Era". *Nature*, 571, 2019, pp. 550–554.

NOBRE, Carlos A., MARENGO, José A., SOARES, W.R. (eds.), *Climate Change Risks in Brazil*. Amsterdam, 2019.

OVERLAND, James E. *et al., "*Nonlinear response of mid-latitude weather to the changing Arctic*" Nature Climate Change*, 6, 26/X/2016.

Potsdam Institute for Climate Impact Research, *Turn Down the Heat: Climate Extremes, Regional Impacts, and the Case for Resilience. A report for the World Bank by the Potsdam Institute for Climate Impact Research and Climate Analytics.* Washington, D.C., 2012.

POULTER, Benjamin *et al.* "Contribution of semi-arid ecosystems to interannual variability of the global carbon cycle". *Nature*, 509, 7.502, 29/V/2014, pp. 600–603.

PRINN, Ron, "400 ppm CO_2? Add other GHGs, and it's equivalent to 478 ppm". *Oceans at MIT*, 6/VI/2013.

RAFTERY, Adrian E. *et al.*, "Less than 2°C warming by 2100 unlikely". *Nature Climate Change*, 31/VII/2017.

RAHMSTORF, Stefan, "Climate Action and Human Wellbeing at a Crossroads: Historical Transformation or Backlash?" Bonn, 2017 https://www.youtube.com/watch?v=io3FI-PLCXA.

RICHARDSON, Katy & BRADSHAW, Catherine, "Assessment of the impacts of climate change on national level food insecurity using the Hunger and Climate Vulnerability Index". *High-End cLimate Impacts and eXtremes* (HELIX), September 2017.

RICKE, Katherine L. & CALDEIRA, Ken, "Maximum warming occurs about one decade after a carbon dioxide emission". *Environmental Research Letters*, 9, 2/XII/2014.

SAMENOW, Jason "Another extreme heat wave strikes the North Pole". *The Washington Post*, 7/V/2018.

SCHLOSSBERG, Tatiana, "2016 Already Shows Record Global Temperatures". *The New York Times*, 19/IV/2016.

SCHURER, Andrew P. *et al.*, "Interpretations of the Paris climate target". *Nature Geoscience*, 11, 19/III/2018.

SENEVIRATNE, Sonia I. *et al.* "Allowable CO_2 emissions based on regional and impact-related climate targets". *Nature*, 529, 28/I/2016.

SIMMS, Andrew, "A cat in hell's chance – why we're losing the battle to keep global warming below 2C", *The Guardian*, 19/I/2017.

SMITH, Joel B. *et al.*, "Assessing dangerous climate change through an update of the Intergovernmental Panel on Climate Change (IPCC) 'reasons for concern'". *PNAS*, 17/III/2009.

SOMERO, George N. *et al.*, *Review of the Federal Ocean Acidification Research and Monitoring Plan*. U.S. National Research Council of the National Academies (Division on Earth and Life Studies), 2013.

STEFFEN, Will *et al.*, "The trajectory of the Anthropocene: The Great Acceleration". *The Anthropocene Review*, 2015, 2(1), pp. 81–98.

STEFFEN, Will *et al.*, "Trajectories of the Earth System in the Anthropocene". *PNAS*, 6/VIII/2018.

STONE, Alvin, "Paris 1.5° C target may be smashed by 2026" *GeoSpace*, 9/V/2017.

TANS, Pieter, "CO_2 levels continue to increase at record rate". *Yale Environment360*, 14/III/2017.

TRENBERTH, Kevin *et al.*, "Insights into Earth's Energy Imbalance from Multiple Sources". *Journal of Climate*, 15/X/2016.

TRENBERTH, Kevin, "Climate change and wildfires – how do we know if there is a link? *phys. org*, 10/VIII/2018.

VICTOR, David G. *et al.* "Prove Paris was more than paper promises", *Nature*, 548, 1/VIII/2017.

WADHAMS, Peter, "Arctic Amplification, Climate Changing, Global Warming. New Challenges from the top of the world" (Lecture). Fondazione Eni Enrico Mattei, Milan, 12/V/2015. https://www.youtube.com/watch?v=aTY9M_ZKk3M.

———. *A Farewell to ice. A Report from the Arctic.* London, 2016.

WALDMAN, Scott, "IPCC. Was the scary report too conservative? E&E News, 11/X/2018.

WATTS, Jonathan, "Arctic warming: scientists alarmed by 'crazy' temperature rises". *The Guardian*, 27/II/2018.

WATTS, Nick *et al.* "The Lancet Countdown on health and climate change: from 25 years of inaction to a global transformation for public health". *Lancet*, 30/X/2017.

WILLEIT, Matteo *et al.*, "Mid-Pleistocene transition in glacial cycles explained by declining CO_2 and regolith removal". *Science Advances*, 5, 4, 3/IV/2019.

WONG, Sam, "So far 2019 has set 35 records for heat and 2 for cold". *New Scientist*, 30/I/2019.

XU, Yangyang & RAMANATHAN, Veerabhadran, "Well below 2°C: Mitigation strategies for avoiding dangerous to catastrophic climate changes". *PNAS*, 14/IX/2017

XU, Yangyang; RAMANATHAN, Veerabhadran & VICTOR, David, "Global Warming will happen faster than we think". *Nature*, 5/XII/2018.

Chapter 8
Climate Feedbacks and Tipping Points

In the preceding chapter, we dealt with facts and projections that reflect, or at least do not differ significantly from, the conclusions of the IPCC reports. The crucial issues discussed in this chapter, including worst-case climate scenarios, have not been widely considered by the IPCC. This is because the IPCC generally does not fully integrate two intimately associated sets of problems that are especially difficult for models to capture: positive climate feedbacks and tipping points.

Let us first define these two points. Positive climate feedbacks are defined as mechanisms of climate warming that are positively influenced by climate itself. As warming increases, so too will the magnitude of these mechanisms, forming a positive feedback loop (Dean et al. 2018). Such positive feedback loops are already at work in the climate system (which includes the oceans, land surface, cryosphere, biosphere, and atmosphere). As Rijsberman and Swart (1990) state, "temperature increases beyond 1°C may elicit rapid, unpredictable, and non-linear responses that could lead to extensive ecosystem damage." Tipping points have been defined as critical thresholds in [climate] forcing[1] or some feature of a system, at which a small perturbation can qualitatively alter the state or development of a system (Lenton and Schellnhuber 2007; Duarte et al. 2012; Lenton et al. 2015, 2019). In *The Seneca Effect: Why Growth is Slow but Collapse is Rapid* (2017), Ugo Bardi describes how, in many processes, the buildup of tensors in a system leads to its collapse: "One way to look at the tendency of complex systems to collapse is in terms of tipping points. This concept indicates that collapse is not a smooth transition; it is a drastic change that takes the system from one state to another, going briefly through an unstable state." The notion of tipping point revives not only Seneca's fascinating dictum that

---

[1] "Climate forcing has to do with the amount of energy we receive from the sun, and the amount of energy we radiate back into space. Variances in climate forcing are determined by physical influences on the atmosphere such as orbital and axial changes as well as the amount of greenhouse gas in our atmosphere." http://ossfoundation.us/projects/environment/global-warming/radiative-climate-forcing.

© Springer Nature Switzerland AG 2020

L. Marques, *Capitalism and Environmental Collapse*,
https://doi.org/10.1007/978-3-030-47527-7_8

gave Bardi's book its title[2] but also the famous Hegelian question of the transition from quantity to quality, a relatively neglected mechanism in accounting for change, either in evolution (Eldredge and Gould 1972/1980) or in large-scale components of the Earth's system, such as climate change, biomass loss, ocean circulation change, ice melting, etc. In his *Encyclopedia of the Philosophical Sciences* (1817), Hegel writes[3]:

> On the one hand, the quantitative determinations of existence can be altered without its quality being thereby affected (…). [O]n the other hand, this process of indiscriminately increasing and decreasing has its limits and the quality is altered by overstepping those limits. (…) When a quantitative change occurs, it appears at first as something quite innocuous, and yet there is still something else hidden behind it and this seemingly innocuous alteration of the quantitative is so to speak a ruse (*List*) through which the qualitative is captured.

In complex systems, it is generally impossible to predict the precise moment when an accumulation of successive quantitative changes will, in Hegel's terms, "overstep a limit." For Antônio Donato Nobre (Rawnsley 2020), we have already crossed the climate red line: "in terms of the Earth's climate, we have gone beyond the point of no return. There's no doubt about this." Many other scientists agree, more or less explicitly, with this perception. Whether or not we have crossed this red line, it can be said that the further one proceeds in a process of small quantitative changes, the greater the risk of crossing a tipping point. Michael Mann uses a suggestive image to speak of this situation: "we are walking out onto a minefield. The farther we walk out onto that minefield, the greater the likelihood that we set off those explosives."[4] Tipping points, therefore, involve analyzing processes that present increasing risks of inflection or nonlinear evolution. Regarding climate change specifically, two hypotheses primarily occupy the debate between experts: the risk that a 2 °C warming could take the Earth system to much higher temperatures (recently dubbed "Hothouse Earth") and the hypothesis contended by James Hansen and colleagues (2016) of an exponential rise in sea level, well above 2.5 meters by 2100 (discussed in the Sect. 8.3 "Higher Rises in Sea Level").

---

[2] See Lucius Annaeus Seneca (4 BCE–65 CE), *Epistolarium Moralium ad Lucilius* 91,6: *Esset aliquod inbecillitatis nostrae sollacium rerumque nostrarum si tam tarde perirent cuncta quant fiunt: nunc incrementa lente exeunt, festinatur in damnun* ("It would be some consolation for the feebleness of our selves and our works if all things should perish as slowly as they come into being; but as it is, increases are of sluggish growth, but the way to ruin is rapid." Translated by Richard Gummere).

[3] Cf. G.W.F. Hegel, *Encyclopedia of the Philosophical Sciences in Basic Outline*. Part I: Science of Logic (1817), par. 108, article: "Measure." Translated by Klaus Brinkmann and Daniel O. Dahlstrom. Cambridge University Press, 2010, p. 170

[4] See "The 'Doomed Earth' Controversy." Kavli Conversations on Science Communication at NYU. Michael Mann in conversation with David Wallace-Wells. Arthur L. Carter Journalism Institute,   30/XI/2017.   https://journalism.nyu.edu/about-us/event/2017-fall/the-doomed-earth-controversy/.

8.1   The "Hothouse Earth" Hypothesis

In a high-impact article, Will Steffen and 14 colleagues (2018) examine the possibility of combined positive feedback loops leading the Earth system to cross tipping points, beyond which irreversible warming dynamics "could prevent stabilization of the climate at intermediate temperature rises and cause continued warming on a 'Hothouse Earth' pathway even as human emissions are reduced." Especially important in this paper is the hypothesis that this tipping point can be crossed at a level of average global warming no higher than 2 °C above the pre-industrial period: "a 2 °C warming could activate important tipping elements,[5] raising the temperature further to activate other tipping elements in a domino-like cascade that could take the Earth System to even higher temperatures." The current trajectory "could lead to conditions that resemble planetary states that were last seen several millions of years ago, conditions that would be inhospitable to current human societies and to many other contemporary species." The Hothouse Earth hypothesis suggests an increasing warming along the following pathways:

(A) Current warming of 1.1 °C–1.5 °C (2015–2025/2030) above the pre-industrial period, here defined as ~200 years BP. The average global temperature is already significantly warmer than the Mid-Holocene Warm Period (~7000 years BP, 0.6 °C–0.9 °C above the pre-industrial period) and is nearly as warm as the Eemian, the last interglacial period (129–116 thousand years ago). "The Earth System has likely departed from the glacial-interglacial limit cycle of the late Quaternary" (Steffen et al. 2018).

(B) Current atmospheric CO_2 concentrations (410–415 ppm) have already reached the lower bound of Mid-Pliocene levels (400–450 ppm). During the Mid-Pliocene (4–3 million years ago), global average temperature was 2 °C–3 °C higher than the pre-industrial period. Although already catastrophic, stabilizing global warming at this level would still be theoretically possible if, and only if, the Paris Agreement were fully complied with, including successive revisions of the GHG emission reduction goals. This best-case scenario is becoming more unrealistic as each day goes by.

(C) Mid-Miocene Climate Optimum (16–11.6 million years ago), with temperatures of as much as 4 °C–5 °C above the pre-industrial period. This is, according to the authors, the most likely scenario by 2100, under current GHG emission levels.

Note that the projections of heating by 2100 proposed by Will Steffen and colleagues are analogous to those of the IPCC's RCP8.5 W/m² scenario. The Hothouse Earth hypothesis distances itself from the IPCC analysis precisely because it

---

[5] The term tipping element, introduced by Timothy Lenton and Hans Joachim Schellnhuber (2007), describes "those components of the Earth System that are at least sub-continental in scale and can be switched — under particular conditions — into a qualitatively different state by small perturbations." See also Lenton et al. 2015.

proposes that this level of warming can become irreversible as soon as a 2 °C warming threshold is crossed.

8.1.1 Heightened Climate Sensitivity

It is possible that the next IPCC Assessment Report (AR6 2021) will incorporate the Hothouse Earth hypothesis because new climate models point to heightened climate sensitivity. One must remember that the various projections of magnitude and speed of global warming depend, in part, on estimates of how the global climate system will respond to a given change in atmospheric concentrations of GHG (expressed in CO_2-eq). The debate revolves around specifying the magnitude of global warming once these concentrations double from 280 ppm in 1880 to 560 ppm. (As seen in the previous chapter, these concentrations exceeded 410 ppm in 2019 and are increasing rapidly.) The climate system immediate response (or transient climate response, TCR) is defined as "the global mean surface warming at the time of doubling of CO_2 in an idealized 1% per year CO_2 increase experiment" (Knutti et al. 2017). The climate system long-term response to the doubling of atmospheric concentrations of GHG is referred to in climate models as Equilibrium Climate Sensitivity (ECS). ECS "has reached almost iconic status as the single number that describes how severe climate change will be" (Knutti et al. 2017). The timescale difference between these two parameters (TCR and ECS) is mainly explained by the fact that the ocean takes centuries or millennia to reach a new thermal equilibrium. In any case, both parameters define the carbon budget, that is, the amount of GHG that can be emitted if we maintain a certain probability that a given level of global surface warming will not be exceeded. Of course, the higher the climate sensitivity, the lower the carbon budget. In turn, the carbon budget is strongly conditioned by the ability of positive climate feedbacks to increase the climate response to a given level of atmospheric GHG concentrations.

   Although the problem of specifying the magnitude of these climate responses to higher atmospheric concentrations of GHG and to positive climate feedbacks dates back to Svante Arrhenius (1896), climate scientists have not been able to decisively narrow the margins of uncertainty on this (Rahmstorf 2008). In its Fourth Assessment Report (AR4, 2007), the IPCC estimated climate sensitivity "likely [> 66% probability] to be in the range 2°C–4.5°C with a best estimate of about 3°C, and is very unlikely to be less than 1.5°C. Values substantially higher than 4.5°C cannot be excluded." This estimate of 3 °C (best estimate) was the consensus some years ago (Knutti and Hegerl 2008). But the current Coupled Model Intercomparison Project Phase 5 (CMIP5) has reported values that are between 2.1 °C and 4.7 °C, slightly higher, therefore, than those of the IPCC (2007), but always with an uncomfortable margin of uncertainty. In its latest report (AR5 2013), for example, the IPCC defined the upper range, with a 10% probability of exceeding 6 °C. That said, the CMIP5 is now giving way to the next generation of climate models, finalized since March 2019, which will make up the sixth Coupled Model Intercomparison Project

(CMIP6). In at least eight of these new models, the ECS has come in at 5 °C or warmer (2.8 °C–5.8 °C) (Belcher et al. 2019). The IPCC scientists who should draw on these new models in their Sixth Assessment Report (AR6 2021) do not yet know how to explain this heightened sensitivity. But, according to Reto Knutti, the trend "is definitely real. There's no question" (Voosen 2019).

In addition to greater climate sensitivity, the Hothouse Earth hypothesis becomes all the more likely given the combined influence of several closely interconnected positive climate feedback mechanisms on the magnitude and speed of warming. We shall take a quick look at five of them:

(1) Anthropogenic aerosols (sulfur dioxide, nitrogen oxides, ammonia, hydrocarbons, black soot, etc.) and their interactions can both cool and warm the atmosphere. As Joyce Penner (2019) said, "these particles remain one of the greatest lingering sources of uncertainty" in climate change predictions. An eventual lower use of fossil fuels, especially coal, will diminish the so-called aerosol masking effect, leading to additional global warming estimated between 0.5 °C and 1.1 °C (Samset et al. 2018). Yangyang Xu et al. (2018) assert that this additional warming is already at work in the climate system, and, therefore, "global warming will happen faster than we think." This is because the recent decline in coal use in China and other countries, however modest, has led to a faster decrease in air pollution than the IPCC, and most climate models have assumed: "lower pollution is better for crops and public health. But aerosols, including sulfates, nitrates, and organic compounds, reflect sunlight. This shield of aerosols has kept the planet cooler, possibly by as much as 0.7°C globally." The urgently needed transition toward low-carbon energy sources will, therefore, lead to an immediate additional increase of at least 0.5 °C in global average temperatures, with relevant regional differences.

(2) Oceans sequester 93% of the current Earth's energy imbalance (EEI). Much of this thermal energy will remain submerged for millennia (with already catastrophic consequences for marine life). But there are recent signals that "the oceans might be starting to release some of that pent-up thermal energy, which could contribute to significant global temperature increases in the coming years. (…) Given the enormity of the ocean's thermal load, even a tiny change has a big impact" (Katz 2015).

(3) Change in oceanic and terrestrial carbon cycles. The more the oceans heat up, the less their ability to absorb and store CO_2 and, hence, the greater the amount of these gases warming the atmosphere. With regard to changes in the carbon cycle specifically on land, the Global Carbon Project (Global Carbon Budget 2016) has warned that:

> Climate change will affect carbon cycle processes in a way that will exacerbate the increase of CO_2 in the atmosphere. Atmospheric CO_2 growth rate was a record high in 2015 in spite of no growth in fossil fuel and industry emissions because of a weaker CO_2 sink on land from hot and dry El Niño conditions.

According to Michael Zika and Karl-Heinz Erb (2009), approximately 2% of the global terrestrial net primary productivity (NPP)—the net amount of carbon captured by vegetation through photosynthesis—are already lost each year "due to dryland degradation, or between 4% and 10% of the potential NPP in drylands. NPP losses amount to 20–40% of the potential NPP on degraded agricultural areas in the global average and above 55% in some world regions."[6] But according to Alessandro Baccini and colleagues (2017), even tropical forests are no longer carbon sinks. Actually, they have become a net carbon source due to deforestation and reductions in carbon density within standing forests (degradation or disturbance), with the latter accounting for 68.9% of overall losses. Based on a 12-year study of MODIS pantropical satellite data, the authors provide evidence that "the world's tropical forests are a net carbon source of 425.2 ± 92.0 teragrams of carbon per year [425 Mt year^{-1}]. This net release of carbon consists of losses of 861.7 ± 80.2 Tg C year^{-1} and gains of 436.5 ± 31.0 Tg C year^{-1}."

(4) The ice–albedo feedback. Albedo is the fraction of solar energy that is reflected from the Earth into space. Because ice and snow are white, they have a high albedo. Ice surface has an albedo of 90%, and, inversely, dark water has an albedo of less than 10%. Peter Wadhams (2015) estimates that "the effect of sea ice retreat in the Arctic alone has been to reduce the average albedo of the Earth from 52% to 48% and this is the same as adding a quarter to the amount of the heating of the planet due to the GHG."

(5) The potentially explosive magnitude and speed of GHG (CO_2, CH_4, and N_2O) release into the atmosphere. This is due to forest fires, the melting of ice in shallow seabeds surrounding the Arctic Ocean (Wadhams 2016), and collapse of land permafrost that is thawing much faster than previously thought, often for the first time since the last glaciation (Turetsky et al. 2019).

8.2 The Arctic Methane Conundrum

These accelerators of global warming cannot be studied separately because they act in synergy. We will discuss, however, only one aspect of these climate feedbacks: the ongoing processes that release carbon, especially methane (a gas that has been gaining visibility in research over the past two decades), into the atmosphere, at high latitudes. Until the early 1970s, methane was considered to have no direct effect on the climate. Since then, many studies evidenced by Gavin Schmidt (2004) have shown the growing importance of methane as a powerful factor in global warming.

Unlike the multi-secular permanence of CO_2 (with a multi-millennial "long tail"; see Archer 2009), methane remains in the atmosphere for only about 9–12 years. However, its global warming potential (GWP) is much higher than that of CO_2. The IPCC Fifth Assessment (2013) updated methane's GWP upward by 20%, making it

---

[6] See also the UNEP GEO5 Assessment, 2012 p. 74. http://www.unep.org/geo/sites/unep.org.geo/files/documents/geo5_report_full_en_0.pdf>.

up to 100 times higher than that of CO_2 over a 5-year horizon, 72–86 times higher over a 20-year horizon, and 34 times higher over a 100-year horizon. Furthermore, according to Maryam Etminan and colleagues (2016), the radiative forcing of methane between 1750 and 2011 is about 25% higher (increasing from 0.48 W/m^2–0.61 W/m^2) than the value found in the IPCC (2013) assessment.[7] In the short term—as the next 20 years are the most important in the threshold situation we find ourselves in—the global warming potential of methane is, therefore, 72–100 times greater than that of CO_2.

It is important to remember, moreover, that the greenhouse effect of methane does not cease completely after 12 years, since it is then oxidized and destroyed by hydroxyl radicals (OH*), being transformed (through various mechanisms in the troposphere and the stratosphere) into CO_2 and water vapor (H_2O), two other greenhouse gases.[8] Finally, and perhaps more importantly, methane and the warming of the atmosphere reinforce each other. As Schmidt (2004) summarizes well, "methane rapidly increases in a warming climate with a small lag behind temperature. Therefore, not only does methane affect climate through greenhouse effects, but it in turn can evidently be affected by climate itself." The importance of this interaction was underlined by Joshua Dean and colleagues (2018), notably in ecosystems characterized by high carbon concentration, such as wetlands, marine and freshwater systems, permafrost, and methane hydrates: "increased CH_4 emissions from these systems would in turn induce further climate change, resulting in a positive climate feedback." In this paper, Joshua Dean and colleagues quantify the direct relationship between heating and the production of methane:

> Experimental temperature rises of 10 °C have caused CH_4 production to increase by 2 orders of magnitude, with 30 °C causing CH_4 production to increase by as much as 6 orders of magnitude; short-term temperature rises of this magnitude above zero are possible during the Arctic growing season and could become more common as regional air temperatures increase.

Methane atmospheric concentrations increased from about 720 ppb in the eighteenth century to 1870 ppb in September 2019, according to NOAA measurements. Figure 8.1 shows continued acceleration since 2007, after a 7-year period of very slow growth.

Since 2014, methane concentrations have been rising faster than at any time in the past two decades, leading us to approach the most greenhouse gas-intensive scenarios (Saunois et al. 2016; Ed Dlugokencky, NOAA/ESRL 2019).[9] An average annual growth in atmospheric methane concentrations of around 5 ppb would be sufficient for the warming produced solely by methane to squash any hope of maintaining a global warming average of 2 °C above the pre-industrial period, the Paris

---

[7] This is for the values of the radiative forcing of methane relative to the global warming potential (GWP) of CO_2, adopted by IPCC/AR4, based on G. Myhre. See "Climate Change 2007: Working Group I: The Physical Science Basis," chapter 2.10.2: Direct Global Warming Potentials.

[8] "The major CH_4 sinks are oxidation by OH* in the troposphere, biological CH_4 oxidation in drier soil, and loss to the stratosphere" (IPCC AR4 7.4.1.1 Biogeochemistry and Budgets of Methane).

[9] See www.esrl.noaa.gov/gmd/ccgg/trends_ch4/.

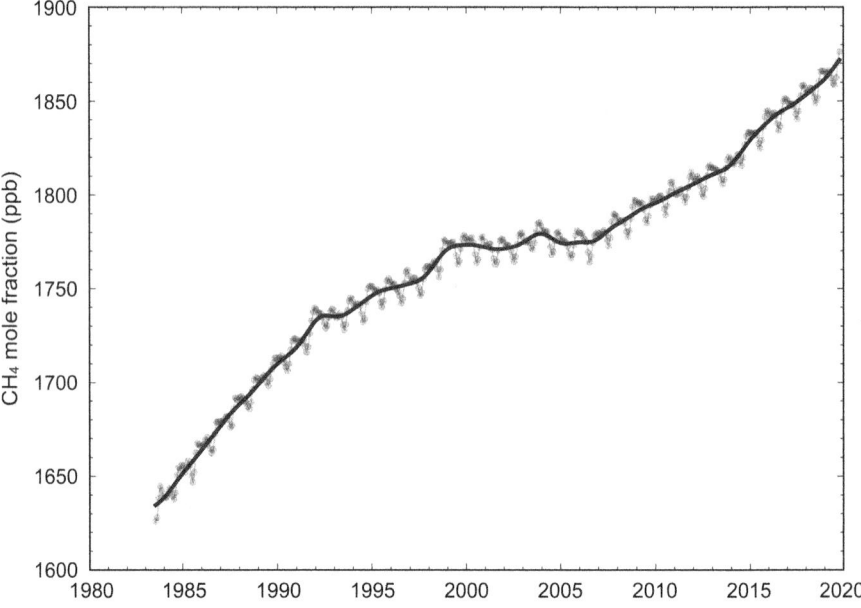

Fig. 8.1 Methane atmospheric concentrations, 1983–February 2020 (parts per billion, ppb). (Source: NOAA, Trends in Atmospheric Methane, 5/II/2020)

Agreement's least ambitious target (Nisbet et al. 2019). According to NOAA data, between 2014 and 2019, methane atmospheric concentrations had an annual average increase of over 9 ppb, against 5.1 ppb in the previous 5 years (2009–2013). This methane escalation was not expected at all in any of the IPCC future greenhouse gas scenarios and clearly raises extreme concern, as anthropogenic methane emissions already caused a warming effect that is about one-third that of CO_2 emissions (Myhre 2017). The current acceleration is probably a result of three complementary causes, identified by Euan Nisbet and colleagues (2019) and then by Sara E. Mikaloff Fletcher and Hinrich Schaefer (2019):

(1) A surge in biogenic methane emissions, whether from wetlands, ruminants, waste, or all of these. The increase in ruminants is certainly an important factor, as "livestock inventories show that ruminant emissions began to rise steeply around 2002 and can account for about half of the CH_4 increase since 2007" (Fletcher and Schaefer 2019). The growth of livestock in Brazil, which accounts for almost 80% of the original vegetation cover removal in the Amazon and the Cerrado (Meirelles 2005, 2014; De Sy 2015; Barbosa et al. 2015; Marques 2019; Azevedo 2019[10]), plays a relevant role in this global increase in methane

---

[10] See Mapbiomas.org 2019 http://mapbiomas.org/map#coverage.

emissions. In 2018, Brazil was the world's largest exporter of beef, providing close to 20% of total global beef exports, and was the second largest commercial cattle herd in the world (232 million head in 2019). Currently, Brazilian livestock occupies approximately 2.2 million km^2, of which 700 thousand km^2 are in the Brazilian Amazon (Barbosa et al. 2015). Methane emissions in Brazil, mainly from enteric fermentation of cattle and from manure deposition in pastures, have grown 163% since 1970 (Azevedo et al. 2018). As is well-known, a slowdown in global warming implies a drastic reduction or abandonment of the carnivorous diet, especially of cattle and sheep (see Chap. 12, Sect. 12.4 "Hypobiosphere: Functional and Nonfunctional Species to Man").

(2) A decline, likely caused by anthropogenic pollution, in the amount of CH_4 destroyed in the atmosphere through CH_4 oxidation.

(3) A strong rise in methane emissions from fossil fuels. This third factor is not only due to the global increase in natural gas, oil, and coal consumption but also, and perhaps most importantly, to methane leaks in all stages of the fossil fuel industry, notably during gas extraction through hydraulic fracturing, as well as methane leaks from active and abandoned coal mines, as discussed in Chap. 5 (Sect. 5.4 "Unconventional Oil and Gas: Maximized Devastation") (Alvarez et al. 2012; Tollefson et al. 2013; Pandey 2019). A crucial point in these processes must be highlighted: the possibility that at least part of the recent increase in atmospheric methane concentrations is already due to feedback loops independent of human action. As noted above, methane rapidly increases in a warming climate (Schmidt 2004; Dean et al. 2018). From this finding, Nisbet and colleagues (2019) advance the hypothesis that the acceleration of global warming since 2014 could be at the root of an increase in wetland CH_4 production. In the words of Fletcher and Schaefer (2019): "If natural wetlands, or changes in atmospheric chemistry, indeed accelerated the CH_4 rise, it may be a climate feedback that humans have little hope of slowing."

8.2.1 The Permafrost Carbon Feedback

In addition to the three sources of anthropogenic methane emission noted above, a fourth factor is the carbon release of rapidly melting permafrost. The process is well detected, but its current magnitude and, above all, its short-term acceleration have been the subject of intense debate in the scientific community. Given the high global warming potential of methane, this uncertainty represents an important hindrance to the development of more reliable climate projections. Permafrost is defined as ground exposed to temperatures equal or less than zero for more than 2 consecutive years; "it is composed of soil, rock or sediment, often with large chunks of ice mixed in" (Turetsky et al. 2019). Permafrost covers 24% of the Northern Hemisphere's land (23 million km^2). Most of these soils have remained frozen since the last glaciation and may have depths of over 700 meters in some regions of northern Siberia and Canada (Schaefer et al. 2012). The estimate of Gustaf Hugelius and colleagues (2014) on the amount of carbon stored globally in permafrost is slightly lower than previous estimates. In their study, soil organic carbon stocks were estimated separately for different depth ranges (0–0.3 m, 0–1 m, 1–2 m, and 2–3 m). According to the authors, total estimated soil organic carbon storage for the permafrost region is around 1300 petagrams (uncertainty range of 1100–1500 Pg or Gt), of which around 500 Gt is in non-permafrost soils, seasonally thawed, while around 800 Gt is perennially frozen. These soils contain approximately twice the amount of carbon currently present in the atmosphere.

Permafrost status is monitored by two networks—Thermal State of Permafrost (TSP) and Circumpolar Active Layer Monitoring (CALM)—coordinated by the International Permafrost Association (IPA). About 7 years ago, a UNEP report, titled *Policy Implications of Warming Permafrost* (Schaefer et al. 2012), warned:

> Overall, these observations [from TSP and CALM] indicate that large-scale thawing of permafrost may have already started. [...] A global temperature increase of 3 °C means a 6 °C increase in the Arctic, resulting in anywhere between 30 to 85% loss of near-surface permafrost. [...] Carbon dioxide (CO_2) and methane emissions from thawing permafrost could amplify warming due to anthropogenic greenhouse gas emissions. This amplification is called the permafrost carbon feedback.

Since 2012, the permafrost carbon feedback has become more and more clear. Given the amplification of Arctic surface air temperature increase, average temperatures in some of its areas have already reached about 3 °C above the 1961–1990 period. As we have seen in the previous chapter, other measurements show a 2.7 °C rise in the whole region's average air temperature since 1971 (3.1 °C during the winter) (Box et al. 2019). This faster Arctic warming (more than double the global average) is causing greater thawing of these soils, activating bacterial decomposition of the organic matter contained in them so that most of their carbon will be released into the atmosphere, either in the form of CO_2 (aerobic decomposition) or CH_4 (anaerobic decomposition), exacerbating climate change. Christian Knoblauch and colleagues (2018) emphasize the importance of methane released in bacterial decomposition in anoxic soils, which may contribute more substantially to global warming than previously assumed. Elizabeth M. Herndon (2018) accepts the results of research done by Knoblauch and colleagues and reiterates that "CH_4 production from anoxic (oxygen-free) systems may account for a higher proportion of global warming potential (GWP) than previously appreciated, surpassing contributions of CO_2." Guillaume Lamarche-Gagnon and colleagues (2019) have provided "evidence from the Greenland ice sheet for the existence of large subglacial methane reserves, where production is not offset by local sinks and there is net export of methane to the atmosphere during the summer melt season."

8.2.2   Methane Hydrates or Clathrates: The Thawing of Subsea Permafrost

Joshua Dean and colleagues (2018) assert that "methane hydrates[11] are not expected to contribute significantly to global CH_4 emissions in the near or long-term future." Although shared by the IPCC (AR4 2007)[12] and by many other scientists,[13] this assumption has been challenged by Natalia Shakhova, Igor Semiletov, Evgeny Chuvilin, Paul Beckwith, John Nissen, Peter Wadhams, Gail Whiteman, and Chris Hope, among many other climate scientists and Arctic specialists, in several articles and public interventions.[14] According to these authors, there is an increasing risk of irreversible destabilization of sediments in the Arctic Ocean's seafloor, with a potentially disastrous release of methane into the atmosphere in the horizon of a few decades or even abruptly. There is also consensus among them that the greatest and most imminent danger comes from the East Siberian Arctic Shelf (ESAS), the most extensive continental shelf in the world's oceans, encompassing the Laptev Sea, the East Siberian Sea, and the Russian part of the Chukchi Sea. Its mean depth is 50 meters, and more than 75% of its area of 2.1 million km^2 is less than 40 meters deep. In the past, sea ice protected the ESAS seabed from solar radiation. But that was the past. The Arctic sea ice is currently 40% smaller than it was only 40 years ago (Urban 2020). Figure 8.2 shows the extent of the Arctic sea ice retreat in October 2019, since 1979.

As can be seen from this chart, monthly sea ice extent reached a record low in October 2019. According to the NSIDC, "the linear rate of sea ice decline for October is 81,400 square kilometers (31,400 square miles) per year, or 9.8% per decade relative to the 1981–2010 average." The 13 lowest sea ice extents all occurred in the last 13 years. Arctic sea ice extent averaged for October 2019 was 5.66 million km^2 (2.19 million square miles), the lowest in the 41-year continuous satellite record. This was 230,000 km^2 (88,800 square miles) below that observed in 2012—the previous record low for the month—and 2.69 million square kilometers (1.04 million square miles) below the 1981–2010 average.

---

[11] Methane hydrates ($CH_4 \cdot 5.75H_2O$ or $4CH_4 \cdot 23H_2O$), also called methane clathrates, are solid structured cage-like substances in which a large amount of methane is trapped. They form a crystal structure of water similar to ice.

[12] See IPCC AR4 Climate Change 2007: Working Group I: The Physical Science Basis. Chapter 4.7.2.4 Subsea Permafrost: "Although the potential release of methane trapped within subsea permafrost may provide a positive feedback to climate warming, available observations do not permit an assessment of changes that might have occurred."

[13] See, for instance, David Armstrong McKay and Rachael Avery, "Fact-check: is an Arctic 'Methane Bomb' about to go off?", *climatetippingpoints.info*, 13/V/2019.

[14] See, for instance, Peter Wadhams, "Arctic Amplification, Climate Changing, Global Warming. New Challenges from the top of the world." Fondazione Eni Enrico Mattei, Milan, 12/V/2015 (YouTube), and Paul Beckwith's many videos on abrupt climate change, mostly in the Arctic. See, for instance, "Jaw-Dropping Methane Levels up to 9 times Global Average Measured Recently Over Arctic Shelf" (two parts), 10/X/2019 (YouTube).

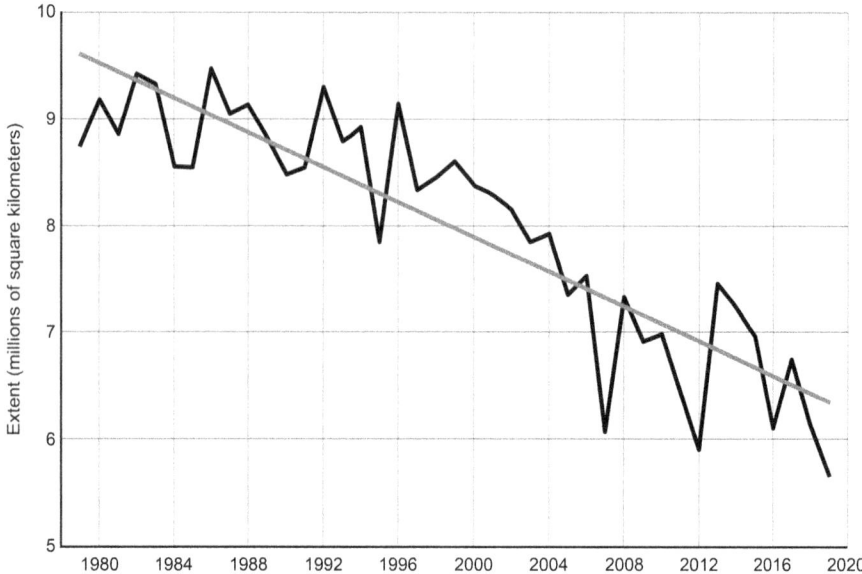

Fig. 8.2 Average monthly Artic sea ice extent, October 1979–2019 (Million km²). (Source: Wild Ride in October. October 2019 compared to previous years. National Snow and Ice Data Center, 5/XI/2019 http://nsidc.org/arcticseaicenews/2019/11/)

During this period, there were sharp declines in the summers of 1989, 1997–1999, 2007, 2012, and 2019.

The date of the first so-called Blue Ocean Event (BOE), that is, the moment when the extent of the Arctic sea ice in September will fall below one million square kilometers, is uncertain. According to Mark Urban (2020), "Arctic summers could become mostly ice-free in 30 years, and possibly sooner if current trends continue." How much sooner is still open to debate. What is certain is that the first BOE looks more and more imminent and that the ESAS has become increasingly exposed, for several months, to solar radiation, which obviously accelerates its heating. It has warmed by up to 17 °C, and the thawing of the subsea permafrost is already releasing methane in annual quantities no smaller than those of terrestrial Arctic ecosystems. The current atmospheric CH_4 emissions from the ESAS are estimated to be between 8 and 17 teragrams (or million tons) annually (Shakhova et al. 2017, 2019). Peter Wadhams never ceases to warn us of the particular importance of this fact, as we can see, for example, in his book from 2016: "We must remember—many scientists, alas, forget— that it is only since 2005 that substantial summer open water has existed on Arctic shelves, so we are in an entirely new situation with a new melt phenomenon taking place."

Already in 2008, a scientific expedition aboard the Russian ship Jacob Smirnitskyi recorded, for the first time, large amounts of methane release from the ESAS seafloor. Orjan Gustafsson wrote about this (Connor 2008):

An extensive area of intense methane release was found. At earlier sites we had found elevated levels of dissolved methane. Yesterday, for the first time, we documented a field where the release was so intense that the methane did not have time to dissolve into the seawater but was rising as methane bubbles to the sea surface. These 'methane chimneys' were documented on echo sounder and with seismic instruments.

Flying over the Arctic at latitudes as far as 82° north, Eric Kort and colleagues (2012) found significant amounts of methane being released from the ocean into the atmosphere over open leads and regions with fractional sea ice cover:

We estimate that sea–air fluxes amount to around 2 mg d^{-1} m^{-2}, comparable to emissions seen on the Siberian shelf. We suggest that the surface waters of the Arctic Ocean represent a potentially important source of methane, which could prove sensitive to changes in sea-ice cover.

The estimate of Eric Kort and his colleagues, reported by Steve Connor (2012), is that the quantities of methane observed could be large enough to affect the global climate: "We were surprised to see these enhanced methane levels at these high latitudes. Our observations really point to the ocean surface as the source, which was not what we had expected." This ebullition of methane is now observed through the Arctic Ocean's water column[15] even in winter, through flaw polynyas (areas of open water surrounded by sea ice) which increased by up to five times during the last decades (Shakhova et al. 2019).

In addition to Eric Kort and his team, many other scientists warn that current amounts of methane incorporated into the atmosphere from the Arctic seabed are already sufficient to strongly interfere with the pace of global warming. According to Peter Wadhams (2016), the global feedback arising from the Arctic snow and ice retreats is already adding 50% to the warming which results from the addition of CO_2:

We have reached the point at which we should no longer simply say that adding CO_2 to the atmosphere is warming our planet. Instead we have to say that the CO_2 which *we have added* to the atmosphere has *already* warmed our planet to the point where ice/snow feedback processes are themselves increasing the effect by a further 50%. We are not far from the moment when the feedbacks will themselves be driving the change – that is, we will not need to add more CO_2 to the atmosphere at all, but will get the warming anyway. This is a stage called *runaway warming*, which is possibly what led to the transformation of Venus into a hot, dry, dead world.

Mark Urban (2020) estimates that "once dark water replaces brilliant ice, Earth could warm substantially, equivalent to the warming triggered by the additional release of a trillion tons of carbon dioxide into the atmosphere." There seem to be strong arguments estimating that the current level of interference from Arctic methane release, and in particular from ESAS, is still, however, very small compared to its immense potential for accelerating global warming. The ESAS became subsea permafrost 12 or 13 millennia ago at the end of the last glaciation, but during the Last Glacial Maximum (from 26.5 to 19 ka), sea level was more than 100 meters

---

[15] A water column is a conceptual column of water from the surface of a sea, river, or lake to the bottom sediment.

lower than it is today (Englander 2014), and the entire shelf area was exposed above sea level, allowing for the accumulation of sediments of high organic content. It is estimated that shallow methane hydrate deposits currently occupy about 57% (1.25 million km^2) of the ESAS' seabed and that they could preserve more than 1400 Gt of methane, which makes this region the largest and most vulnerable store of subsea CH_4 in the world (Wadhams 2016; Shakhova et al. 2019).

8.2.3   A Slow-Motion Time Bomb or a Methane and CO_2 Burst?

If all these observations suggest that the release of CO_2 and methane will continue to increase in the Arctic, exacerbating warming, the rate of increase is still uncertain, especially on the horizon of this century. Ted Schuur created the metaphor "slow-motion time bomb," exploding initially in a way that is imperceptible (Borenstein 2006). The expression was reused in the film *Death Spiral and the Methane Time Bomb* (2012). Slow motion because, according to Schuur and colleagues (2015):

> At the proposed rates, the observed and projected emissions of CH_4 and CO_2 from thawing permafrost are unlikely to cause abrupt climate change over a period of a few years to a decade. Instead, permafrost carbon emissions are likely to be felt over decades to centuries as northern regions warm, making climate change happen faster than we would expect on the basis of projected emissions from human activities alone.

But, some studies in recent years project a melting of the permafrost (land and subsea) that is faster or much faster than previously supposed. Kevin Schaefer and colleagues (2011), for instance:

> predict that the permafrost carbon feedback will change the Arctic from a carbon sink to a source after the mid-2020s and is strong enough to cancel 42–88% of the total global land sink. The thaw and decay of permafrost carbon is irreversible and accounting for the permafrost carbon feedback (PCF) will require larger reductions in fossil fuel emissions to reach a target atmospheric CO_2 concentration.

An abrupt thaw in terrestrial permafrost could be initiated or exacerbated, for example, by wildfires or through thermokarst processes.[16] Fires in the boreal regions (above 50°N) reduce the albedo by covering the reflective white snow with black soot that absorbs sunlight, accelerating permafrost melting, bacterial activity, and the subsequent release of CO_2 and methane into the atmosphere. James L. Partain Jr. and colleagues (2015) estimate that climate change in Alaska has increased the risk of a fire year as severe as 2015 by 34%–60%. Alaska wildfires have increased dramatically since 1990 (Sanford et al. 2015), and fires in this state have spread for more than five thousand square kilometers only in June and July 2019, the hottest 2

---

[16] Thermokarst is "the process by which characteristic landforms result from the thawing of ice-rich permafrost or the melting of massive ice" (Shakhova et al. 2019).

months on record. In these 2 months, satellites detected more than 100 wildfires raging above the Arctic Circle. Clare Nullis from the WMO declared that in June 2019 alone, Arctic fires emitted 50 Mt of CO_2 into the atmosphere: "this is more than was released by Arctic fires in the same month between 2010 and 2018 combined" (WMO 2019). The magnitude of these wildfires is unprecedented, and, more importantly, they have ignited peat soils, which burn deeper in the ground and release huge quantities of methane and CO_2.

Regarding thermokarst processes and the formation of thaw lakes in which seeps and ebullition (bubbling) of methane are created, more than 150,000 seeps have already been identified (2019). Already in the beginning of the century, measurements of methane emissions from thaw lakes in northern Siberia, especially through ebullition, were much higher than those recorded previously. Katey Walter and colleagues (2006) showed that "ebullition accounts for 95% of methane emissions from these lakes, and that methane flux from thaw lakes in our study region may be five times higher than previously estimated." Walter warned that the effects of those emissions "can be huge. It's coming out a lot and there's a lot more to come out" (Borenstein 2006). Merritt Turetsky and colleagues (2019) claim that current models of GHG release are mistaken in assuming that the permafrost thaws gradually from the surface downward:

> Frozen soil doesn't just lock up carbon — it physically holds the landscape together. Across the Arctic and Boreal regions, permafrost is collapsing suddenly as pockets of ice within it melt. Instead of a few centimetres of soil thawing each year, several metres of soil can become destabilized within days or weeks. The land can sink and be inundated by swelling lakes and wetlands. (…) Permafrost is thawing much more quickly than models have predicted, with unknown consequences for greenhouse-gas release.

The authors estimate that abrupt permafrost thawing will occur in less than 20% of frozen land and could release between 60 and 100 Gt of carbon by 2300. They also project that another 200 Gt of carbon will be released in other regions that will thaw gradually. However, given that abrupt processes release more carbon per square meter—and particularly more methane—than does gradual thaw, "the climate impacts of the two processes will be similar. So, together, the impacts of thawing permafrost on Earth's climate could be twice that expected from current models."

Let us come back to the ESAS subsea permafrost. Shakhova et al. (2019) observe that there is "a potential for possible massive/abrupt release of CH_4, whether from destabilizing hydrates or from free gas accumulations beneath permafrost; such a release requires only a trigger." This trigger can occur at any moment because, as the authors remind us:

> The ESAS is a tectonically and seismically active area of the world ocean. During seismic events, a large amount of over-pressurized gas can be delivered to the water column, not only via existing gas migration pathways, but also through permafrost breaks.

Although ESAS is possibly the biggest problem, acceleration in the speed of methane release has also been observed from other Arctic continental shelves, such as the Kara Sea shelf. Irina Streletskaya and colleagues (2018) studied the methane content in ground ice and sediments of the Kara seacoast. The study states that "permafrost degradation due to climate change will be exacerbated along the coasts where declining sea ice is likely to result in accelerated rates of coastal erosion (…), further releasing the methane which is not yet accounted for in the models." Likewise, Alexey Portnov and colleagues (2013, 2014) measured methane release in the South Kara Sea shelf and in the West Yamal shelf, in northeast Siberia. In their 2013 report, they show that "this Arctic shelf region where seafloor gas release is widespread suggests that permafrost has degraded more significantly than previously thought." Their studies provide an example of another Arctic marine shelf where seafloor gas release is widespread and where permafrost degradation is an ongoing process. Commenting on both of these studies, Brian Stallard cites the following projection proposed by Portnov: "If the temperature of the oceans increases by two degrees as suggested by some reports, it will accelerate the thawing to the extreme. A warming climate could lead to an explosive gas release from the shallow areas."[17]

Natalia Shakhova and Igor Semiletov fear there will be an escape of 50 Gt of methane into the atmosphere in the short-term only from ESAS. For both of them, a vast methane belch is "highly possible at any time" (Pearce 2013; Mascarelli 2009). Whiteman et al. (2013) reinforce this perception: "a 50-gigatonne (Gt) reservoir of methane, stored in the form of hydrates, exists on the East Siberian Arctic Shelf. It is likely to be emitted as the seabed warms, either steadily over 50 years or suddenly." Peter Wadhams calculates that in case of a decade-long pulse of 50 $GtCH_4$, "the extra temperature due to the methane by 2040 is 0.6°C, a substantial extra contribution." And he adds: "although the peak of 0.6 °C is reached 25 years after emissions begin, a rise of 0.3–0.4°C occurs within a very few years." The likelihood that such a methane pulse will occur is considerable, according to Wadhams. If it does, it will accelerate positive feedback loops that can release even more methane into the atmosphere at an exponential rate of progression, leading to warming that is completely outside even the most pessimistic projections. This is, in any case, the reason for the various interventions of the Arctic Methane Emergency Group (AMEG), a UK-based group of scientists, who affirm (2012) that[18]:

> The tendency among scientists and the media has been to ignore or understate the seriousness of the situation in the Arctic. AMEG is using best available evidence and best available explanations for the processes at work. These processes include a number of vicious cycles which are growing in power exponentially, involving ocean, atmosphere, sea ice, snow, permafrost and methane. If these cycles are allowed to continue, the end result will be runaway global warming.

The "runaway global warming" conjecture, feared by the AMEG scientists, but rejected by the IPCC,[19] would be able to lead the Earth toward conditions that

---

[17] Cf. "Siberian Methane Release is on the Rise, and That's VERY Frightening." *Nature World News*, 31/XII/2014

[18] See AMEG Strategic Plan 12/IV/2012 http://a-m-e-g.blogspot.com/.

[19] See IPCC 31st Session, Bali 26–29 October 2009, p. 90: "Some thresholds that all would consider dangerous have no support in the literature as having a non-negligible chance of occurring. For instance, a 'runaway greenhouse effect'—analogous to Venus— appears to have virtually no

prevail today in Venus. This conjecture may be interesting from a strictly scientific point of view, but it is totally useless from the point of view of the fate of vertebrates and forests, because both would cease to exist under conditions that are much less extreme. The scenario of warming above 3 °C over the next few decades, defined as "catastrophic" (Xu and Ramanathan 2017), and the collapse of biodiversity already underway, addressed in Chaps. 10 and 11, will possibly be enough to cross the tipping points conducive to a "Hothouse Earth" pathway, as described above by Will Steffen and colleagues (2018). When Gaia Vince (2011) asked what the chances of survival of our young species would be in such circumstances, Chris Stringer, a paleoanthropologist at London's Natural History Museum, affirmed:

> One of the most worrying things is permafrost melting. If it continues to melt as we think it's already doing in some regions, we may well have a runaway greenhouse effect. We're also very dependent on a few staple crops, such as wheat and rice. If they get hit by climate change, we're in trouble. We're medium- to large-size mammals, we take a long time to grow up, we only produce one child at a time and we're demanding of our environment — this type of mammal is the most vulnerable. So, no, we're not immune from extinction. (...) The danger is that climate change will drive us into pockets with the best remaining environments. The worst-case scenario could be that everyone disappears except those who survive near the North and South poles — maybe a few hundred million, if the environment still supports them — and those will be the remaining humans. The problem is that once you've got humans isolated in small areas, they are much more vulnerable to extinction through chance events.

Perhaps the most privileged part of the human species will be able to adapt to the consequences of a drastic shrinking of the cryosphere, with global average warming above 5 °C and widespread degradation of the biosphere. At bay, in high latitudes, however, they will live in a terribly hostile world, one that is tragically depleted of animal and plant life and certainly unrelated to the organized societies of our times.

8.3   Higher Rises in Sea Level

"I don't think 10 years ago scientists realized just how quickly the potential for rapid sea level rise was," affirmed Maureen Raymo, Director of the Lamont–Doherty Core Repository at the Lamont–Doherty Earth Observatory of Columbia University (Lieberman 2016). The rise in sea level is another central factor in the destabilization of contemporary societies. It threatens coastal ecosystems, urban and transportation infrastructure, and many nuclear power plants, in addition to flooding and salinizing aquifers and deltas that are suitable for agriculture. As we shall see, the IPCC AR5 (2013) projections have not captured the increasing speed of this process because their predictions of sea level rise do not include the contribution of melting ice sheets, which has emerged as the main driver of this. Sea level rise is linked to two major factors driven by climate change; both are in acceleration: (1) thermal

---

chance of being induced by anthropogenic activities."

expansion and (2) melting glaciers and the loss of Greenland's and Antarctica's ice sheets. According to GISS/NASA, measurements derived from coastal tide gauge data and, since 1993, from satellite altimeter data indicate that between 1880 and 2013, global mean sea level (GMSL) rose by 22.6 cm, or an average of 1.6 mm per year over 134 years. Between 1993 and 2017 alone, GMSL rose by more than 7 centimeters. "This rate of sea-level rise is expected to accelerate as the melting of the ice sheets and ocean heat content increases as GHG concentrations rise" (Nerem et al. 2018). And indeed, according to Habib-Boubacar Dieng and colleagues (2017), GMSL rise since the mid-2000s shows significant increase compared to the 1993–2004 time span. Since at least 2007 (Livina and Lenton 2013), the thawing of the cryosphere reached a tipping point, entering into a phase of acceleration and irreversibility.

8.3.1 An Average Rise of 5 mm per Year That Is Accelerating

The rise in sea level doubled or tripled in the years following 1993, when compared to the rise observed during most of the twentieth century. Whether it doubled or tripled depends on the somewhat uncertain estimates of the pace of sea level rise in the last century. For Sönke Dangendorf and colleagues (2017), GMSL rise before 1990 was 1.1 (+/−0.3) mm per year, while from 1993 to 2012, it was 3.1 (+/−1.4) mm per year. In this case, the speed of sea level rise almost tripled when compared to that of the twentieth century. According to NOAA, "the pace of global sea level rise doubled from 1.7 mm/year throughout most of the twentieth century to 3.4 mm/year since 1993" (Lindsey 2017). Steven Nerem and colleagues (2018) estimate the climate change-driven acceleration of GMSL rise over the last 25 years to be 0.084 ± 0.025 mm/y^2. Finally, Fig. 8.3 shows an increase by a factor of more than seven in the pace of mean sea level rise between 1900–1930 (0.6 mm/year) and 2010–2015 (4.4 mm/year).

And John Englander (2019) finally shows a mean rise of 5 mm per year between 2012 and 2017.

8.3.2 Greenland

"Since the early 1990s, mass loss from the Greenland Ice Sheet has contributed approximately 10% of the observed global mean sea level rise" (McMillan et al. 2016). The second-largest ice sheet in the world is melting at a breakneck speed. At its highest point, this ice sheet still stands more than 3000 meters above sea level. GMSL would rise to 7.4 meters, according to the IMBIE team (2019), should it all melt and run off into the ocean. This could occur within a millennium in the absence of drastic reductions in GHG emissions (Aschwanden et al. 2019). Greenland lost about 3.8 trillion tons of ice (1 trillion ton = 400,000,000 Olympic Pools) between

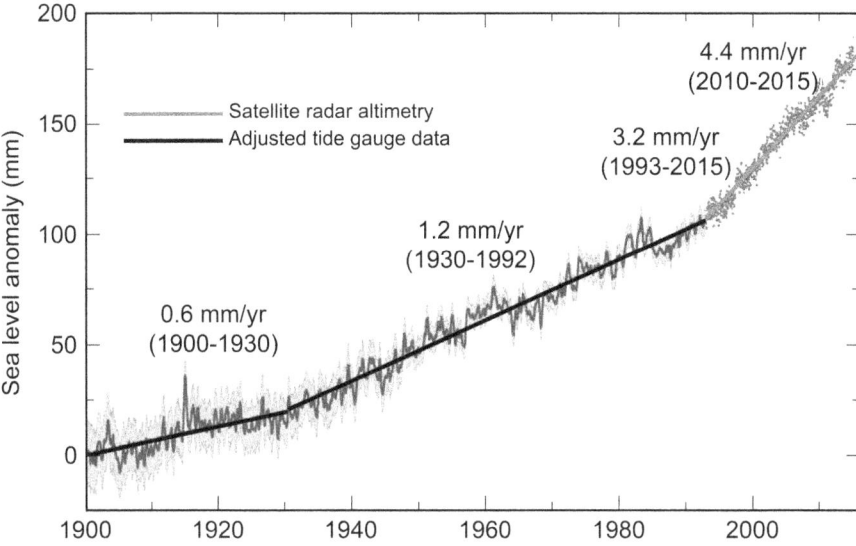

Fig. 8.3 Global mean sea level change (1900–2015). (Source: Hawai'i Climate Adaptation Portal http://climateadaptation.hawaii.gov/sea-level-rise/)

1992 and 2018, causing GMSL to rise by 10–11 mm (it takes about 360 Gt of ice loss to raise GMSL 1 mm). This loss has raised GMSL by 13.7 mm since 1972, half during the last 8 years (Mouginot et al. 2019), and will likely raise GMSL between 2.5 cm and 10 cm by 2050 (Merzdorf 2019).The pace of ice loss is much higher than predicted by the models, and the observed acceleration is stunning. It has risen from 33 Gt a year in the 1990s to 254 Gt a year in the 2010s, according to the IMBIE team (2019). There are two other slightly larger estimates of ice loss (Bevis et al. 2019; Mouginot et al. 2019). In any case, Greenland lost almost 200 Gt of ice in July 2019 alone in the wake of one of the largest heat waves in recorded climate history. Isabella Velicogna and colleagues (2020) recorded "a mass loss of 600 Gt by the end of August 2019, which is comparable to the mass loss in the warm summer of 2012." The island crossed one or more tipping points around 2002 or 2003. Rain is becoming more frequent, and it is melting Greenland's ice even in the dead of winter (Fox 2019). Mostly important, Andy Aschwanden and colleagues (2019) have shown that warmer waters along the west coast of Greenland "led to a disintegration of buttressing floating ice tongues, which triggered a positive feedback between retreat, thinning, and outlet glacier acceleration," a process they call outlet glacier-acceleration feedback.

8.3.3 Antarctica

As is well-known, the Antarctic thaw has a potential, in the long run, to cause a rise in sea level of about 58 meters. The continent lost 2720 (+/− 1390) Gt of ice between 1992 and 2017, leading to a rise in sea level of 7.6 (+/− 3.9) millimeters, with a marked increase in ice loss in recent years. In fact, the pace of glacial loss in Antarctica almost tripled since 2012, jumping from 76 Gt in 2012 to 219 Gt in 2017. In that period alone (2012–2017), Antarctic thaw increased global sea levels by 3 millimeters (Shepherd and the IMBIE team 2018). According to the World Meteorological Organization, about 87% of the glaciers along Antarctica's west coast have retreated over the past 50 years. Almost a quarter of its glaciers, among them the huge Thwaites and Pine Island Glaciers, can now be considered unstable. More recently, the glaciers spanning an eighth of the East Antarctica coastline are also being melted by warming seas. In February 2020, Antarctica logged its hottest temperatures on record, 18.3 °C and 20.75 °C (at Seymour Island), beating the previous record (2015) by more than 1 °C. If confirmed by the World Meteorological Organization, this temperature anomaly will be the first 20 °C-plus temperature measurement for the entire region south of 60° latitude (on July 2019, the Arctic region also hit its own record temperature of 21 °C). As pointed out by James Renwick, "it's a sign of the warming that has been happening there that's much faster than the global average (Readfern 2020). Thawing in parts of the West Antarctic Ice Sheet has already put it in a stage of collapse and the retreat is unstoppable (Joughin et al. 2014; Paolo et al. 2015; DeConto and Pollard 2016). Pietro Milillo and colleagues (2019) measured the evolution of ice velocity, ice thinning, and grounding line retreat of the huge Thwaites Glacier (around 170 thousand km^2) from 1992 to 2017. The Thwaites Glacier is melting very fast and is currently responsible for approximately 4% of global sea level rise. It contains enough ice to raise sea levels by about 60 cm. When it breaks off into the Amundsen Sea Embayment, it will probably trigger the collapse of five other nearby glaciers that are also melting (Pine Island, Haynes, Pope, Smith and Kohler). Together, they "have three meters of sea level locked up in them" (Englander 2019).

8.3.4 Projections of 2 Meters or More by 2100

According to Qin Dahe, Co-Chair of the IPCC-AR5 (2013), "as the ocean warms, and glaciers and ice sheets reduce, global mean sea level will continue to rise, but at a faster rate than we have experienced over the past 40 years." How much faster this rise will be is still uncertain. As stated above, the IPCC (2013) predictions about sea level rise until 2100 (26–98 cm) have long been considered too conservative (Rahmstorf et al. 2012). Anders Levermann, from the Potsdam Institute for Climate Impact Research (2013), calculates an impact of 2.3 meters for every 1 °C rise in

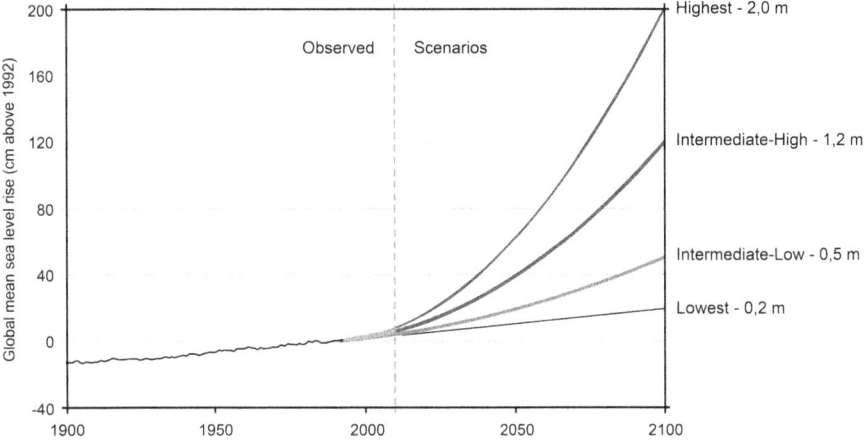

Fig. 8.4 Global mean sea level (GMSL) rise, observed and projected (1900–2100), 1992 = 0. (Source: NOAA, 6/XII/2012)

average global temperatures.[20] For its part, NOAA stated in 2012 that GMSL should increase by at least 20 centimeters and not more than 2 meters by 2100, considering the 1992 average level as a starting point.[21] Figure 8.4 shows these two extreme possibilities and their two intermediary variables.

In 2013, a GMSL of more than 1 meter throughout the century was no longer considered unlikely, given the acceleration of the thawing of Greenland and West Antarctica. But in 2017, NOAA redid its estimates and raised the upper limit of estimated mean sea level rise to 2.5 meters and the lower limit to 30 centimeters by 2100. Table 8.1 discriminates the median values of GMSL in each decade up to 2100 and successively in 2120, 2150, and 2200, according to six scenarios.

Five Observations:

(1) The Low and Intermediate-Low scenarios are no longer plausible because they assume no acceleration in GMSL rise (constant rise at rates of 3 and 5 millimeters per year by 2010, respectively, as per Table 8.2).
(2) The High scenario is the most likely, maintaining the current pace of RCP8.5 W/ m² as defined by the IPCC.
(3) "New evidence regarding the Antarctic ice sheet, if sustained, may significantly increase the probability of the Intermediate-High, High, and Extreme scenarios, particularly for RCP8.5 projections" (Sweet et al. 2017, p. 21).
(4) By 2030, there should be a GMSL rise between 16 cm and 24 cm above 2000 levels. This is an average of 20 cm, almost equivalent to the 22.6 cm rise observed between 1880 and 2013, as seen above. In the 2020s alone, GMSL is

---

[20] Cf. "Seas may rise 2.3 metres per degree of global warming: report." *World Bulletin*, 13/VII/2013
[21] Cf. NOAA, *Global Sea level Rise Scenarios for the U.S. National Climate Assessment*, 6/ XII/2012

Table 8.1 The GMSL rise scenario heights in meters composed of 19-year average estimates (1993–2012) reported on a decadal basis from 2000 through 2100 (and subsequently for 2120, 2150, and 2200)

GMSL Scenario (meters)	2010	2020	2030	2040	2050	2060	2070	2080	2090	2100	2120	2150	2200
Low	0.03	0.06	0.09	0.13	0.16	0.19	0.22	0.25	0.28	0.30	0.34	0.37	0.39
Intermediate-Low	0.04	0.08	0.13	0.18	0.24	0.29	0.35	0.4	0.45	0.50	0.60	0.73	0.95
Intermediate	0.04	0.10	0.16	0.25	0.34	0.45	0.57	0.71	0.85	1.0	1.3	1.8	2.8
Intermediate-High	0.05	0.10	0.19	0.30	0.44	0.60	0.79	1.0	1.2	1.5	2.0	3.1	5.1
High	0.05	0.11	0.21	0.36	0.54	0.77	1.0	1.3	1.7	2.0	2.8	4.3	7.5
Extreme	0.04	0.11	0.24	0.41	0.63	0.90	1.2	1.6	2.0	2.5	3.6	5.5	9.7

From Sweet et al. 2017, NOAA, Table 5, p. 23 (the Low and Extreme scenarios are scientifically plausible; the remaining four scenarios correspond to different levels under RCP2.6, RCP4.5, RCP 6.0, and RCP8.5)

Table 8.2   Rise rates in millimeters per year for 19-year averages (1993–2012) centered on decade associated with the median GMSL scenario heights this century

GMSL scenario rates (mm/year)	2010	2020	2030	2040	2050	2060	2070	2080	2090
Low	3	3	3	3	3	3	3	3	3
Intermediate-Low	4	5	5	5	5	5	5	5	5
Intermediate	5	6	7	9	10	12	13	14	15
Intermediate-High	5	7	10	13	15	18	20	22	24
High	6	8	13	16	20	24	28	31	35
Extreme	6	10	15	20	25	30	35	40	44

From Sweet et al. 2017, NOAA, Table 6, p. 23

expected to rise between 6 and 13 cm, which is enough to cause recurrent and potentially catastrophic flooding in various cities around the world.

(5) As shown in the table below, the speed of GMSL rise is multiplied by factors between 3 (intermediate) and 7.3 (extreme) by 2100.

In certain regions, these rises are much higher than the global average and are already equivalent to the IPCC projections for the 2081–2100 period. In New York, the US Federal Emergency Management Agency (FEMA) predicts elevations of 25 centimeters by 2020, 76 centimeters in the 2050s, 147 centimeters in the 2080s, and 191 cm in the first decade of the twenty-second century.[22] According to Andrew Shepherd, "around Brooklyn you get flooding once a year or so, but if you raise sea level by 15 cm then that's going to happen 20 times a year" (Pierre-Louis 2018).

8.3.5   A New Projection: "Several Meters Over a Timescale of 50–150 Years"

Scary as they may be, the above reported NOAA projections (Sweet et al. 2017) were largely surpassed by a new analysis (Hansen et al. 2016), based primarily on two arguments:

(1) Information provided by paleoclimatology: "Earth is now as warm as it was during the prior interglacial period (Eemian), when sea level reached 6–9 m higher than today" (Hansen et al. 2016).
(2) The amplifying feedbacks caused by the ongoing thawing in Antarctica and Greenland, which should increase subsurface ocean warming and mass loss of ice sheets. According to the authors, because of these feedbacks, the ice sheets that come into contact with the ocean are vulnerable to accelerating disintegration, causing much bigger GMSL increases on a much smaller timescale:

We hypothesize that ice mass loss from the most vulnerable ice, sufficient to raise sea level several meters, is better approximated as exponential than by a more linear response.

---

[22] Cf. "Trend watch," *Nature*, 518, 26/II/2015, p. 461

Doubling times of 10, 20 or 40 years yield multi-meter sea level rise in about 50, 100 or 200 years. (…) These climate feedbacks aid interpretation of events late in the prior inter-glacial [Eemian], when sea level rose to +6–9 m with evidence of extreme storms while Earth was less than 1 °C warmer than today.

This projection, put forth by 18 scientists and coordinated by James Hansen, is very similar to and is reinforced by the Hothouse Earth hypothesis (Steffen et al. 2018) discussed above. Both studies consider an average global warming of "only" 2 °C above the pre-industrial period as a planetary threshold that, if crossed, can trigger climate feedback loops and nonlinear climate responses. In fact, James Hansen and colleagues warn that "the modeling, paleoclimate evidence, and ongo-ing observations together imply that 2°C global warming above the pre-industrial level could be dangerous." They predict that this level of warming will cause:

(1) Cooling of the Southern Ocean, especially in the Western Hemisphere
(2) Slowing of the Southern Ocean overturning circulation, warming of ice shelves, and growing mass loss of ice sheets
(3) Slowdown and eventual shutdown of the Atlantic overturning circulation with cooling of the North Atlantic region
(4) Increasingly powerful storms
(5) Nonlinearly growing sea level rise, reaching several meters over a timescale of 50–150 years

8.3.6   Climate Refugees

In a video about this study, James Hansen discusses some implications of climate change and rising sea levels. He underlines the interactions between ocean warming and the melting of the Antarctic and Greenland ice sheets, in addition to analyzing the positive feedback loops at play in global warming, sea level rise, and extreme weather events, such as "superstorms stronger than any in modern times" and "pow-erful enough for giant waves to toss 1000 ton megaboulders onto the shore in the Bahamas." These feedbacks, he states:

raise questions about how soon we will pass points of no return in which we lock in conse-quences that cannot be reversed on any timescale that people care about. Consequences include sea level rise of several meters, which we estimate could occur this century, or at latest next century, if fossil fuel emissions continue at a high level. (…) The IPCC does not report these effects for two reasons. First, most models used by IPCC exclude ice melt. Second, we conclude that most models, ours included, are less sensitive than the real world to the added fresh water.

There is no need to stress the obvious impacts of these combined processes of sea level rise, larger and more extreme hurricanes, and superstorms capable of flooding thousands of square kilometers in the most densely populated coastal regions. The first of these impacts is the emergence of hundreds of millions of people who may fall into a new category of so-called "natural" disaster victims, the climate refugees, a term that does not exist yet in international law. According to a Norwegian Refugee Council (NRC) report, between 2008 and 2013, "natural" disasters caused, on average, the forced displacement of 27.5 million people *per year*. In 2013 alone, about 22 million people in at least 119 countries were displaced from their homes because of these disasters; this is triple the number of displaced people due to armed conflict in the same period (Yonetani et al. 2014). In the past two decades, the global population of displaced people jumped from 33.9 million in 1997 to 65.6 million in 2016, the third consecutive year in which the number of refugees has broken the all-time high, according to the United Nations High Commissioner for Refugees (UNHCR). We present further data from the UNHCR report, *Global Trends: Forced Displacements in 2016.* In 2016, 20 people were forced to leave their homes every minute. These forced displacements resulted in 22.5 million refugees (50% of them under the age of 18), the highest number in historical records, with 40.3 million moving involuntarily within the borders of their own countries. Just a decade ago, 1 in 160 people was a refugee. In 2016, 1 in 113 people was sentenced to this condition. In 2016, asylum seekers totaled 2.8 million people in 164 countries.[23] The UNHCR's report *Climate Change and Disaster Displacement in 2017* states that "between 2008 and 2015, 203.4 million people were displaced by disasters, and the likelihood of being displaced by disasters has doubled since the 1970s."

Not all these more than 203 million displaced people can be considered climate refugees, of course, but, as stated in the same document, "climate change is also a threat multiplier, and may exacerbate conflict over depleted resources." Indeed, given that many wars and armed conflicts are fueled and/or intensified by climate problems, such as droughts, water depletion, food shortages, floods, and hurricanes, one must consider the "hidden" weight of environmental crises in intensifying conflicts that are predominantly political, ethnic, and religious in nature. In any case, the 2017 "Lancet Countdown" states that "annual weather-related disasters have increased by 46% from 2000 to 2013." Referring to the next three decades, a press release titled "Sustainability, Stability, Security," published in 2016 by the UN Convention to Combat Desertification (UNCCD) for an African inter-ministerial meeting at COP22, reminds us that more than 90% of Africa's economy depends on a climate-sensitive natural resource, like rain-fed, subsistence agriculture, and issues a very clear warning: "unless we change the way we manage our land, in the next 30 years we may leave a billion or more vulnerable poor people with little choice but to fight or flee."

---

[23] Cf. *Global Trend. Forced Displacements in 2016.* UNHCR http://www.unhcr.org/globaltrends2016/.

8.3.7   Consequences of a Rise in Sea Level of Up to 1 Meter

(1) A GMSL rise of only about 50 cm by 2050 (Sweet et al. 2017: Intermediate-High = 44 cm; High = 54 cm) relative to 2000 will cause the forced migration of over 40 million people, according to simulations proposed by the NGO *GlobalFloodMap.org*. Several points on the coast of Africa (almost five million people), Europe (almost six million people), and Asia (14 million people) will be particularly affected. The case of Bangladesh, with a population of 153 million concentrated in just 144,000 km^2, is one of the most serious, since two-thirds of its land is less than 5 meters above the current sea level. According to UN projections, by 2050 Bangladesh could lose 40% of its arable land. Hasan Mahmud, the country's Minister for Environment, Forest and Climate Change, told *Le Monde* that "the sea level in the Gulf of Bengal has already risen and if scientists' predictions are confirmed, 30 million people are expected to flee their lands by the end of the century." (Bouissou 2013).

(2) The next scenario entails a rise in sea level between 60 cm and 1.3 m between 2060 and 2080 (Sweet et al. 2017: 2060 Intermediate-High = 60 cm; High = 77 cm; 2080 Intermediate-High = 1 m; High = 1.3 m). This rise in sea level will be enough to cause flooding of deltas; changes in shorelines; complete submergence of islands, lowlands, and arable lands; as well as destruction of coastal ecosystems and displacement of populations that today live near the coast. This, in turn, will produce waves of new refugees with social traumas and colossal losses of urban infrastructure, as well as an overload of territories, some of which are already saturated due to human occupation. In this second scenario, a rise in sea level will directly reach almost 150 million people (Anthoff et al. 2006). Among the 20 most populous cities in the world, 13 are sea or river ports in coastal areas. A study coordinated by Susan Hanson (2011) identified 136 port cities over one million people most exposed to the impacts of rising sea levels and climate extremes by the 2070s. Of these 136 cities, 52 are in Asia, 17 in the USA, and 14 in South America.

"The Ocean Conference" promoted by the United Nations (2017) estimates that people currently living in coastal communities represent 37% of the global population. Immense human contingents and entire populations of countless other species will be increasingly affected by more and more aggressive and frequent flooding, gradually being condemned to the condition of climate refugees. According to CoastalDEM, a new digital elevation model produced by Climate Central (2019), land currently home to 300 million people will fall below the elevation of an average annual coastal flood by 2050. Scott Kulp and Benjamin Strauss (2019) estimate that "one billion people now occupy land less than 10 meters above current high tide lines, including 230 million below 1 meter." Under high emissions, the authors add, "CoastalDEM indicates up to 630 million people live on land below projected annual flood levels for 2100."

But let us focus on the catastrophic current situation. The future is now in South Florida, for instance, where some 2.4 million people are at risk of flooding from even a moderate hurricane-driven storm surge. The odds of a catastrophic 100-year flood by 2030 are now 2.6 times higher than they would have been without global warming (Lemonick 2012). The same can be said of the 52 nations called SIDS (*Small Island Developing States*), where environmental collapse is already an ongoing reality, as the oceans are about to wipe out these small paradises which are home to a huge amount of diverse cultures, to biodiversity, and to almost 1% of humanity. For terrestrial species, including humans, living in the Pacific Islands made up of coral atolls, this collapse has a scheduled date, if the projections of Curt D. Storlazzi and colleagues (2018) are confirmed:

> We show that, on the basis of current greenhouse gas emission rates, the nonlinear interactions between sea-level rise and wave dynamics over reefs will lead to the annual wave-driven overwash of most atoll islands by the mid-twenty-first century. This annual flooding will result in the islands becoming uninhabitable because of frequent damage to infrastructure and the inability of their freshwater aquifers to recover between overwash events.

8.3.8   Cyclones, Hurricanes, Typhoons, Tornadoes… and Nuclear Power Plants

Extreme weather events are thermodynamic phenomena that are classified in five categories on the Saffir-Simpson scale, based on the increasing speed of their winds, starting with the weakest of them, which can reach a speed of over 117 km/h. One of the conditions that induce these events is when surface layers of the ocean, up to 50 meters deep, reach temperatures above 26 °C. There is evidence that the number of Category 4 and 5 hurricanes on the Saffir-Simpson scale has been rising. According to Jeffrey St. Clair (2019), in the last 169 years, only 35 Atlantic hurricanes have attained Category 5 status, and 5 of them occurred in the last 4 years: Matthew (2016), Irma and Maria (2017), Michael (2018), and Dorian (2019). Humans are increasingly exposed to extreme flooding by tropical cyclones, and "there is also growing evidence for a future shift in the average global intensity of tropical cyclones towards stronger storms" (Woodruff et al. 2013). NOAA's Geophysical Fluid Dynamics Laboratory (GFDL) assessment, titled *Global Warming and Hurricanes: An Overview of Current Research Results* (2019), summarizes its main conclusions for a 2 °C global warming scenario in three points:

(1) Tropical cyclone rainfall rates will likely increase "on the order of 10–15% for rainfall rates averaged within about 100 km of the storm."
(2) Tropical cyclone intensities globally will likely increase on average by 1% to 10%. "This change would imply an even larger percentage increase in the destructive potential per storm, assuming no reduction in storm size."
(3) The global proportion of tropical cyclones that reach very intense levels (Categories 4 and 5) will likely increase over the twenty-first century.

The growing vulnerability of nuclear power plants is perhaps the most worrying consequence of these processes. Since they are cooled by water, their reactors are always next to rivers and estuaries, or by the sea. Although caused by a tsunami, the 2011 Daiichi disaster in Fukushima showed that electrical installations, indispensable for the functioning of the reactor cooling system, are vulnerable to flooding, whatever may be its cause: hurricanes, torrential rains, dam failure, or rise in sea level (Kopytko 2011). The vulnerability of nuclear plants tends to increase because their design, in most cases dating from the 1970s to 1990s, calculates safety margins for such phenomena on the basis of historical limits, which are now being exceeded due to climate change, in particular the intensification of hurricanes and rising sea levels.

Examples of increased risks abound. In 1999, the Blayais Nuclear Power Plant in the Gironde estuary in France was flooded by a tide and unprecedented winds, damaging two of its reactors. In June 2011, a flood in the Missouri River endangered the Fort Calhoun Nuclear Generating Station in Nebraska, and in October 2012 Hurricane Sandy seriously threatened the safety of the Salem and Oyster Creek Nuclear Plants in New Jersey.[24] In England, "12 of Britain's 19 civil nuclear sites are at risk of flooding and coastal erosion because of climate change," 9 of which "have been assessed [...] as being vulnerable now" (Edwards 2012). In Brazil, there are growing risks of this phenomenon for the reactors in the Angra dos Reis power plants (Pereira 2012).

8.3.9   A Call to Arms

This is the title of the last chapter of Peter Wadhams book, *A Farewell to Ice: A Report from the Arctic* (2016). Its conclusion is one of brutal simplicity:

> We have destroyed our planet's life support system by mindless development and misuse of technology. A mindful development of technology, first for geoengineering, then for carbon removal, is now necessary to save us. It is the most serious and important activity in which the human race can now be involved, and it must begin immediately.

Wadhams does not ignore the costs, risks, and limitations of his proposal at the present stage of human technology, hence his appeal for an agenda to improve science, so as to make viable, technically and financially, a geoengineering with risks that are less imponderable. Yes, it is now necessary to contemplate all technological alternatives that can alleviate the present and future impacts of the climate emergency. None of them, however, will be able to prevent a global collapse, at the same time environmental and social. Technology, the economy, and universal carbon taxation (proposed since the Toronto Conference on Climate Change in 1988!) are all important and necessary. But only politics can significantly mitigate this ongoing global collapse. This perception that politics is at the center of our survival possibilities as organized societies does not yet seem clear enough among economists and scientists. The paradigm that has enabled the immense development of our societies

---

[24] Cf. "Flood Risk at Nuclear Power Plants." Union of Concerned Scientists

is based on an energy system built on burning fossil fuels and on an intoxicating, animal protein-based, globalized food system that destroys forests and the foundations of life on the planet. This paradigm has exhausted itself. A paradigm exhausts itself when its harms far outweigh its benefits. This is precisely what is happening with increasing clarity, especially since the last quarter of the twentieth century. There are still those who think—especially among economists—that the benefits continue to outweigh the harms. But they ignore science, for science never ceases to warn us that the worst evils are yet to come and that they are already inevitable, even if the worst-case scenarios do not materialize.

The human species has, therefore, come to the most important crossroad in its history. If the decision on strategic investments in energy, food, and transportation systems is left in the hands of state–corporations, there will be no technology to save us. In this case, we will be condemned, within the horizon of current generations, to a warming of 2 °C, with increasing probabilities of warming equal or above 5 °C by 2100, due to the action of positive climate feedbacks. A warming of 3 °C or 4 °C is already considered beyond adaptation and implies unspeakable suffering to all, even the tiny minority who consider themselves invulnerable. An average global warming of 5 °C above the pre-industrial period implies extreme precariousness and, ultimately, the extinction of our species and of countless others.

This realization should not and cannot feed useless despair because another way is still possible. The way forward is one of full political mobilization to keep warming *as low as is still possible*. Maintaining global warming at the levels set by the IPCC and pursued by the Paris Agreement already is, in all likelihood, a sociophysical impossibility, as seen in the preceding chapter. But politics is not the art of the possible, as those who resign themselves to the status quo like to claim. To long for the impossible in order to broaden the horizon of the possible is the fundamental teaching of political reason. Is maintaining global warming "well below" 2 °C—to remember the aim of the Paris Agreement—impossible? "Of course, it is impossible, but we have to try," exhorts Piers Forster (Lawton 2018). And Graham Lawton adds, wisely: "Even if we go over 1.5°C, every bit of extra warming we shave off makes the world more liveable." The whole political program of our society can now be summed up by this attempt.

References

ALVAREZ, R. A. *et al.*"Greater focus needed on methane leakage from natural gas infrastructure". *PNAS*, 109, 17, 2012, pp. 6.435–6.440

ANTHOFF, R.J *et al.*, "Global and regional exposure to large rises in sea-level: a sensitivity analysis". *Working Paper* 96, 2006. Tyndall Centre for Climate Change Research, Norwic. UNEP/Grid-Arendal.

ARCHER, David *et al.*, "Atmospheric Lifetime of Fossil Fuel Carbon Dioxide". *Annual Review of Earth and Planetary Sciences*, 37, 2009, pp. 117–134.

ASCHWANDEN, Andy *et al.*, "Contribution of the Greenland Ice Sheet to sea level over the next millennium". *Science Advances*, 5, 6, 19/VI/2019.

AZEVEDO, Tasso R., ANGELO, Claudio & RITTL, Carlos, *Emissões de GEE no Brasil e suas implicações para políticas públicas e a contribuição brasileira para o Acordo de Paris. Documento de análise.* Observatório do Clima/SEEG, 2018.

BACCINI, Alessandro *et al.*, "Tropical forests are a net carbon source based on above ground measurements of gain and loss". *Science*, 28/IX/2017.

BARBOSA, Fabiano A. *et al.*, *Cenários para a pecuária de corte amazônica*, Belo Horizonte, IGC/UFMG 2015.

BARDI, Ugo, *The Seneca Effect. Why Growth is Slow but Collapse is Rapid.* Springer, 2017.

BELCHER, Steven, BOUCHER, Olivier & SUTTON, Rowan, "Why Results from the Next Generation of Climate Models Matter". *CarbonBrief*, 21/III/2019.

BEVIS, Michael *et al.*, "Accelerating changes in ice mass within Greenland, and the ice sheet's sensitivity to atmospheric forcing". *PNAS*, 5/II/2019.

BORENSTEIN, Seth, "Study Says Methane a New Climate Threat". *The Washington Post*, 6/IX/2006.

BOUISSOU, Julien, "Au Bangladesh, survivre avec le changement climatique". *Le Monde*, 12/II/2013.

BOX, Jason *et al.*, "Key indicators of Arctic climate change: 1971–2017". *Environmental Research Letters*, 14, 2019.

CLIMATE CENTRAL, *Flooded Future: Global Vulnerability to Sea Level Rise Worse Than Previously Understood.* 29/X/2019.

CONNOR, Steve, "Exclusive. The methane time bomb", *The Independent*, 23/IX/2008.

———. "Danger from the deep: New climate threat as methane rises from cracks in Arctic ice". *The Independent*, 23/IV/2012.

DANGENDORF, Sönke *et al.* "Reassessment of 20th century global mean sea level rise". *PNAS*, 114, 23, 22/V/2017.

DEAN, Joshua F. *et al.*, "Methane Feedbacks to the Global Climate System in a Warmer World". *Reviews of Geophysics*, 56, 15/II/2018.

DeCONTO, Robert M. & POLLARD, David. "Contribution of Antarctica to past and future sea-level rise". *Nature*, 531, 31/III/2016.

DE SY, Veronique *et al.*, "Land use patterns and related carbon losses following deforestation in South America". *Environmental Research Letters*, 10, 2015.

DIENG, Habib-Boubacar *et al.*, "New estimate of the current rate of sea level rise from a sea level budget approach". *Geophysical Research Letters*, 22/IV/2017.

DUARTE, Carlos *et al.*, "Abrupt Climate Change in the Arctic". *Nature. Climate Change.* 27/I/2012, 2, 60–62.

EDWARDS, Rob, "UK nuclear sites at risk of flooding, report shows". *The Guardian*, 7/III/2012.

ELDREDGE, Niles & GOULD, Stephen Jay, "Punctuated equilibria: an alternative to phyletic gradualism. *Models in Paleobiology*, 1972, pp. 82–115 (republished in S. Jay Gould, *Panda's thumb. More reflections in natural history*, New York, 1980, chapter 17: The episodic nature of evolutionary change).

ENGLANDER, John, *High Tide on Main Street. Rising Sea Level and the Coming Coastal Crisis* (2012), The Science Bookshelf, 2nd edition, 2014.

———, "Sea Level Rise Can No Longer Be Stopped, What Next?" (Lecture at the Royal Institution, uploaded on 29/V/2019) https://www.youtube.com/watch?v=MvqY2NcBWI8.

ETMINAN, Maryam *et al.*, "Radiative forcing of carbon dioxide, methane and nitrous oxide: a significant revision of the methane radiative forcing." *Geophysical Research Letters,* 27/XII/2016.

FLETCHER, Sara E. Mikaloff & SCHAEFER, Hinrich, "Rising methane: A new climate challenge". *Science*, 364, 6444, 7/VI/2019, pp. 932–933.

FOX, Alex, "Rain is melting Greenland's ice, even in winter, raising fears about sea level rise". *Science*, 7/III/2019.

Global Carbon Budget 2016, see LE QUÉRÉ, Corinne et al. Global Carbon Budget, Earth System Science Data, 8, 2016, pp. 605–649.

HANSEN, James *et al.* "Ice melt, sea level rise and superstorms: evidence from paleoclimate data, climate modeling, and modern observations that 2°C global warming could be dangerous". *Atmospheric Chemistry and Physics. An interactive open-access journal of the European Geosciences Union*, 16, 22/III/2016.

HANSON, Susan *et al.*, "A global ranking of port cities with high exposure to climate extremes". *Climatic Change*, 104, 2011, PP. 89–111.

HERNDON, E. M., "Permafrost slowly exhales methane". *Nature Climate Change*, 8, 4, April 2018, pp. 273–274.

HUGELIUS, G. *et al.*, "Estimated stocks of circumpolar permafrost carbon with quantified uncertainty ranges and identified data gaps". *Biogeosciences* 11, 2014, pp. 6573–6593.

IMBIE team (Ice sheet Mass Balance Inter-comparison Exercise), "Mass balance of the Greenland Ice Sheet from 1992 to 2018". *Nature*, 10/XII/2019.

JOUGHIN, Ian; SMITH, Benjamin E. & MEDLEY, Brooke. "Marine Ice Sheet Collapse Potentially Underway for the Thwaites Glacier Basin, West Antarctica". *Science*, 12/V/2014.

KATZ, Cheryl, "How long can oceans continue to absorb Earth's excess heat?". *Yale360*, 18/III/2015.

KNOBLAUCH, Christian *et al.*, "Methane production as key to the greenhouse gas budget of thawing permafrost". *Nature Climate Change*, 8, 19/III/2018, pp. 309–312.

KNUTTI, Reto & HEGERL, Gabriele C., "The equilibrium sensitivity of the Earth's temperature to radiation changes". *Nature Geosciences*, 1, November 2008, pp. 735–43.

KNUTTI, Reto, RUGENSTEIN, Maria A.A. & HEGERL, Gabriele C., "Beyond Equilibrium Climate Sensitivity". *Nature Geosciences*, 10, 4/XI/2017, pp. 727–736.

KOPYTKO, N., "The climate change threat to nuclear power". *New Scientist*, 2813, 24/V/2011.

KORT, Eric *et al.*, "Atmospheric observations of Arctic Ocean methane emissions up to 82° north". *Nature Geoscience*, 5, 22/IV/2012, pp. 318–321.

KULP, Scott A. & STRAUSS, Benjamin, "New elevation data triple estimates of global vulnerability to sea-level rise and coastal flooding". *Nature Communications*, 29/X/2019.

LAMARCHE-GAGNON, Guillaume *et al.*, "Greenland melt drives continuous export of methane from the ice-sheet bed". *Nature*, 565, 2/I/2019, pp. 73–77.

LAWTON, Graham, "Hitting 1.5oC". *New Scientist*, 8/XII/2018.

LEMONICK, Michael D., "The Future is Now for Sea Level Rise in South Florida". *Climate Central*, 6/IV/2012.

LENTON, Timothy M. & SCHELLNHUBER, Hans Joachim, "Tipping the scales". *Nature Climate Change*, 1, 2007, pp. 97–98.

LENTON, Timothy M. *et al.*, "Tipping elements in the Earth's climate system". *PNAS*, 105, 6, 2015, pp. 1786–1793.

LENTON, Timothy *et al.*, "Climate tipping points – too risky to bet against". *Nature*, 575, 29/XI/2019.

LIEBERMAN, Amy, "Preparing for the Inevitable Sea-Level Rise". *The Atlantic*, 29/II/2016.

LINDSEY, Rebecca, "Climate Change: Global Sea Level". NOAA, 11/IX/2017.

LIVINA, Valerie N. & LENTON, Timothy M., "A recent tipping point in the Arctic sea-ice cover: abrupt and persistent increase in the seasonal cycle since 2007". *The Cryosphere*, 7, 1, 2013, pp. 275–286

MEIRELLES Fº, João. "Você já comeu a Amazônia hoje?", Instituto Peabiru, Belém, 2005

———. "É possível superar a herança da ditadura brasileira (1964–1985) e controlar o desmatamento na Amazônia? Não, enquanto a pecuária bovina prosseguir como principal vetor de desmatamento". *Boletim do Museu Par. Emílio Goeldi*, 9, 1, 2014, pp. 219–241.

MARQUES, Luiz, "Abandonar a carne ou a esperança". *Jornal da Unicamp*, 10/VII/2019.

MASCARELLI, Amanda Leigh, "A sleeping giant?" *Nature Reports Climate Change*, 5/III/2009.

McMILLAN, Malcom *et al.*, "A high-resolution record of Greenland mass balance". *Geophysical Research Letters*, 16/VI/2016.

MERZDORF, Jessica, "Study Predicts More Long-Term Sea Level Rise from Greenland Ice". NASA's Goddard Institute for Space Studies, 20/VI/2019.

MILILLO, Pietro *et al.*, "Heterogeneous retreat and ice melt of Thwaites Glacier, West Antarctica". *Science Advances, 5, 1,* 30/I/2019.

MOUGINOT, Jérémie *et al.*, "Forty-six years of Greenland Ice Sheet mass balance from 1972 to 2018". *PNAS*, 7/V/2019.

MYHRE, Gunnar, "Effect of methane on climate change could be 25% greater than we thought". *Cicero*, 12/I/2017.

NEREM, R.S *et al.*, "Climate-change–driven accelerated sea-level rise detected in the altimeter era". *PNAS*, 12/II/2018.

NISBET, Euan G. *et al.*, "Very Strong Atmospheric Methane Growth in the 4 Years 2014–2017: Implications for the Paris Agreement". *Global Biogeochemical Cycles*, 5/II/2019.

PANDEY, Sudhanshu *et al.*, "Satellite observations reveal extreme methane leakage from a natural gas well blowout". *PNAS*, 116, 52, 26/XII/2019, pp. 26376–26381.

PAOLO, Fernando S., FRICKER, Helen A. & PADMAN, Laurie, "Volume loss from Antarctic ice shelves is accelerating". *Science*, 348, 6232, 17/IV/2015, pp. 327–331.

PARTAIN Jr., James *et al.* "An Assessment of the Role of Anthropogenic Climate Change in the Alaska Fire Season of 2015". In HERRING, S. C. *et al.* (eds.), "Explaining Extreme Events of 2013 from a Climate Perspective". *Bulletin of the American Meteorological Society*, 97, 12, 2015.

PEARCE, Fred, "Vast methane belch possible any time". *New Scientist*, 27/VII/2013.

PENNER, Joyce E., "Soot, sulfate, dust and the climate – three ways through the fog". *Nature*, 11/VI/2019.

PEREIRA, A., "Glub, glub, glub". *Folha de São Paulo*, 6/XII/2012.

PIERRE-LOUIS, Kendra, "Antarctica Is Melting Three Times as Fast as a Decade Ago". *The New York Times*, 13/VI/2018.

PORTNOV, Alexey *et al.* "Offshore permafrost decay and massive seabed methane escape in water depths >20 m at the South Kara Sea shelf". *Geophysical Research Letters*, 40, 15, 1/VIII/2013, pp. 3.962–3.967.

PORTNOV, Alexey *et al.* "Modeling the evolution of climate-sensitive Arctic subsea permafrost in regions of extensive gas expulsion at the West Yamal shelf". *Journal of Geophysical Research*, 119, 11, 17/XI/2014, pp. 2.082–2.094.

RAHMSTORF, Stefan *et al.* "Projected sea-level rise may be underestimated". *Potsdam Institute of Climate Impact Research*, 28/XI/2012.

RAHMSTORF, Stefan, "Anthropogenic Climate Change: Revisiting the Facts". In Zedillo, E. *Global Warming: Looking Beyond Kyoto*, 2008, pp. 34–53.).

RAWNSLEY, Jessica, "Amazon rainforest reaches point of no return". *Climate News Network*, 16/III/2020.

READFERN, Graham, "Antarctica logs hottest temperature on record with a reading of 18.3C". *The Guardian*, 7/II/2020.

RIJSBERMAN, F.R. & SWART, R.J. (ed.), "Targets and Indicators of Climatic Change". The Stockholm Environment Institute, 1990.

St. CLAIR, Jeffrey, "Roaming Charges: Blood in the Eye of the Storm". *Counterpunch*, 6/IX/2019.

SAMSET, Bjorn H. *et al.*, "Climate Impacts From a Removal of Anthropogenic Aerosol Emissions". *Geophysical Research Letters*, 8/I/2018.

SANFORD, Todd, WANG, Regina & KENWARD, Alyson, "The age of Alaskan Wildfires". *Climate Central*, 2015.

SAUNOIS, M., *et al.*, "The growing role of methane in anthropogenic climate change". *Environmental Research Letters*, 11, 12, 12/XII/2016

SCHAEFER, Kevin *et al.*, "Amount and timing of permafrost carbon release in response to climate warming". *Tellus B. Chemical and Physical Meteorology*, 63, 2, 2011, pp. 168–180.

SCHAEFER, Kevin, LANTUIT, Hugues; ROMANOVSKY, Vladimir & SCHUUR, Edward A. G. *Policy Implications of Warming Permafrost*. UNEP, 2012.

SCHMIDT, Gavin, "Methane: A Scientific Journey from Obscurity to Climate Super-Stardom". *Goddard Institute for Space Studies*, September 2004.

SCHUUR, Edward A. G. *et al.* "Climate change and the permafrost carbon feedback". *Nature*, 520, 9/IV/2015, pp. 171–179.

SHAKHOVA, Natalia *et al.*, "Current rates and mechanisms of subsea permafrost degradation in the East Siberian Arctic Shelf". *Nature Communications*, 22/VII/2017.

SHAKHOVA, Natalia, SEMILETOV, Igor & CHUVILIN, Evgeny, "Understanding the Permafrost–Hydrate System and Associated Methane Releases in the East Siberian Arctic Shelf". *Geosciences*, 9, 6, 251, 2019.

SHEPHERD, Andrew (and the IMBIE team), "Mass balance of the Antarctic Ice Sheet from 1992 to 2017". *Nature*, 13/VI/2018.

STEFFEN, Will *et al.*, "Trajectories of the Earth System in the Anthropocene". *PNAS*, 6/VIII/2018.

STORLAZZI, Curt D. *et al.* "Most atolls will be uninhabitable by the mid-21st century because of sea-level rise exacerbating wave-driven flooding". *Science Advances*, 4, 4, 25/IV/2018.

STRELETSKAYA, Irina D. *et al.*, "Methane Content in Ground Ice and Sediments of the Kara Sea Coast". *Geosciences*, 8, 12, 434, 2018.

SWEET, William V. *et al.*, *Global and Regional Sea Level Rise Scenarios for the United States*. NOAA, January 2017.

TOLLEFSON, Jeff *et al.* "Methane leaks erode green credentials of natural gas". *Nature*, 493, 7.430, 2/I/2013.

TURETSKY, Merritt R. *et al.*, "Permafrost collapse is accelerating carbon release". *Nature*, 30/IV/2019.

URBAN, Mark U., "Life without ice" (Editorial). *Science*, 14/II/2020.

VELICOGNA, Isabella *et al.*, "Continuity of ice sheet mass loss in Greenland and 2 Antarctica from the GRACE and GRACE Follow-On 3 missions". *Geophysical Research Letters*, 2020 (first version).

VINCE, Gaia, "A human perspective. Interview with Chris Stringer". *Nature Climate Change*, 1, September 2011.

VOOSEN, Paul, "New climate models predict a warming surge". *Science*, 16/IV/2019.

WADHAMS, Peter, "Arctic Amplification, Climate Changing, Global Warming. New Challenges from the top of the world" (Lecture). Fondazione Eni Enrico Mattei, Milan, 12/V/2015. <https://www.youtube.com/watch?v=aTY9M_ZKk3M>.

———. *A Farewell to ice. A Report from the Arctic*. London, 2016.

WALTER, Katey *et al.* "Methane bubbling from Siberian thaw lakes as a positive feedback to climate warming". *Nature*, 443, 7/IX/2006, pp. 71–75.

WHITEMAN, Gail, HOPE, Chris & WADHAMS, Peter, "Vast costs of Arctic change". *Nature*, 499, 24/VII/2013, pp. 401–403.

WMO, "Unprecedented wildfires in the Arctic", 12/VII/2019.

WOODRUFF, Jonathan D., IRISH, Jennifer L. & CAMARGO, Suzana J., "Coastal flooding tropical cyclones and sea-level rise" (Review). *Nature*, 504, 5/XII/2013, pp. 44–52.

XU, Yangyang & RAMANATHAN, Veerabhadran, "Well below 2°C: Mitigation strategies for avoiding dangerous to catastrophic climate changes". *PNAS,* 14/IX/2017.

XU, Yangyang; RAMANATHAN, Veerabhadran & VICTOR, David, "Global Warming will happen faster than we think". *Nature*, 5/XII/2018.

YONETANI, Michelle *et al.*, *Global Estimates 2014. People displaced by disasters.* Norwegian Refugee Council, September 2014.

ZIKA, Michael & ERB, Karl-Heinz, "The global loss of net primary production resulting from human-induced soil degradation in drylands". *Ecological Economics*, 14/VII/2009.

Chapter 9
Demography and Democracy

"The long-lasting debate involving population, development and environment appears to have become an insoluble equation." Such is the conclusion of George Martine and José Eustáquio Diniz Alves (2019) about the apparently insurmountable difficulties in harmonizing the three terms of the demographic equation. That said, one can at least safely assert that the number of people currently inhabiting the planet is not, in and of itself, a *fundamental* stressor of ecosystems as an isolated factor. It is not possible to subscribe to the opinion of Sir David King, former science adviser to the UK government, for whom "the massive growth in the human population through the twentieth century has had more impact on biodiversity than any other single factor" (Campbell et al. 2007). In reality, the cumulative model of contemporary societies—not population growth itself—"has had more impact on biodiversity than any other single factor." The expansive, environmentally destructive, and socially exploitative dynamics of global capitalism, its energy system based on the burning of fossil fuels, and the new globalized food system centered on consumption of animal protein would continue to drive humanity and biodiversity to collapse, *even if the human population were reduced to less than half, that is, to its 1970 levels.* Population increase is, undoubtedly, an aggravating factor, but it is not the engine of this dynamic of collapse. As George Martine states (2016):

> Obviously the more people who are consuming, the faster the rate of degradation under the present system. Reducing population size is part of any long-term solution. Yet, it is disingenuous to cite population size and growth as the main culprit of environmental degradation or to suggest that family planning programs could provide a quick fix. (…) What really matters is who has access to "development", such as we know it. Of the 7.3 billion people currently on Earth, only a third can be minimally construed as "middle class" consumers and the remainder contribute marginally to insoluble global environmental threats.

There is a clear correspondence, for example, between environmental degradation and anthropogenic CO_2 emissions. These emissions do not increase simply as a result of population growth, but as a result of a per capita increase in emissions. Per capita anthropogenic CO_2 emissions did not cease to increase year after year in the period from 1950 to 2010. On average, each individual on the planet emitted more

© Springer Nature Switzerland AG 2020

L. Marques, *Capitalism and Environmental Collapse*,

https://doi.org/10.1007/978-3-030-47527-7_9

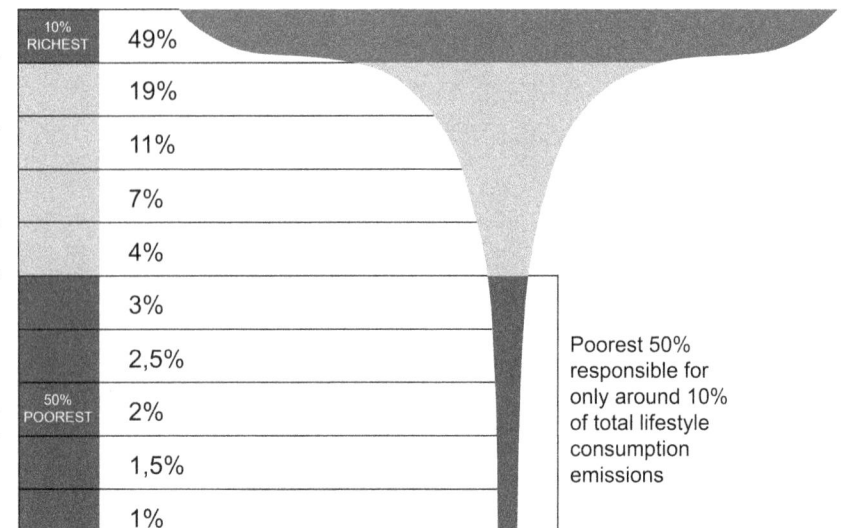

Fig. 9.1 Global income deciles and associated lifestyle consumption CO_2 emissions. (Source: Oxfam, "Extreme Carbon Inequality", 2/XII/2015)

than twice as much CO_2 as an individual in 1950, according to the following progression in million tons per capita: 1950, 0.64; 2001, 1.12; and 2010, 1.33 (Boden & Andres 2013). Moreover, the increase in per capita consumption should not obscure who is largely responsible for GHG emissions, the richest 10% of humanity, since these GHGs are emitted as a function of their consumption levels. Figure 9.1 shows how these emissions are distributed in the global income pyramid.

The richest 10% of humanity produce nearly half of anthropogenic CO_2 emissions, and the richest 30%, which George Martine classifies as "middle class" in the aforementioned quote, produce almost 80% of them, while the poorest half of the planet's population produce only 10%. Yangyang Xu and Veerabhadran Ramanathan (2017) bring more data to Oxfam's analysis. Focusing on the poorest three billion people, they find that:

Their contribution to CO_2 pollution is roughly 5% compared with the 50% contribution by the wealthiest 1 billion. This bottom 3 billion population comprises mostly subsistent farmers, whose livelihood will be severely impacted, if not destroyed, with a one- to five-year megadrought, heat waves, or heavy floods.

Hannah Ritchie and Max Rosen (2017, revised in 2019) also highlight this extreme inequality[1]:

The world's poorest have contribute less than 1% of emissions. (…) When aggregated in terms of income, (…) the richest half (high and upper-middle income countries) emit 86%

---

[1] I thank Dr. José Eustáquio Diniz Alves for having kindly called my attention to this reference.

of global CO_2 emissions. The bottom half (low and lower-middle income) only 14%. The very poorest countries (home to 9% of the global population) are responsible for just 0.5%. This provides a strong indication of the relative sensitivity of global emissions to income versus population. Even several billion additional people in low-income countries — where fertility rates and population growth is already highest — would leave global emissions almost unchanged. 3 or 4 billion low income individuals would only account for a few percent of global CO_2. At the other end of the distribution however, adding only one billion high income individuals would increase global emissions by almost one-third.

Furthermore, it is not overpopulation that is causing migratory waves from Central America toward the United States. Climate change, drought, an unsustainable agricultural model, and other environmental degradation factors are the real culprits for the growing agricultural crisis in these countries. All in all, the demographic factor plays only a secondary role in the causal link between globalized capitalism and the ongoing socio-environmental collapse. That said, population size is not an irrelevant factor, far from it. According to the 2019 UN *World Population Prospects* (WPP 2019):

> The global population is expected to reach 8.5 billion in 2030, 9.7 billion in 2050 and 10.9 billion in 2100, according to the medium-variant projection, which assumes a decline of fertility for countries where large families are still prevalent, a slight increase of fertility in several countries where women have fewer than two live births on average over a lifetime, and continued reductions in mortality at all ages.

We know, however, that there is inherent uncertainty in demographic projections. Thereby, it would suffice for the decline in current high-fertility countries to be a little slower than expected for demography to return to the forefront of the socioenvironmental crises. Moreover, even if population growth remains within the bounds of the medium-variant projection, the WPP 2019 concludes that "with a certainty of 95%, the size of the global population will stand between 8.5 and 8.6 billion in 2030, between 9.4 and 10.1 billion in 2050, and between 9.4 and 12.7 billion in 2100." The difference between the two ends of the projection for 2030 (100 million people) is not very relevant, since it only represents a difference of about 1.2% (between the highest and lowest predictions). However, by 2050, this uncertainty range jumps to 700 million people, a difference of about 7.5% relative to the lower value of 9.4 billion, and by 2100 there is a potential difference of 3.3 billion, that is, of about 35% between the lowest and highest projection. This is a wide gap. Just to measure its extent, imagine that by mid-2020 the planet was home to 11 billion humans and not the current 7.8 billion. In spite of these uncertainties, what is proposed here can be summarized in two points:

(1) 2019 data from the WPP confirms trends that have been well-detected for over a decade:

> The rate of population growth remains especially high in the group of 47 countries designated by the United Nations as least developed, including 32 countries in sub-Saharan Africa. With an average growth of 2.3% annually from 2015 to 2020, the total population of the least developed countries (LDCs) as a group is growing 2.5 times faster than the total population of the rest of the world.

According to 2019 data from the WPP, there is a population of 1.066 billion in sub-Saharan African countries, 1.37 billion in India, 271 million in Indonesia, and 204 million in Pakistan. Together, these countries are home to 2.911 billion people, or 39% of the planet's current human population, and their demographic growth rate is still very high. By 2050, the population of these countries is expected to reach 2.118 billion, 1.643 billion, 327 million, and 403 million, respectively. By 2050, therefore, these countries will have nearly 4.5 billion people or about 47% of the total estimated mid-century global population of 9.7 billion. Moreover, in keeping with their current trajectory, India, Indonesia, and a number of African countries will tend to lead the global capitalist accumulation (if there is no global environmental collapse by 2030, which is increasingly likely) and provide commodities for that accumulation, with increasing impacts on global ecosystems and on biodiversity. Economic globalization, coupled with a four-decade decline food self-sufficiency at the country level in a growing number of countries (Schramski et al. 2019[2]), as well as a diet increasingly based on meat, will lead many commodity-supplying countries, including Brazil and other Latin American and African countries, to destroy their forests—large carbon stockpiles and also rich in biodiversity—and to replace them with agriculture. It is not the population and local economy of tropical countries that drive the destruction of wildlife habitats, but the globalization of the food and energy system. Only a radical change in the animal-based protein diet and a dismantling of economic globalization have the power today to reduce or even reverse environmental degradation.

(2) Furthermore, waste and overconsumption of humanity's richest 30% must radically decrease so that the rising socioeconomic inequality can be reversed, safeguarding the ability of planetary ecosystems to support human population at current levels. Even if the most optimistic projections of rapid deceleration in population growth are confirmed, the planet's richest 30% will always have the greatest environmental impact—in absolute terms and per capita (measured by indicators such as the ecological footprint)—should we persist with the capitalist model of higher energy production and expansion of surplus and consumption. The capitalist supermarket cannot be universalized, even in an imaginary scenario of zero or negative demographic growth. As George Martine (2016) states, "under the current paradigm, it is simply absurd to imagine that the current living standards of the richer minority can be adopted by the entire world population — whether of 8 or 15 billion — without drastically overstepping planetary boundaries."

The crux of the demographic problem is not just knowing how much the human population will increase by 2050 and 2100, but, above all, what the economic system's impact on the biosphere and the climate system will be. And, not unlike the

---

[2] See Schramski et al. (2019): "The number of countries producing enough food to meet the caloric requirements of their populations decreased by 35%, from 101 to 66, over the 45-year period [1965–2010]; on average, curiously persistent, three countries fell into food production deficit every 4 years."

other socio-environmental crises, the magnitude of this impact will depend on societies' capacity for democratic governance, whence the title of this chapter.

The Ehrlich Formula I = PAT

Paul and Anne Ehrlich (1974/1990, p. 58) conceptualized—correctly, according to our understanding—the demographic impact on the biosphere:

> The impact of any human group on the environment can be usefully viewed as the product of three different factors. The first is the number of people. The second is some measure of the average person's consumption of resources (which is also an index of affluence). Finally, the product of those two factors–the population and its per-capita consumption–is multiplied by an index of the environmental disruptiveness of the technologies that provide the goods consumed. (…) In short, Impact = Population x Affluence x Technology, or I = PAT.

Paul Ehrlich and John Holdren (1971) wrote, always correctly in my view, that "the total negative impact" (I) of a given society on the environment "can be expressed, in the simplest terms, by the relation $I = P \times F$, where P is the population, and F is a function which measures the per capita impact." However, the authors overestimate the population factor. So, in the same article, they add: "The per capita consumption of energy and resources, and the associated per capita impact on the environment, are themselves functions of the population size. Our previous equation is more accurately written $I = P \times F(P)$, displaying the fact that impact can increase faster than linearly with population." This perception was understandable during the years of the publication of *The Population Bomb* (1968) and *The Population Explosion* (1974). In fact, during 1965–1970, the growth rate of the world's population was increasing by 2.1%. In 2015–2020, it is growing by less than 1.1% per year. It is, of course, still too much, but it is estimated that this growth rate will slow down by the end of the century.

The concerns brought forth by these classics by Paul and Anne Ehrlich began to prove themselves outdated in the last two decades of the twentieth century. Today, more than ever, it is the energy and food systems inseparable from economic globalization that are the true drivers of deforestation, loss of wildlife habitats, ongoing collapse of biodiversity, and pollution of soil, air, and water by emissions of particulate matter, pesticides, fertilizers, plastic waste, and other pollutants, as discussed in Chap. 4.

If we return to the equation I = PAT (Impact = Population × Affluence × Technology), it is true that the *stricto* sensu demographic pressure on ecosystems is getting worse because in many populous countries, the increase in population (P Index) is decelerating too slowly. But, of extreme importance is the fact that the two other factors are definitely not diminishing but growing rapidly.

The second environmental impact factor (Affluence)—measured by GHG emissions, consumption of energy and natural resources, production of commodities, and generation of waste—is increasing at a rapid pace. In many cases, technology, the third factor, is becoming even more destructive, such as in the production of unconventional oil and the even more destructive regression to coal (Chaps. 5 and 6). Although there is scientific and technological knowledge to significantly lower the T index, the choices made by the corporate–state network, which retains control of global investment flows, have not helped diminish this destructiveness. On the contrary, the observed trend, illustrated a thousand times in this book, is always the same: the less abundant natural resources become (fish schools, forests, soils, freshwater resources, liquid oil, and the potential for hydropower generation), the more invasive and destructive become the technologies used to obtain them.

9.1 Demographic Choices and Democracy: Reciprocal Conditioning

It is absolutely necessary to accelerate the demographic transition. But demographic transition is, above all, a function of democracy, without which societies will not be endowed with the five pillars of demographic rationality:

(1) A socioeconomic system that is environmentally friendly, understanding the economy as a subsystem of the biosphere
(2) Lower consumption, less waste, and less waste generation by the richest 30%, in order to reduce inequality in wealth and income
(3) Education for all, but especially for girls/women
(4) Female sexual freedom
(5) Secularism

Democracy is obviously the *conditio* sine qua non for the existence of these five pillars of demographic rationality. As is well known, items three to five are "spontaneous" promoters of a successful demographic transition. With regard to secularism, in particular, it must be stressed that, without democracy, societies will continue to suffer a blockade from the three great monotheistic religions on family planning, various contraceptive practices, and clinically assisted and state-guaranteed abortion. The belief in the existence of an immaterial entity, ontologically independent of the body, which would a priori constitute the individual—*nefesh* or *neshamah* in Judaism, *anima* in Christianity, and *nafs* in Islam—leads monotheistic religions to postulate that the fetus is already endowed with a supposed "essence" in the intrauterine phase before the formation of its central nervous system, anatomical structures, and physiology. Abortion, therefore, would be prohibited according to these faiths. Religious beliefs became constant in human imagination, possibly from the moment when awareness of one's own mortality became a central fact of our ancestors' awareness. The problem is not, of course, religiosity. The problem is the ability of religions to become a power system capable of obstructing protective legislation for women.

According to the Guttmacher Institute, pregnancy is often unwanted by women (Engelman 2013). Therefore, as long as women do not conquer, as they have done in some countries, the first and most elementary of democratic rights—the right over themselves, their bodies, their sexuality, and their procreation—and, in addition to this, as long as they do not have access to legal, medical, and financial means for family planning and abortion in the event of an unwanted pregnancy, all the UN's calculations on probabilities will be simple statistical exercises on the evolution of fertility rates, disregarding, at their own risk, a factor whose absence may distort all others: reproductive democracy. It should be feared, therefore, that Asia and Africa, continents that house 76% of the world's population, and in which religious extremism, theocracies, and state religions are spreading, may not be able to evolve into secularism. As a result, they will suffer from calamitous demographic increases for themselves and for the whole world.

That said, not only Asia and Africa but also the Americas, and notably Brazil and the United States, are victims of a strong offensive against secularism. In the pulpits and in the Brazilian National Congress, the Catholic, Protestant, and neo-Pentecostal churches are united when it comes to barring the right to state-assisted abortion. The same is true in the United States where the freedom to have an abortion in the first 3 months of pregnancy, stemming from a landmark 1973 Supreme Court ruling (*Roe v. Wade*), is eroding, with brutal setbacks leading to the criminalization of abortion in several states.

This is not to say that demography is a simple function of democracy, because the opposite is also true. The capacity for democratic governance also depends, to a large extent, on the pace and scale of population growth, since fertility rates above the replacement level hinder a minimum of political stability in the medium and long term, nullify government efforts for better education and sanitation, and leave the population at the mercy of religious obscurantism.

9.2 Beyond the Arithmetic Addition: Urbanization, Tourism, and Automobiles

Returning to the Ehrlich formula ($I = PAT$), the environmental impact of population growth is strongly conditioned by Affluence (A), that is, the average per capita consumption multiplied by the environmental destruction index of technologies (T) that provide energy, production assets, and consumer goods. Thus, for various reasons (GHGs, waste production, use or consumption of energy, water, soils, meat, minerals, wood, etc.), the environmental impact of an American or an European is, on average, obviously much greater than that of an African, Asian, or Latin American who does not belong to the economic elite.

The association of the affluence index with the phenomenon of intense urbanization is a supplementary factor of anthropic pressure, since the urban footprint is larger than that of the population as a whole. According to the UN's *World Urbanization Prospects* (*The 2003 Revision*), the urban population reached one billion in 1960, two billion in 1985, and three billion in 2002, and it should reach five billion in 2030. According to a 2014 review of the World Urbanization Prospects from the United Nations Population Division, the world's urban population is expected to exceed six billion by 2045. In many cases, the urbanization process is extreme, with the formation of gigantic urban and suburban patches that further enhance the environmental impact, especially in the new megacities in poor and infrastructure-deficient countries. In 1950, New York and Tokyo were the only cities with more than ten million inhabitants. In 1990, there were ten megacities with more than ten million inhabitants. In 2012, there were 23 megacities of this caliber, four of which were in China. In 2014, there were 28 megacities in the world, 16 of which were in Asia, 4 in Latin America, 3 in Africa, 3 in Europe, and 2 in North America. By 2025, there will be 37 megacities with more than ten million inhabitants in the world, seven of them in China. By 2030, there will probably be 41 megacities with ten million inhabitants or more, with the 9 largest located in Asia and Africa (in descending order, Tokyo, Delhi, Shanghai, Mumbai, Beijing, Dhaka, Karachi, Cairo, and Lagos).

This process of mega-urbanization is spontaneous and seemingly inexorable within today's state–corporations, committed to the dynamics of the global market and unable to carry out an agenda of urban planning and decentralization. This urbanization is sometimes even encouraged by governments. In China, for example, between 1982 and 2012, the urban population went from 200 million to over 700 million. In the next 15 years, another 300 million Chinese, equivalent to the US population, will migrate to the cities.

Far from pondering the negative effects of this process, the Chinese government plans to accelerate it by merging nine major cities in the Pearl River Delta in the South of the country into a single urban sprawl of 50 million. This urbanization was considered by the Chinese leaders and their 5-year plan (2011–2015) as the "essential engine" of economic growth. Thus, in 2010, there were 94 cities in China with over one million inhabitants. And according to Beijing's plans, by 2025, there will be 143 cities of this scale. According to Peiyue Li, Hui Qian, and Jianhua Wu (2014), in Lanzhou:

> 700 mountains are being levelled to create more than 250 km^2 of flat land. (…) Land-creation projects are already causing air and water pollution, soil erosion and geological hazards such as subsidence. They destroy forests and farmlands and endanger wild animals and plants.

Now, simply because they will be concentrated predominantly in these consumer *hotspots*—cities of more than one million inhabitants or megacities of more than ten million—the additional two billion people who will be added around 2043, or even earlier, will tend to produce more heat radiation, more air pollution, more municipal solid waste, more industrial waste, more CO_2, more methane, and more tropospheric ozone and will consume more energy and natural resources per capita than the two billion people that were added to humanity between 1987 and 2012.

9.2.1 Tourism

The tourism industry, among the largest in the world, promotes increasing pressure on the environment, as highlighted by UNEP: deforestation and degradation of forests and soils, loss of natural habitats, increased pressure on native species and increase in invasive species, increased forest vulnerability to fire, water scarcity, more land and sea pollution, and increased greenhouse gas emissions from more travel. The aviation sector accounts for more than 1 GtCO$_2$, or about 2% of global emissions per year, and tourism now accounts for 60% of air transport (UNEP). Emissions from this sector are expected to triple by 2050 or more than double if planes become more fuel-efficient. According to the World Tourism Organization (UNWTO), in 1995, 540 million tourists traveled outside their countries. In 2010, that number reached 940 million. In 2018, international tourist arrivals worldwide reach 1.4 billion, a mark which was reached 2 years ahead of UNWTO's long-term forecast issued in 2010.[3] In 2000, the number of Chinese tourists traveling the world was ten million, and in 2013, it rose to 98.19 million. In the first 11 months of 2014 alone, this number reached 100 million. The China National Tourism Administration informed that "Chinese tourists traveled overseas on 131 million occasions in 2017, an increase of 7% from the previous year,"[4] and the *China Daily* (5/VIII/2019) reported that Chinese tourists made 149 million overseas trips in 2018.

9.2.2 Automotive Vehicles

A third classic example of how the factors "Affluence" and "Technological Destructiveness" enhance demographic impact is the amount and the per capita increase of oil-powered vehicles in the world. Table 9.1 gives a picture of this evolution:

In 40 years (1970–2010), the number of vehicles in operation (cars and light and heavy commercial vehicles) more than quadrupled, while the population did not double. In the next 10 years (2021–2030), it is estimated that this fleet will reach two billion vehicles, for a population that is about 20% greater than in 2010. The auto industry resumed its expansion globally since 2009 and, in Europe, since 2014. Global sales of the automotive industry have maintained a rate of between 77 and 79 million vehicles per year since 2016, as shown in Fig. 9.2:

Table 9.2 shows the number of motor vehicles by country between 2015 and 2019.

---

[3] See UNWTO, World Tourism Barometer, 17, 1, January 2019.

[4] See Nielsen Holdings, "2017 Outbound Chinese Tourism and Consumption Trends."
    https://www.nielsen.com/wp-content/uploads/sites/3/2019/05/outbound-chinese-tourism-and-consumption-trends.pdf

Table 9.1 Increase of oil-powered vehicles in the world (1950–2030) compared with world population

Year	Millions of units	World population in billions
1970	250	3.6
2010	1.015	6.9
2030	2.000	8.5

Sources: (1) *Wardsauto*; (2) *International Transport Forum. Meeting the Needs of 9 Billion People.* OECD, 2011 and (3) Deborah Gordon and Daniel Sperling, "Surviving Two Billion Cars". *Environmental360*

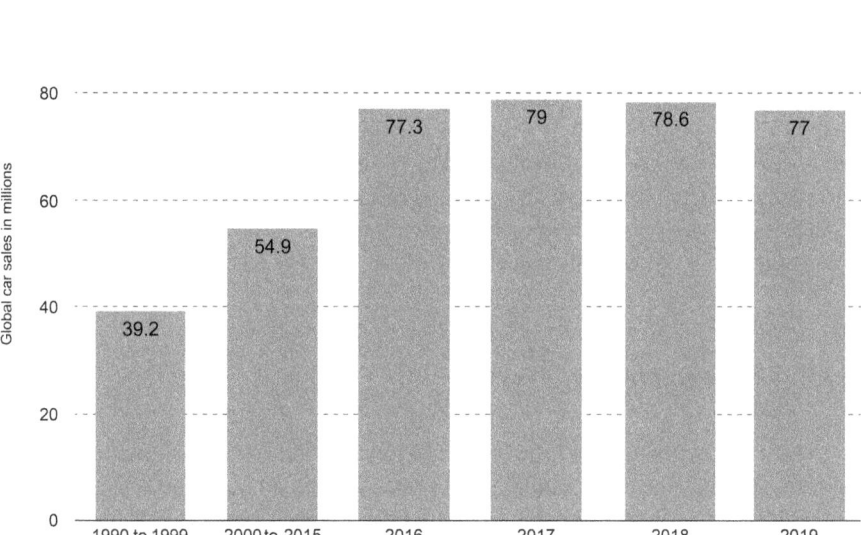

Fig. 9.2 Number of cars sold worldwide from 1990 to 2019 (in million units). (Source: http://www.statista.com/statistics/200002/international-car-sales-since-1990/)

Table 9.2 Number of motor vehicles by country between 2015 and 2019

Country	Motor vehicles per 1000 people
United States	811 (2017)
Italy	695 (2015)
Spain	591 (2015)
Germany	589 (2016)
United Kingdom	579 (2016)
France	569 (2016)
Brazil	350 (2019)
China	179 (2019)
India	22 (2015)

Source: List of Countries by Vehicle per capita (Wikipedia)

Furthermore, the increase in consumption of oil for transport should continue to increase in the foreseeable future. According to the EIA's International Energy Outlook 2017:

> Because of the increase in electric vehicle penetration, the share of petroleum-based fuel for light-duty vehicle use decreases from 98% in 2015, to 90% by 2040. But, liquid fuels consumption is still expected to increase by almost 20% between 2015 and 2040 as more petroleum-based cars are still being added to the stock and other uses of liquid fuels continue to grow.

The projected increase in this fleet obviously depends on the supply elasticity of oil, gas, ethanol, and batteries for electric vehicles. According to Daniel Sperling (2010), maintaining current fuel consumption conditions per kilometer driven, a fleet of two billion vehicles—which should be reached by 2030—would consume 120 million barrels of oil per day, almost 20% more than today's *total* daily oil consumption.

9.3 The Destructiveness of Technology (the T Index)

The transport sector, in general, includes products that are derived from very destructive technology, whether it be the materials used, the energy consumed, or its direct contribution to global CO_2 emissions. According to the Transport and Climate Change Global Status Report (2018), produced by the Partnership on Sustainable, Low Carbon Transport (SLoCaT), "transport sector CO_2 direct emissions increased 29% (from 5.8 Gt to 7.5 Gt) between 2000 and 2016, at which point transport produced about 23% of global energy-related CO_2 emissions, and (as of 2014) 14% of global GHG emissions." These numbers will increase not only because of the expected increase in fossil fuel-powered vehicles over the next decade but also because of the higher proportion of fuel (particularly in the United States) coming from unconventional oil, whose extraction emits more CO_2. According to data from SLoCaT 2018, GHG emissions from rail transport accounted for only 3% of the total emissions in 2015, while emissions from individual automobiles and long-distance transport for tourism and trade accounted for 88% of global CO_2 emissions in the transport sector, as shown in Table 9.3.

The decarbonization of technologies in the transportation, cement, and steel industries is extremely difficult if we maintain the paradigms of economic growth and of today's globalized economy. As Steven Davis et al. (2018) affirm, "these difficult-to-decarbonize energy services include aviation, long-distance transport, and shipping; production of carbon-intensive structural materials such as steel and cement; and provision of a reliable electricity supply that meets varying demand."

Table 9.3 Global CO_2 emissions in the transport sector (in %)

Cars (light-duty vehicles)	45%
Trucks	21%
Aviation	11%
Shipping	11%

According to estimates from the OECD International Transport Forum, by 2050, CO_2 emissions from oil-powered vehicles can multiply 2.5–3 times relative to 2000 (with likely improvements in efficiency during that period already included in this estimate).

This set of findings and projections brings us back to the same juncture that at once motivates and runs through this book: either societies advance rapidly to overcome capitalism—and by overcoming capitalism we mean, and repeat exhaustively, to radically redefine man's position in the biosphere and deepen democracy (starting with reproductive democracy)—or the factors that make up the demographic impact on natural resources and ecosystems ($I = PAT$) will, together, overcome capitalism in their own way, that is, through social and environmental collapse.

9.3.1   A Fragile Premise

The realization of any of the demographic impact scenarios discussed above depends, of course, on an implicit assumption that "the other variables" remain relatively unchanged. The cumulative apparatus of global capitalism is already leading to a collapse of water resources, soils, biodiversity, and ecosystems in general, as well as ever more severe changes in climate coordinates. Occurring in synergy, such phenomena will imply more or less brutal demographic contractions, squarely contradicting UN Population Division projections which are based mainly on variations in fertility rates. This methodological condition of a ceteris paribus—that is, the condition that ecosystems remain functional and societies relatively organized—is increasingly questionable.

References

BODEN, Tom & ANDRES, Bob. "Global CO_2 Emissions from Fossil-Fuel Burning, Cement Manufacture, and Gas Flaring: 1751–2010". Carbon Dioxide Information Analysis Center. *Oak Ridge National Laboratory*, 30/VII/2013.

CAMPBELL, Martha et al., "Public Health. Return of the Population Growth Factor". *Science*, 315c 5818, March 2007.

DAVIS, Steven et al., "Net-zero emissions energy systems". *Science*, 360, 6396, 29/VI/2018.

EHRLICH, Paul R. *The Population Bomb*. Sierra Club Ballantine Books, 1968.

EHRLICH, Paul R. & EHRLICH, Anne H. *The Population Explosion*. New York, Simon and Schuster (1974), 1990.

EHRLICH, Paul R. & HOLDREN, John Paul. "Impact of population growth". *Science*, 171, 1971, pp. 1.212–1.217.

ENGELMAN, Robert, "Our Overcrowded Planet: A Failure of Family Planning". *Yale environment360*, 24/VI/2013.

LI, Peiyue; QIAN, Hui & WU, Jianhua. "Accelerate research on land creation". *Nature*, 7503, 510, 2014, pp. 29–31.

MARTINE, George, "Sustainability and the missing links in global governance". *News of the International Union for the Scientific Study of Population* (N-IUSSP), 14/III/2016.

MARTINE, George & ALVES, José Eustáquio Diniz, "Disarray in global governance and climate change chaos". *Revista Brasileira de Estudos Populacionais*, 36, 2019, pp. 1–30.

RITCHIE, Hannah & ROSER, Max, "CO_2 and Greenhouse Gas Emissions". *Our World in Data*, May 2017, last revised in December 2019.

SCHRAMSKI, John R. et al. "Declining Country-Level Food Self-Sufficiency Suggests Future Food Insecurities". *BioPhysical Economics and Resource Quality*, 4, 12, 2019.

SPERLING, Daniel, *Two Billion Cars, Transforming Transportation*. Chicago, 2010.

XU, Yangyang & RAMANATHAN, Veerabhadran, "Well below 2°C: Mitigation strategies for avoiding dangerous to catastrophic climate changes". *PNAS*, 14/IX/2017.

Chapter 10
Collapse of Terrestrial Biodiversity

Numerous scholars from various fields of science today are concerned with the ongoing collapse of biodiversity. The first Global Assessment of the Intergovernmental Science-Policy Platform on Biodiversity and Ecosystem Services (IPBES),[1] published in 2019, estimates that:

> The rate of global change in nature during the past 50 years is unprecedented in human history (…) Human actions threaten more species with global extinction now than ever before. (…) An average of around 25% of species in assessed animal and plant groups are threatened, suggesting that around 1 million species already face extinction, many within decades, unless action is taken to reduce the intensity of drivers of biodiversity loss.

Societies' very survival depends on their ability to avert the impending threat of biological annihilation via the ongoing sixth mass extinction of species, triggered or intensified by the globalization of capitalism over the last 50 years. Sir Robert Watson, Chair of IPBES (2016), doesn't mince his words to say what is at stake: "We are eroding the very foundations of our economies, livelihoods, food security, health and quality of life worldwide." There is no hyperbole in the claim that the collapse of biodiversity and the acceleration of global warming, two processes that interact in synergy, entail an increasing risk of extinction for the *Homo sapiens*. As pointed out by Cristiana Paşca Palmer, Executive Secretary of the Convention on Biodiversity (2018), "I hope we aren't the first species to document our own extinction." Julia Marton-Lefèvre, former Director General of the International Union for Conservation of Nature (IUCN), reiterates this warning for the umpteenth time in a statement to delegations meeting at Rio+20 in 2012:

> Sustainability is a matter of life and death for people on the planet. A sustainable future cannot be achieved without conserving biological diversity—animal and plant species, their habitats and their genes—not only for nature itself, but also for all 7 billion people who depend on it.

---

[1] The IPBES was established in 2010 by the UN to improve the interface between science and policy on issues of biodiversity and ecosystem services.

© Springer Nature Switzerland AG 2020
L. Marques, *Capitalism and Environmental Collapse*,
https://doi.org/10.1007/978-3-030-47527-7_10

10.1   Defaunation and Biological Annihilation

Rodolfo Dirzo, Mauro Galetti, Ben Collen, and other co-authors of a review titled "Defaunation in the Anthropocene" (2014) conceptualize one of the central aspects of the current sixth mass extinction of species: the term defaunation is used to denote the loss of both species and populations of wildlife, as well as local declines in abundance of individuals. The defaunation process is in full swing:

> In the past 500 years, humans have triggered a wave of extinction, threat, and local population declines that may be comparable in both rate and magnitude with the five previous mass extinctions of Earth's history. Similar to other mass extinction events, the effects of this "sixth extinction wave" extend across taxonomic groups, but they are also selective, with some taxonomic groups and regions being particularly affected. (…) So profound is this problem that we have applied the term "defaunation" to describe it.

In a 2017 article, Gerardo Ceballos, Paul Ehrlich, and, again, Rodolfo Dirzo warn about the false impression that the threat of biological annihilation is not imminent:

> The strong focus on species extinctions, a critical aspect of the contemporary pulse of biological extinction, leads to a common misimpression that Earth's biota is not immediately threatened, just slowly entering an episode of major biodiversity loss. This view overlooks the current trends of population declines and extinctions. Using a sample of 27,600 terrestrial vertebrate species, and a more detailed analysis of 177 mammal species, we show the extremely high degree of population decay in vertebrates, even in common "species of low concern." Dwindling population sizes and range shrinkages amount to a massive anthropogenic erosion of biodiversity and of the ecosystem services essential to civilization. This "biological annihilation" underlines the seriousness for humanity of Earth's ongoing sixth mass extinction event.

10.2   The 1992 Convention on Biological Biodiversity

This process of mass extinction of plant and animal species has accelerated despite diplomatic and other efforts. In 1992, 1 day after the United Nations Framework Convention on Climate Change (UNFCCC) was signed, 194 states[2] signed the Convention on Biological Diversity (CBD). In this robust 30-page document, the signatories solemnly affirm "that the conservation of biological diversity is a common concern of humankind," declare to be "concerned that biological diversity is being significantly reduced by certain human activities," and claim they are "determined to conserve and sustainably use biological diversity for the benefit of present and future generations." In contrast to these moving intentions, this is the reality: while opening the third edition of the UN Global Biodiversity Outlook (GBO-3) in 2010, Ban Ki-moon, then UN Secretary-General, highlighted the accelerating decline in biodiversity in the first decade of the twenty-first century—"the principal

---

[2] See Convention on Biological Diversity. http://www.cbd.int/convention/parties/list/. The USA is the only country that has not ratified this treaty.

pressures leading to biodiversity loss are not just constant but are, in some cases, intensifying." This third report was presented at the Nagoya meeting (COP10) of the Convention on Biological Diversity. In this meeting, nearly 200 parties subscribed to the Strategic Plan for Biodiversity (the so-called Aichi Biodiversity Targets), establishing 20 Targets (subdivided into 56 goals, grouped under 5 strategic goals) for the conservation of biodiversity between 2011 and 2020, with successive stages until 2050. After the first 4 years of this first decade, the fourth edition of the UN Global Biodiversity Outlook (GBO-4), presented in October 2014 at CBD COP 12 in South Korea, admits that "pressures on biodiversity will continue to increase at least until 2020, and that the status of biodiversity will continue to decline."[3] With regard to Target 12 (Reducing risk of extinction), the report concludes that:

> Multiple lines of evidence give high confidence that based on our current trajectory, this target would not be met by 2020, as the trend towards greater extinction risk for several taxonomic groups has not decelerated since 2010. Despite individual success stories, the average risk of extinction for birds, mammals, amphibians and corals shows no sign of decreasing.

Using 55 indicator datasets, Derek Tittensor and 50 colleagues (2014) reiterate this same assessment of progress on the Aichi Targets: "On current trajectories, results suggest that despite accelerating policy and management responses to the biodiversity crisis, the impacts of these efforts are unlikely to be reflected in improved trends in the state of biodiversity by 2020." For Richard Gregory, one of the authors of this report, "world leaders are currently grappling with many crises affecting our future. But this study shows there is a collective failure to address the loss of biodiversity, which is arguably one of the greatest crises facing humanity" (Vaughan 2014). In December 2016, the 13th meeting of the CBD in Cancun, Mexico, revealed the miserable failure of the plans established by signatory countries to achieve the Aichi Targets by 2020. No less than 90% of these targets, including halting habitat loss, saving endangered species, reducing pollution, and making fishing and agriculture more sustainable, was then expected to be unmet by 2020 (Coghlan 2018).

Having failed to honor their Aichi commitments, countries will meet in October 2020, in Kunming, China, for the 15th Conference of the Parties (CBD COP15). A new landmark global pact, titled 2050 CBD Vision "Living in harmony with nature", is the expected outcome of this meeting. Unfortunately, according to Friends of the Earth, the draft that will be discussed in China is not ambitious enough. It fails to recognize the negative impacts of monoculture agribusiness and pesticides, nor does it call for divestment from other destructive projects:[4]

> Corporations have a vested interest in avoiding strict regulation and any attempts to scale down their profit-driven activities. As long as they have a seat at the negotiation table, no measures will be taken to live within planetary boundaries. Yet instead of seeking to reduce corporate conflicts of interest—a controversial issue in the CBD—the Draft repeatedly

---

[3] See *Global Biodiversity Outlook 4*, 2014, p. 87 e p. 10.
[4] See "Time to tackle biodiversity loss: Draft post-2020 UN Framework not ambitious enough." Friends of the Earth, 17/I/2010.

promotes closer collaboration with the private sector, and states that increased production will be necessary.

10.3 The Biodiversity of the Holocene

There is no unanimity on the number of species that make up today's biodiversity. In 1988, Robert M. May suggested—on the assumption that the number of species increased inversely to the proportion of their size—that there were an estimated 10 million to 50 million terrestrial species. He stressed, however, the uncertainties about this, especially from a theoretical point of view. In 1996, Richard Leakey and Roger Lewin considered that the planet hosted perhaps 50 million species. For James P. Collins and Martha L. Crump (2009), the number of species of organisms living on Earth today would range from 10 million to 100 million, and Richard Monastersky (2014) accepts estimates—that include animal, plant, and fungal species—oscillating between 2 million and over 50 million.

When it comes to the domain of microorganisms, which control major nutrient cycles and directly influence plant, animal, and human health, we enter further into terra incognita. As Diana R. Nemergut and co-authors (2010) stated in a paper on bacterial biogeography, "we know relatively little about the forces shaping their large-scale ecological ranges." Thus, if global biodiversity quantification estimates include the domains of bacteria and archaea, our planet could contain nearly one trillion species (Kenneth and Lennona 2016). That said, at least for eukaryotic species (organisms whose cells have a nucleus enclosed within membranes), most estimates have shrunk to less than 10 million species. According to Nigel E. Stork et al. (2015), there are about 1.5 million beetle species (estimates ranging from 0.9 to 2.1 million), 5.5 million insect species (2.6–7.8 million), and 6.8 million species of terrestrial arthropods (range 5.9–7.8 million), "which suggest that estimates for the world's insects and their relatives are narrowing considerably." Camilo Mora and his team (2011) have proposed a new way of estimating the number of species. According to them:

> the higher taxonomic classification of species (i.e., the assignment of species to phylum, class, order, family, and genus) follows a consistent and predictable pattern from which the total number of species in a taxonomic group can be estimated. This approach was validated against well-known taxa, and when applied to all domains of life, it predicts ~8.7 million (±1.3 million Standard Error) eukaryotic species globally, of which ~2.2 million (±0.18 million SE) are marine.

This study elicited diverse reactions, including criticism of the methodology used, which would be unable to account for a global biodiversity. But Sina Adl, one of the co-authors of this work, is the first to admit that "the estimate in the manuscript is a gross under-estimate" (as cited in Mathiesen 2013). In any case, a consensus seems to be outlined regarding the magnitude of global biodiversity, at least in regard to eukaryotic species. Thus, Rodolfo Dirzo, Mauro Galetti, and Ben Collen (2014) implicitly consolidate this proposal by stating: "of a conservatively estimated 5 million to 9 million animal species on the planet, we are likely losing ~11,000 to 58,000 species annually." IPBES also works with this scale. As stated by Andy Purvis (2019), "in the [Global] Assessment, we have used a recent mid-low estimate of 8.1 million animal and plant species, of which an estimated 5.5 million are insects (i.e., 75%) and 2.6 million are not."

10.4   The Sixth Extinction

Species extinction is an inherent fact of evolution; "some thirty billion species are estimated to have lived since multicellular creatures first evolved, in the Cambrian explosion" (Leakey and Lewin 1996), which gives us an idea of their transience, since the number of multicellular species existing today does not go beyond the millions, as we have seen. Therefore, at least 99% of all species that have ever existed are extinct. This is mainly due to the large mass extinctions of species that abruptly disrupted the increasing trend of biodiversity. One can speak of mass extinctions when the Earth loses more than three-quarters of its species in a geologically short interval (Barnosky et al. 2011). Five major extinctions fall into this category: (1) in the end of the Ordovician period (440 million years ago), (2) in the Late Devonian Period (365 million years ago), (3) in the Permian–Triassic Period (251 million years ago), (4) in the end of the Triassic Period (210 million years ago), and (5) in the end of the Cretaceous Period (65 million years ago). Jun-xuan Fan and colleagues (2010) showed no evidence of a discrete mass extinction near the end of the Devonian period, the so-called the Late Devonian Frasnian–Famennian extinction. There was instead "a protracted diversity decline that began in the Eifelian (392.72 Ma) and continued to ~368.24 Ma with no subsequent major recovery." The fifth extinction (or the fourth one, according to this study) opened the Cenozoic Era, generally called the Age of Mammals, but one that might more appropriately be called the Age of Arthropods, according to Richard Leakey and Roger Lewin (1996), since they are the largest phylum in existence, encompassing over 80% of known animal species.

In the late 1970s, a team of researchers led by Luis Alvarez of the University of California advanced the now widely accepted hypothesis that the impact of a large asteroid on Mexico's Yucatan Peninsula (or of an asteroid rain) unleashed or dealt the final blow to the chain of events known as the fifth extinction. The hypothesis that external factors also caused other extinctions has gained traction since 1984, when David Raup and Joseph John Sepkoski from the University of Chicago propounded the occurrence of about 12 extinction events (including the five largest), with a mean interval between events of 26 million years. The statistical analysis advanced by these scholars leads to the conclusion that 60% of all extinctions throughout the Phanerozoic Eon were at least triggered (if not caused) by the impact of asteroids or comets, which would act as first strikes, making the biotas vulnerable to other debilitating and destructive processes (Leakey and Lewin 1996).

Whatever may be the current number of existing eukaryotic species—around eight million now being the most accepted number—since the 1980s, many scientists have been warning that the Earth is in the midst of a massive biodiversity extinction crisis caused by human activities, a process called the sixth extinction. This threatens to be as or even more annihilating than the previous five, given three characteristics that are peculiar to it:

(1) It is not triggered by an exceptional, external *event*, but by a *process internal* to the biosphere—the increasing destructiveness of our globalized economic activity—a conscious, announced, increasingly well-known, and hitherto unstoppable process. The dynamics of the sixth extinction is not like the irradiation of waves caused by the impact of a stone on the surface of water, which tends to wane as its range extends into space and time; it is a process that is intensified in direct relation to the expansion of the commodity market, especially after 1950 and in the ensuing wave of capitalism's extreme globalization, causing pollution, degradation, and destruction of non-anthropized habitats. Figure 10.1 clearly shows the brutal acceleration of species extinctions from the 1980s onward, closely related to increase in human population, following the degradation of natural habitats by the expansion of global agribusiness and other economic activities.

(2) The second characteristic is that, far from implying the dominance of one taxonomic group over the others, the sixth extinction endangers the allegedly "dominant" species by destroying the web of biological sustenance that allows it to exist (Jones 2011). We will return to this point in the last chapter as we analyze

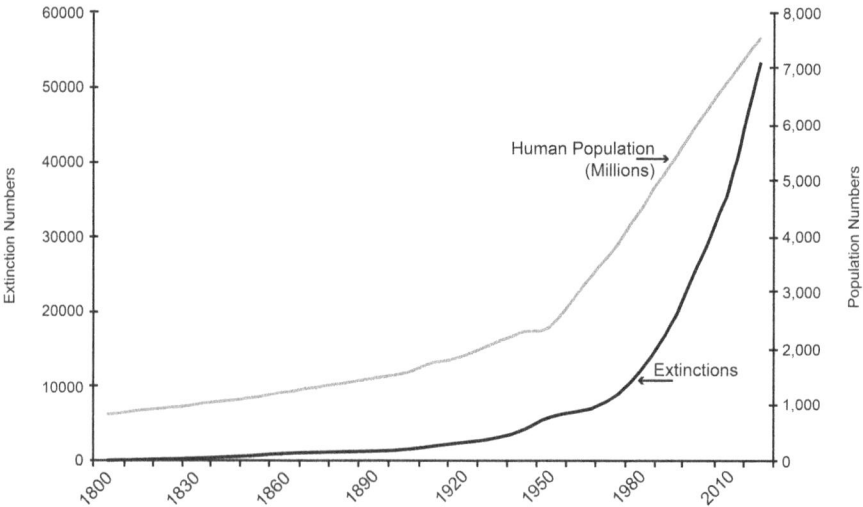

Fig. 10.1 Population and species extinctions. Source: Scott, J.M. 2008. *Threats to Biological Diversity: Global, Continental, Local*. US Geological Survey, Idaho Cooperative Fish and Wildlife, Research Unit, University of Idaho

the illusions of anthropocentrism, but it is clear that the web of life that sustains our global society and, ultimately, our species is becoming weaker and weaker, as a multitude of scientists and diplomats never cease to repeat. In 2011, prefacing the *Global Biodiversity Outlook 3* report, Achim Steiner, former Executive Director of the UNEP (2006–2016), warned of this compelling fact: "The arrogance of humanity is that somehow we imagine we can get by without biodiversity or that it is somehow peripheral: the truth is we need it more than ever on a planet of six billion heading to over nine billion people by 2050."

(3) The third characteristic of the sixth extinction is its overwhelming speed. The current rate of extinction is speeding up. As pointed out by Andy Purvis (2019), "while it is true that the very highest rate of *confirmed* extinctions was in the late 19th century, as flightless island birds fell prey to cats and rats introduced by Europeans, the rate was as high again by the end of the 20th century." And over the past two decades, it has accelerated even further. Speed is perhaps the most important feature of the sixth extinction because this almost sudden characteristic suppresses a crucial evolutionary variable: the time it takes for species to adapt to and survive environmental changes. Based on extrapolations from a sample of 200 snail species, Claire Régnier and colleagues (2015) showed that we may have already lost 7% of the previously described nonmarine animal species. In other words, 130,000 of these nonmarine species may have been lost forever since their taxonomic record (see also Pearce 2015). Unlike the previous ones, the sixth extinction is not measurable on a geological scale, but on a historical scale, and the unit of time by which this scale is measured is shortening. In 1900, it occurred on the scale of centuries. Fifty years ago, the most appropriate observation scale would be the decade. Today, the unit of measurement for the advancement of the sixth extinction is the year or even the day. In June 2010, UNEP's *The State of the Planet's Biodiversity* document estimated that 150 to 200 species become extinct every 24 hours. This speed of the sixth extinction—I must repeat—cannot be compared to the previous five mass extinctions. More than 20 years ago, Peter M. Vitousek et al. (1997), reporting on then-recent calculations, stated that "rates of species extinction are now on the order of 100 to 1000 times those before humanity's dominance of Earth." According to a global survey of plant extinctions, "since 1900, nearly 3 species of seed-bearing plants have disappeared per year, which is up to 500 times faster than they would naturally" (Ledford 2019). In 2005, the *Ecosystems and Human Well-being: Synthesis* (2005), a report established by the Millennium Ecosystem Assessment, stated:

> The rate of known extinctions of species in the past century is roughly 50–500 times greater than the extinction rate calculated from the fossil record of 0.1–1 extinctions per 1,000 species per 1,000 years. The rate is up to 1,000 times higher than the background extinction rates if possibly extinct species are included.

And the same document estimates that "projected future extinction rate is more than ten times higher than current rate." In line with this projection, Chris Thomas et al. (2004) "predict, on the basis of mid-range climate-warming scenarios for 2050, that 15–37% of species in our sample of regions and taxa will be committed to extinction."

10.5   The IUCN Red List of Threatened Species

The number species threatened with extinction has grown rapidly over the past 50 years, as the International Union for Conservation of Nature (IUCN) records show. Today, this assessment divides species into nine categories: Not Evaluated, Data Deficient, Least Concern, Near Threatened, Vulnerable, Endangered, Critically Endangered, Extinct in the Wild, and Extinct. In 1964, the first IUCN's Red List of Threatened Species was published, which periodically lists the species evaluated as Critically Endangered, Endangered, or Vulnerable.[5] Table 10.1 gives us an idea of the speed of the process of species extinction that is underway.

To date, more than 105,700 species have been assessed for the IUCN Red List, and more than 28,000 species are threatened with extinction (27%). Between 2000 and 2013, with the increase in the number of species evaluated having quadrupled, the number of threatened species nearly doubled, from just over 10,000 in 2000 to 21,286 in 2013, as shown in Fig. 10.2.

Table 10.2 gives a picture of the evolution of the relationship between the evaluated species and those threatened with extinction over the last 10 years.

In the 2015 assessment, 14 species were shifted from the "Endangered" category to the "Critically Endangered (Possibly Extinct)" category.

Table 10.1 Threatened species among mammals, amphibians, and birds in 2009 and in 2019 (Threatened Species = Vulnerable, Endangered, or Critically Endangered)

	2009 (%)	2019 (%)
Mammals	21	25
Amphibians	30	40
Birds	12	14

Sources: IUCN. http://www.iucnredlist.org/about/summary-statistics#How_many_threatened; https://www.iucnredlist.org

---

[5] A species is "Critically Endangered" when (a) it has less than 250 mature individuals, or (b) it has lost at least 90% of its population for more than 10 years or for three generations (whichever is longer), if this reduction is considered reversible, or (c) it has lost 80% if such loss—observed or projected—is considered irreversible, or (d) its geographic extent is too small or fragmented, or (e) if a quantitative analysis shows that the probability of extinction in the wild is at least 50% within 10 years or three generations (whichever is longer). See IUCN Red List Categories and Criteria
http://www.iucnredlist.org/technical-documents/categories-and-criteria/2001-categories-criteria#critical

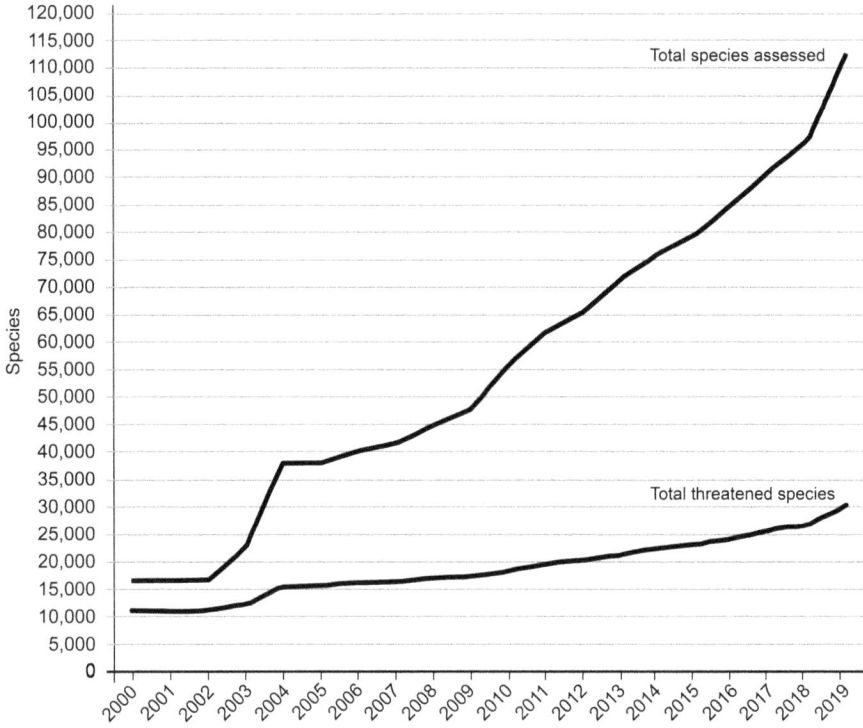

Fig. 10.2 Assessed species and threatened species (2000 and 2013). (Source: UICN http://www.iucnredlist.org/about/summary-statistics#Fig_1)

	Species assessed	Threatened species
2009	47,677	17,291
2012	63,837	19,817
2013	71,576	21,286
2014	73,686	22,103
2015	77,340	22,784
2019	105,732	>28,000

Table 10.2 Evolution of the relationship between the evaluated species and those threatened with extinction over the last 10 years

10.6   The Living Planet Index (LPI) and the Decline in Terrestrial Vertebrate Populations

In the important 2017 study, quoted at the beginning of this chapter, Gerardo Ceballos, Paul Ehrlich, and Rodolfo Dirzo emphasize the global extent and speed of the recent decline in vertebrates:

Considering all land vertebrates, our spatially explicit analyses indicate a massive pulse of population losses, with a global epidemic of species declines. Those analyses support the

view that the decay of vertebrate animal life is widespread geographically, crosses phyloge-
netic lineages, and involves species ranging in abundance from common to rare. (…) In the
last few decades, habitat loss, overexploitation, invasive organisms, pollution, toxification,
and more recently climate disruption, as well as the interactions among these factors, have
led to the catastrophic declines in both the numbers and sizes of populations of both common and rare vertebrate species. For example, several species of mammals that were relatively safe one or two decades ago are now endangered.

In this paper, the authors' objective differs from that of the IUCN, which is mainly
concerned with risks of extinction and, therefore, with species whose populations
are already scarce or rare and that have less territorial distribution. Unlike the IUCN,
the three authors cited here also take into account the decline in "common" or abundant species. Thus, they propose a picture of the present process of "biological
annihilation" of terrestrial vertebrates, perhaps even more dramatic than the one
proposed by the IUCN's Red List. Starting from the IUCN assessments, they divide
this picture into two large groups. On one side, there are the species placed by the
IUCN under the category "Low Concern" or "Data Deficient." On the other side,
there are the "Threatened Species," according to their various threat levels
("Critically Endangered," "Endangered," "Vulnerable," and "Near Threatened").
The result is captured in Fig. 10.3.

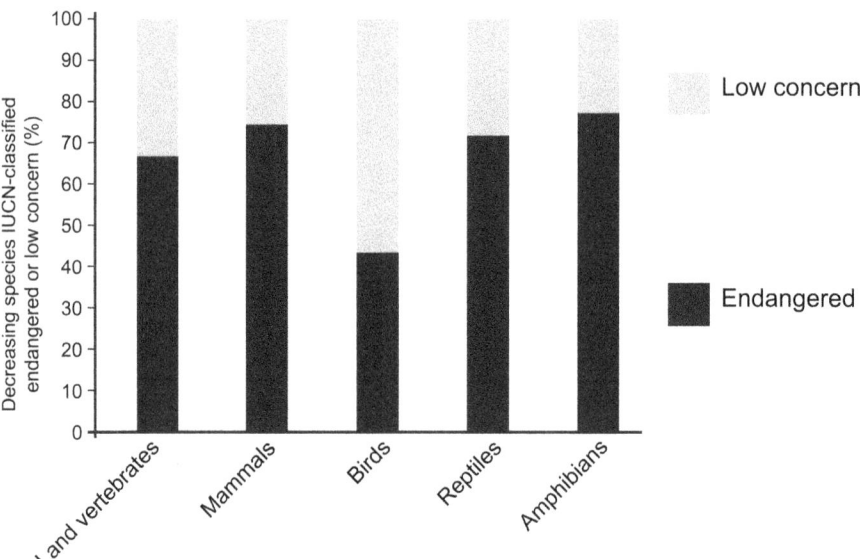

Fig. 10.3 Decreasing species classified by IUCN as "Endangered" (including "Critically
Endangered," "Endangered," "Vulnerable," and "Near Threatened") or "Low Concern" (including
"Low Concern" and "Data Deficient") in terrestrial vertebrates, in %. Source: Gerardo Ceballos,
Paul R. Ehrlich, Rodolfo Dirzo, "Biological annihilation via the ongoing sixth mass extinction
signaled by vertebrate population losses and declines." *PNAS*, 25/VII/2017, Fig. 4

The first column (Land vertebrates), which summarizes the four successive ones (mammals, birds, reptiles, and amphibians), shows that populations of about two thirds of terrestrial vertebrates are in decline; "this figure emphasizes," the authors conclude, "that even species that have not yet been classified as endangered (roughly 30% in the case of all vertebrates) are declining. This situation is exacerbated in the case of birds, for which close to 55% of the decreasing species are still classified as 'low concern.'"

This evaluation converges with the Living Planet Report 2018 (twelfth edition) of the Living Planet Index (WWF/ZSL), which measured the state of 16,704 populations of 4005 species. The results are frightening; on average, vertebrate animal populations in 2014 were well under half their 1970 levels. Between 1970 and 2014, 60% of worldwide vertebrate animal populations had been wiped out. In freshwater habitats, populations have collapsed by 83%, and in South and Central America, the worst affected region, there has been an 89% total drop in vertebrate animal populations. The key drivers of biodiversity decline, and the primary threats to these populations remain overexploitation (37%), habitat degradation (31.4%), and habitat loss (13.4%). As stressed by the authors of the Living Planet Report 2018, "of all the plant, amphibian, reptile, bird and mammal species that have gone extinct since AD 1500, 75% were harmed by overexploitation or agricultural activity or both."

10.7   The Two Extinction Pathways

Global capitalism tears the web of life in two complementary ways: as a direct and immediate consequence of its activities—legal or illegal—and as a reflexive and systemic mode of impact on habitats. In deforestation, timber trade, hunting, fishing, and trafficking of wild species, especially in the tropical world and in particular in Brazil, legal activity which is profoundly destructive conceals illegality and is inextricably intertwined with it. In June 2014, a joint report from UNEP and Interpol looked at illegal wildlife trafficking, assessing the magnitude of the monetary values involved (Nellemann 2009):

> In the international community, there is now growing recognition that the issue of the illegal wildlife trade has reached significant global proportions. Illegal wildlife trade and environmental crime involve a wide range of flora and fauna across all continents, estimated to be worth USD 70–213 billion annually.

In Colombia, wildlife trafficking financed the FARC until the so-called pacification; in Uganda, it financed the Lord's Resistance Army (LRA); in Somalia, Al-Shabaab; and in Darfur, the Janjaweed. The same symbiosis between trafficking and militias is established in the Congo, the Central African Republic, Mali, and Niger. The Russian mafia has long profited from the trafficking of sturgeons and caviar. According to Interpol's David Higgins, environmental crimes of this nature have increased by a factor of five in the past 20 years (Barroux 2014; Baudet and Michel

2015). According to the Brazilian *I Relatório Nacional Sobre o Tráfico de Animais Silvestres [1st National Report on the Trafficking of Wild Animals]* (2001), there are three types of illegal wildlife trade:

(a) Animals for zoos and private collectors. This type of wildlife trafficking prioritizes the most endangered species, also because the rarer the animal, the greater its demand and, thus, the higher its market value.
(b) Biopiracy for scientific, pharmaceutical, and industrial use. This type of trafficking focuses on species used in animal experiments and those that supply chemicals. It involves high monetary values, mostly coming from the pharmaceutical industry. In 2001, the value of one gram of poison from a brown spider (*Loxosceles* sp.) and from a coral snake (*Micrurus frontalis*), for example, could be as high as US$ 24,000 and US$ 30,000, respectively.
(c) Animals for pet shops. According to the report, this is the process that most encourages wildlife trafficking, at least in Brazil.

This report also notes that countries which are among the major wildlife exporters include Brazil, Peru, Argentina, Guyana, Venezuela, Paraguay, Bolivia, Colombia, South Africa, Zaire, Tanzania, Kenya, Senegal, Cameroon, Madagascar, India, Vietnam, Malaysia, Indonesia, China, and Russia. And among the countries that import them, the most are: the USA, Germany, Holland, Belgium, France, England, Switzerland, Greece, Bulgaria, Saudi Arabia, and Japan. According to the Animal Rights News Agency (ANDA), Brazil accounts for 15% of wildlife trafficking worldwide.[6] According to the *I Relatório Nacional Sobre Gestão e Uso Sustentável da Fauna Silvestre [1st National Report on Use and Management of Wildlife in Brazil]* (2016) by Renctas, defaunation rates in Brazil are colossal:

> At least 60 million vertebrates (mainly birds, mammals, and some reptiles) are hunted illegally each year in the Brazilian Amazon. There are no data for other regions of Brazil but poaching still occurs in all ecosystems. (…) It is estimated that Brazil illegally exports 1.5 billion wild animals per year, an amount that globally reaches 10 billion reais [US$ 2 billion], losing in revenue only to drug and arms trafficking.

Like all other types of trafficking—of drugs, timber, electronic waste, weapons, and people (prostitution, organs, and tissues)—this one is also exceptionally profitable. The governments of exporting and importing countries are complicit in this ecocide by not equipping their inspecting agencies with budgets that are consistent with their surveillance and repression duties. These agencies remain vulnerable to corruption and have been unable to thwart gangs and deter criminals.

---

[6] See "Brasil responde por 15% do tráfico de animais silvestres." *ANDA*, 16/XI/2017.

10.8   The International Financial System

The international financial system is a key element in trafficking. Banks not only profit from crime, but—through a tangle of mechanisms that prevent the tracking of wealth from trafficking and that legalize this wealth—they make the line between legal and illegal economies invisible. The gains from wildlife trafficking end up sustaining the financial system through a complex network of resource transfusions among the various mafias, whether they are involved in environmental crimes or other types of crimes. Debbie Banks et al. (2008) are authors of a study by the London-based Environmental Investigation Agency (EIA) on money laundering in illegal timber trafficking in Papua. The study reports that:

> An expert in forging shipping documents working from Singapore boasted to EIA investigators that he was "timber mafia" and that the trade was "better than drug smuggling." The vast profits from this illegal trade accrued in bank accounts in Singapore and Hong Kong.

In one of the cases investigated, a corrupt police officer received $120,000, paid through 16 bank transfers from individuals linked to companies accused of illegal logging. These were the very companies this police officer was supposed to be investigating. Similar mechanisms occur in deforestation, a major factor in the decline of tropical biodiversity, as well as in animal trafficking. In the case of deforestation, we must remember, for example, that over 90% of deforestation in Brazil is illegal (Wilkinson 2019; MapBiomas http://mapbiomas.org/). Gains from environmental crimes (deforestation, hunting, illegal animal trafficking, etc.) navigate the network of international financial corporations and are in symbiosis with them: crime brings in wealth and banks legalize it and profit from it. This was evidenced by the collusion between organized crime and HSBC, which, according to Charles Ferguson, is not unique to that bank (Ferguson 2012). As Neil Barofsky (2012) emphasizes, for the "Judicial system": "HSBC is not only too big to fail, but is also too big to jail." In fact, when reported, their directors paid a fine corresponding to a few weeks' earnings but were never defendants in a criminal lawsuit. The high-profit–low-risk mechanism reinforces the tacit symbiosis between the financial system and organized environmental crime.

10.9   Systemic Destruction: 70% of Biodiversity Loss Comes from Agriculture and Livestock

Even when the destruction of plant and animal species is not illegal or is not an immediate business focus, global capitalism or, more precisely, the globalized production of soft commodities is *systemically* the main cause of biodiversity collapse. Agribusiness is certainly one of the most important causes of population decline and, ultimately, of the extinction of vertebrates and invertebrates. As the 2016 *I*

Relatório Nacional Sobre Gestão e Uso Sustentável da Fauna Silvestre [*1st National Report on Use and Management of Wildlife in Brazil*] produced by Renctas states:

> The destruction of natural environments eliminates thousands of animals locally. In the Amazon, between 5 and 10 thousand square kilometers of forest are cleared every year. In forests where there is no hunting, each square kilometer can house up to 95 medium and large mammals, depending on the forest. Therefore, deforestation may be eliminating between 475 to 950 thousand animals per year in the Amazon.

In addition to advancing over native vegetation cover, agribusiness intoxicates the soil, causes eutrophication of the aquatic environment through the use of chemical fertilizers, and poisons the environment with systemic pesticides. David Gibbons, Christy Morrissey, and Pierre Mineau (2014) present the results of a review of 150 studies on the direct (toxic) and indirect (food chain) action of three insecticides—fipronil and two neonicotinoid-class insecticides (imidacloprid and clothianidin)—on mammals, birds, fish, amphibians, and reptiles:

> All three insecticides exert sub-lethal effects, ranging from genotoxic and cytotoxic effects, and impaired immune function, to reduced growth and reproductive success, often at concentrations well below those associated with mortality. Use of imidacloprid and clothianidin as seed treatments on some crops poses risks to small birds, and ingestion of even a few treated seeds could cause mortality or reproductive impairment to sensitive bird species.

Caspar Hallmann et al. (2014) also showed that regions with concentrations of more than 20 nanograms of imidacloprid per liter of water present a rapid decline in insectivorous bird population:

> Recent studies have shown that neonicotinoid insecticides have adverse effects on non-target invertebrate species. Invertebrates constitute a substantial part of the diet of many bird species during the breeding season and are indispensable for raising offspring. We investigated the hypothesis that the most widely used neonicotinoid insecticide, imidacloprid, has a negative impact on insectivorous bird populations. Here we show that, in the Netherlands, local population trends were significantly more negative in areas with higher surface-water concentrations of imidacloprid. At imidacloprid concentrations of more than 20 nanograms per litre, bird populations tended to decline by 3.5 per cent on average annually.

According to the fourth edition of the UN Global Biodiversity Outlook (GBO-4, 2014), "drivers [of biodiversity loss] linked to agriculture account for 70% of the projected loss of terrestrial biodiversity." Indeed, Manfred Lenzen et al. (2012) show that about one third of species threatened with extinction in "developing nations" are in this condition as a result of the extreme globalization of commodities trade. The study is the first to detect and quantify the cause and effect relationship between the 15,000 commodities produced for international trade in 187 countries and the threat to 25,000 animal species listed in the 2012 IUCN Red List. Orangutans, elephants, and tigers in Sumatra are on this list, for example, as victims of habitat degradation due to palm oil and pulpwood plantations. Within the past 25 years, agribusiness has destroyed 69% of the habitat of the Sumatran elephants (*Elephas maximus sumatranus*). Its population had declined by at least 80% during the past three generations, estimated to be about 75 years, and is now reduced to between 2400 and 2800 individuals. The IUCN has just reclassified the situation of these animals, no longer placing them in the "Endangered" species category, but in the "Critically Endangered" category. Also, in Brazil or, perhaps, especially in Brazil, agribusiness (mostly livestock) is the main culprit of biodiversity loss. The *Red Book of Endangered Brazilian Fauna* (ICMBio, Vol., 2018) clearly points to the major role that agribusiness plays in the destruction of Brazil's immense natural heritage:

> Throughout the country, the main pressure factors on continental species relate to the consequences of agriculture and livestock activities, either by fragmentation and decrease in habitat quality in areas where the activity is consolidated or by the ongoing process of habitat loss in areas where the activity is expanding. These activities affect 58% of the 1014 endangered continental species.

Figure 10.4 quantifies these pressure factors on Brazilian biodiversity.

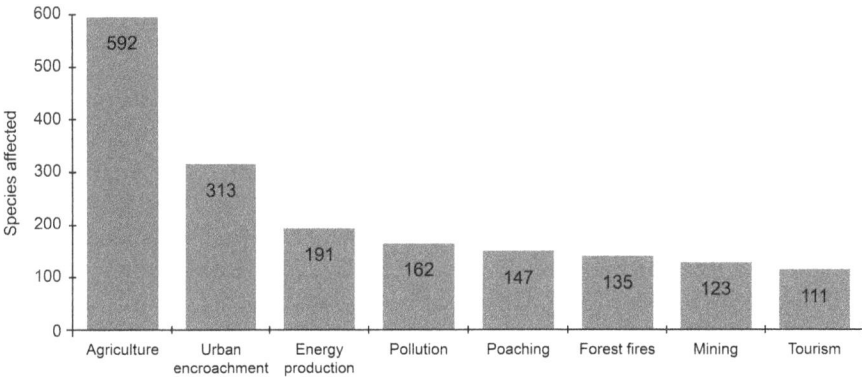

Fig. 10.4 Continental species affected by the major threats in Brazil. (Source: *Livro Vermelho da Fauna Brasileira Ameaçada de Extinção* [Brazil Red Book of Threatened Species of Fauna] ICMBio, 2018)

Of the 1014 continental species considered threatened, 592 are in this condition as a direct result of agriculture and livestock activities. Furthermore, 162 species are threatened by pollution, in particular by the use of pesticides, which affect "mainly invertebrates (river crabs, limnic mollusks, butterflies and springboards), but also fish, birds, amphibians, reptiles, and mammals." Finally, 135 species are threatened by forest fires, the vast majority of which are caused by the agribusiness sector. This means that of the 1014 species evaluated in Brazil, close to 880, or about 87% of them, are directly threatened by agribusiness.

10.10   Mammals

Regarding mammal species that are threatened (i.e., Critically Endangered, Endangered, and Vulnerable), the exacerbation of the phenomenon can be established with a high degree of reliability, since the 5506 species evaluated by IUCN coincide with the number of species described (a number that is certainly close to that of existing mammal species) and were calculated as follows, 21% in 2009 against 25% in 2019, a huge difference in the span of just 10 years. Population decline in non-synanthropic mammals is widespread, yet it varies according to their trophic level and is most acute among species at the top of the food pyramid: 77% of 31 large (over 15 pounds) carnivore species are declining, and more than half of these 31 species have already declined by 50% from their historical records (Ripple et al. 2014). In another paper, William Ripple et al. (2015) warn that large terrestrial herbivorous mammals, a group of about 4000 species, are also experiencing alarming rates of population decline:

> Large herbivores are generally facing dramatic population declines and range contractions, such that ~60% are threatened with extinction. Nearly all threatened species are in developing countries, where major threats include hunting, land-use change, and resource depression by livestock. Loss of large herbivores can have cascading effects on other species including large carnivores, scavengers, mesoherbivores, small mammals, and ecological processes involving vegetation, hydrology, nutrient cycling, and fire regimes. The rate of large herbivore decline suggests that ever-larger swaths of the world will soon lack many of the vital ecological services these animals provide, resulting in enormous ecological and social costs.

An example of the decline of large terrestrial herbivores (those with a body mass of ≥ 100 kg) also in developed countries: according to the 2018 Arctic Report Card (NOAA), since the mid-1990s, the size of reindeer and caribou herds has declined by 56%. That's a drop from an estimated 4.7 million animals to 2.1 million, a loss of 2.6 million (Sullivan 2018). Boreal caribou are disappearing across Canada as their original habitat has been cut in half. Only 14 of 51 herds are considered self-sustaining, and another third of the remaining boreal caribou could disappear in the next 15 years (Berman 2019).

Within the mammalian class, the primate order is declining particularly rapidly. The 2013 IUCN Red List reported that of the 420 species evaluated, 49% were threatened globally. The 2018 Red List shows that of the 450 species evaluated, 59.8% are now threatened. This race toward extinction was brought to light by a comprehensive study in 2017 involving 31 scientists and coordinated by Alejandro Estrada (UNAM):

> Current information shows the existence of 504 species in 79 genera distributed in the Neotropics, mainland Africa, Madagascar, and Asia. Alarmingly, ~60% of primate species are now threatened with extinction and ~ 75% have declining populations. This situation is the result of escalating anthropogenic pressures on primates and their habitats—mainly global and local market demands, leading to extensive habitat loss through the expansion of industrial agriculture, large-scale cattle ranching, logging, oil and gas drilling, mining, dam building, and the construction of new road networks in primate range regions.

10.11 Birds

The situation of birds has worsened at least since 1988, the date of the first full IUCN assessment. In the 2012 BirdLife International assessment (accredited by the IUCN), there were 1313 bird species threatened with extinction (Vulnerable, Endangered, and Critically Endangered), representing 13% of the 10,064 bird species described in the world. In BirdLife International's *State of the World's Birds 2018*, this percentage rises to 13.3%. In absolute terms, this means a huge increase in just 5 years, since now an additional 156 bird species are threatened with extinction, or 1469 species out of a universe of 11,000 species which have been described. The biggest cause of this decline, according to the document, continues to be agribusiness.

Europe lost 400 million birds from 1980 to 2009, and around 90% of this decline can be attributed to the 36 most common species, such as sparrows, starlings, and skylarks (Inger et al. 2014). Kenneth Rosenberg et al. (2019) reported population losses across much of the North American avifauna over 48 years, including once-common species and from most biomes: "Integration of range-wide population trajectories and size estimates indicates a net loss approaching 3 billion birds, or 29% of 1970 abundance." The *State of India's Birds 2020* assessed the status of 867 species. Of the 261 species for which long-term trends could be determined, 52% have declined since the year 2000, with 22% declining strongly. Current annual trends could be estimated for 146 species. Of these, 79% have declined over the last 5 years, with almost 50% declining strongly. Just over 6% are stable and 14% increasing. According to a study by the Brazilian government conducted in 2016, of the 1924 bird species described, 233 species and subspecies were threatened with extinction. Of these, 42 were critically endangered. In the Amazon, BirdLife International's 2012 IUCN Red List shows that almost 100 bird species are now closer to extinction.[7] According to Leon Bennun, Director of science, policy, and information at BirdLife: "We have previously underestimated the risk of extinction that many of Amazonia's bird species are facing. However, given the recent weakening of Brazilian forest law, the situation may be even worse than recent studies have predicted" (Harvey 2012).

Despite the international Convention on Migratory Species (1979), migratory birds continue to suffer from massive hunting. In Northeast India, 120,000 to 140,000 Amur falcons, or Eastern red-footed falcons (*Falco amurensis*), are hunted each year, mainly with nets as they pass through the Nagaland region on their annual migration from Siberia and northern China to Africa (Dalvi and Sreenivasan 2012). In Africa, the practice of catching migratory birds from northern latitudes dates back to the time of the pharaohs. But it is now performed on an infinitely larger scale, with equipment imported from China, such as huge plastic nets and recordings that attract the birds. Such techniques allow as many as 140 million birds to be killed each migratory season. According to Brian Finch, author of a migratory bird

---

[7] See http://www.mma.gov.br/mma-em-numeros/biodiversidade.

census in Kenya, "there's a massive die-off of birds. We should be aware of the implications of taking all these insectivores out of the ecosystem" (Mangat 2013).

This awareness is critical. The abominable extermination of birds deprives the biosphere of fantastically beautiful, sensitive, and intelligent creatures. Birds are pollinators. According to WWF, the seeds of over 90% of all woody tree species are moved around by birds. Moreover, they are indispensable for the ecological balance and, therefore, for humanity. They control the numbers of pests that would otherwise plague our natural environments and our crops. We must remember the terrible consequences of the loss of these and other insectivores. By causing a terrible proliferation of insects, the near-extermination in China of the Eurasian tree sparrow (*Passer montanus*) under the Great Sparrow Campaign in 1958 was one of the causes of the so-called Great Chinese Famine (1958–1961), in which more than 20 million people starved to death. Today, climate change and the destruction of forest habitats, particularly by agriculture and livestock activities, are often cited as factors that explain the spread and sudden increase in populations of *Aedes aegypti* and *Aedes albopictus*, especially in the tropical and subtropical regions of the planet. But, as a supplementary factor, one might point out the drastic worldwide decline in bird populations and in other natural insect predators, such as reptiles and amphibians which are even more threatened. The Great Sparrow Campaign of 1958, ordered by Mao Zedong, was an act of ecological dementia. But the current decimation of birds through loss of wildlife, deforestation, increased use of pesticides, hunting, and trafficking is no less insane. This equation was suggested by Philip Lymbery and Isabel Oakeshott (2014):

> The effect of agricultural policy in Europe and the Americas in the past few decades has been almost exactly the same as Mao's purge. Tree sparrows—the same species that Mao targeted—have declined in Britain by 97% over the last forty years, largely due to the intensification of agriculture. The figures for other well-loved birds like turtle doves and corn buntings are no less alarming. Modern farming has become so "efficient" that the countryside is now too sterile to support native farmland birds.

Drugs and pesticides also kill animals on a large scale. Diclofenac injected into cattle is leading to kidney failure in five out of eight vulture species and in one eagle species (*Aquila nipalensis*) from India, which feed on their carcasses. This is an evil that now also affects 13 other eagle species in Asia, Africa, and Europe.[8] Similarly, the US EPA has published a paper assessing the lethality of nine rodenticides on birds (especially owls, hawks, eagles, and crows) that feed on poisoned rats and mice.[9]

---

[8] See "Drug may poison Europe's Eagles." *New Scientist*, 2972, 7/VI/2014, p. 7.
[9] See "Potential Risks of Nine Rodenticides to Birds and Nontarget Mammals: a Comparative Approach." EPA, 2004.

10.12   Terrestrial Arthropods and the Decline in Pollinators

According to Nico Eisenhauer, Aletta Bonn, and Carlos Guerra (2019), the Red List of Threatened Species of the IUCN "is still heavily biased towards vertebrates, with invertebrates being particularly underrepresented." Invertebrates include arthropods (insects, arachnids, myriapods, crustaceans), which, as seen above (Leakey and Lewin 1996), account for over 80% of all known living animal species. They are crucial to the functioning of ecosystems. Insects, for instance, pollinate 80% of wild plants and provide a food source for 60% of birds (Hallmann et al. 2017). Yet, a comprehensive review of 73 historical reports of insect declines from across the globe (Sanchez-Bayo and Wyckhuys 2019) reveals that the current proportion of insect species in decline is twice as high as that of vertebrates. Every year about 1% of all insect species are added to the list of threatened species, "with such biodiversity declines resulting in an annual 2.5% loss of biomass worldwide." These dramatic rates of decline "may lead to the extinction of 40% of the world's insect species over the next few decades." According to the authors:

> Habitat change and pollution are the main drivers of such declines. In particular, the intensification of agriculture over the past six decades stands as the root cause of the problem, and within it the widespread, relentless use of synthetic pesticides is a major driver of insect losses in recent times.

Mikhail Beketov et al. (2013) examined 23 streams in Germany, 16 in France, and 24 in Australia. They classified streams according to different levels of pesticide contamination: uncontaminated, slightly contaminated, and highly contaminated. Highly contaminated streams in Australia showed a decrease of up to 27% in the number of invertebrate families when compared to uncontaminated streams (Oosthoek 2013):

> Pesticides caused statistically significant effects on both the species and family richness in both regions [Europe and Australia], with losses in taxa up to 42% of the recorded taxonomic pools. Furthermore, the effects in Europe were detected at concentrations that current legislation considers environmentally protective.

In 2014, the Task Force on Systemic Pesticides (TFSP), bringing together 29 researchers, stated in its findings that systemic pesticides pose an unmistakable and growing threat to both agriculture and ecosystems. Jean-Marc Bonmatin, a CNRS researcher in this working group, summarized these results (as cited by Carrington 2014):

> The evidence is very clear. We are witnessing a threat to the productivity of our natural and farmed environment equivalent to that posed by organophosphates or DDT. Far from protecting food production, the use of neonicotinoid insecticides is threatening the very infrastructure which enables it.

10.13   The Ongoing Collapse of Flying Insect Biomass

Caspar A. Hallmann et al. (2017) shed new light on the current scale of the decrease in flying insects in Europe; no longer using indicators of population abundance for specific species or taxonomic groups, they look at changes in insect biomass. From observations conducted during 27 years in 63 nature reserves in Germany, the authors estimate that the average flying insect biomass declined 76% (up to 82% in midsummer) in just 27 years in these locations. Tyson Wepprich et al. (2019) estimated the rate of change in total butterfly abundance and the population trends for 81 species using 21 years of systematic monitoring in Ohio, USA. They showed that "total abundance is declining at 2% per year, resulting in a cumulative 33% reduction in butterfly abundance."

The ongoing collapse of flying insects in Europe and in the USA is the most visible part of a much broader phenomenon which entomologists are characterizing as "an accelerated decline in all insect species since the 1990s" (Foucart 2014). Insecticides are the main reason for this decline in insects, as pointed out by hundreds of scientific papers on insecticides that are called systemic. The order Lepidoptera, which represents about 10% of the known insect species and is one of the most studied, is not among the most severely affected. Nevertheless, several species of butterflies and moths are declining because of the destruction of their habitats and the elimination, by herbicides, of the plants that caterpillars feed on. According to Rodolfo Dirzo et al. (2014):

> Globally, long-term monitoring data on a sample of 452 invertebrate species indicate that there has been an overall decline in abundance of individuals since 1970. Focusing on just the Lepidoptera (butterflies and moths), for which the best data are available, there is strong evidence of declines in abundance globally (35% over 40 years) Non-Lepidopteran invertebrates declined considerably more, indicating that estimates of decline of invertebrates based on Lepidoptera data alone are conservative.

Bill Freese and Martha Crouch (2015) showed that spraying Monsanto's glyphosate herbicide on Roundup Ready corn and soybean seeds (genetically engineered to tolerate this herbicide) eradicated milkweeds, the only food source for the monarch butterfly (*Danaus plexippus*). As a result, its existence is threatened. In the state of Florida, five species of butterflies were considered extinct in May 2013 by entomologist Marc Minno, and two more (*Epargyreus zestos oberon* and *Hesperia meskei pinocayo*) became extinct in July 2013 in the same region. Another species which is among the most beautiful, the so-called Madeiran large white (*Pieris brassicae wollastoni*), in Madeira, a Portuguese island, was declared extinct in 2006, falling victim to what IUCN's Red List calls "natural system modification."

The European Grassland Butterfly Indicator (EEA) estimates that in just over 20 years (1990–2011), half of the field butterflies disappeared from the landscape of 19 European countries. In Brazil, about 50 species of butterflies are threatened with extinction and listed in the IBAMA Red List.[10] This decline is all the more worrying

---

[10] "Pesquisadores da Unicamp elaboram plano contra extinção de borboletas". *Globo.com G1*, 12/IX/2012.

because butterflies are important bioindicators: they are considered indicators that represent trends observed in most terrestrial insects. Moreover, they are key to the preservation of these ecosystems, being a source of food for birds and, above all, because of their role as pollinators. This terrible decline in invertebrate biodiversity is not restricted to insects. The *Earth's Endangered Creatures* lists 30 endangered arachnid species, stressing that the list is not exhaustive.

10.14   Pollinators and the Crisis in Pollination

As pointed out by the IPBES (2016), over 40% of invertebrate pollinators and 16.5% of vertebrate pollinators are threatened with global extinction (increasing to 30% for island species). Since at least the end of the twentieth century, the decline in invertebrate pollinators has attracted public attention (Allen-Wardell et al. 2008). Pollination is a vital service for plant reproduction and biodiversity maintenance. Around 100,000 invertebrate species, especially insects, are involved in pollination, and "in extreme cases, their decline may lead to the extinction of plants and animals," since "of the estimated 250,000 species of modern angiosperms, approximately 90% are pollinated by animals, especially insects" (Rocha 2012). Our food dependence on pollinators is, therefore, immense, and the consequences of this crisis can be catastrophic. According to a UNEP report (2010), based on FAO's estimates, "out of some 100 crop species which provide 90% of food worldwide, 71 of these are bee-pollinated." In the UK, the Living with Environmental Change Network (LWEC 2014) warned that:

> Over three-quarters of wild flowering plant species in temperate regions need pollination by animals like insects to develop their fruits and seeds fully. This is important for the long-term survival of wild plant populations and provides food for birds and mammals. Pollinators improve or stabilize the yield of three-quarters of all crop types globally; these pollinated crops represent around one third of global crop production by volume.

In addition, 90% of the vitamin C we need comes from insect-pollinated fruits, vegetables, oils, and seeds. This support network for plant reproduction and biodiversity is weakening. The 2016 IPBES assessment on pollinators reaffirmed and updated these estimates: "75% of our food crops and nearly 90% of wild flowering plants depend at least to some extent on animal pollination." Furthermore, the assessment concluded that "a high diversity of wild pollinators is critical to pollination even when managed bees are present in high numbers." A study by Berry Brosi and Heather M. Briggs (2013) revealed the importance of this diversity:

> We show that loss of a single pollinator species reduces floral fidelity (short-term specialization) in the remaining pollinators, with significant implications for ecosystem functioning in terms of reduced plant reproduction, even when potentially effective pollinators remained in the system. Our results suggest that ongoing pollinator declines may have more serious negative implications for plant communities than is currently assumed.

10.15   Colony Collapse Disorder (CCD) and the Pesticides

The threat of extinction that hovers over bees (*Apis mellifera*) affects both wild and managed bees. David Hackenberg, an American commercial beekeeper, coined the term colony collapse disorder (CCD) in 2006: healthy hives have been deserted by the vast majority of worker bees who abandon the queen bee, a few nurse bees, the larvae, and the food reserves. This phenomenon currently strikes apiaries in the USA, Europe, Africa, China, Taiwan, Japan, the Middle East, and Brazil. It has manifested itself since the 1990s, but most markedly since 2006 in the USA, with losses reaching 30–90% of the hive population in Europe and in the USA. The winter 2018–2019 saw the highest honeybee colony losses on record in the USA (Reiley 2019).

Three groups of scientists—the French NGO Pollinis (Réseau des Conservatoires Abeilles et Pollinisateurs); the working group on systemic pesticides (TFSP, or Task Force on Systemic Pesticides), already mentioned in Chap. 4 (Sect. 4.10 "Industrial Pesticides"); and the IPBES—have accumulated evidence showing that the decline in bees is caused by habitat loss and systemic neurotoxic pesticides made from substances such as fipronil or neonicotinoids (clothianidin, imidacloprid, and thiamethoxam). The IPBES (2016) assessment concludes that "recent evidence shows impacts of neonicotinoids on wild pollinator survival and reproduction at actual field exposure." These molecules, commercialized under the names of Gaucho, Cruiser, Poncho, Nuprid, Argento, etc., comprise about 40% of the world's agricultural insecticide market, representing a value of over 2.6 billion dollars. They are distinguished from previous generations of insecticides by their toxicity, which is 5000–10,000 times greater than the famous DDT, for example (Foucart 2014). In addition to disrupting the neurological orientation system of bees, these pesticides weaken them and make them more vulnerable to viruses, mites (Varroa destructor), and other pathogens. In a study published by Jeffery Pettis et al. (2013), in which healthy bees were fed beehive pollen collected from 7 pesticide-contaminated agricultural crops, 35 pesticides (with high doses of fungicides) were detected in those samples, and contact with the samples made bees more susceptible to intestinal parasites. Ten categories of pesticides (carbamates, cyclodienes, formamidines, neonicotinoids, organophosphates, oxadiazines, pyrethroids, etc.) were considered:

> Insecticides and fungicides can alter insect and spider enzyme activity, development, oviposition behavior, offspring sex ratios, mobility, navigation and orientation, feeding behavior, learning and immune function. Reduced immune functioning is of particular interest because of recent disease-related declines of bees including honey bees. Pesticide and toxin exposure increases susceptibility to and mortality from diseases including the gut parasite *Nosema* spp.. These increases may be linked to insecticide-induced alterations to immune system pathways, which have been found for several insects, including honey bees.

Tomasz Kiljaneck et al. (2016) published a new method for detecting the presence of 200 insecticides in bees at concentrations of 10 ng/g or less. They conclude, with regard to European bees, that "in total, 57 pesticides and metabolites were determined in poisoned honeybee samples." Furthermore, these pesticides "even at very low levels at environmental doses and by interaction could weaken bees defense systems allowing parasites or viruses to kill the colony." The lethality of pesticides is sufficiently demonstrated for the subgroup of species that includes bumblebees and *Apis mellifera*, which accounts for 80% of insect pollination. Juliet Osborne (2012) stressed the importance of a study conducted by Richard Gill, Oscar Ramos-Rodriguez, and Nigel Raine (2012) on bumblebees. The authors examined bumblebee responses to two pesticides:

> We show that chronic exposure of bumblebees to two pesticides (neonicotinoid and pyrethroid) at concentrations that could approximate field-level exposure impairs natural foraging behaviour and increases worker mortality leading to significant reductions in brood development and colony success. We found that worker foraging performance, particularly pollen collecting efficiency, was significantly reduced with observed knock-on effects for forager recruitment, worker losses and overall worker productivity. Moreover, we provide evidence that combinatorial exposure to pesticides increases the propensity of colonies to fail.

The *EPILOBEE* study in 17 European countries found a widespread incidence of CCD, albeit more severe in northern European countries, with beehive losses ranging from 27.7% in France to 42.5% in Belgium (Foucart 2014). According to the USDA and the Apiary Inspectors of America, hives have been decimated in this country at rates between 21% and 34% per year since the winter of 2006/2007.[11] In Brazil, the world champion in pesticide use, bee mortality has already reached 14 states. In the state of Sao Paulo, 55% of bees have disappeared, according to Lionel Segui Gonçalves (2017). According to Oscar Malaspina, there is a clear link between these occurrences and the use of pesticides, especially when launched by airplanes, since the wind carries the sprayed product to areas adjacent to those of the target plantation.[12]

10.16   Three Certainties Among So Many Uncertainties

During the Pleistocene, our planet was inhabited by gigantic and spectacular animals. By the end of this geological age, between 2.5 million and 11,700 years BP, many of these species, such as mammoths, ground sloths, and giant tortoises, in addition to the various species belonging to the sabertooth genera, had been extinct, many of them probably by our species. According to William Ripple and Blaire Van Valkenburgh (2010), "humans may have contributed to the Pleistocene megafaunal extinctions. The arrival of the first humans, as hunters and scavengers, through top-down forcing, could have triggered a population collapse of large herbivores and their predators." Since then, the still rich animal diversity of the Holocene has continued to become impoverished, with hundreds of species officially becoming extinct in the last 500 years and, especially, in the last half century (1970–2020). We are now suffering more and more from the so-called empty forest syndrome, a term

---

[11] See Kim Kaplan, "Usada/AIA Survey Reports 2010/2011 Winter Honey Bee Losses", Agricultural Research Center. 23/V/2012; S. Goldenberg, "Rate of U.S. honeybee deaths 'too high for long-term survival'". *The Guardian*, 15/V/2014.

[12] Quoted in "Nova pesquisa acirra polêmica sobre uso de agrotóxicos". *Fapesp*, 30/VI/2019

coined by Kent Redford in 1992 to refer to a forest already depleted of its fauna by humans, even before it would be completely destroyed by them. Defaunation, in fact, precedes the final devastation of tropical forests by globalized capitalism. The final triumph of globalization in recent decades has greatly accelerated and continues to accelerate the ongoing collapse of biodiversity.

Chinachem, Bayer-Monsanto, DowDuPont and BASF are the largest suppliers of pesticides in the world. They are definitely the main culprits for the collapse of biodiversity, especially among invertebrates. But climate change, habitat degradation, and landscape homogeneity combined can aggravate effects of pesticides in nature. Climate change has accelerated range losses among many species. Peter Soroye, Tim Newbold, and Jeremy Kerr (2020) evaluated how climate change affects bumblebee species in Europe and North America. Their measurements provided evidence of rapid and widespread declines of bumblebees across Europe and North America: "bumble bees richness declined in areas where increasing frequencies of climatic conditions exceed species' historically observed tolerances in both Europe and North America." Thus, the current approach to regulatory environmental risk assessment of pesticides must be updated. "The overall picture is of a need to move to a more holistic view" (Topping et al. 2020). In fact, we are only beginning to understand the scale, dynamics, and consequences of the ongoing sixth mass extinction. In 1992, in his classic *The Diversity of Life*, Edward O. Wilson stated: "extinction is the most obscure and local of all biological processes." As the IUCN warns, extinction risk has been assessed for only a small percentage of the total number of species described, which are, in turn, a small fraction of the biodiversity universe, recently estimated, as seen above, at 8 million eukaryotic species. Among so many uncertainties, three certainties, however, stand out:

1. Defaunation and species extinctions in all taxonomic groups are undoubtedly accelerating, as evidenced by successive assessments from the IUCN, the Living Planet Index, and many other scientific assessments.
2. Every species is precious not only for its own sake but also for its multiple interactions with other species in the web of life. As Sacha Vignieri (2014) rightfully reminds us:

   Though for emotional or aesthetic reasons we may lament the loss of large charismatic species, such as tigers, rhinos, and pandas, we now know that loss of animals, from the largest elephant to the smallest beetle, will also fundamentally alter the form and function of the ecosystems upon which we all depend.

3. With this increasing fragmentation of the web of life, we are approaching critical points. This is the fundamental message both of the IPBES Global Assessment Report (2019) and of the study by Rodolfo Dirzo et al. (2014), so often cited in this chapter: "Cumulatively, systematic defaunation clearly threatens to fundamentally alter basic ecological functions and is contributing to push us toward global-scale 'tipping points' from which we may not be able to return." And Sacha Vignieri (2014) completes: if we are unable to reverse this ongoing collapse, "it will mean more for our own future than a broken heart or an empty forest."

References

ALLEN-WARDELL, G. *et al.* "The Potential Consequences of Pollinator Declines on the Conservation of Biodiversity and Stability of Food Crop Yields". *Conservation Biology*, vol. 12, 1, 2008, pp. 8–17.

BANKS, Debbie *et al.*, *Environmental Crime. A threat to our future.* EIA – Environmental Investigation Agency, London, Emmerson Press, 2008.

BARNOSKY, Anthony *et al.*, "Has the Earth's Sixth Mass Extinction Already Arrived?" *Nature*, 471, 7336, February 2011, pp. 51–57.

BAROFSKY, Neil, "Too Big to Jail: Our Banking System's Latest Disgrace". *The New Republic*, 12/XII/2012.

BAUDET, Marie-Béatrice, MICHEL, Serge, "Sur la piste des mafias de l'environnement". *Le Monde*, 25/I/2015.

BARROUX, Rémi, "La criminalité environnementale explose". *Le Monde*, 25/VI/2014.

BEKETOV, Mikhail A. *et al.*, "Pesticides reduce regional biodiversity of stream invertebrates". *PNAS*, 17/VI/2013.

BERMAN, Tzeporah, "Canada clearcuts one million acres of boreal forest every year … a lot of it for toilet paper". *The Narwhal*, 21/III/2019.

BROSI, Berry J. & BRIGGS, Heather M. "Single pollinator species losses reduce floral fidelity and plant reproductive function". *Proceedings of National Academy of Sciences*, 20/VI/2013.

CARRINGTON, Damian, "Insecticides put world food supplies at risk, say scientists". *The Guardian*, 24/VI/2014.

CEBALLOS, Gerardo, EHRLICH, Paul R., DIRZO, Rodolfo, "Biological annihilation via the ongoing sixth mass extinction signaled by vertebrate population losses and declines". *PNAS*, 25/VII/2017.

COGHLAN, Andy, "Biodiversity betrayal as nations fail miserably on conservation. *New Scientist*, 8/XII/2018.

COLLINS, J. P. & CRUMP, M. L. *Extinction in our times: global amphibian decline.* Oxford University Press, 2009.

DALVI, S. & SREENIVASAN, R., "Shocking Amur falcon massacre in Nagaland". *Conservation India*, X/2012.

DIRZO, Rodolfo; GALETTI, Mauro; COLLEN, Ben *et al.* "Defaunation in the Anthropocene". *Science*, 345, 6195, 25/VII/2014, pp. 401–406.

EISENHAUER, Nico, BONN, Aletta & GUERRA, Carlos A., "Recognizing the quiet extinction of invertebrates". *Nature Communications*, 10, 50, 2019.

ESTRADA, Alejandro *et al.* "Impending extinction crisis of the world's primates: Why primates matter". *Science Advances*, 18/I/2017.

FAN, Jun-xuan *et al.*, "A high-resolution summary of Cambrian to Early Triassic marine invertebrate biodiversity". *Science*, 367, 6475, 17/I/2010, pp. 272–277.

FERGUSON, Charles, "Bare-faced bankers should be treated as criminals: prosecuted and imprisoned". *The Guardian*, 20/VII/2012.

FOUCART, Stéphane, "Le déclin massif des insects menace l'agriculture". *Le Monde*, 25/VI/2014.
_____. "En Europe, le déclin des abeilles frappe lourdement les pays du Nord". *Le Monde*, 8/IV/2014.

FREESE, Bill & CROUCH, Martha. "Monarchs in peril. Herbicide-Resistant crops and the decline of Monarch butterflies in North America". *Center for Food Safety*, February 2015.

GIBBONS, David; MORRISSEY, Christy & MINEAU, Pierre. "A review of the direct and indirect effects of neonicotinoids and fipronil on vertebrate wildlife". *Environmental Science Pollution Research*, 18/VI/2014.

GILL, Richard J.; RAMOS-RODRIGUEZ, Oscar & RAINE, Nigel E. "Combined pesticide exposure severely affects individual- and colony-level traits in bees". *Nature*, 21/X/2012.

GONÇALVES, Lionel Segui, "O massacre das abelhas pelo agrotóxico". *Revista IHU*, 4/IX/2017.

HALLMANN, Caspar. A. *et al.*, "Declines in insectivorous birds are associated with high neonic-otinoid concentrations". *Nature*, 9/VII/2014.

HALLMANN, Caspar A. *et al.*, "More than 75 percent decline over 27 years in total flying insect biomass in protected areas". *Plos One*, 18/X/2017.

HARVEY, Fiona, "Nearly 100 bird species face increased risk of extinction in the Amazon. *The Guardian*, 7/VI/2012.

INGER, Richard *et al.*, "Common European birds are declining rapidly while less abundant species' numbers are rising". *Ecology Letters*, 18, 1, 2/XI/2014.

IPBES (2016). The assessment report on Biodiversity and Ecosystem Services on pollinators, pollination and food production. S.G. Potts, V. L. Imperatriz-Fonseca, and H. T. Ngo (eds). Bonn, 2016.

IPBES (2019), Global Assessment Report on Biodiversity and Ecosystem Services, 2019.

JONES, Andrew. "The Next Mass Extinction: Human Evolution or Human Eradication". *In*: SHIELD, Rudolph (ed.). *Extinctions. History, Origins, Causes and Future of Mass Extinctions. Journal of Cosmology*. Cambridge, Harvard-Smithsonian, 2011, pp. 259–277.

KENNETH, J. Loceya & LENNONA, Jay T., "Scaling laws predict global microbial diversity". *PNAS*, 2016.

KILJANECK, Tomasz *et al.* "Multi-residue method for the determination of pesticides and pesticide metabolites in honeybees by liquid and gas chromatography coupled with tandem mass spectrometry – Honeybee poisoning incidents". *Journal of Chromatography A*, 2016.

LEAKEY, Richard & LEWIN, Roger. *The Sixth Extinction. Biodiversity and its Survival*. London, Weidenfel and Nicolson, 1996.

LEDFORD, Heidi, "World largest plant survey reveals alarming extinction rate". *Nature*, 10/VI/2019.

LENZEN, Manfred *et al.* "International trade drives biodiversity threats in developing nations". *Nature*, 486, 7/VI/2012, pp. 109–112.

LWEC (2014), *Living with Environmental Change Policy and Practice Notes*, 9, April 2014.

LYMBERY, Philip & OAKESHOTT, Isabel, *Farmageddon: the true cost of cheap meat*. London, Bloomsbury, 2014.

MANGAT, Rupi. "Horror: Egypt's Bird killing fields". *Africa Review*, 4/XII/2013.

MATHIESEN, Karl, "How many more new species are left to discover?" *The Guardian*, 30/X/2013.

MAY, R. M. "How Many Species are there on Earth?". *Science*, 241, 4.872, 16/IX/1988, pp. 1441–1449.

MONASTERSKY, Richard, "Biodiversity: Life – a status report". *Nature*, 10/XII/2014, pp. 158–161.

MORA, Camilo *et al.* "How Many Species Are There on Earth and in the Ocean?". *Plos One Biology*, 23/VIII/2011.

NELLEMANN, Christian (ed.). *The environmental food crisis – the environment's role in averting future food crises. A UNEP rapid response assessment*. UNEP, Grid-Arendal, 2009.

————. *The Environmental Crime Crisis. Threats to Sustainable Development from Illegal Exploitation and Trade in Wildlife and Forest Resources. A UNEP Rapid Response Assessment*. UNEP, Grid-Arendal, 2014.

NEMERGUT, Diana R. *et al.*, "Global patterns in the biogeography of bacterial taxa". *Environmental Microbiology*, 1/VIII/2010.

OOSTHOEK, Sharon, "Pesticides spark broad biodiversity loss". *Nature*, 17/VI/2013.

OSBORNE, Juliet L. "Bumblebees and pesticides". *Nature*, 21/X/2012.

PEARCE, Fred, "Snail's demise suggests sixth mass extinction is under way". *New Scientist*, 8/VI/2015.

PETTIS, Jeffery S. *et al.* "Crop Pollination Exposes Honey Bees to Pesticides Which Alters Their Susceptibility to the Gut Pathogen Nosema ceranae". *Plos one*, 24/VII/2013.

PURVIS, Andy. "A million threatened species? Thirteen questions and answers". IPBES, 2019.

RAUP, David M. & SEPKOSKI, Jr., Joseph John, "Periodicity of extinctions in the geologic past". *PNAS*, 81, 3, 1984, pp. 801–805.

RÉGNIER, Claire *et al.* "Mass extinction in poorly known taxa". *PNAS*, 112, 25, 23/VI/2015.

REILEY, Laura, "This past winter saw the highest honeybee colony losses on record". *The Washington Post*, 5/VII/2019.

RIPPLE, William J. & VALKENBURGH, Blaire Van, "Linking Top-down Forces to the Pleistocene Megafaunal Extinctions', *BioScience*, 60, 7, July/August 2010, pp. 516–526.

RIPPLE, William J. *et al.* "Status and Ecological Effects of the World's Largest Carnivores". *Science*, 10/I/2014.

RIPPLE, William, J. *et al.*, "Collapse of the world's largest herbivores". *Science Advances*, 1, 1/V/2015.

_____. "Ruminants, climate change and climate policy". *Nature Climate Change*, 4, 2–5, 2014.

ROCHA, Maria Cecília de Lima e Sá de Alencar. *Efeitos dos agrotóxicos sobre as abelhas silvestres no Brasil. Proposta metodológica de acompanhamento.* Brasília, Ibama-MMA, 2012.

ROSENBERG, Kenneth V. *et al.*, "Decline of the North American avifauna". *Science*, 366, 6461, 4/X/2019, pp. 120–124.

SANCHEZ-BAYO, Francisco & WYCKHUYS, Kris A.G., "Worldwide decline of the entomofauna: A review of its drivers". *Biological Conservation*, 232, 2019, pp. 8–27.

SOROYE, Peter, NEWBOLD, Tim & KERR, Jeremy, "Climate change contributes to widespread declines among bumble bees across continents". *Science*, 367, 6478, 7/II/2020, pp. 685–688.

STORK, Nigel E. *et al.* "New approaches narrow global species estimates for beetles, insects, and terrestrial arthropods". *PNAS*, 112, 24, 16/VI/2015.

SULLIVAN, Cody, "2018 Arctic Report Card: Reindeer and caribou populations continue to decline". NOAA, 11/XII/2018.

THOMAS, Chris D. *et al.* "Extinction risk from climate change". *Nature*, 427, 8/I/2004, pp. 145–148.

TITTENSOR, Derek P. *et al.* "A mid-term analysis of progress toward international biodiversity targets". *Science*, 2/X/2014.

TOPPING, Christopher John, ALDRICH, Annette, BERNY, Philippe, "Overhaul environmental risk assessment for pesticides". *Science*, 367, 6476, 24/I/2020, pp. 360–363.

UNEP 2010, *Global Honey Bee Colony Disorder and Other Threats to Insect Pollinators.* 2010.

VAUGHAN, Adam, "Global biodiversity targets won't be met by 2020, scientists say". *The Guardian*, 3/X/2014.

VIGNIERI, Sacha. "Vanishing fauna". *Science*, 345, 6.195, 25/VII/2014.

VITOUSEK, Peter M. *et al.*, "Human Domination of Earth's Ecosystems". *Science*, 27, 5325, 25/VII/1997, pp. 494–499.

WEPPRICH, Tyson *et al.*, "Butterfly abundance declines over 20 years of systematic monitoring in Ohio, USA". *Plos One*, 9/VII/2019.

WILKINSON, Allie, "Brazilian Amazon still plagued by illegal use of natural resources". *Nature*, 17/X/2019.

Chapter 11
Collapse of Biodiversity in the Aquatic Environment

For Samuel Iglésias (Santi 2015), "we are probably in the century that will see the extinction of marine fish." In freshwater, where about 35,000 species of fish are known, the collapse of biodiversity has definitely begun, as hundreds of these species have already disappeared because of human activities. Ten years ago, Charles J. Vörösmarty et al. (2010) reported estimates, based on the IUCN, that "at least 10,000–20,000 freshwater species are extinct or at risk." The ongoing death toll in freshwater ecosystems is obviously more overwhelming because of the smaller scale of these ecosystems and their greater exposure and vulnerability to fishing, to large dams which interfere with fish spawning, and to human pollution. Oceans cover 71% of the Earth's surface, and in these huge ecosystems the impacts of humans are more diffuse and delayed. But already in 2011 (almost simultaneously, therefore, with the work coordinated by Vörösmarty on the extinction of freshwater species), a press release signed by the International Programme on the State of the Ocean (IPSO), the IUCN, and the World Commission on Protected Areas (WCPA) summarized the conclusions of an interdisciplinary and international workshop on the state of marine biodiversity in one sentence: "The world's ocean is at high risk of entering a phase of extinction of marine species unprecedented in human history" (IPSO, IUCN, WCPA 2011). This scientific panel examined the combined effects of overfishing, pollution, ocean warming, acidification, and deoxygenation and concluded that:

> The combination of stressors on the ocean is creating the conditions associated with every previous major extinction of species in Earth history. The speed and rate of degeneration in the ocean is far faster than anyone has predicted. Many of the negative impacts previously identified are greater than the worst predictions. Although difficult to assess because of the unprecedented speed of change, the first steps to globally significant extinction may have begun with a rise in the extinction threat to marine species such as reef forming corals.

In fact, over the past three decades, marine defaunation has been developing with increasing brutality and speed, as Douglas McCauley et al. (2015) warn:

© Springer Nature Switzerland AG 2020
L. Marques, *Capitalism and Environmental Collapse*,
https://doi.org/10.1007/978-3-030-47527-7_11

> Although defaunation has been less severe in the oceans than on land, our effects on marine animals are increasing in pace and impact. Humans have caused few complete extinctions in the sea, but we are responsible for many ecological, commercial, and local extinctions. Despite our late start, humans have already powerfully changed virtually all major marine ecosystems.

Green sea turtles (*Chelonia mydas*), for example, might become extinct within the timespan of a generation or so. This is because their sex determination depends on the incubation temperature during embryonic development. The higher the temperature of the sand where the eggs are incubated, the higher the incidence of females. This warming also causes high mortality rates in their offspring. Michael Jensen et al. (2018) state that:

> Turtles originating from warmer northern Great Barrier Reef nesting beaches were extremely female-biased (99.1% of juvenile, 99.8% of subadult, and 86.8% of adult-sized turtles). (…) The complete feminization of this population is possible in the near future.

Extermination of aquatic life is caused by a wide range of anthropogenic impacts, including overfishing, pollution, river damming, warming of the aquatic environment, aquaculture, decline in phytoplankton, eutrophication, deoxygenation, acidification, death, and destruction of corals. Each endangered species in the aquatic environment is vulnerable to at least one of these factors. Just to name two examples:

1. Sturgeon (the common name for the 27 species of fish belonging to the family Acipenseridae): this magnificent water giant that has been able to evolve and adapt to so many transformations of the Earth's system since the Triassic period, over 200 million years ago, is now succumbing to overfishing and to anthropic interference in its habitats. No less than 24 of its 27 species are endangered, and, according to IUCN, sturgeons are "more critically endangered than any other group of species."
2. In the 1950s, thousands of baiji (*Lipotes vexillifer*), a species that had evolved over 20 million years, lived in the waters of the Yangtze; by 1994, fewer than 100 individuals remained, and by 2006, the dolphin had become extinct by pollution, dam building, and reckless navigation (Bosshard 2015).

11.1   Mammals

After the Baiji extinction, there are, according to WWF, only six remaining species of freshwater dolphins in the world, three in South America, and three in Asia, five of them vulnerable, endangered, or critically endangered: (1) the Amazon river dolphin (*Inia geoffrensis*), also known as *boto* or pink river dolphin; (2) the Bolivian river dolphin (*Inia boliviensis*); (3) the tucuxi (*Sotalia fluviatilis*), found throughout the Amazon and Orinoco river basins; (4) the Ganges river dolphin (*Platanista gangetica*), also known as the Ganga or the soons; (5) the Indus river dolphin (*Platanista gangetica minor*); and (6) the Irrawaddy dolphin (*Orcaella brevirostris*). The population of the Irrawaddy dolphin in Laos has been decimated by the fishing industry, and the fate of this species is definitely sealed by the construction of the Don Sahong Dam on the Mekong River in southern Laos (Russo 2015). Two Amazonian freshwater dolphin species, the pink river dolphin and the tucuxi, are critically endangered if the reduction rate of their populations in the observed area (the Mamirauá Sustainable Development Reserve) is the same in all of their habitats. Both species are now in steep decline, with their populations halving every 10 years (botos) and 9 years (tucuxis) at current rates, according to a study conducted between 1994 and 2017 (Silva et al. 2018). The extinction of the soons, the river dolphins of India, is imminent, despite their status as the National Aquatic Animal since 2009. Between the end of the nineteenth and early twentieth centuries, its population was estimated at 50,000 in the Ganges alone. It had fallen to 5000 in 1982 and to 1200–1800 in 2012 on the Brahmaputra and Ganges rivers and their tributaries. The proposed Tipaimukh dam on the Barak River in northeast India may sound the death knell for the Ganges river dolphin in Assam's Barak River system (Ghosh 2019).

Similar existential threats also loom over many marine mammals, and species extinctions have started to occur in recent years. The Japanese sea lion (*Nihon ashika*, *Zalophus japonicus*) was last observed in 1974 and is considered extinct. The Caribbean monk seal (*Monachus tropicalis*), a mammal over 2 m long, was officially declared extinct by the IUCN in 2008. The vaquita (*Phocoena sinus*), a small cetacean endemic to the northern part of the Gulf of California, is on the brink of extinction. According to IUCN, "the vaquita is the most endangered marine mammal species in the world. (…) Its status has continued to deteriorate [since 1958] because of unrelenting pressure from incidental mortality in fisheries." In fact, the vaquita usually dies entangled in nets used to catch totoaba (*Totoaba macdonaldi*), a species that is also "critically endangered" (IUCN), as its swim bladders are in high demand, especially in China. In the 1990s, there were already only 600 to 800 vaquitas and in 2014 only 100. Its extinction is imminent because in 2018 its population was reduced to no more than 15. The South Island Hector's and Māui dolphins are subspecies of the Hector's dolphin *Cephalorhynchus hectori*, whose only habitat is New Zealand's coastal waters, which is critically endangered. Both subspecies are threatened with extinction. The Māui dolphin (*Cephalorhynchus hectori maui*) or popoto is on the edge of extinction. Its population is estimated at 55 individuals, and this species, the smallest of the world's 32 dolphin species, may be extinct in a near future (Rotman 2019).

According to the Marine Mammal Commission of the USA, the population of the Florida manatee (*Trichechus manatus latirostris*) is listed as threatened under the Endangered Species Act and designated as depleted (8810 individuals in 2016) under the Marine Mammal Protection Act. An estimated 824 Florida manatees died in 2018, "a nearly 50% increase over the number of deaths in 2017 and the second-highest death count ever" (Platt and Kadaba 2019). According to Jacqueline Miller (2016):

[t]he historical population of the blue whale [*Balaenoptera musculus*] has been estimated at somewhere between 300,000 and 350,000. Between 1909 and 1965, the reported kill of the blue whales in the Antarctic alone (…) was 328,177. By 1966 this number had increased to an estimated 346,000 blue whales killed. When commercial whaling of the blue whale was

finally banned, it was estimated they had suffered up to a 99% population loss. Present estimates are between 5,000 and 10,000 in the Southern Hemisphere, and between 3,000 and 4,000 in the Northern Hemisphere.

Currently, the IUCN considers the blue whale the largest animal ever known to have existed (24–30 m), and other 20 marine mammals to be endangered, and the following whale species to be critically endangered (Findlay and Child 2016): southern blue whale (*Balaenoptera musculus intermedia*), western gray whale (*Eschrichtius robustus*), Svalbard-Barents sea bowhead whale (*Balaena mysticetus*), North Pacific right whale (*Eubalaena japonica*), Chile-Peru southern right whale (*Eubalaena australis*), and Cook Inlet beluga whale (*Delphinapterus leucas*).

Just like the Japanese sea lion (*Nihon ashika, Zalophus japonicus*) which was wiped out by hunting, whales can also suffer the same fate, given the fact that Japan, Iceland, and Norway, which are currently considered pirate whaling nations, continue to hunt and kill fin, minke, and sei whales every year. Japan alone has killed more than 8000 whales under the guise of scientific research since the 1986 Moratorium on Commercial Whaling established by the International Whaling Commission (IWC). On the other hand, since 2002, over 6000 whales have been saved from the Japanese whalers by the Sea Shepherd Conservation Society (SSCS), the heroic and combative NGO, created by Paul Watson and severely repressed by the governments of the USA, Australia, and New Zealand, in addition to being deemed an eco-terrorist organization by the Japanese government. Due to such pressures, Paul Watson declared in 2017 that the SSCS was abandoning the pursuit of Japanese whalers.

11.2   Noise Pollution

One of the biggest stressors of marine life is man-made noise pollution. "Hydrophones anchored to the continental slope off California, for instance, have recorded a doubling of background noise in the ocean every two decades since the 1960s" (Brannen 2014). Alison Stimpert et al. (2014) documented the susceptibility of Baird's beaked whales (*Berardius bairdii*) to mid-frequency active military sonar. There is a well-documented link between the military sound pulses and mass strandings in which dozens of the mammals have died (Batchelor 2019). The Apache Alaska Corporation deafened Beluga whales by detonating—every 10 or 12 seconds for 3–5 years—air guns for underwater oil and gas prospecting in the only habitat in which they still survive (Broad 2012). A 2012 *New York Times* editorial underlines that:[1]

> Sound travels much faster through water than it does through air, magnifying its impact, and many of the sounds the Navy plans to generate fall in the frequencies most damaging to marine mammals. More than five million of them may suffer ruptured eardrums and

---

[1] See "Marine Mammals and the Navy's 5-Year Plan." *The New York Times*, 12/X/2012.

temporary hearing loss, in turn disrupting normal behavioral patterns. As many as 1800 may be killed outright, either by testing or by ship strikes.

Between 2014 and 2019, the US Navy conducted firing exercises in the Atlantic and Pacific Oceans and in the Gulf of Mexico.

11.3 Overfishing and Aquaculture

Dirk Zeller and Daniel Pauly (2019) highlighted the two factors that contributed to the Great Acceleration after 1950, referring specifically to industrial scale fishing: "(1) the reliance on and availability of cheap fossil fuels; and (2) the gradual incorporation of technologies developed for warfare (e.g., radar, echo sounders, satellite positioning, etc.)." Due to these two factors, worldwide per capita fish consumption more than doubled in the last 60 years. According to FAO (2018), between 1961 and 2016, the average annual increase in global food fish consumption (3.2%) outpaced population growth (1.6%). In per capita terms, this consumption grew from 9 kg in 1961 to 20.2 kg in 2015, at an average rate of about 1.5% per year. Already in 2013, industrialized countries consumed 26.8 kg of fish per capita. Consumption of fish in the European Union increased for nearly all of the main commercial species. It reached 24.33 kg per capita in 2016, 3% more than in 2015. The top five species eaten in the EU—tuna, cod, salmon, Alaska pollock, and shrimp—amounted to 43% of the market in 2016. These species were mostly imported from non-EU countries (EUMOFA 2018).

Global seafood production by fisheries and aquaculture was estimated at 167 million tons in 2014 and at 171 million tons in 2016, of which 88% was utilized for direct human consumption (FAO 2018). In 2012, 15 countries accounted for 92.7% of all aquaculture production, with 88% of global production coming from Asia. Twenty years ago, Rosamond Naylor et al. (2000) had already announced the unsustainable character of aquaculture:

> Many people believe that such growth [of aquaculture] relieves pressure on ocean fisheries, but the opposite is true for some types of aquaculture. Farming carnivorous species requires large inputs of wild fish for feed. Some aquaculture systems also reduce wild fish supplies through habitat modification, wild seedstock collection and other ecological impacts. On balance, global aquaculture production still adds to world fish supplies; however, if the growing aquaculture industry is to sustain its contribution to world fish supplies, it must reduce wild fish inputs in feed and adopt more ecologically sound management practices.

China's aquatic farms provide more fish than the country's fishing activities, and the interaction between sewage, industrial and agricultural waste, and pollution caused by aquaculture itself has devastated Chinese marine habitats (Barboza 2007).

In 2016, lakes, rivers, and oceans were scoured by 4.6 million boats and fishing vessels, 3.5 million of which were Asian (75%), searching for declining shoals of fish (FAO 2018). In fact, the decline in shoals explains why fishery output (excluding discarded fish) reached 93.8 million tons in 1996 and stabilized thereafter to around 90–92 million tons. Since then, the growth in seafood consumption has been increasingly supplied by aquaculture, which more than doubled during the 1990s with a 10% annual growth rate, although its average annual growth fell to 6% between 2000 and 2014 and to 5.8% between 2010 and 2016 (FAO 2018). According to a Rabobank report (2017), "aquaculture is expected to continue its growth in 2018, albeit at a gradually declining rate. Growth in both 2017 and 2018 is estimated to be in the 3% to 4% in volume terms." This growth rate decline could be irreversible as aquaculture is susceptible to climate change and environmental extremes (Plagányi 2019; FAO 2018). Figure 11.1 differentiates between wild capture and aquaculture.

With regard to fishing, Daniel Pauly and Dirk Zeller (2016) show that this graph does not reflect the actual world fishing curve because 30% of global fish catch go unreported every year. According to these two experts on seafood depletion, this curve has peaked at a much higher number (130 million tons) and has since been declining three times faster than the FAO curve suggests, with a loss of one million tons a year due to overfishing (see also Carrington 2016). Figure 11.2 provides a clearer picture of both overfishing and the decline in marine fish stocks beginning in the 1990s:

The graph above shows that this decline is undoubtable and fishing corporations, a sector that is heavily controlled by oligopolies, bear sole responsibility for it. In fact, according to Henrik Österblom et al. (2015):

> Thirteen corporations control 11–16% of the global marine catch (9–13 million tons) and 19–40% of the largest and most valuable stocks, including species that play important roles in their respective ecosystem. They dominate all segments of seafood production, operate through an extensive global network of subsidiaries and are profoundly involved in fisheries and aquaculture decision-making.

Of these 13 multinationals, 4 have headquarters in Norway, 3 in Japan, 2 in Thailand, 1 in Spain, 1 in South Korea, 1 in China, and 1 in the USA. The revenues of these

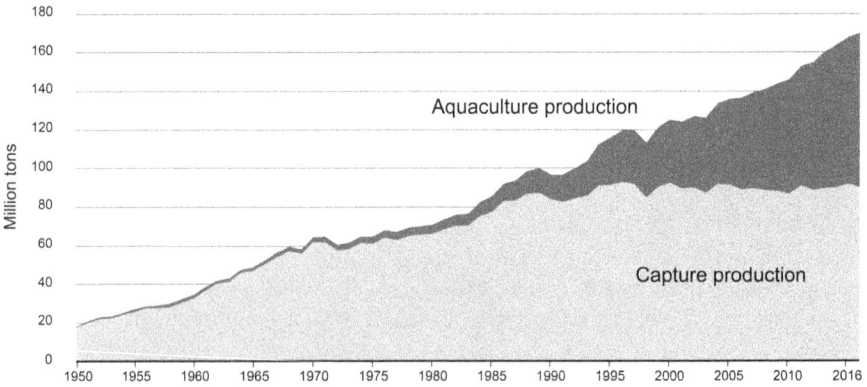

Fig. 11.1 World fish production in millions of tons (1950–2013). (Source: FAO, *The State of World Fisheries and Aquaculture*. Rome, 2018)

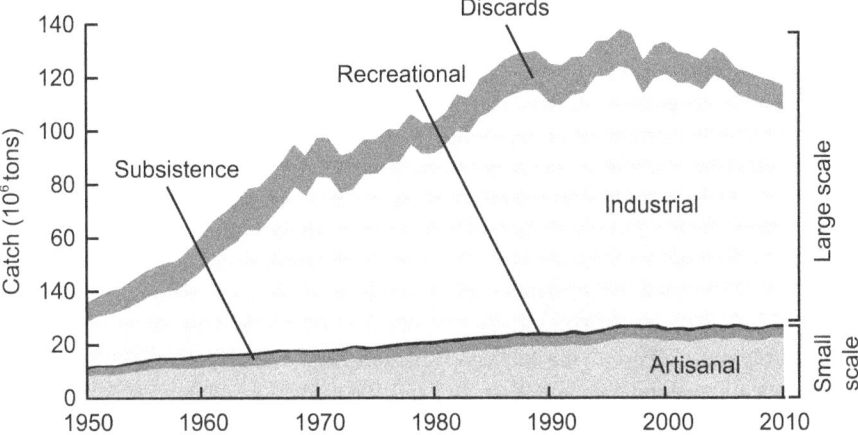

Fig. 11.2 Peak and decline of industrial sea fishing (1950–2010). (Source: Daniel Pauly & Dirk Zeller, "Catch reconstructions reveal that global marine fisheries catches are higher than reported and declining". *Nature Communications*, 7, 19/I/2016)

13 corporations (0.5% of the 2250 fishing and aquaculture companies worldwide) corresponded to 18% of the value of seafood in 2012 (USD 252 billion).

11.4   Fish Stocks on the Verge of Collapse

Back in 2006 Boris Worm and many other ocean scientists published an alarming report on the state of fisheries:

> There is no reasonable doubt that most major declines in stock biomass and in corresponding yields are due to unsustainable levels of fishing pressure. Global assessments of stock biomass support our conclusions that a large fraction of the world's fished biodiversity is overexploited or depleted (24% of assessed stocks in 2003), that this fraction is increasing (from 10% in 1974), and that recovery of depleted stocks under intense management is still an exception (1% in 2003).

Graham Edgar et al. (2014) showed that "total fish biomass has declined about two-thirds from historical baselines as a result of fishing." Assessing 5829 populations of 1234 marine vertebrate species, the Living Blue Planet Report (WWF-ZSL 2015) provides a comprehensive picture of this situation until 2012:

> Marine vertebrate populations declined 49% between 1970 and 2012. Populations of fish species utilized by humans have fallen by half, with some of the most important species experiencing even greater declines. Around one in four species of sharks, rays and skates is now threatened with extinction, due primarily to overfishing. Tropical reefs have lost more than half their reef-building corals over the last 30 years. Worldwide, nearly 20% of mangrove cover was lost between 1980 and 2005. 29% of marine fisheries are overfished. If current rates of temperature rise continue, the ocean will become too warm for coral reefs by 2050. (…) Just 3.4% of the ocean is protected, and only part of this is effectively managed.

As Ken Norris, Director of Science at the Zoological Society of London (ZSL), said in a statement (Doyle 2015), "this report suggests that billions of animals have been lost from the world's oceans in my lifetime alone. (…) This is a terrible and dangerous legacy to leave to our grandchildren." This harsh reality, one that our grandchildren will have no means of coping with, is one that even we are now beginning to feel. Populations of some commercial fish stocks, such as tuna, mackerel, and bonito, fell by almost 75% between 1970 and 2012. Other less iconic and smaller species face similar situations (Zakaib 2011). Over the past 20 years, stocks of hake, flounder, sole, halibut, mackerel, mackerel scad (*Decapterus macarellus*), horse mackerel, and other fish belonging to the Carangidae family have gone from 30 million tons to just 3 million. To continue to supply the market, fishing must now extend over 6000 km from Peru to the Antarctic boundary and cover 120° of longitude, half the distance between Chile and New Zealand. The stocks of Chilean jack mackerel (*Trachurus murphyi* of the family *Carangidae*) decreased by 63% between 2006 and 2011 alone and, according to Daniel Pauly, "when Chilean jack mackerel disappears, everything else will be gone."[2] Data from Sea Around Us on fishing corporations allow Pauly to state that "the catch is going down by 2% a year, but it looks more stable than it is. What they do is deplete one stock and then move onto another, which means they're going to run out of fish in a few decades" (Pala 2016). Boris Worm et al. (2006) project the global collapse of all taxa currently fished by the mid-twenty-first century. Also for Paul Watson (Facebook 8/XII/2019), founder of the Sea Shepherd Conservation Society, "by 2048 there will be no commercial fishing industry because there will be no fish. (…) If the fishes die out, the Ocean dies and when the Ocean dies, humanity dies with it."

The current decline in fish populations explains why "aquaculture has recently superseded wild-capture fisheries as the main source of seafood for human consumption" (Guillen et al. 2019). In fact, industrial fishing would already be unprofitable if it were not supported by subsidies that encourage overfishing, mostly in developed countries. These subsidies "are worth an estimated US$14-35 billion— even though the global fishing fleet is 2–3 times larger than the ocean can sustainably support" (WWF-ZSL 2015).

11.5   Regional Declines, Some Irreversible

The decline in marine life has, in many cases, reached irreversible levels. In the USA, for example, the Sustainable Fisheries Act (SFA) of 1996 set specific quotas for the 44 species protected by law. Even so, eight of them were no longer recoverable. Moreover, these restrictive fishing quotas in US waters may have an even greater impact on oceans that are distant from the country, as 62–65% of seafood consumed in the USA is imported, according to Jessica Gephart, Halley Froehlich,

---

[2] Quoted in "Pacifique sud. La ruée sur un poisson menace tous les autres." *Le Monde*, 29/I/2012.

and Trevor Branch (2019). Richard Conniff (2014) adds a factor that aggravates this picture: one-third of US fish imports hail from trafficking, that is, they are considered IUU fishing, the technical term for illegal, unreported, and unregulated fishing. On the west coast of the USA, shoals of sardines have decreased by about 90% since 2007, from an estimated 1.4 million tons (2007) to roughly 100,000 tons in 2015. In light of this critical situation, the Pacific Fishery Management Council decided to ban fishing of sardines from Mexico to Canada, starting on July 1, 2015 (Fimrite 2015). This was too late, though. Four years after this moratorium, a new stock assessment released by the National Marine Fisheries Service in July 2019 revealed that the population of Pacific sardines amounted to only 27,547 tons. The Pacific sardine population has plummeted by 98.5% since 2006. The impacts of this collapse are echoing in the entire marine ecosystem. According to Elizabeth Grossman (2015):

> Sardines, which are exceptionally nutritious because of their high oil content, are vital to mother sea lions feeding their pups and to nesting brown pelicans. (…) Starving pups have been seen on California's Channel Islands (…). In addition, California brown pelicans have been experiencing high rates of nesting failure and thousands have been dying.

In January 2015, fishing corporations in Brazil obtained, from the government, amendments to the legislation that regulates the capture of 409 fish species and 66 aquatic invertebrates (Annex I list of IUCN). Alexander Lees (2015) warns that such a change may represent "a serious setback for conservation and for the sustainable management of fisheries in Brazil." In 2010, according to a survey conducted in the seas of Brazil, overfishing devastated 80% of marine species. One hundred thousand tons of sardines were fished, for example, between the coast of Rio de Janeiro and Santa Catarina alone.[3] The increasing pressure of overfishing on Brazilian shoals has been ignored because the last survey on fishing in Brazil dates from 2011 and, since then, there has been no quantitative or qualitative data. In the Western and Eastern Mediterranean, 96% and 91% of fish stocks are overfished, respectively. European fishermen are catching fish on average six times more than the so-called maximum sustainable yield (MSY), a limit beyond which the reproduction of these species becomes impossible. This limit is respected by only 4% of European fisheries.

11.6   Bottom Trawling: Fishing as Mining

With industrial fishing, we observe once again the most defining law of global capitalism: as natural resources become scarce, capitalism generates new technologies and more radical means of exploration and increasing devastation. Therefore, the first declines or collapses of shoals in the 1990s typically led to the use of an even

---

[3] See "Overfishing prompts Brazil sardine conservation curbs," UPI, 9/III/2010.

more brutal method of exploitation than those that caused the scarcity: bottom trawling. According to the Living Blue Planet Report (WWF-ZSL 2015):

> Only a few decades ago it was virtually impossible to fish deeper than 500 meters: now, with technological improvements in vessels, gear and fish-finding equipment, bottom trawling is occurring at depths of up to 2,000 meters. Most deep-sea fisheries considered unsustainable have started to target fish populations that are low in productivity, with long lifespans, slow growth and late maturity. This leads to rapid declines in the population and even slower recovery once the stock has collapsed.

Trawling can refer to bottom trawling in search of demersal species or to midwater trawling, which involves the use of probes and GPS to fix the exact depth of the net. In both cases, the result is catastrophic for marine populations. As Elliott A. Norse, President of the Marine Conservation Institute in Washington, and colleagues (2011) reveal how bottom trawling, which scrape the seafloor, are a form of mining for marine life:

> As coastal fisheries around the world have collapsed, industrial fishing has spread seaward and deeper in pursuit of the last economically attractive concentrations of fishable biomass. (…) The deep sea is by far the largest but least productive part of the oceans, although in very limited places fish biomass can be very high. (…) Many deep-sea fisheries use bottom trawls, which often have high impacts on nontarget fishes (e.g., sharks) and invertebrates (e.g., corals), and can often proceed only because they receive massive government subsidies. The combination of very low target population productivity, nonselective fishing gear, economics that favor population liquidation and a very weak regulatory regime makes deep-sea fisheries unsustainable with very few exceptions. Rather, deep-sea fisheries more closely resemble mining operations that serially eliminate fishable populations and move on.

According to the FAO and the authors of this study, bottom trawling increased sevenfold between 1960 and 2004. It has also expanded since the 1950s toward the southern seas at an average rate of 1° latitude per year.

11.7   Hypoxia and Anoxia

"There is no environmental variable of such ecological importance to marine ecosystems that has changed so drastically in such a short period of time as a result of human activities as dissolved oxygen" (Laffoley and Baxter 2019). The loss of oxygen from the world's ocean is a growing threat. The oxygen concentrations have already begun to decline in the twentieth century, both in coastal and offshore areas. They declined by roughly 2% between 1960 and 2010 (Laffoley and Baxter 2019), and the 2013 IPCC Assessment Report (AR5) predicts that they will decrease by 3–6% during this century, only in response to water warming.[4] Multidecadal trends and variability of dissolved oxygen (O_2) in the ocean from 1958 to 2015 were quantified by Takamitsu Ito et al. (2017). The authors conclude that:

---

[4] See Ocean Deoxygenation, in Ocean Scientists for Informed Policy (OSIP).

A widespread negative O_2 trend is beginning to emerge from the envelope of interannual variability. The global ocean O_2 inventory is negatively correlated with the global ocean heat content. (…) The trend of O_2 falling is about 2 to 3 times faster than what we predicted from the decrease of solubility associated with the ocean warming.

The primary causes of ocean deoxygenation are eutrophication, nitrogen deposition from the burning of fossil fuels, and ocean warming. Eutrophication in coastal areas, a phenomenon described by Richard Vollenweider in 1968, is the degenerative response of an ecosystem to the abnormal accumulation of nitrogen and phosphate in water, which stimulates the proliferation of algae and results in the formation of toxic material; release of hydrogen sulfide (H_2S), an equally toxic gas; and obstruction of solar light by algae. When the algae die, they sink and are broken down by microorganisms in a process called bacterial respiration, one that consumes oxygen. This results in hypoxia or, ultimately, anoxia (insufficient or no oxygen concentrations in the water), which leads to the death and decomposition of aquatic organisms and, consequently, to more bacterial activity, in a snowball effect of environmental intoxication and decrease in biodiversity. Although eutrophication may occur for natural reasons, this process has recently been the result of anthropogenic activities, such as agriculture, industry, and sewage disposal. According to a new IUCN report (Laffoley and Baxter 2019), "over 900 areas of the ocean around the world have already been identified as experiencing the effects of eutrophication. Of these, over 700 have problems with hypoxia."

There is a cause and effect relationship between the discharge of municipal effluents, nitrogenated fertilizers, and other phytostimulation compounds composed of nitrogen (N), phosphorus (P), and phosphate (K)—a sector dominated by ten global corporations— and the pollution of soil, atmosphere, and water. By prefacing the 2013 report of *Our Nutrient World* (Sutton et al. 2013), Achim Steiner summarizes the problem:

Excessive use of phosphorus is not only depleting finite supplies, but triggering water pollution locally and beyond while excessive use of nitrogen and the production of nitrogen compounds is triggering threats not only to freshwaters but to the air and soils with consequences for climate change and biodiversity.

11.8 Industrial Fertilizers

Since the second half of the twentieth century, there has been a growing consumption of fertilizers, one that is occurring at much higher rates than population growth. According to Denise Breitburg et al. (2018):

The human population has nearly tripled since 1950. Agricultural production has greatly increased to feed this growing population and meet demands for increased consumption of animal protein, resulting in a 10-fold increase in global fertilizer use over the same period. Nitrogen discharges from rivers to coastal waters increased by 43% in just 30 years from 1970 to 2000, with more than three times as much nitrogen derived from agriculture as from sewage.

Figure 11.3 shows this growth curve by one order of magnitude. World consumption of industrial fertilizers jumped from about 18 million tons in 1950 to 180 million tons in 2013.

In 1998, the world produced 137 million tons of chemical fertilizers, 15% of which were consumed in the USA. Between 1950 and 1998, worldwide use of petrochemical fertilizers increased more than four times per capita (Horrigan et al. 2002). In addition, fertilizer use per hectare of plowed land is increasing, from 110.2 kg/ha in 2003 to 122.4 kg/ha in 2009, according to data from the World Bank. This overconsumption of fertilizers is a consequence of soil depletion, but it is also induced by the dictate of profit maximization. We can speak of hyper-consumption because most of the nitrogen and/or phosphorus contained in these fertilizers is not absorbed by plants. David Tilman (1998) estimates that agriculture absorbs only 33% to 50% of the nitrogen contained in it. For its part, *Our Nutrient World*, the report mentioned above, states that 80% of nitrogen and between 25% and 75% of phosphorus from fertilizers are not incorporated into plants and are dispersed into the environment. Part of this excess penetrates the water tables, and part of it is carried by rain to rivers, lakes, and the sea (Sutton et al. 2013).

There is no recent assessment of the current degree of water eutrophication worldwide. A survey of water eutrophication at the end of twentieth century by the UNEP showed that, globally, 30% to 40% of lakes and reservoirs were then affected

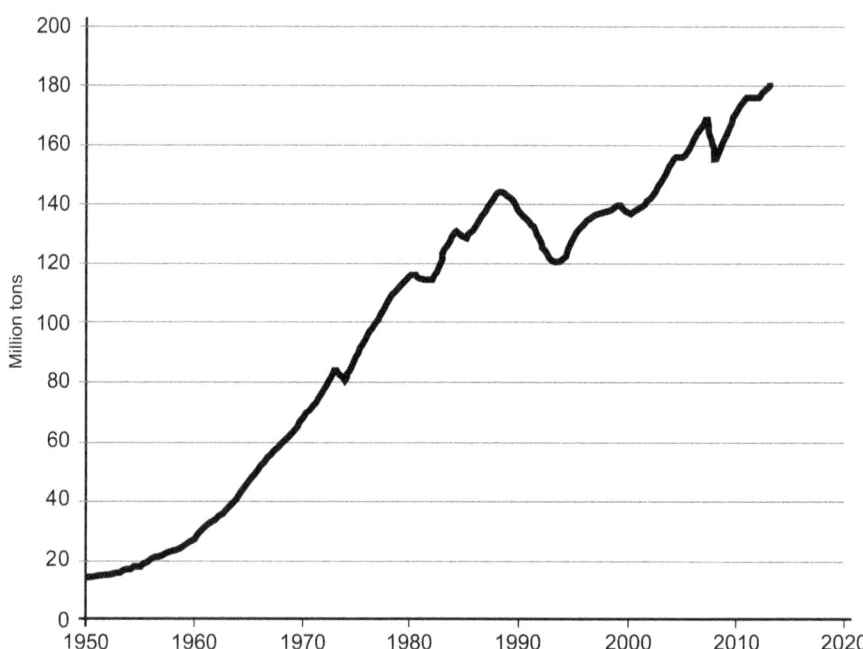

Fig. 11.3 World fertilizer consumption in millions of tons (1950–2013). (Source: Lester R. Brown, "Many Countries Reaching Diminishing Returns in Fertilizer Use". Earth Policy Institute, 8/I/2014)

to varying degrees of eutrophication (Zhang et al. 2019). In 1988, 1990, 1991, 1992, and 1994, the International Lake Environment Committee (ILEC) published interim reports, titled *Survey of the State of World Lakes*. They indicated that eutrophication affected 54% of lakes in Asia, 53% of lakes in Europe, 48% in North America, 41% in South America, and 28% in Africa. In the USA, the EPA's first *National Rivers and Streams Assessment 2008–2009* (NRSA), published in 2013, examined 1924 sites in rivers and streams in the country. The scope of the NRSA was to determine to what extent US rivers and streams provide healthy biological conditions, as well as the magnitude of chemical stressors that affect them:

> NRSA reports on four chemical stressors: total phosphorus, total nitrogen, salinity and acidification. Of these, phosphorus and nitrogen are by far the most widespread. Forty-six percent of the nation's river and stream length has high levels of phosphorus, and 41% has high levels of nitrogen.

In summary, only 21% of the country's rivers and streams are in "good biological condition," and 23% of them are in "fair condition," while 55% of them are in poor condition." Eastern rivers face an even worse fate, with 67.2% of rivers and streams in poor biological condition.

11.9 Ocean Dead Zones: An Expanding Marine Cemetery

In seawater, oxygen exists in different concentrations, varying between 6 ml/l and 8.5 ml/l, depending on the water's temperature and salinity. The term hypoxia applies to when the oxygen concentration is less than 2 ml/l and the term anoxia to when this concentration is less than 0.5 ml/l. Under these conditions, fish that cannot flee in time tend to become disoriented, or they faint and die from asphyxia. Organisms that cannot move with speed, such as crustaceans, and those that live fixed on other structures die in their entirety and their putrefaction retroactively feeds (through bacterial respiration) hypoxia and anoxia.

  The phenomena of hypoxia or anoxia may occur naturally, but they are rare, small-scale, and only seasonal. The anthropogenic factors listed above turn them into frequent and growing phenomena that are large scale and sometimes permanent. An anoxic zone, thus, becomes a marine cemetery where there is no place for vertebrates and other multicellular species. Ocean dead zones with almost zero oxygen have quadrupled in size since 1950, but between 2003 and 2011 alone they more than tripled. "Currently [2016], the total surface area of Oxygen Minimum Zones (OMZ) is estimated to be ~30×10^6 km² (~8% of the ocean's surface area)."[5] According to Robert J. Diaz of the Virginia Institute of Marine Science, these areas

---

[5] See *A Regular Process for Global Reporting and Assessment of the State of the Marine Environment*. Oceans & Law of the Sea United Nations, 21/I/2016: Chapter 3.2.3 Coastal Eutrophication and "Dead Zones."

have roughly doubled every decade (Perlman 2008). Figure 11.4 shows the 10-year increase in the number of dead zones between the early twentieth century and 2006.

In 2011, the WRI and the Virginia Institute of Marine Science identified 530 dead zones and 228 zones displaying signs of marine eutrophication.[6] Some of the most acute cases of dead zones are the North Adriatic Sea; the Chesapeake Bay in the USA, an area of over 18,000 km[2] off the northern shores of the Gulf of Mexico; the mouth of the Mississippi River; Tokyo Bay; certain sea areas that bathe China, Japan, Southeast Australia, and New Zealand; as well as Venezuela's Gulf of Cariaco. Other larger ocean dead zones are found in parts of the North Sea, the Baltic Sea, and the Kattegat Strait between Denmark and Sweden. In 2012, Osvaldo Ulloa from the Center for Oceanographic Research at the University of Concepción in Chile noted the emergence of new anoxic zones off northern Chile's Iquique coast. According to Ulloa, prior to this study, it was not thought that there could be completely oxygen-free areas in the open sea, let alone at levels so close to the surface: "fish lose their habitat and die or move away because they are unable to survive. Only microorganisms, especially bacteria and archaea, can survive" (Gutiérrez 2012).

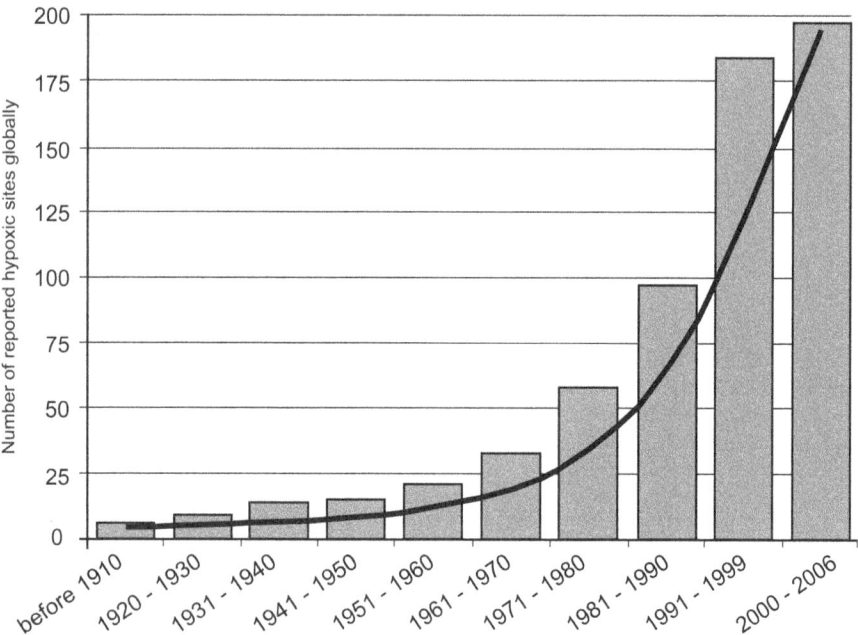

Fig. 11.4 Global increase in hypoxia zones in the sea. (Source: *GEO Year Book, 2013*, from Diaz and Rosenberg (2008))

---

[6] See "New Web-Based Map Tracks Marine 'Dead Zones' Worldwide." World Resources Institute, 20/I/2011.

11.10   Up to 170% More Ocean Acidification by 2100

"The surface ocean currently absorbs approximately one-third of the excess CO_2 injected into the atmosphere from human fossil fuel use and deforestation, which leads to a reduction in pH and wholesale shifts in seawater carbonate chemistry" (Donney et al. 2009). Since 1870 oceans have absorbed 150 Gigatons (Gt) of CO_2, and between 2004 and 2013 alone, they absorbed on average 2.6 Gt of this gas per year. Oceans have a huge ability to absorb cumulative impacts. The dynamic of ocean response to acidification is very slow and is not easily detected within the timeframe of scientific experiments. For example, the increase in ocean CO_2 concentrations resulting from human activities over the past 50–100 years has so far only reached a depth of 3000 m below the surface. But this absorption increases as CO_2 concentrations in the atmosphere increase. This causes chemical changes in the aqueous environment. One of these changes is acidification, that is, a change in ocean pH, or ionic hydrogen potential, which is a measure of hydrogen ion levels (H+) on a scale that indicates the acidity (low pH), neutrality, or alkalinity (high pH) of an environment. A lower pH means that the ocean has become more acidic.

Ocean acidification has been called "the other CO_2 problem" (Donney et al. 2009) and Carol Turley of Plymouth University's Marine Laboratory called it its "evil twin." Rising atmospheric CO_2 concentrations are causing changes in the ocean's carbonate chemistry system. The dissolution of CO_2 in water produces carbonic acid (H_2CO_3), and this process of oceanic acidification decreases the concentrations of calcium carbonate, such as calcite ($CaCO_3$), high-magnesium calcite (HMC), and aragonite, which makes it more difficult for a vast group of marine organisms—corals, crustaceans, urchins, mollusks, etc.—to use these minerals to turn them into their shells or exoskeletal structures. The deficit and/or weakening of these protections slows embryo growth, prevents them from fully forming, or makes their calcareous protections less adherent to stone, less dense, more brittle, and more vulnerable to predators and pathogens. The rapidity of ocean acidification has surprised scientists. In 1999, it was predicted that changing ocean chemistry could affect corals in the mid-twenty-first century. Robert H. Byme (2010) showed that pH changes in the northern Pacific reached a depth of 500 m between 1991 and 2006. Extrapolating from laboratory tests, A. Kleypas et al. (2006) predict that:

> Calcification rates will decrease up to 60% within the 21st century. We know that such extrapolations are oversimplified and do not fully consider other environmental and biological effects (e.g., rising water temperature, biological adaptation); nor do they address effects on organism fitness, community structure, and ecosystem functioning. Any of these factors could increase or decrease the laboratory-based estimates, but it is certain that net production of $CaCO_3$ will decrease in the future.

According to Carbon Brief projections (2018), if GHG emissions continue unabated (RCP8.5 W/m^2), CO_2 atmospheric concentrations will increase to 550 ppm by 2050 and to about 950 ppm by 2100. Richard Feely et al. (2008) state that "this increase [500 ppm by 2050 and 800 ppm by 2100] would result in a decrease in surface-water pH of ~0.4 by the end of the century, and a corresponding 50% decrease in carbonate ion concentration." According to an assessment conducted in 2013 by a US National Academy of Sciences (NAS) committee on the worsening of acidification:[7]

> The rate of release of CO_2 is the greatest for at least the past 55 million years. Since the start of the Industrial Revolution in the middle of the 18th century, atmospheric CO_2 levels have risen by ~40% and the pH of seawater has decreased by ~0.12 pH units, which corresponds to an approximately 30% rise in acidity. By the end of this century, models based on "business as usual" scenarios for CO_2 release predict a further decrease in pH that would lead to an approximately 100–150% rise in ocean acidity relative to the mid-18th century.

A report from the International Geosphere-Biosphere Program (IGBP) presented in 2013 at the 19th United Nations Conference on Climate Change (COP 19) in Warsaw went beyond the estimate of a 100% to 150% increase in ocean acidification, stating that water acidification can increase by up to 170% by 2100, as a result of human activity, leading to the likely disappearance of around 30% of ocean species.

11.11   Ongoing Effects

Ocean acidification is already affecting, for example, the reproductive capacity of oysters grown in the northern Pacific coastal regions. In fact, between 2005 and 2008, the death of millions of oyster larvae on the US Pacific coast was recorded (Welch 2009). According to a study by Nina Bednarsek et al. (2014), acidification is also dismantling the protection of pteropods in California's sea:

> We show a strong positive correlation between the proportion of pteropod individuals with severe shell dissolution damage and the percentage of undersaturated water in the top 100 m with respect to aragonite. We found 53% of onshore individuals and 24% of offshore individuals on average to have severe dissolution damage. (…) Relative to pre-industrial CO_2 concentrations, the extent of undersaturated waters in the top 100 m of the water column has increased over sixfold along the California Current Ecosystem (CCE). We estimate that the incidence of severe pteropod shell dissolution owing to anthropogenic Oceanic Acidification has doubled in near shore habitats since pre-industrial conditions across this region and is on track to triple by 2050.

One of the authors of this study, William Peterson, from the NOAA, stated: "We did not expect to see pteropods being affected to this extent in our coastal region for several decades."[8] Two species of edible bivalve shellfish (*Mercenaria mercenaria*

---

[7] *Review of the Federal Ocean Acidification Research and Monitoring Plan*, The National Academy Press, 2013, p. 10. This is an evaluation of the report commissioned to the National Research Council by the US Congress: *Ocean Acidification: The National Strategy to Meet the Challenges of a Changing Ocean*, NRC, 2010, and the document jointly written by a working group, the Interagency Working Group on Ocean Acidification (Iwgoa), and titled *Strategic Plan for Federal Research and Monitoring of Ocean Acidification*, 2012.

[8] See "NOAA-led researches discover ocean acidity is dissolving shells of tiny snails off the U.S. West Coast." NOAA, 30/IV/2014.

and *Argopecten irradians*) were affected by exposure to varying levels of acidity in the aquatic environment, as shown by Stephanie C. Talmage and Christopher J. Gobler (2010):

> Larvae grown under near preindustrial CO_2 concentrations (250 ppm) displayed significantly faster growth and metamorphosis as well as higher survival and lipid accumulation rates compared with individuals reared under modern day CO_2 levels. Bivalves grown under near preindustrial CO_2 levels displayed thicker, more robust shells than individuals grown at present CO_2 concentrations, whereas bivalves exposed to CO_2 levels expected later this century had shells that were malformed and eroded. These results suggest that the ocean acidification that has occurred during the past two centuries may be inhibiting the development and survival of larval shellfish and contributing to global declines of some bivalve populations.

The decline of these organisms has a chain of repercussions, given their multiple functions in the marine biosphere. Many of them, such as pteropods, are critical to the feeding of other species and to water filtering, making water less toxic to marine organisms. Finally, a lower pH in the blood and cellular fluids of some marine organisms may decrease their ability to capture oxygen, impairing their metabolic and cellular processes. Research is beginning to outline the broad range of these harms to marine life, which include photosynthesis, respiration, nutrient acquisition, behavior, growth, reproduction, and survivability.

11.12   Corals Die and Jellyfish Proliferate

Corals are cnidarian animals that secrete an exoskeleton which is formed of limestone or organic matter. Corals live in colonies considered "superorganisms" which are incomparable repositories of sea life. More than a quarter of all marine species are known to spend part of their lives in coral reefs (Caldeira 2012). Corals are anchors of various marine systems and their disappearance may be lethal to underwater life. Warming waters are the most important *causa mortis* of corals. Warming causes corals to expel the microscopic algae (*zooxanthellae*) living in endosymbiosis with them from their tissues. This, in turn, occasions a process called coral bleaching, from which they do not always recover. The extent and frequency of these bleaching phenomena are increasing. Sixty major mass bleaching events occurred between 1979 and 1990 worldwide. In Australia, scientists from AIMS (Australian Institute of Marine Science) have been monitoring mass bleaching events on the Great Barrier Reef since the early 1980s. Mass bleaching events in 1998, 2002, 2006, 2016, 2017 (Vaughan 2019), and most recently 2020 were caused by unusually warm sea surface temperatures during the summer season. On March 26, 2020, David Wachenfeld, Chief Scientist of the Great Barrier Reef Marine Park Authority, stated in a video posted on its website: "We can confirm that the Great Barrier Reef is experiencing its third mass bleaching event in five years." Some reefs that did not bleach in 2016 or 2017 are now experiencing moderate or severe bleaching. Normally, mass bleaching events are associated with the El Niño climate phenomenon, but the 2017 and 2020 bleaching episodes took place without one. Terry Hughes (dubbed "Reef Sentinel" by the journal *Nature*) and co-authors published an article in *Science* (2018) on a survey of 100 coral reefs in 54 countries; the conclusions could not be more alarming:

> Tropical reef systems are transitioning to a new era in which the interval between recurrent bouts of coral bleaching is too short for a full recovery of mature assemblages. We analyzed bleaching records at 100 globally distributed reef locations from 1980 to 2016. The median return time between pairs of severe bleaching events has diminished steadily since 1980 and is now only 6 years. (…) As we transition to the Anthropocene, coral bleaching is occurring more frequently in all El Niño–Southern Oscillation phases, increasing the likelihood of annual bleaching in the coming decades.

The authors also show that the percentage of bleached corals per year increased from 8% in the 1980s to 31% in 2016. In a subsequent publication, Terry Hughes et al. (2019) show that the reproductive capacities of corals are decreasing after unprecedented back-to-back mass bleaching events caused by global warming: "As a consequence of mass mortality of adult brood stock in 2016 and 2017 owing to heat stress, the amount of larval recruitment declined in 2018 by 89% compared to historical levels." Pollution also plays an important role in coral decline, as tiny plastic particles found in polyps alter their feeding capacity (Hall et al. 2015). Other causes of coral reef death are acidification of water by increasing CO_2 absorption, direct discharge of sewage and domestic and industrial effluents, and the use of dynamite or cyanide to kill fish, which also destroys or poisons reefs. According to the *ReefBase: A Global Information System for Coral Reefs* (with data from the World Atlas of Coral Reefs from the UNEP-World Conservation Monitoring Center), fishing with dynamite and/or cyanide continues to be practiced, although it has been prohibited since 1985. This is the case, for example, in Indonesia, a nation that, together with Australia, is home to the largest coral reefs in the world.

Up to 90% of coral reefs in the seas off the Maldives in the southern Atlantic and off the Seychelles in the Indian Ocean have been killed by global warming. While coral reef systems have existed for over 500 million years, "the Great Barrier Reef is relatively young at 500,000 years of age, with the most recent addition developing only 8,000 years ago" (Esposito 2020). Since 1985, the 2000-km Great Barrier Reef off Australia's east coast—a crucial reservoir for the survival of 400 coral species, 1500 species of fish, and 4000 species of mollusks—has been destroyed by 50%, falling victim to industrial ports, global warming, ocean acidification, pollution, and, additionally, acanthaster, a starfish whose proliferation is due mainly to the agribusiness sector which dumps increasing quantities of industrial fertilizers in the ocean.

An assessment of the Caribbean coral decline between 1970 and 2012, promoted by the IUCN (Jackson, Donovan, Cramer, Lam 2014), states:

> Caribbean coral reefs have suffered massive losses of corals since the early 1980s due to a wide range of human impacts including explosive human population growth, overfishing, coastal pollution, global warming, and invasive species. The consequences include widespread collapse of coral populations, increases in large seaweeds (macroalgae), outbreaks of coral bleaching and disease, and failure of corals to recover from natural disturbances

such as hurricanes. Alarm bells were set off by the 2003 publication in the journal *Science* that live coral cover had been reduced from more than 50% in the 1970s to just 10% today [2012].

Finally, on July 13, 2012, Roger Bradbury, a Specialist on Corals from the Australian National University, wrote an understandably outspoken article in the *New York Times*:

> It's past time to tell the truth about the state of the world's coral reefs, the nurseries of tropical coastal fish stocks. They have become zombie ecosystems, neither dead nor truly alive in any functional sense, and on a trajectory to collapse within a human generation. There will be remnants here and there, but the global coral reef ecosystem—with its storehouse of biodiversity and fisheries supporting millions of the world's poor—will cease to be. Overfishing, ocean acidification and pollution are pushing coral reefs into oblivion. Each of those forces alone is fully capable of causing the global collapse of coral reefs; together, they assure it. The scientific evidence for this is compelling and unequivocal, but there seems to be a collective reluctance to accept the logical conclusion—that there is no hope of saving the global coral reef ecosystem. (…) Coral reefs will be the first, but certainly not the last, major ecosystem to succumb to the Anthropocene—the new geological epoch now emerging.

11.13 Jellyfish

While corals die, jellyfish thrive. In 2010, scientists at the University of British Columbia established that global warming was causing a proliferation and an earlier yearly appearance of numerous species of jellyfish (McVeigh 2011). Once kept in balance by the self-regulating mechanisms of ecosystems, the jellyfish belonging to the subphylum Medusozoa of the phylum Cnidaria, which dates back half a billion years, now proliferate wildly in the oceans, benefiting from (1) warming waters; (2) their transportation to all ports of the world in the ballasts of ships; (3) the multiplication of hard surfaces at sea—piers, boat hulls, oil rigs, and garbage—all ideal nurseries for their eggs; (4) the decline of predatory species, such as sharks, tuna, and turtles (that die by eating pieces of plastic, thinking they are jellyfish); (5) the extinction of competing species caused by overfishing, pollution, fertilizers, and habitat destruction; and (6) lower concentrations of diluted oxygen in the sea. Jellyfish devour huge amounts of plankton, depriving small fish of food, impacting the entire food chain. Tim Flannery (2013) illustrates well the behavior of one jellyfish species:

> *Mnemiopsis* acts like a fox in a henhouse. After they gorge themselves, they continue to collect and kill prey. As far as the ecosystem goes, the result is the same whether the jellyfish digest the food or not: they go on killing until there is nothing left. That can happen quickly.

By seizing niches deserted by locally extinct species or declining populations, jellyfish complete the work of man. Moreover, they exterminate the eggs and larvae of other species. They have also proved to be creatures that have enormous adaptive advantages to the ocean's new environmental conditions, characterized by waters that are more polluted, warmer, more acidic, and less oxygenated, since their metabolism is exceptionally efficient. Lisa-ann Gershwin (2013), Director of the Australian Marine Stinger Advisory Services, issues a terrible warning:

> We are creating a world more like the late Precambrian than the late 1800—a world where jellyfish ruled the seas and organisms with shells didn't exist. We are creating a world where we humans may be soon unable to survive, or want to.

11.14   Plastic Kills and Travels Through the Food Web

Globally, it is estimated that over 100 million marine animals are killed each year by plastic waste. Plastic damages habitats, entangles wildlife, and causes injury via ingestion. Sarah Gall and Richard Thompson (2015) reviewed and updated the current knowledge on the effects of plastic marine debris on marine organisms:

> A total of 340 original publications were identified documenting encounters between marine debris and marine organisms. These [publications] reported encounters for a total of 693 species. 76.5% of all reports listed plastic amongst the debris types encountered by organisms making it the most commonly reported debris type. 92% of all encounters between individual organisms and debris were with plastic.

According to the IUCN, marine plastic pollution has affected at least 267 species and at least 17% of species affected by entanglement and ingestion were listed as threatened or near threatened. Moreover, leachate from plastics has been shown to cause acute toxicity in some species of phytoplankton, such as the fresh species *Daphnia magna* and *Nitocra spinipes* (Bejgarn et al. 2015). It is now a well-established fact that, along with overfishing, ocean warming, acidification, eutrophication, and pollution caused by aquaculture, marine plastic pollution, predicted to increase by an order of magnitude by 2025 (Jambeck et al. 2015), is among the relevant factors responsible for the current annihilation of marine life. Let us return for a moment to this problem, already covered in Chap. 4 (Sect. 4.9 Plastic in the Oceans), to examine its impact on biodiversity in the oceans. The French Research Institute for Exploration of the Sea (IFREMER) estimates that there are 150 million pieces of macrowaste (*macrodéchets*) in the bottom of the North Sea and more than 175 million of them in the northwest Mediterranean basin, with a total of about 750 million floating in the Mediterranean Sea as a whole (Rescan and Valo 2014). Like oil, its raw material, plastic floats, at least in most of its forms. As Andrés Cózar et al. (2014) claim:

> Exposure of plastic objects on the surface waters to solar radiation results in their photodegradation, embrittlement, and fragmentation by wave action. (…) Persistent nano-scale particles may be generated during the weathering of plastic debris, although their abundance has not been quantified in ocean waters.

Each of these particles in this plastic soup retains its chemical characteristics and toxicity. Mammals, fish, birds, and mollusks cannot digest these fragments or particles, which they involuntarily ingest or confuse with plankton, jellyfish, or other food sources. A study published in the *Bulletin of Marine Pollution* (Lusher et al. 2012) detected the presence of microplastics in the gastrointestinal tract of 36.5% of the 504 fish examined from ten different species in the English Channel. The fish were collected at a distance of 10 km from Plymouth and at a depth of 55 m. Alina Wieczorek et al. (2018) studied the presence of plastics in Mesopelagic fishes from the Northwest Atlantic. The results are even more alarming: "Overall 73% of fish contained plastics in their stomachs with *Gonostoma denudatum* having the highest frequency of occurrence (100%), followed by *Serrivomer beanii* (93%) and *Lampanyctus macdonaldi* (75%)." The ingestion of plastic by fish and other marine organisms may cause asphyxiation, blockage of the digestive tract, perforation of the intestines, loss of nutrition, or a false sense of satiety. Philipp Schwabl et al. (2018) explain that:

> Microplastic may harm via bioaccumulation (especially when the intestinal barrier is damaged) and can serve as a vector for toxic chemicals or pathogens. Moreover, ingested plastic may affect intestinal villi, nutritional uptake and can induce hepatic stress.

Microplastics (between 5 mm and < 1 μm in size) are now more numerous than larger fragments. They are infiltrating ecosystems and transferring their toxic components to tissues, as well as to the tiny organisms that ingest them. Mark Anthony Browne et al. (2013) showed the impacts of these components on lugworms (*Arenicola marina*), a marine species of the phylum Annelida, exposed to sand containing a 5% concentration of microplastics presorbed with pollutants (nonylphenol and phenanthrene) and additive chemicals (Triclosan and PBDE-47). The authors warn, more generally, that "as global microplastic contamination accelerates, our findings indicate that large concentrations of microplastic and additives can harm ecophysiological functions performed by organisms." Stephanie Wright et al. (2012) also underline that there are many other species at risk, as their eating behavior is similar to that of *A. marina*, and conclude that "high concentrations of microplastics could induce suppressed feeding activity, prolonged gut residence times, inflammation and reduced energy reserves, impacting on growth, reproduction and ultimately survival." Microplastics can ultimately carry bacteria and algae to other regions of the ocean, causing invasions of species and imbalances with unknown consequences to marine ecosystems.

Since plastic travels the food chain and has become ubiquitous in contemporary societies, it is not surprising that its components are now in human stool, as first detected by Philipp Schwabl et al. (2018). In a pilot study, they identified the presence of 11 types of plastic in human stools: polypropylene (PP), polyurethane, polyethylene, polyvinyl chloride, polyethylene terephthalate (PET), polyamide (nylon), polycarbonate, polymethylmethacrylate (acrylic), polyoxymethylene (acetal), and melamine formaldehyde. PP and PET were detected in 100% of the samples. From this pilot study, the authors conclude that "more than 50% of the world population might have microplastics in their stool." For the authors, the two main sources of plastic ingestion are (1) ingestion of seafood correlated with microplastics content and (2) food contact materials (packaging and processing).

11.15   Ocean Warming, Die-Offs, and Decline of Phytoplankton

More than 90% of the warming that has happened on Earth over the past 50 years has occurred in the ocean (Dahlman and Lindsey 2018). As seen in Chap. 7 (item 7.3 between 1.1 °C and 1.5 °C and accelerating, Fig. 7.6), compared to the 1880–1920 period, global ocean warming reached 1 °C in 2016 and 0.8 °C in 2018. The linear trend of Ocean Heat Content (OHC) at a depth of 0–700 m during 1992–2015 shows a warming trend four times stronger than the 1960–1991 period (Cheng et al. 2017). The staggering scale of ocean warming and the more frequent and intense marine heatwaves already have devastating consequences on ecosystems. For example, this warming is driving shoals away from the equator at a rate of about 50 km per decade. Thus, human populations in tropical countries, whose diet is particularly dependent on fish, will be the most quickly affected. As Daniel Pauly (Davidson 2017) states:

> In temperate areas you will have the fish coming from a warmer area, and another one leaving. You'll have a lot of transformation but they will actually—at least in terms of fishery—adapt. In the tropics you don't have the replacement, you have only fish leaving.

But the problem also presents itself for fish in high latitudes, for which there are no colder places to go. In general, ocean warming is predicted to affect marine species through physiological stress, making them more susceptible to diseases. Ocean warming has also caused mass die-offs of fish and other marine species. Blobs of warm water are killing clams in the South Atlantic (Mooney and Muyskens 2019). Since 2013, a sea star wasting disease, tied to ocean warming, has affected more than 20 sea star species from Mexico to Alaska (Wagner 2014; Harvell et al. 2019; Herscher 2019). About 62,000 dead or dying common murres (*Uria aalge*) washed ashore between summer 2015 and spring 2016 on beaches from California to Alaska. John Piatt et al. (2020) estimate that total mortality approached 1 million of these sea birds. According to the authors, these events were ultimately caused by the most powerful marine heatwave on record that persisted through 2014–2016 and created an enormous volume of ocean water (the "Blob") from California to Alaska with temperatures that exceeded average by 2–3 standard deviations. Overall, die-offs or mass mortality events (MMEs) are becoming increasingly common among species of fish and marine invertebrates over the last 70 years, although it remains unclear whether the increase in the occurrence of MMEs represents a true pattern or simply a perceived increase (Fey et al. 2015).

In addition, rising sea levels lead to the progressive decline of beaches, endangering the species that live on or depend on them to reproduce. According to Shaye Wolf of the Center for Biological Diversity in San Francisco, the loss of beaches threatens 233 species protected by the Endangered Species Act (ESA) in the USA. These include the large loggerhead turtle (*Caretta caretta*), which lays its eggs on Florida's beaches, and the Hawaiian monk seal (*Neomonachus schauin-*

slandi) (Schneider 2015). The same fate is reserved for five turtle species that lay eggs on Brazilian beaches, two of which are endangered and all of which are considered vulnerable by the IUCN.

We saw in the preceding section that plastic pollution is poisoning some species of phytoplankton, such as *Daphnia magna* and *Nitocra spinipes*. But warming waters, along with other factors, also threaten phytoplankton. Daniel Boyce, Marlon Lewis, and Boris Worm (2010) were the first to warn of a long-term downward trend in phytoplankton, a decline that is now occurring at alarming rates:

> We observe declines in eight out of ten ocean regions, and estimate a global rate of decline of approximately 1% of the global median per year. Our analyses further reveal interannual to decadal phytoplankton fluctuations superimposed on long-term trends. These fluctuations are strongly correlated with basin-scale climate indices, whereas long-term declining trends are related to increasing sea surface temperatures. We conclude that global phytoplankton concentration has declined over the past century.

The decline in phytoplankton observed by the authors is so threatening to life on the planet that this finding gave rise to intense discussion, summarized by David Cohen (2010). According to Kevin Friedland from NOAA, new measurements taken in 2013 confirm the lowest ever recorded levels of these organisms in the North Atlantic (Fischetti 2013; Spross 2013). Looking at satellite data collected between 1997 and 2009, a study coordinated by Mati Kahru (2011) links Arctic warming and melting to the hastening of phytoplankton's maximum spring bloom by up to 50 days in 11% of the Arctic ocean's area. This puts the phytoplankton out of sync with the reproductive cycles of various marine mammals that feed on it. In addition, a study published by Alexandra Lewandowska et al. (2014) reaffirms the phenomenon of phytoplankton decline:

> Recently, there has been growing evidence that global phytoplankton biomass and productivity are changing over time. Despite increasing trends in some regions, most observations and physical models suggest that at large scales, average phytoplankton biomass and productivity are declining, a trend which is predicted to continue over the coming century. While the exact magnitude of past and possible future phytoplankton declines is uncertain, there is broad agreement across these studies that marine phytoplankton biomass declines constitute a major component of global change in the oceans. Multiple lines of evidence suggest that changes in phytoplankton biomass and productivity are related to ocean warming.

In September 2015, Cecile S. Rousseaux and Watson W. Gregg published an analysis from data obtained through NASA's satellite observations. The authors confirm the decrease in diatoms, the largest and most common form of phytoplankton: "We assessed the trends in phytoplankton composition (diatoms, cyanobacteria, coccolithophores and chlorophytes) at a 16 global scale for the period 1998–2012. (…) We found a significant global decline in diatoms (-1.22% y^{-1})." If this decline persists, it could be one of the most destructive factors, reducing biodiversity to levels never faced by our species, levels that could produce what we call, in the next chapter, hypobiosphere.

References

BARBOZA, David, "In China, Farming Fish in Toxic Waters". *The New York Times*, 15/XII/2007.
BATCHELOR, Tom, "Scientists demand military sonar ban to end mass whale strandings". *The Independent*, 30/I/2019.
BEDNARSEK, Nina *et al.* "Limacina helicina shell dissolution as an indicator of declining habitat suitability owing to ocean acidification in the California Current Ecosystem". *Proceedings of the Royal Society. Biological Sciences*, 30/IV/2014.
BEJGARN, Sofia et al., "Toxicity of leachate from weathering plastics: An exploratory screening study with Nitocra spinipes". Chemosphere, 132, August 2015.
BOYCE, Daniel G.; LEWIS, Marlon R. & WORM, Boris, "Global phytoplankton decline over the past century". *Nature*, 466, 29/VII/2010, pp. 591–596.
BRADBURY, Roger, "A World without Coral Reefs". *The New York Times*, 13/VII/2012.
BYME, Robert H., "Direct observations of basin-wide acidification of the North Pacific Ocean". *Geophysical Research Letters*, 20/I/2010.
BOSSHARD, Peter, "A Health Check-up for Our Environment — Ignored at Our Own Risk". *The Huffington Post*, 11/I/2015.
BRANNEN, Peter, "Sound off". *Aeon*, 14/X/2014.
BREITBURG, Denise *et al.*, "Declining oxygen in the global ocean and coastal waters". *Science*, 359. 6371, 5/I/2018.
BROAD, William J., "A Rising Tide of Noise Is Now Easy to See". *The New York Times*, 10/XII/2012
BROWN, Lester R., "Many Countries Reaching Diminishing Returns in Fertilizer Use". Earth Policy Institute, 8/I/2014.
BROWNE, Mark Anthony *et al.* "Microplastic Moves Pollutants and Additives to Worms, Reducing Functions Linked to Health and Biodiversity". *Current Biology*, 23, December 2013, pp. 2.388–2.392.
CALDEIRA, Ken, "The Great Climate Experiment. How far can we push the planet?" *Scientific American*, September 2012.
CARBON BRIEF (2018), "Rising CO_2 levels could push 'hundreds of millions' into malnutrition by 2050". 27/VIII/2018.
CARRINGTON, Damian, "Overfishing causing global catches to fall three times faster than estimated". *The Guardian*, 19/I/2016.
CHENG, Lijing et al., "Improved estimates of ocean heat content from 1960 to 2015". Science Advances, 10/III/2017.
COHEN, David, "Reactions to the Phytoplankton Crisis". *The decline of the empire*, 8/V/2010.
CONNIFF, Richard, "Unsustainable Seafood: A New Crackdown on Illegal Fishing". *Yale Environment360*, 22/IV/2014.
CÓZAR, Andrés *et al.* "Plastic debris in the open ocean". *PNAS*, 30/VI/2014, pp. 1–6.
DAHLMAN, LuAnn & LINDSEY, Rebecca, "Climate Change: Ocean Heat Content". NOAA, 1/VIII/2018.
DAVIDSON, Helen, "Marine expert warns of climate emergency as fish abandon tropical waters". *The Guardian*, 15/VI/2017.
DIAZ, Robert J. & ROSENBERG, R. "Spreading Dead Zones and Consequences for Marine Ecosystems". *Science*, 321, 5891, 15/VIII/2008, pp. 926–929.
DONNEY, Scott C. *et al.*, *Ocean Acidification, the other CO_2 problem*, Annual Review of Marine Science. January 2009.
DOYLE, Alister, "Ocean fish numbers on 'brink of collapse' – WWF". *Reuters*, 16/IX/2015.
EDGAR, Graham J. *et al.* "Global conservation outcomes depend on marine protected areas with five key features". *Nature*, 506, 13/II/2014, pp. 216–220.
ESPOSITO, Lara, "Will the Great Barrier Reef Survive Climate Change?". A Climate Institute Publication, January 2020.
EUMOFA, The European Market Observatory for Fisheries and Aquaculture Products, 2018.

FAO, *The State of World Fisheries and Aquaculture. Meeting the sustainable development goals.* Rome, 2018.

FEELY, Richard A. *et al.,* "Evidence for upwelling of corrosive 'acidified' water onto the Continental Shelf". *Science,* 320, 5882, 2008, pp. 1490–1492.

FEY, Samuel B. *et al.,* "Recent shifts in the occurrence, cause, and magnitude of animal mass mortality events". *PNAS,* 12/I/2015.

FIMRITE, Peter, "Sardine population collapses, prompting ban on commercial fishing". *SFGate,* 14/IV/2015.

FINDLAY, K, CHILD, M.F., "A conservation assessment of Balaenoptera musculus". In Child, M. F., Roxburgh, L., Do Linh San, E., Raimondo, D., Davies-Mostert, H.T. (eds.), *The Red List of Mammals of South Africa, Swaziland and Lesotho.* South African National Biodiversity Institute and Endangered Wildlife Trust, South Africa, 2016.

FISCHETTI, Mark, "Sweeping Change in Phytoplankton Populations Could Remake Oceans". *Scientific American,* 8/VIII/2013.

FLANNERY, Tim, "They're Taking Over!". *The New York Review of Books,* 26/IX/2013.

GALL, Sarah C. & THOMPSON, Richard C. "The impact of debris on marine life". *Marine Pollution Bulletin,* March 15, 2015, pp. 170–179.

GEPHART, Jessica A., FROELICH, Halley E. & BRANCH, Trevor A., "Opinion: To create sustainable seafood industries, the United States needs a better accounting of imports and exports". *PNAS,* 7/V/2019, pp. 9142–9146.

GERSHWIN, Lisa-ann. *Stung! On Jellyfish Blooms and the Future of the Ocean.* University of Chicago Press, 2013.

GHOSH, Sahana, "Is the Endgame Inevitable for the Ganges River Dolphin in Assam's Barak River?" *The Wire,* 3/V/2019.

GROSSMAN, Elizabeth, "A Little Fish with Big Impact in Trouble on U.S. West Coast". *Yale Environment360,* 18/VI/2015.

GUILLEN, Jordi *et al.,* "Global seafood consumption footprint". *Ambio,* February 2019.

GUTIÉRREZ, E., "Aparecen zonas anóxicas en la costa de Chile". *La Jornada. Ciências,* 7/XI/2012.

HALL, N. M.; BERRY, K. L. E.; RINTOUL, L. & HOOGENBOOM, M. O. "Microplastic ingestion by scleractinian corals". *Marine Biology,* 162, 3, 2015, p. 725.

HARVELL, C. Drew *et al.,* "Disease epidemic and a marine heat wave are associated with the continental-scale collapse of a pivotal predator (*Pycnopodia helianthoides*)" *Science Advances,* 5, 1, 30/I/2019.

HERSCHER, Rebecca, "Massive Starfish Die-Off is Tied to Global Warming". *NPR,* 30/I/2019.

HORRIGAN, L.; LAWRENCE, R. S. & WALKER P. "How Sustainable Agriculture Can Address the Environmental and Human Health Harms of Industrial Agriculture". *Environ Health Perspectives,* 110, 2002, pp. 445–456.

HUGHES, Terry P. *et al.,* "Spatial and temporal patterns of mass bleaching of corals in the Anthropocene". *Science,* 6371, 5/I/2018, pp. 80–83.

HUGHES, Terry P. *et al.,* "Global warming impairs stock–recruitment dynamics of corals". *Nature,* 568, 3/IV/2019, pp. 387–390.

IPSO, IUCN, WCPA (2011), "Multiple ocean stresses threaten 'globally significant' marine extinction". 20/VI/2011.

ITO, Takamitsu *et al.* "Upper ocean O_2 trends: 1958–2015", *Geophysical Research Letters,* 9/V2017.

JAMBECK, Jenna R. *et al.* "Plastic waste inputs from land into the ocean". *Science,* 347, 6223, 13/II/2015, pp. 768–771.

JENSEN, Michael P. *et al.* "Environmental Warming and Feminization of One of the Largest Sea Turtle Populations in the World". *Current Biology,* 28, 1, 8/I/2018, pp. 154–159.

KAHRU, Mati *et al.* "Are phytoplankton blooms occurring earlier in the Arctic?" *Global Change Biology,* 17, 4, 2011, pp. 1.733–1.739.

KLEYPAS, J. A. *et al.*, "Impacts of Ocean Acidification on Coral Reefs and Other Marine Calcifiers: A Guide for Future Research". NSF, NOAA, USGS, 2006.

LAFFOLEY, Dan & BAXTER, John M. (eds.), *Ocean deoxygenation: everyone's problem. Causes, impacts, consequences and solutions*. IUCN, 2019.

LEES, Alexander C., "Leave Brazil's Red List alone". Nature, 518, 11/II/2015, p. 167.

LEWANDOWSKA, Alexandra *et al.* "Effects of sea surface warming on marine plankton". *Ecology Letters*, 2014, pp. 1–10.

LUSHER, A. L.; McHUGH, M. & THOMPSON, R. C. "Occurrence of microplastics in the gastrointestinal tract of pelagic and demersal fish from the English Channel". *Bulletin of Marine Pollution*, 26/XII/2012.

McCAULEY, Douglas J. *et al.* "Marine defaunation: Animal loss in the global ocean". *Science*, 347, 6.219, 16/I/2015.

McVEIGH, Tracy, "Explosion in jellyfish numbers may lead to ecological disaster". *The Guardian*, 12/VI/2011.

MILLER, Jacqueline, "Blue Whale Research". *Royal Ontario Museum,* 2/VIII/2016.

MOONEY, Chris & MUYSKENS, John, "Dangerous new hot zones are spreading around the world". *The Washington Post*, 11/IX/2019.

NAYLOR, R. L. *et al.* "Effect of aquaculture on world fish supplies", *Nature*, 405, June 2000, pp. 1.017–1.024.

NORSE, Elliott *et al.* "Sustainability of deep-sea fisheries". *Marine Policy*, 25/VI/2011.

ÖSTERBLOM, Henrik *et al.*, "Transnational Corporations as 'Keystone Actors' in Marine Ecosystems". *Plos One*, 10, 5, 2015.

PALA, Christopher, "Oceans running out of fish as undeclared catches add a third to official figures". *The Ecologist*, 19/I/2016.

PAULY, Daniel & ZELLER, Dirk. "Catch reconstructions reveal that global marine fisheries catches are higher than reported and declining". *Nature Communications*, 7, 19/I/2016.

PERLMAN, David, "Scientists alarmed by ocean dead-zone growth". *SFGate*, 15/VIII/2008.

PIATT, John F. *et al.*, "Extreme mortality and reproductive failure of common murres resulting from the northeast Pacific marine heatwave of 2014–2016". *Plos One*, 15/I/2020.

PLAGÁNYI, Éva, "Climate change impacts on fisheries". *Science*, 363, 6430, 1/III/2019, pp. 930–931.

PLATT, John R. & KADABA, Dibika, "13 Percent of Florida Manatees Died Last Year". *The Revelator*, 31/I/2019.

RABOBANK, *Global Animal Protein 2018*. Rabo Research Food & Agribusiness, November 2017.

report (2018), "aquaculture is expected to continue its growth in 2018

RESCAN, Manon & VALO, Martine, "10 chiffres qui vous ignoriez (peut-être) sur l'environnement en France". *Le Monde*, 27/XII/2014.

ROTMAN, Veronica, "Maui's dolphin: Going, going, gone?". The University of Auckland, 1/I/2019.

ROUSSEAUX, Cecile S. & GREGG, Watson W., "Recent decadal trends in global phytoplankton composition". *Global Biogeochemical Cycles*, 23/IX/2015.

RUSSO, Christina, "Dolphin's death leaves only 5 of her kind on the planet". *The Dodo*, 8/IV/2015.

SANTI, Pascale, "On verra probablement, au XXIᵉ siècle, l'extinction des poissons marins". *Le Monde*, 27/VII/2015.

SCHNEIDER, Amy, "The Changing Tide". *Eugeneweekly.com*, 16/IV/2015

SCHWABL, Philipp *et al.*, "Assessment of microplastic concentrations in human stool". *United European Gastroenterology (UEG)*, 23/X/2018.

SILVA, Vera Maria Ferreira da *et al.*, "Both cetaceans in the Brazilian Amazon show sustained, profound population declines over two decades". *Plos One*, 13, 5, 2/V/2018.

SPROSS, Jeff, "Rapid Plankton Decline Puts the Ocean's Food Web in Peril". *ClimateProgress*, 26/XI/2013.

STIMPERT, A. K. *et al.* "Acoustic and foraging behavior of a Baird's beaked whale, *Berardius bairdii*, exposed to simulated sonar". *Nature*, 7.031, 14/V/2014.

SUTTON, M. A. *et al. Our Nutrient World: The challenge to produce more food and energy with less pollution.* Edinburgh, Centre for Ecology and Hydrology, 2013.

TALMAGE, Stephanie C. & GOBLER, Christopher J., "Effects of past, present, and future ocean carbon dioxide concentrations on the growth and survival of larval shellfish". PNAS, October 5, 107, 40, 2010, pp. 17246–17251.

TILMAN, David *et al.* "The greening of the green revolution". *Nature*, 396, 1998, pp. 211–212.

VAUGHAN, Adam, "The Great Barrier Reef is losing its ability to recover from bleaching". *New Scientist*, 2/IV/2019.

VÖRÖSMARTY, Charles *et al.* "Global threats to human water security and river biodiversity". *Nature*, 467, 30/IX/2010, pp. 555–561.

WAGNER, Eric, "Scientists Look for Causes of Baffling Die-off of Sea Stars". *Yale Environment 360*, 17/VII/2014.

WELCH, Craig, "Oysters in deep trouble: Is Pacific Ocean's chemistry killing sea life?" *Seattle Times*, 14/VI/2009.

WIECZOREK, Alina *et al.*, "Frequency of Microplastics in Mesopelagic Fishes from the Northwest Atlantic". *Frontiers in Marine Science,* 5, 19/II/2018.

WORM, Boris *et al.* "Impacts of Biodiversity Loss on Ocean Ecosystem Services". *Science*, 314, 787, 3/XI/2006.

WRIGHT, Stephanie L.; ROWE, Darren; THOMPSON, Richard C. & GALLOWAY, Tamara S. "Microplastic ingestion decreases energy reserves in marine worms". *Current Biology*, December, 23, 2012, pp. 2.388–2.392.

WWF/ZSL, *Living Blue Planet Report. Species, habitats and human well-being*, London, 2015.

ZAKAIB, G. Dickey. "Overfishing hits all creatures great and small". *Nature*, 3/V/2011.

ZELLER, Dirk & PAULY, Daniel, "Viewpoint: Back to the future for fisheries, where will we choose to go?" *Global Sustainability* 2, 11, 2019, pp. 1–8.

ZHANG, Y., LIANG, J., ZENG G., *et al.*, "How climate change and eutrophication interact with microplastic pollution and sediment resuspension in shallow lakes: A review". *Science of the Total Environment*, 2019.

Chapter 12
Genesis of the Idea of the Anthropocene and the New Man–Nature Relationship

The Anthropocene Working Group of the Subcommission on Quaternary Stratigraphy defines the Anthropocene as[1]:

> […] the present geological time interval, in which many conditions and processes on Earth are profoundly altered by human impact. This impact has intensified significantly since the onset of industrialization, taking us out of the Earth System state typical of the Holocene Epoch that post-dates the last glaciation. (…) Phenomena associated with the **Anthropocene** include: an order-of-magnitude increase in erosion and sediment transport associated with urbanization and agriculture; marked and abrupt anthropogenic perturbations of the cycles of elements such as carbon, nitrogen, phosphorus and various metals together with new chemical compounds; environmental changes generated by these perturbations, including global warming, sea-level rise, ocean acidification and spreading oceanic 'dead zones'; rapid changes in the biosphere both on land and in the sea, as a result of habitat loss, predation, explosion of domestic animal populations and species invasions; and the proliferation and global dispersion of many new 'minerals' and 'rocks' including concrete, fly ash and plastics, and the myriad 'technofossils' produced from these and other materials.

James Syvitski (2012) rightly observed that "the concept of the Anthropocene has manifested itself in the scientific literature for over a century under various guises." In fact, the underlying idea that man's actions shape the Earth system more decisively than non-anthropic forces is more than two centuries old. It dates back to the late eighteenth century, a time of lively reflection on the relationship between man and nature. Its history must be remembered here with its essential milestones; otherwise, we run the risk of not understanding the intellectual foundation of the concept of the Anthropocene. This concept culminates in one of the richest chapters in the history of ideas of the contemporary age (Lorius and Carpentier 2010; Bonneuil and Fressoz 2013) and, at the same time, in the most crucially decisive chapter of human experience on this planet.

In 1780, in his book, *Époques de la nature* and, more precisely, in the seventh and last part of the book titled *Lorsque la puissance de l'homme a fécondé celle de*

---

[1] See <http://quaternary.stratigraphy.org/workinggroups/anthropocene/>.

© Springer Nature Switzerland AG 2020
L. Marques, *Capitalism and Environmental Collapse*,
https://doi.org/10.1007/978-3-030-47527-7_12

la Nature (II, 184–186), Buffon notes that "the whole face of the Earth today bears the mark of man's power." But in claiming the superiority of a nature "fertilized" by man over a "brute" nature, he still understands human omnipresence as a benevolent force:

> It is only about thirty centuries ago that man's power merged with that of nature and extended over most of the Earth; the treasures of its fertility, hitherto hidden, were revealed by man. (...) Finally, the whole face of the Earth today bears the mark of man's power, which, although subordinate to that of nature, has often done more than she did, or has, at least, made her wonderfully fruitful, for it is with the help of our hands that nature has developed in all its extension (…) Compare brute nature with cultivated nature (…).

This idea of the superiority of cultivated nature over brute nature rests on a philosophical tradition that was well described a century before Buffon in the conclusion of Leibniz's *De rerum originatione radicali* (On the Ultimate Origination of Things, 1697, paragraph 16), where one reads:

> It must be recognized that there is a perpetual and most free progress of the whole universe towards a consummation of the universal beauty and perfection of the works of God, so that it is always advancing towards greater cultivation. Just as now a large part of our earth has received cultivation, and will receive it more and more.

But already in the end of the eighteenth century and early nineteenth century, diagnoses quite different from those of Leibniz and Buffon, especially on man's disastrous impact on forests, were beginning to emerge from the pen of naturalists, such as Lamarck (1820), José Bonifácio de Andrada e Silva (Padua 2002), Alexander von Humboldt, Dietrich Brandis, and George Perkins Marsh, as well as Gifford Pinchot in the early twentieth century. Due to the immense breadth of his intellectual capacity, Humboldt is perhaps the most appropriate starting point for a prehistory of the Anthropocene, at least in the realm of modern science. As Andrea Wulf (2015) clearly points out, he seems to have been the first to scientifically conceive of the Earth as a "natural whole" animated by inward forces to the point that he had thought (more than a century before James Lovelock) to give the title *Gäa* to the volumes that he would eventually title *Cosmos*. Of even greater importance to the Anthropocene issue, Humboldt was, along with Lamarck, the first to realize that human activity had a decisive impact on ecosystems. From his analysis of the devastation of forests by colonial plantations at Lake Valencia in Venezuela in 1800 (*Personal Narrative* 1814–1829, vol. IV, p. 140), "Humboldt became the first scientist to talk about harmful human-induced climate change" (Wulf 2015). As Wulf also states, "he warned that humans were meddling with the climate and that this could have an unforeseeable impact on future generations."

A brief historical retrospect of the ideas that underlie the Anthropocene concept since the nineteenth century was proposed by Steffen et al. (2011). The authors' starting point is neither Lamarck nor Humboldt, but Marsh who was, in fact, among the first to realize that human action on the planet had become a threat to life. Hence, in *The Earth as Modified by Human Action* (1868/1874), a largely rewritten version of *Man and Nature: Or, Physical Geography as Modified by Human Action* (1864), he had a goal that was diametrically opposed to Buffon's *défense et illustration* of human power over nature. Starting from the preface to the first edition, Marsh spells out his concerns:

> The object of the present volume is: to indicate the character and, approximately, the extent of the changes produced by human action in the physical conditions of the globe we inhabit; to point out the dangers of imprudence and the necessity of caution in all operations which, on a large scale, interfere with the spontaneous arrangements of the organic or the inorganic world; to suggest the possibility and the importance of the restoration of disturbed harmonies and the material improvement of waste and exhausted regions; and, incidentally, to illustrate the doctrine that man is, in both kind and degree, a power of a higher order than any of the other forms of animated life, which, like him, are nourished at the table of bounteous nature.

In the same period, that is, between 1871 and 1873, abbot Antonio Stoppani (1824–1891) coined the term Anthropozoic; although not devoid of religious connotations, it clearly referred to human interference in the geological structures of the planet:

> It is in this sense, precisely, that I do not hesitate in proclaiming the Anthropozoic era. The creation of man constitutes the introduction into nature of a new element with a strength by no means known to ancient worlds. And, mind this, that I am talking about physical worlds, since geology is the history of the planet and not, indeed, of intellect and morality. (…) This creature, absolutely new in itself, is, to the physical world, a new element, a new telluric force that for its strength and universality does not pale in the face of the greatest forces of the globe.

In 1896, Svante Arrhenius (1859–1927) first confronted the difficult question of the doubling of CO_2, calculating by how much a given change in the atmospheric concentrations of gases that trap infrared radiation would alter the planet's average temperature. He states, for example, that a 40% increase or decrease in the atmospheric abundance of the trace gas CO_2 might trigger the glacial advances and retreats, an estimate similar to that of the IPCC (Le Treut and Somerville 2007 IPCC AR4). He also stated that "the temperature of the Arctic regions would rise about 8 °C or 9 °C, if the [atmospheric concentrations of] CO_2 increased 2.5 to 3 times its present value." This estimate is also probably not far from the truth, given the phenomenon known as Arctic amplification, briefly discussed in Chaps. 7 and 8. In 1917, Alexander Graham Bell finally coins the term "greenhouse effect" to describe the impact of CO_2 concentrations on planetary temperatures. In the early 1920s, Vladimir I. Vernadsky (1926) introduced the idea that just as the biosphere had transformed the geosphere, the emergence of human knowledge (which Teilhard de Chardin (1923) and Édouard Le Roy would name noosphere) was transforming the biosphere (Clark et al. 2005).

These pioneering scientific contributions, from 1860 to 1920, go hand in hand (especially in England and the USA) with the first philosophical and moral reactions to industrialization and urbanization by artists and intellectuals such as John Ruskin, George Bernard Shaw (Dukore 2014), Henry Thoreau, and John Muir.[2]

---

[2] John Muir (1838–1914) founded the Sierra Club, one of the oldest and most active environmental NGOs in the world, in 1892.

They also go hand in hand with the first legal initiatives and environmental organizations, such as the Sea Birds Preservation Act (1869), considered the first conservationist law in England, the Plumage League (1889) in defense of birds and their habitats, the Coal Smoke Abatement Society (1898), the Sierra Club (1892), the Rainforest Action Network (1895), the Ecological Society of America (1915), the Committee for the Preservation of Natural Conditions (1917), and the Save the Redwoods League (1918), which were mobilized especially to safeguard sequoias.

The environmental awareness strongly emerging in this period will wane with Europe sinking into World War I and with the extreme political tensions that would lead to World War II. The idea that man's power could be equated to the forces of nature takes on—with the carnage of World War I—the characteristics that Freud had attributed to it: humans could, thenceforth, use their growing technological control over these forces of nature to give free rein to their inherent aggressive drives,[3] which now threaten to destroy us. In 1930, in the conclusion of his *Civilization and Its Discontents*, Freud expresses this fear:

> The fateful question for the human species seems to me to be whether and to what extent their cultural development will succeed in mastering the disturbance of their communal life by the human instinct of aggression and self-destruction. It may be that in this respect precisely the present time deserves a special interest. Men have gained control over the forces of nature to such an extent that with their help they would have no difficulty in exterminating one another to the last man. They know this, and hence comes a large part of their current unrest, their unhappiness and their mood of anxiety.

12.1 Cold War, the Atomic Age, and the Industrial and Agrochemical Machine

After World War II, industrial capitalism reaches its Golden Age (1945–1973). In France, this period is called *Les Trente Glorieuses* (The Glorious Thirty): the "economic miracles" of Germany, Italy, and Japan take place, the USA achieves complete economic hegemony, the socialist bloc under Soviet leadership maintains an equally strong economic performance, and economic growth rates remain generally very high, even in some countries with late and incipient industrialization. But this widespread rush of economic activity is precisely what produces its inevitable counterpart: the first major pollution crises. The tension between economic benefits and environmental harms—of a civilization model based on the accumulation of surplus and capital—gradually becomes inevitable. The perception of an environmental threat begins to rival the threat of total destruction by war that was forecast by Freud well before Hiroshima, as seen.

---

[3] Human aggression is an unfailing constant of our species and the technology of war precedes that of agriculture. A mass grave in northern Sudan with 59 human skeletons showing signs of violent deaths (having spears and arrowheads inserted between their bones) dates back to about 13,000 BC and is the first archaeological evidence of a war (Cf. Jan and Mat Zalasiewicz (2015)).

In 1973, at the end of his life, Arnold Toynbee would attempt, in *Mankind and Mother Earth* (1976, posthumous), a new synthesis of the history of civilization: new because he would now symptomatically place it under the man–biosphere antinomy. Indeed, the British historian's approach is not distant from the Freudian analysis of humanity's (self)destructive drives. It only emphasizes, half a century after Freud, the environmental consequences of this destructiveness, anticipating the prospect of an uninhabitable Earth which is later taken up in David Wallace-Wells' famous article and book (2017, 2019):

> Mankind's material power has now increased to a degree at which it could make the biosphere uninhabitable and will, in fact, produce this suicidal result within a foreseeable period of time if the human population of the globe does not now take prompt and vigorous concerted action to check the pollution and the spoliation that are being inflicted on the biosphere by short-sighted human greed.

Between Freud's late work (1930) and Toynbee's posthumous (1976) one, the great expansion of capitalism had unleashed, alongside the big Cold War crises, immense impacts on nature, which were beginning to generate counter-impacts. The impacts of World War II and then the Cold War on the degradation of the biosphere were considerable. Jan and Mat Zalasiewicz (2015) provide some examples of this:

> After both world wars, the exhausted armies were left with millions of unused bombs, including chemical weapons. There was neither the time nor the resources to make them safe; most were simply shipped out to sea and thrown overboard. There are over a hundred known weapons dumps in the seas around north-west Europe alone. (…) Even larger dumping grounds were improvised elsewhere in the world. (…) Around a trillion [bullets] have been fired since the beginning of the second world war—that's a couple of thousand for every square kilometre of Earth's surface, both land and sea.

According to the same authors, between 1965 and 1971, Vietnam, a country smaller than Germany, was bombarded with twice as much high explosives as the US forces used during the entire World War II: "The region was pulverized by some 26 million explosions, with the green Mekong delta turning into 'grey porridge,' as one soldier put it." Vietnam and other Southeast Asian countries are undoubtedly among the main victims of the Cold War. But its most systemic victim was the biosphere, obviously including humanity as a whole, primarily as a result of the arms race. Between 1945 and 2013, there were 2421 atomic tests of which over 500 occurred in the atmosphere in the 1950s and 1960s (Jamieson 2014).[4] In the 1950s, scientists began warning of their impacts. Hence, after October 10, 1963, atomic tests generally happened underground, as established by the Treaty Banning Nuclear Weapon Tests in the Atmosphere, in Outer Space, and Under Water, which entered into force on that day. Nevertheless, France and China continued to carry out nuclear tests in the atmosphere until 1974 (France) and 1980 (China).

---

[4] Of the 2421 atomic tests performed between 1945 and 2013, 1123 were conducted by the USA between 1945 and 1992; 982 by the Soviet Union between 1949 and 1991; 210 by France between 1960 and 1996; 45 by the UK between 1952 and 1991; 45 by China between 1952 and 1996; and 16 by other countries between 1974 and 2013. See also (YouTube) Isao Hashimoto, *A Time-Lapse Map of Every Nuclear Explosion Since 1945* (2003).

The full assessment of the effects of these tests on life on the planet will only be achieved by future generations. Among its most feared and delayed effects is the impact of 30 tons of plutonium-239 waste with a half-life of 24,000 years, abandoned by the Americans on the Pacific Islands. No less than "67 nuclear and atmospheric bombs were detonated on Enewetak and Bikini between 1946 and 1958—an explosive yield equivalent to 1.6 Hiroshima bombs detonated every day over the course of 12 years," writes Jose et al. (2015). With rising sea levels, this radioactive waste, stored since 1979 in the Marshall Islands in a concrete dome called Runit Dome, will be submerged, likely causing it to leak into the ocean. In fact, as these three authors warn, "underground, radioactive waste has already started to leach out of the crater: according to a 2013 report by the US Department of Energy, soil around the dome is already more contaminated than its contents." In this regard, they quote Michael Gerrard, Director of the Sabin Center for Climate Change Law at Columbia University, who visited the dome in 2010:

> Runit Dome represents a tragic confluence of nuclear testing and climate change. It resulted from US nuclear testing and the leaving behind of large quantities of plutonium. Now it has been gradually submerged as result of sea level rise from greenhouse gas emissions by industrial countries led by the United States.

In short, if there are many historical reasons for the birth of the term Anthropocene, the main one is the fact that the geological strata and the Earth system in general have been profoundly shaped—and will be even more so in the twenty-first century—by numerous consequences of the world wars that ravaged the twentieth century. These range from the atomic weapon and the various regional wars of the Cold War to the scorched-earth war against the biosphere, driven by the deadliest of weapons: the fossil fuel and agrochemical machine of globalized capitalism.

12.2 Anthropization: A Large-Scale Geophysical Experiment

In 1939, in an article titled "The Composition of the Atmosphere through the Ages," Guy Stewart Callendar bore witness to the emerging awareness of the anthropogenic character of climate change. As James Fleming (2007) notes, the article "contains an early statement of the now-familiar claim that humanity has become an agent of global change." Although they date back 80 years, Callendar's words would be quite appropriate for a text from 2020 on the Anthropocene:

> It is a commonplace that man is able to speed up the processes of Nature, and he has now plunged heavily into her slow-moving carbon dioxide into the air each minute. This great stream of gas results from the combustion of fossil carbon (coal, oil, peat, etc.), and it appears to be much greater than the natural rate of fixation.

Callendar's article later gave rise to the term Callendar effect, which refers to the perception of the direct and well-quantified correlation between anthropogenic GHG emissions and structural changes in the planet's climate. But Arrhenius and Callendar still thought that this planetary warming would be beneficial, delaying a "return of the deadly glaciers." In addition, this article came out on the eve of the war, and its wording naturally went unnoticed in the face of much more immediate and tangible dangers. War, killing on an industrial scale, the so-called trivialization of evil on a scale never before imagined by man, concentration camps, and Hiroshima submerged everything. In the 1950s, two articles by Gilbert Plass (1956, 1959) brought up the problem of global warming once more. Curiously, the impact of CO_2 atmospheric concentrations on the climate was still considered no more than a scientific "theory." In fact, one of Plass' papers (1956) was titled "The Carbon Dioxide Theory of Climatic Change." Plass calculated in this article that "the average surface temperature of the earth increases 3.6°C if the CO_2 concentration in the atmosphere is doubled." In any case, the relationship between the CO_2 concentration in the atmosphere and the rise of the Earth's surface temperature began to be seen in a different light (see also Chap. 8). In the 1950s, the perception of man's potentially dangerous interference in the climate merges with a new awareness of widespread pollution, destruction of nature, and the ills of industrial society. Many environmental disasters mobilize people's awareness in England and in the USA. These disasters include the Great Smog of London, the Cuyahoga River Fire in Ohio (both in 1952), the return to deforestation caused by the US housing boom, projects to flood and build dams in the Grand Canyon (1963), and the explosion of a Union Oil platform off California's coast in 1969 (polluting its sea and beaches with 16 million liters of oil, the largest event of its kind at that time). Climate change also seems to take public perception by storm. In 1953, the famous magazines, *Popular Mechanics* and *Time*, make room for this topic. In a *Time* article titled "The invisible blanket," Gilbert Plass predicts that:

> At its present rate of increase, the CO_2 in the atmosphere will raise the earth's average temperature 1.5° Fahrenheit [0.9°C] each 100 years (...) For centuries to come, if man's industrial growth continues, the earth's climate will continue to grow warmer.

In 1957, Roger Revelle and Hans Suess conclude an article on the increase in atmospheric CO_2 concentrations with a paragraph that has become, perhaps, the best known hallmark in the contemporary history of global warming science:

> Thus human beings are now carrying out a large scale geophysical experiment of a kind that could not have happened in the past nor be reproduced in the future. Within a few centuries we are returning to the atmosphere and oceans the concentrated organic carbon stored in sedimentary rocks over hundreds of millions of years.

The following year, in 1958, an educational film called *The Unchained Goddess*, produced by Frank Capra, the great Italian-American Filmmaker who was also a Chemical Engineer, predicts that atmospheric warming and glacial melting, both caused by human activity, would be calamitous:

> Due to our release through factories and automobiles every year of more than six billion tons of carbon dioxide, our atmosphere seems to be getting warmer. (...) A few degrees rise in the Earth's temperature would melt the polar ice caps. And if this happens, an inland sea would fill a good portion of the Mississippi valley.

This rising awareness of the harmful effects of pollution and GHG emissions is reflected in the establishment (between 1947 and 1971) of the eight most influential US environmental NGOs: Defenders of Wildlife (1947), The Nature Conservancy (1950), WWF (1961), Environmental Defense Fund (1967), Friends of the Earth (1969), International Fund for Animal Welfare (IFAW) (1969), Natural Resources Defense Council (NRDC) (1970), and Greenpeace (1971) (Hunter 2002). Another reflection of this awakening is the publication of *Silent Spring* by Rachel Carson (1907–1964) in 1962, a book that is, as we know, a major milestone in the history of environmental awareness. Linda Lear, her biographer, recalls the context in which the book appears[5]:

> *Silent Spring* contained the kernel of social revolution. Carson wrote at a time of new affluence and intense social conformity. The cold war, with its climate of suspicion and intolerance, was at its zenith. The chemical industry, one of the chief beneficiaries of postwar technology, was also one of the chief authors of the nation's prosperity. DDT enabled the conquest of insect pests in agriculture and of insect-borne disease just as surely as the atomic bomb destroyed America's military enemies and dramatically altered the balance of power between humans and nature.

Silent Spring, in fact, has the impact of "another" bomb, this time affecting American and even global consciousness. It is the first book on environmental science to be discussed at a press conference by President John F. Kennedy and to remain on the bestseller list for a long time. It is not surprising, thus, that the establishment of the Clean Air Act, a federal air pollution monitoring and control program, happened in 1963. Between 1962 and 1966, the book was translated (in chronological order) into German, French, Swedish, Danish, Dutch, Finnish, Italian, Spanish, Portuguese, Japanese, Icelandic, and Norwegian. It was later translated into Chinese (1979), Thai (1982), Korean (1995), and Turkish (2004). By warning of the death of birds and other animals caused by the pesticide DDT, Carson emphasized—in the very same year as the Cuban missile crisis—that the risks of human annihilation no longer stemmed only from a nuclear winter, but also from a silent spring. More than nuclear war, one should, henceforth, fear the less noisy, but no less destructive, war against nature. This is because (notwithstanding the fear expressed by Freud in 1930, 15 years before the atomic bomb) a nuclear winter could be avoided, but not a spring without birds—an allusion to a dead nature—if man did not learn to contain his (self)destructiveness in relation to nature.

In retrospect, it is becoming clearer that the 1960s can conveniently be called the Rachel Carson decade. At the end of the decade, René Dubos, a French Biologist naturalized in the USA, launches another book that is emblematic of those years, *So Human an Animal*. It begins with a statement that reflects the indignation felt in that decade:

> This book should have been written with anger. I should be expressing in the strongest possible terms my anguish at seeing so many human and natural values spoiled or destroyed in affluent societies, as well as my indignation at the failure of the scientific community to organize a systematic effort against the desecration of life and nature. (…) The most hopeful

---

[5] See Linda Lear, "Introduction" to the 2002 edition of *Silent Spring*.

sign for the future is the attempt by the rebellious to reject our social values. (…) As long
as there are rebels in our midst, there is reason to hope that our societies can be saved.

Perhaps influenced by Carson and Dubos, and certainly under the impact of the
Vietnam War and the 1968 *annus mirabilis*, the Union of Concerned Scientists
(UCS) was launched at MIT. Its inaugural document was initially signed by 50
senior scientists, including the heads of the biology, chemistry, and physics depart-
ments, and then endorsed by other scientists from this renowned center of scientific
activity in the USA. The document redefines the meaning of scientific activity. It is
no longer just about seeking to broaden knowledge on nature; it is about understanding science now as a critical activity. It proposes, thenceforth, "to devise means
for turning research applications away from the present emphasis on military technology toward the solution of pressing environmental and social problems."[6]

12.2.1   *"The Masters of the Apocalypse"*

Although it occurred simultaneously on both sides of the northern Atlantic, the
awareness of human arrogance and human impact on the environment had less
momentum in continental Europe. As Hicham-Stéphane Afeissa (2009) points out,
it is not coincidental that, until recently, European research on environmental issues
seemed to suffer from a significant deficit in relation to the USA. But in addition to
having less momentum, the awakening of an environmental consciousness in
Europe had different characteristics. Contrary to the US autistic complex of being
the embodiment of good, twentieth-century Europe gasped under the weight of bad
conscience: the self-destruction caused by the two wars; the unconditional political,
economic, and ideological capitulation to the USA; the atrocities of colonization
and decolonization; and genocides. Moreover, it was the main zone of friction
between the spheres of influence of the so-called superpowers (and, thus, the most
plausible scenario of a nuclear hecatomb in the event of a cold war slippage), making European public opinion and intelligentsia more sensitive to Hiroshima than to
environmental disasters. Therefore, reflection on the ecological question in the Old
World emerges slowly from a meditation on the retreat of thought in the post-
genocide era and the new precariousness of the human condition in the nuclear age.
No one better than Michel Serres (*Éclaircissements* 1992) expresses the advent of
this seismic tremor in the consciousness of European scientists and philosophers:
"Since the atomic bomb, it had become urgent to rethink scientific optimism. I ask
my readers to hear the explosion of this problem in every page of my books.
Hiroshima remains the sole object of my philosophy." In fact, the famous sentence
from the *Bhagavad-Gita*, "Now I am become Death, the destroyer of worlds," mut-
tered by J. Robert Oppenheimer on July 16, 1945, in the face of the explosion of
"his" bomb in the Los Alamos desert echoed paradoxically more in Europe than in

---

[6] See <http://www.ucsusa.org/about/founding-document-1968.html>.

the USA: in the writings of Bertrand Russell, Einstein, Karl Jaspers, Friedrich Dürrenmatt (in his grim comedy, *Die Physiker*, 1962), and, especially, Günther Anders, whose later work is strongly devoted to reflecting on the atomic bomb. In "The Bomb and the Roots of Our Blindness toward the Apocalypse" (essay published in *The Obsolescence of Humankind* 1956), Günther Anders spells out the essence of man's "existential situation" (to put it in the language of that time):

> If something in the consciousness of men today has the value of Absolute or Infinity, it is no longer the power of God or the power of nature, not even the alleged powers of morality or culture: it is our own power. Creation *ex nihilo*, once a manifestation of omnipotence, has been replaced by the opposite power: the *power to annihilate*, to reduce to nothing—and that power is in our hands. We really have gained the omnipotence that we had been yearning for so long, with Promethean spirit, albeit in a different form to what we hoped for. Given that we possess the strength to prepare each other's end, we are *the masters of the Apocalypse. We are infinity*.

12.2.2   From the Nuclear to the Ecological

Today, this paragraph refers not only to the nearly 23 thousand nuclear warheads in the hands of 8 countries, an arsenal of destructive power 200 thousand times larger than the bomb dropped in Hiroshima (Sidel and Levy 2014),[7] but to the continuous and increasing destructive action of man over ecosystems in the Anthropocene. It could appear as an epigraph in the Red List of Threatened Species first published in 1963 by the International Union for Conservation of Nature (IUCN), an institution born in Europe immediately after the war.

The transition from nuclear to ecological does not mean, however, just a change in subject or a broadening of its scope. Unlike a nuclear catastrophe, the ecological threat does not become a catastrophe because of a war event or a *hamartia*, a tragic flaw, but due to a (relatively) peaceful economic process, generally regarded (even today) as beneficial and even essential; one in which the shifts in phases toward collapse were still, in those years, almost imperceptible in their overall configuration. This difference between nuclear and ecological catastrophes was highlighted by Hans Jonas in 1979 in *The Imperative of Responsibility* and described precisely in a preface to the English edition of this work (1984):

> Lately, the other side of the triumphal advance [of modern technology] has begun to show its face, disturbing the euphoria of success with threats that are as novel as its welcomed fruits. Not counting the insanity of a sudden, suicidal atomic holocaust, which sane fear can avoid with relative ease, it is the slow, long-term, cumulative—the peaceful and constructive use of worldwide technological power, a use in which all of us collaborate as captive beneficiaries through rising production, consumption, and sheer population growth—that poses threats much harder to counter.

---

[7] Data from the Federation of American Scientists: Status of World Nuclear Forces (October 2009).

12.2.3   1972–2002: The Final Formulation
of the Multiauthored Concept of the Anthropocene

In the abovementioned text, Hans Jonas summarizes the slow awareness of the advent of the Anthropocene in Europe. To a large extent, the notion of the Anthropocene, despite its technicality and its pertaining strictly to the realm of stratigraphy, would not have been conceivable outside of the sphere of critical thinking which gained momentum in the 1960s. This period was rife with philosophical insurgency and political action, and it also saw the emergence of figures such as Barbara Ward and René Dubos, both who drafted the report commissioned by Maurice Strong for the seminal 1972 United Nations Conference on Human Development. This report, titled *Only One Earth: The Care and Maintenance of a Small Planet*, brought together the contributions of 152 experts from 58 countries, resulting in the 26 principles that make up the Stockholm Declaration and in the creation of UNEP. The first of these principles clearly advances the central idea of a new geological epoch that is conditioned by human action as much as (or more than) by non-anthropic variables:

> In the long and tortuous evolution of the human race on this planet a stage has been reached when, through the rapid acceleration of science and technology, man has acquired the power to transform his environment in countless ways and on an unprecedented scale.

These precedents must be kept in mind so as not to confine the concept of the Anthropocene to the narrow limits of scientific terminology. Indeed, while the International Commission on Stratigraphy (ICS) is still debating its official adoption, this concept is not limited to a proposed revision of stratigraphic nomenclature. It is a collectively formulated idea, circulating through the *Zeitgeist* of the 1970s–1990s. As James Syvitski (2012) states, "in the 1990s the term Anthroposphere was widely used in the Chinese science literature under the influence of Chen Zhirong of the Institute of Geology and Geophysics at the Chinese Academy of Sciences in Beijing." In the West, the birth of the term Anthropocene is credited mainly to two biologists, Eugene F. Stoermer and Andrew C. Revkin, as seen in Revkin (2011) and in a historical study of the term produced by Steffen et al. (2011):

> Eugene F. Stoermer wrote: 'I began using the term "anthropocene" in the 1980s, but never formalized it until Paul contacted me'. About this time other authors were exploring the concept of the Anthropocene, although not using the term. More curiously, a popular book about *Global Warming*, published in 1992 by Andrew C. Revkin, contained the following prophetic words: 'Perhaps earth scientists of the future will name this new post-Holocene period for its causative element—for us. We are entering an age that might someday be referred to as, say, the Anthrocene [sic]. After all, it is a geological age of our own making'. Perhaps many readers ignored the minor linguistic difference and have read the new term as Anthro(po)cene!

Indeed, it is from the seminal ideas of Humboldt, Lamarck, Marsh, Stoppani, Vernadsky, and Teilhard de Chardin, but not least from the emergent reflection of biologists, chemists, meteorologists, environmentalists, anthropologists, and philosophers of their generation,[8] that Crutzen and Stoermer proposed, in the 2000 congress of the International Geosphere–Biosphere Program (IGBP) in Cuernavaca and then in a 2002 article by Crutzen alone, the recognition of a new geological epoch, the Anthropocene. Considering the combination of biogeophysical forces that shape the Earth system, this epoch is characterized by the fact that anthropic action prevails over forces generated by nonhuman factors. "It seems appropriate," Crutzen wrote in 2002, "to assign the term 'Anthropocene' to the present, in many ways human-dominated, geological epoch, supplementing the Holocene—the warm period of the past 10–12 millennia."

12.2.4   The Great Acceleration and Other Markers of Anthropic Interference

According to Paul Crutzen, the birth date of this new geological epoch could be conventionally fixed in 1784, the year of James Watt's steam engine patent and also the birth of the atmospheric carbonization era. Jan Zalasiewicz (2014), Director of the Anthropocene Working Group of the International Commission on Stratigraphy (ICS), prefers to date the beginning of the Anthropocene from the second half of the twentieth century, choosing as his criteria the increase in greenhouse gas emissions and pollution, as well as the inscription on rocks of the radioactivity emitted by the detonation of atomic bombs in open air, among other factors. The proposal put forth by 25 researchers and coordinated by Zalasiewicz to date the Anthropocene from 1950 onward was reported by Richard Monastersky in 2015:

> These radionuclides, such as long-lived plutonium-239, appeared at much the same time as many other large-scale changes wrought by humans in the years immediately following the Second World War. Fertilizer started to be mass produced, for instance, which doubled the amount of reactive nitrogen in the environment, and the amount of carbon dioxide in the atmosphere started to surge. New plastics spread around the globe, and a rise in global commerce carried invasive animal and plant species between continents. Furthermore, people were increasingly migrating from rural areas to urban centres, feeding the growth of megacities. This time has been called the Great Acceleration.

The iconic 24 graphics created by Will Steffen and colleagues in 2015 coined the idea of Great Acceleration, which was forever fixed by the International Geosphere–Biosphere Programme (IGBP—Global Change, 1986–2015) in these terms:

> The second half of the 20th Century is unique in the history of human existence. Many human activities reached take-off points sometime in the 20th Century and sharply accelerated towards the end of the century. The last 60 years have without doubt seen the most

---

[8] In his autobiography (on the Nobel Prize website), Crutzen recounts how many researchers and friends at the Stockholm University Meteorological Institute, where he made his career, became heavily involved in issues such as acid rain and the carbon cycle, which attracted considerable political interest in the first United Nations Conference on the Human Environment, held in Stockholm in 1972.

profound transformation of the human relationship with the natural world in the history of humankind. The effects of the accelerating human changes are now clearly discernible at the Earth system level. Many key indicators of the functioning of the Earth system are now showing responses that are, at least in part, driven by the changing human imprint on the planet. The human imprint influences all components of the global environment—oceans, coastal zone, atmosphere, and land.

The Anthropocene concept expresses the preponderance of anthropic forces in relation to the other forces that intervene in shaping the Earth's system. In the previous 11 chapters, we have seen how these forces are causing deforestation, deep imbalances in the climate system, and mass extinctions of plant and animal species on a growing scale, one that is already comparable to the previous five major extinctions. Future paleontologists—assuming that a future remains for us—will notice the sudden disappearance of fossil records of an uncountable number of species. Instead of plant and animal fossils, the traces left by anthropic forces on terrestrial and underwater rocks will be the signatures of isotopes such as plutonium-239, mentioned by Zalasiewicz, as well as numerous other markers, such as the various forms of pollution by fossil fuels, different explosives, POPs, concrete, plastic, aluminum, fertilizers, pesticides, and other industrial waste.

Ugo Bardi's book, *Extracted: How the Quest for Mineral Wealth Is Plundering the Planet* (2014), provides some insight into this. Mining operations extract 2 billion tons of iron and 15 million tons of copper globally per year. The USA alone extracts 3 billion tons of ore a year from its territory, and according to the US Geological Survey, the removal of sand and gravel for construction on a global scale can exceed 15 billion tons a year. In relation to rocks and earth alone, humans remove, per year, the equivalent of two Mount Fujis (with an altitude of 3776 m, it is the highest mountain in the Japanese archipelago). According to Jan Zalasiewicz, mining operations dig up the Earth's crust to depths that reach several thousand meters (e.g., 5 km deep in a gold mine in South Africa). And in its extraction processes alone, the oil industry drilled about 5 million kilometers underground, which, as the scholar points out, is equivalent to the total length of man-made highways on the planet.[9]

In 2000, the burning of fossil fuels emitted about 160 Tg per year of sulfur dioxide (SO_2) into the atmosphere, which is more than the sum emitted by all natural sources, and more synthetic nitrogen for fertilizers was produced and applied to agriculture than is naturally fixed by all the other terrestrial processes added together. Furthermore, more than half of all accessible freshwater resources had already been used by humans, and 35% of the mangroves in coastal areas had been lost. At least 50% of non-iced land surface had already been transformed by human action in 2000, and, over the last century, the extent of land occupied by agriculture has doubled, to the detriment of forests. Anthropic action interferes decisively not only "externally" in vegetation cover, in the behavior of physical forces, and in the extinction of species but also inside organisms, infiltrating into the cellular tissues

---

[9] Cf. Jan Zalasiewicz, "The Anthropocene as a potential new unit of the Geological Time Scale." <https://www.youtube.com/watch?v=y_FbbXlgkgE>

of countless species and altering their metabolism, hormones, and chemical balance, as discussed in Chaps. 4, 10, and 11. According to Erle C. Ellis and Navin Ramankutty (2008), biomes have been so altered by humans that it would be better to call them "Anthromes" or "Anthropogenic biomes," terms that provide "in many ways a more accurate description of broad ecological patterns within the current terrestrial biosphere than are conventional biome systems that describe vegetation patterns based on variations in climate and geology."

12.3   The New Man–Nature Relationship

In his conversations with Bruno Latour (*Éclaircissements* 1992), Michel Serres states: "A global subject, the Earth, emerges. A global subject, on the other hand, is constituted. We should, therefore, think about the global relations between these two globalities. We do not yet have a theory that allows us to think about them." There is, in fact, something unthought in the concept of the Anthropocene. Its importance is, first of all, philosophical. This concept abolishes the separation—foremost in man's consciousness of himself—between the human and the nonhuman spheres. In the Anthropocene, nature ceased to be a variable that was independent of man and, ultimately, became a dependent variable. Therefore, nature has ultimately become a social relationship. But the inverse is equally true: relations between men in the broadest sense—from the economic to the symbolic sphere—lose their autonomy and gradually become functions that are dependent on environmental variables.

Similar to the concept of culture, we know that the concept of nature cannot be defined. Its polysemy allows for irreconcilable, perhaps even contradictory, meanings to coexist in it, especially since the subject who defines nature is himself a part of nature. The only thing we can say is that during the Holocene, nature presented itself to human experience in two fundamental and contradictory ways: (a) as other than him, nature appeared to man's consciousness as that which is not human, as something essentially different from him and in opposition to how he defined himself, and (b) as a totality, nature was *physis* and, as such, it encompassed and unified everything, including man and the gods. As Pierre Hadot teaches in a remarkable book, *Le Voile d'Isis* (2004), this duality was already present in Greece. It arises there in the form of tension between two ideas on nature, which imply opposite attitudes. As early as 1989, Hadot outlined this idea in a lesson at the Collège de France:

> One can distinguish two fundamental attitudes of ancient man toward nature. The first can be symbolized by the figure of Prometheus: this represents the ruse that steals the secrets of nature from the gods who hide them from mortals, the violence that seeks to overcome nature in order to improve the lives of humans. The theme already appears in medicine (*Corpus hippocraticum*, XII, 3), but especially in mechanics [...]. The word *mêchanê* refers, additionally, to ruse. Opposed to this 'promethean' attitude, which puts ruse at the service of men's needs, there is a totally different kind of relation to nature in Antiquity that could

be described as poetic and philosophical [that is, orphic], where 'physics' is conceived of as a spiritual exercise.

Christian Godin (2012) notices quite well that these two attitudes of the Greeks toward the "secrets" of nature "may characterize more generally (and not just in Greece) the two opposite attitudes that man can adopt toward nature: one of fusion and one of conquest."

The Promethean attitude gradually reduces nature to the "object" of the subject until it becomes completely alienated in the modern age by its conversion into quantity, vector force, and *res extensa*. It can be said that the whole history of philosophy in the modern and contemporary age is strongly dominated by the unfinished double enterprise of determining the ontology of this object and the epistemological status of the subject's relationship with it. Alexandre Koyré's definition of this object quite accurately specifies the point that we have reached along this path: "Nobody knows what nature is except that it is whatever it is that falsifies our hypotheses" (as cited by Crombie 1987). In the nineteenth century, Hegel's logic would still attempt to restore unity between subject and object through a dialectical identification between the spirit and the world or the reciprocal subsumption of one by the other (Beiser 1993). But modern science would simply ignore Hegel as much as he ignored the two categories that have progressively come to characterize science in the modern age: the central role of experiment and the mathematization of knowledge. Such a restoration actually occurs, in Hegel's century, only in the aesthetic and existential experience of *Stimmung*, an empathetic convergence and communion of spirit with nature in the lyrical instant or in the spirit's submission to the sublime. In the lyrical instant, to stick to emblematic examples, we have Goethe's *Über allen Gipfeln ist Ruh*, certain small landscapes of the French or Italian countryside by the likes of Valenciennes to Corot, and the second movement of Beethoven's *Pastoral*; in the second case, that of the sublime, we have Leopardi's *L'infinito* or the fourth movement of Beethoven's *Pastoral*.

It is superfluous to remember that as a living organism, man is, objectively, nature. But the very idea of anthropogenesis, or hominization, has always been perceived, at least in the West, as a slow and gradual process of distancing and differentiation of the human species from other species and from nature in general. In this process, nature meant, at the same time, that which is not human, that which surrounds man (his *Umwelt*), and that which is the origin of man. Whatever the meaning—biological, utilitarian, phenomenological, or symbolic—of the word origin, man was, in short, the effect of that origin.

12.3.1   The Powerlessness of Our Power

In the Anthropocene, by contrast, it is nature that becomes an effect of man. Wherever he goes, from the stratosphere to the deep sea, man now finds—objectively, and no longer merely as a projection of his consciousness—the effects of himself, of his action, and of his industrial pollution. *La Terre, jadis notre mère, est devenue notre fille*[10]: proposed by Michel Serres, this metaphor of the ancestral mother turned into our daughter illustrates perfectly the concept of the Anthropocene. It urges us to take care of the Earth just as we would do for our child.

But this supposed parental responsibility should not mislead us: we have not acquired any parental power over "our child." If the Earth has become a variable that is dependent on anthropic action, this does not mean that man has greater control over it. On the contrary, since the mother could, occasionally, be a stepmother, the degraded daughter is systematically unsubmissive and "vindictive," to use James Lovelock's metaphor in *The Revenge of Gaia* (2006). Henceforth, societies will be increasingly governed by boomerang effects, that is, by negative effects on man of the imbalances that he has wrought upon ecosystems, as can be best seen in Chap. 15. In its own way, the Anthropocene achieved the ideal unity of science—gradually abandoned since the nineteenth century—because by abolishing the separation between the human and nonhuman spheres, it abolished *ipso facto* the boundaries between the natural sciences and the "human sciences." As Michel Serres (2009) also states, today "human and social sciences have become a kind of subsection of life and earth sciences. And the reciprocal is also true."

Today, more than ever, we are existentially vulnerable to that which has become vulnerable to us. The Anthropocene is, in short, the revelation of the powerlessness of our power. This impotence is precisely our inability to halt the effects of that which we have caused and our inability to act economically according to what science tells us about the limits of the Earth system and its growing imbalances. On a more fundamental and also more concrete level, it is our inability to free ourselves psychically from the quantitative, compulsively expansive, and anthropocentric paradigm of capitalist economy. As we will discuss in Chaps. 14 and 15, this inability is the *causa causans* of the environmental collapse that looms on our horizon. Rachel Carson was already aware of this when she made the following statement in an American documentary aired on CBS in April 1963 (Gilding 2011): "man's attitude towards nature is today critically important simply because we have now acquired a fateful power to destroy nature. But man is part of nature and his war against nature is inevitably a war against himself."

12.3.2   A New World, Biologically, Especially in the Tropics

The current radical reduction of vertebrate and invertebrate life-forms is a central feature of the environmental collapse we are facing in the Anthropocene. In fact, we are headed for "a new world, biologically." A collective synthesis of research done over the past two decades and published in June 2012 in the journal *Nature* suggests

---

[10] Serres returns to this image more than once. He also mentions it in *Éclaircissements. Entretiens avec Bruno Latour* (Paris, 1992, pp. 170–171).

this conclusion. It shows that "within a few generations," the planet can transition to a new biospheric state never known to *Homo sapiens*. The main author of this study, Anthony Barnosky of the University of California, declared to Robert Sanders (2012)[11]:

> "It really will be a new world, biologically. (…) The data suggests that there will be a reduction in biodiversity and severe impacts on much of what we depend on to sustain our quality of life, including, for example, fisheries, agriculture, forest products and clean water. This could happen within just a few generations.

The forms of this "new world" are beginning to emerge, insofar as expanding economic activity destroys ecosystems and alters the physical, chemical, and biological parameters of the planet. Compared to the lush biodiversity of the Holocene, the Anthropocene will be almost unrecognizable.

The contrast between the former exuberance and the forthcoming destitution will be more acute in the tropics because the greatest biodiversity is still concentrated there and because such latitudes, which are already warmer, will be more rapidly and profoundly affected by global warming and other biospheric degradation factors, as shown by three studies published in 2011, 2012, and 2013. Diffenbaugh and Scherer (2011) state that:

> In contrast to the common perception that high-latitude areas face the most accelerated response to global warming, our results demonstrate that in fact tropical areas exhibit the most immediate and robust emergence of unprecedented heat, with many tropical areas exhibiting a 50% likelihood of permanently moving into a novel seasonal heat regime in the next two decades. We also find that global climate models are able to capture the observed intensification of seasonal hot conditions, increasing confidence in the projection of imminent, permanent emergence of unprecedented heat.

Camilo Mora et al. (2013) also predict that "unprecedented climates will occur earliest in the tropics." For their part, Oliver Wearn et al. (2012) warn that the extinction rate of vertebrate species in the Brazilian Amazon should expand enormously in the future:

> […] local extinctions of forest-dependent vertebrate species have thus far been minimal (1% of species by 2008), with more than 80% of extinctions expected to be incurred from historical habitat loss still to come. Realistic deforestation scenarios suggest that local regions will lose an average of nine vertebrate species and have a further 16 committed to extinction by 2050.

As seen in the previous chapter, there will also be a radical reduction in most forms of marine life, including (possibly) phytoplankton. Therefore, as the biosphere regresses, both on land and in water, this new world of the Anthropocene will move toward a reduced biosphere that could perhaps be called hypobiosphere.

---

[11] See also Barnosky (2012) and Cardinale et al. (2012).

12.4   Hypobiosphere: Functional and Nonfunctional Species to Man

We propose this neologism, hypobiosphere, to refer to the growing areas of the planet in which deforestation and defaunation will deprive the biosphere of the vast majority of highly evolved multicellular forms of animal and plant life still present in nature. The first 11 chapters of this book, and, in particular, the preceding two chapters, offer a display of partial prefigurations of the hypobiosphere. In fact, to some extent, we are already living in a hypobiosphere. According to Yinon Bar-On, Rob Phillips, and Ron Milo (2018), humans currently account for about 36% of the biomass of all mammals, livestock and domesticated animals for 60%, and wild mammals for only 4%:

> Today, the biomass of humans ($\approx$0.06 Gigatons of Carbon, Gt C) and the biomass of livestock ($\approx$0.1 Gt C, dominated by cattle and pigs) far surpass that of wild mammals, which has a mass of $\approx$0.007 Gt C. This is also true for wild and domesticated birds, for which the biomass of domesticated poultry ($\approx$0.005 Gt C, dominated by chickens) is about threefold higher than that of wild birds ($\approx$0.002 Gt C). (…) Intense whaling and exploitation of other marine mammals have resulted in an approximately fivefold decrease in marine mammal global biomass (from $\approx$0.02 Gt C to $\approx$0.004 Gt C). While the total biomass of wild mammals (both marine and terrestrial) decreased by a factor of $\approx$6, the total mass of mammals increased approximately fourfold from $\approx$0.04 Gt C to $\approx$0.17 Gt C due to the vast increase of the biomass of humanity and its associated livestock.

Given these numbers, it does not seem arbitrary to say that the Anthropocene is reducing the biosphere into two groups. On one side are the species controlled by man and on the other side are those uncontrolled but able to withstand anthropogenic impacts, either due to their minimal contact with humans (such as species living in the deep ocean) or because they thrive in the anthromes, thanks to rubbish or other disturbances in the ecosystems. If this is the case, we should note the gradual prevalence of ten categories of life on the planet:

(1) Plants intended for human consumption and animal husbandry
(2) Plant inputs for industry (cellulose, ethanol, etc.)
(3) Domestic animals
(4) Animals raised for human consumption
(5) Animals raised for scientific experiments
(6) Plant and animal species unaffected by pesticides and human pollutants
(7) Species that benefit from anthropogenic environmental imbalances
(8) Species that feed on our food and our waste
(9) Species that live in remaining remote areas with little contact with humans (e.g., in deep ocean trenches)
(10) Fungi, worms, microorganisms (viruses,[12] bacteria, mites, etc.)

---

[12] As pointed out by Luis P. Villarreal, "Are viruses alive?" *Scientific American*, 8/VIII/2008: "Regardless of whether or not we consider viruses to be alive, it is time to acknowledge and study them in their natural context—within the web of life."

Due to its apparent arbitrariness, this classification may be part of an array of absurd taxonomies, like the one imagined by Jorge Luis Borges (1952). However, this one has a rigorous logic, since, as might be expected in the Anthropocene, it divides species into those deeply manipulated by humans (1–5) and those that live outside the sphere of human domination (6–10). The last five categories of this classification, especially the last one, encompass billions or trillions of species; so, this new equilibrium of the biota will not necessarily be hostile to most life-forms. But it will be hostile—perhaps at the point of extermination—to the vast majority of approximately eight million eukaryote species. This is especially the case for vertebrates, a phylum (or subphylum) that contains, according to the 2004 IUCN report, 57,739 described species. Within this phylum, the class of mammals (endowed with a neocortex) includes about 5500 described species, a number that is probably very close to that of existing mammal species.

12.4.1   The Increase in Meat Consumption

Many factors in globalized capitalism combine and strengthen each other to push us toward the hypobiosphere, this impoverished state of the biosphere. Many of these factors have already been examined in previous chapters: the widespread pollution of soil, water, and the atmosphere; the increasing use of pesticides; extreme heat waves in the atmosphere and oceans, which eliminate coral reefs and many other forms of terrestrial and aquatic life; the disappearance of beaches, essential for the spawning of some species, through rising sea levels; destruction (by man and by extreme weather events) of mangroves which are large breeding grounds; the damming of rivers by hydroelectric dams; eutrophication of waters due to overuse of industrial fertilizers; ocean acidification; overfishing; expansion of urban areas and road networks; mining, etc. But in the short term, the most powerful factor for the impoverishment of the biosphere is, undoubtedly, the elimination of wild habitats, especially tropical ones, due to deforestation and fires caused by agribusiness and, in particular, by the livestock industry. According to FAO's Food Production document:

> Livestock is the world's largest user of land resources, with grazing land and cropland dedicated to the production of feed representing almost 80% of all agricultural land. Feed crops are grown in one-third of total cropland, while the total land area occupied by pasture is equivalent to 26% of the ice-free terrestrial surface.

In other words, the most important factor currently driving us toward the hypobiosphere is the increasing human consumption of meat. The combined per capita consumption of meat, eggs, and milk in developing countries grew by about 50% from the early 1970s to the early 1990s. Based on this finding, a team of researchers from FAO, the International Food Policy Research Institute (IFPRI), and the International Livestock Research Institute (ILRI) produced a document 20 years ago titled

Livestock to 2020: The Next Food Revolution (Delgado et al. 1999). Christopher Delgado and his colleagues state in this document:

> It is not inappropriate to use the term 'Livestock Revolution' to describe the course of these events in world agriculture over the next 20 years. Like the well-known Green Revolution, the label is a simple and convenient expression that summarizes a complex series of inter-related processes and outcomes in production, consumption, and economic growth.

The authors were obviously correct in their projections to 2020, since the transition to animal protein consumption that began in the 1970s has continued unabated for the past 20 years. But what seemed to them to be a "Livestock Revolution" turned out to be a "Livestock Apocalypse" or, as Philip Lymbery and Isabel Oakeshott (2014) prefer to call it, a "Farmageddon." Animal breeding for human consumption which, until the nineteenth century, was an almost artisanal activity has notably become, in recent decades, a highly industrialized activity, led by factory farms, first for chickens, then pigs, and more recently cattle. From 2000 to 2014, Chinese production of meat, eggs, and milk has rapidly increased, and this will continue—especially for pork. Its production has jumped from around 40 million tons in 2000 to approximately 56 million tons in 2014 (Yang et al. 2019). The meat industry's fast food distribution chains are among the largest in the world. McDonald's, for example, which in the 1950s was still a simple diner in California, has become the world's largest international meat chain in the past half century, serving about 70 million customers a day in over 100 countries.

The consumption of meat on an industrial scale by today's highly urbanized societies is, first and foremost, a moral problem that increasingly mobilizes philosophical, anthropological, and biological arguments: "Every child should visit an abattoir. If you disagree, ask yourself why. What can you say about a society whose food production must be hidden from public view?" asks George Monbiot (2014). The meat industry and the consumption of its products infringe on the first right of animals: the right to a life without avoidable suffering. And it is not just a threat to the rights of animals directly victimized by agribusiness, but the gigantic increase in herds from the second half of the twentieth century threatens, as stated above, the survival of tropical wild habitats where 80% of terrestrial biodiversity is present.

Indeed, meat consumption by humans has reached insane levels, and there is no indication that we have reached the top of this escalation in animal cruelty, one that is driven by the logic of maximum productive efficiency at the lowest cost. Actual numbers surpass all efforts at dystopian imagination. According to Mia MacDonald (2012), who bases herself on 2010 data:

> Each year, more than 60 billion land animals are used in meat, egg, and dairy production around the world. If current trends continue, by 2050 the global livestock population could exceed 100 billion—more than 10 times the expected human population at that point.

In 2010, this meant about ten animals for each human. According to Alex Thornton (2019), basing himself on data from 2014, "an estimated 50 billion chickens are slaughtered for food every year—a figure that excludes male chicks and unproductive hens killed in egg production. The number of larger livestock, particularly pigs, slaughtered is also growing," as Fig. 12.1 shows.

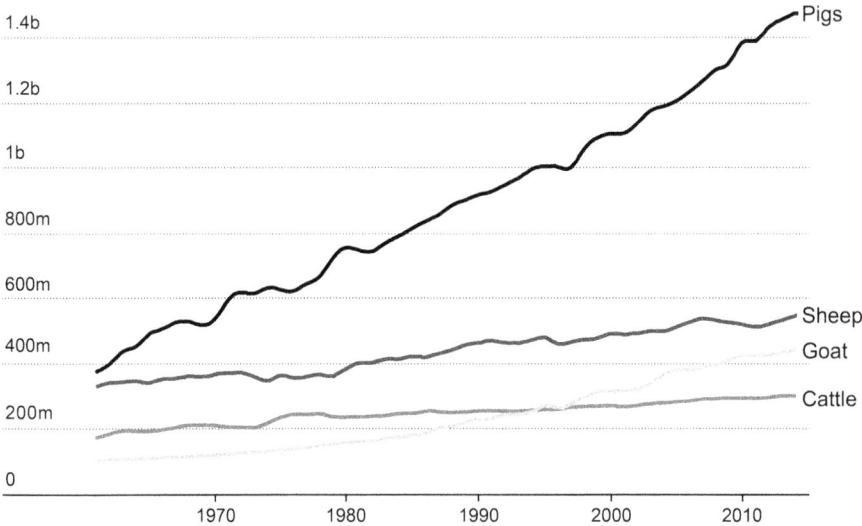

Fig. 12.1 Evolution of animals slaughtered for food per year (1961–2014). Source: Alex Thornton, "This is how many animals we eat each year." *World Economic Forum*, 8/II/2019

Table 12.1 Animals killed by the food industry each year (in millions)

	1961	2014
Pigs	376	1470
Sheep	331	545
Goat	103	444
Cattle	173	300
Total	983	2759

The graph above translates into Table 12.1 which shows the amounts of animals killed by the food industry each year (in millions), often after a life of terrible suffering.

Adding these numbers to the 50 billion chickens mentioned above, the estimates are of nearly 53 billion animals killed for food, which amounts to 7.4 animals killed for each human on the planet in 2014 alone when the population had reached 7.2 billion. Gary Dagorn (2015) reports slightly higher numbers:

> We never produced and consumed as much meat as we do today. In 2014, the world produced 312 million tons of meat, which is equivalent to an average of 43 pounds per person per year. Each year, 65 billion animals are killed (that is, almost two thousand animals … per second) to end up on our plates.

The stock of animals raised for human consumption is clearly larger, as Lymbery and Oakeshott (2014) show in their book:

> Some 70 billion farm animals are produced worldwide every year, two-thirds of them now factory-farmed. They are kept permanently indoors and treated like production machines, pushed ever further beyond their natural limits, selectively bred to produce milk or eggs, or to grow fat enough for slaughter at a younger and younger age.

In fact, since 1925, at least in the USA, chickens have had their lives shortened from 112 to 48 days, while their weight at the time of slaughter has gone from 2.5 pounds to 6.2 (Elam 2015), due to immobility and the administration of additives and antibiotics in their feeding. Brazilian agribusiness is the largest exporter of beef in the world. In 2018, there were 213.5 million head of cattle, more than one animal per person in the Brazilian territory. Most of the herds are concentrated in the North and Midwest of the country, with pastures expanding rapidly over the Cerrado and the Amazon rainforest, which are among the richest areas of planetary biodiversity. Along with China and some countries in Southeast Asia and in Africa, Brazil is the largest laboratory of the planetary hypobiosphere.

12.4.2   Doubling of Per Capita Consumption Between 2000 and 2050

The world population is expected to increase by just over 50% between 2000 and 2050 (from 6.1 to about 9.7 billion), while world meat production is expected to increase by just over 100% in that same period. According to FAO (2006):

> Global production of meat is projected to more than double from 229 million tonnes in 1999/01 [312 million tonnes in 2014 and 317 million tonnes in 2016] to 465 million tonnes in 2050, and that of milk to grow from 580 to 1,043 million tonnes. The environmental impact per unit of livestock production must be cut by half, just to avoid increasing the level of damage beyond its present level.

Assuming the increasingly unlikely hypothesis that the global food system will not collapse in the next few decades of the century, meat consumption in Europe's richest countries will tend to stabilize at a level of 80 kg per year per inhabitant (about 220 grams per day). This demand will be met by ever-increasing meat imports from the so-called "developing" countries, such as Brazil, where demand for meat continues to grow at high rates. Resilient as they may be, ecosystems will not be able to resist such increases. Indeed, for several reasons, the mounted knights of the livestock apocalypse are riding species that are bred for human consumption and, especially, cattle.

12.4.3   Meat = Climate Change

Two reports by FAO (2006, 2013) estimate that CO_2e (CO_2 equivalent) emissions from agribusiness account for about 15% to 18% of total anthropogenic emissions per year. According to FAO's *Livestock's Long Shadow* (2006), livestock production generates more GHG, measured in CO_2e, than the transport sector. But these percentages are not accepted by Robert Goodland and Jeff Anhang of the Worldwatch Institute. In 2009, they issued a very well-documented article on the responsibility of carnivorism in the environmental crises ("Livestock and Climate Change. What if the key actors in climate change are cows, pigs, and chickens?"). It states:

> *Livestock's Long Shadow*, the widely-cited 2006 report by FAO, estimates that 7,516 M per year of CO_2 equivalents (CO_2e), or 18% of annual worldwide GHG emissions, are attributable to cattle, buffalo, sheep, goats, camels, horses, pigs, and poultry. That amount would easily qualify livestock for a hard look indeed in the search for ways to address climate change. But our analysis shows that livestock and their byproducts actually account for at least 32,564 million tons of CO_2e per year, or 51% of annual worldwide GHG emissions.

Whatever may be the extent of its role in anthropogenic greenhouse gas emissions, in considering the environmental liability of meat-eating, we should include the impacts of the use of antibiotics, hormones, and pesticides on animals, as well as the two next equations:

12.4.4 Meat = Deforestation, Land-Use Change, and Soil Degradation

We present some data on this equation, taken from FAO's 2006 study. In that year, pasture land occupied 34 million km^2 or 26% of dry land. This is more than the total area of Africa (30.2 million km^2). The study states that 20% of these lands are degraded. Globally, about 24,000 km^2 of forest are replaced by pasture each year, and, according to data from another report (*Climate Focus*, 2017), an average of 27,600 km^2 of forests are replaced by pasture each year (Whyte 2018). Globally, soy is responsible for the deforestation of 6000 km^2 per year, but a major part of its cultivation goes toward animal feed. Thus, the deforestation caused by soybean production is also largely due to carnivorism. Two texts by João Meirelles (2005, 2014) show the responsibility of meat-eating in the destruction of the Amazon rainforest, either through domestic or international consumption. David Pimentel (1997) estimates that "nearly 40% of world grain is being fed to livestock rather than being consumed directly by humans." Much more recent estimates are very close to those of Pimentel. For Jonathan Foley (2014), "today only 55% of the world's crop calories feed people directly; the rest are fed to livestock (about 36%) or turned into biofuels and industrial products (roughly 9%). In 1997, Pimentel estimated that "the 7 billion livestock animals in the United States consume five times as much grain as is consumed directly by the entire American population." In addition, Pimentel continued, "if all the grain currently fed to livestock in the United States were consumed directly by people, the number of people who could be fed would be nearly 800 million." This means that a vegetarian diet would allow us to feed more than twice the current US population and almost all of the world's starving population.

After deforestation, overgrazing exterminates the remaining biodiversity through the effect of trampling, as well as that of feces and urine in excess. Also, according to Pimentel (1997), 54% of US pasture land is being overgrazed.

12.4.5   Meat = Water Depletion

This second equation also makes meat consumption a major contributor to the depletion of water resources, as shown in Fig. 12.2:

Since the emergence of the concept of "water footprint," coined by Arjen Hoekstra (from John Anthony Allan's concept of "virtual water"), we know how to calculate how much water is used or polluted in the development of each agriculture and livestock product. The production of 1 kg of meat requires the use of 20,000 liters of water. Although the amount of water consumed varies depending on the region where the crop is grown, the irrigation system, and the type of cattle feed, the proportion indicated in the figure above remains substantially the same. The use of water by cattle is even greater when the animals are fed fodder and grain. According to calculations made by David Pimentel (1997), US agriculture accounts for 87% of all freshwater consumed per year. Only 1.3% of this water is used directly by cattle. But when we include water used for fodder and grain, the amount of water used increases dramatically. All in all, each pound of beef consumes 100,000 liters of water.

Due to the multiplicity of these impacts, the process of extreme "carnivoration" of the human diet from the 1960s onward is one of the most emblematic characteristics of the Anthropocene. It drives the tendency toward biodiversity collapse and the advent of the hypobiosphere.

Fig. 12.2 Liters of water necessary to produce 1 k of agriculture and livestock products. Based on data from the Brazilian National Water Agency—Agência Nacional de Águas (ANA—MMA)

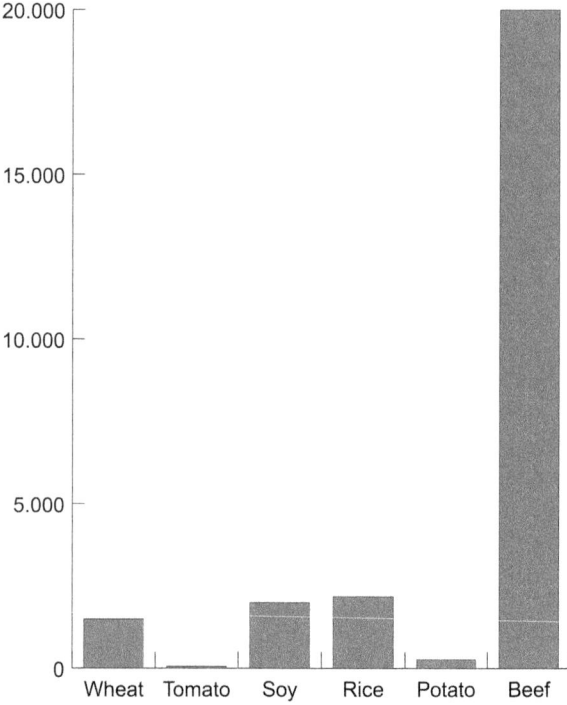

References

AFEISSA, Hicham-Stéphane, *Qu'est-ce que l'écologie?*. Paris, Vrin, 2009.

ANDERS, Günter, *Die Antiquiertheit des Menschen*. French translation: *L'obsolescence de l'homme*. I: *Sur l'âme à l'époque de la deuxième revolution industrielle* (1956); II: *Sur la destruction de la vie à l'époque de la troisième revolution industrielle*. Paris, 2002 and 2011.

BARDI, Ugo. *Extracted. How the Quest for Mineral Wealth is Plundering the Planet. A Report to the Club of Rome* (2013). Vermont, Chelsea Green Publisher, 2014.

BARNOSKY, Anthony *et al.* "Approaching a state shift in Earth's biosphere". *Nature* 486, 7/VI/2012, pp. 52–58.

BAR-ON, Yinon M., PHILLIPS, Rob & MILO, Ron, "The biomass distribution on Earth". *PNAS*, 19/VI/2018.

BEISER, Frederick C. "Hegel and the problem of metaphysics". *The Cambridge Companion to Hegel*. Ed. por F. C. Beiser. Cambridge Un. Press, 1993, pp. 1–24.

BONNEUIL, Christophe & FRESSOZ, Jean-Baptiste, *L'événement anthropocène*. Paris, 2013.

BORGES, Jorge Luis, "El idioma analítico de John Wilkins". *Otras inquisiciones* (1952). Obras Completas, Barcelona, 1989, II, pp. 84–90.

CALLENDAR, Guy S., "The Composition of the Atmosphere Through the Ages". *Meteorological Magazine*, 74, 878 March 1939, pp. 33–39.

CARDINALE, Bradley J. et al., "Biodiversity loss and its impact on humanity". *Nature*, 486, 6/VI/2012, pp. 59–67.

CARSON, Rachel. *A Silent Spring* (1962). Penguin Classics, 2000.

CLARK, W.C., CRUTZEN, P.J. & SCHELLNHUBER, H.J., *Science for Global Sustainability. Toward a New Paradigm*. Working Paper, 120, March 2005. Center for International Development at Harvard University and MIT Press.

CROMBIE, A.C., "Alexandre Koyré and Great Britain: Galileo and Mersenne". *History and Technology*, 4, 1987, pp. 81–91.

CRUTZEN, Paul J. "Geology of mankind: the Anthropocene". *Nature*, 415, 6867, 2002, pp. 23–25.

DAGORN, Gary, "Avant d'être cancerigène, la viande est polluante pour la planète". *Le Monde*, 29/X/2015.

DELGADO, Christopher *et al.*, *Livestock to 2020. The Next Food Revolution*. FAO, IFPRI, & ILRI, 1999.

DIFFENBAUGH, Noah S. & SCHERER, Martin, "Observational and model evidence of global emergence of permanent, unprecedented heat in the 20th and 21st centuries. A letter". *Climatic Change*, 107, 2011, pp. 615–624.

DUBOS, René. *So human an animal*. New York, 1968.

DUKORE, Bernard F. "Environmental Shaw". *SHAW. The Annual of Bernard Shaw Studies*, 34, 2014, pp. 16–45.

ELAM, Thomas E., *Live Chicken Production Trends*, FarmEcon LLC, 26/IV/2015.

ELLIS, Erle C. & RAMANKUTTY, Navin. "Putting people in the map: anthropogenic biomes of the world". *Frontiers in Ecology and the Environment*, 6, October 2008, pp. 439–447.

FAO, *Livestock's Long Shadow. Environmental Issues and Options*, Rome, 2006.

FAO, *Tackling Climate Change Through Livestock*, Rome, 2013.

FLEMING, James R., *The Callendar Effect: The Life and Work of Guy Stewart Callendar (1898-1964)*. American Meteorological Society, 2007.

FOLEY, Jonathan, "The future of food". *National Geographic*, 2014.

FREUD, Sigmund, *Das Unbehagen in der Kultur* (1930). English translation *Civilization and its Discontents*. In S. Freud, The Standard Edition of the Complete Psychological Works, translated by James Strachey, London, 2001.

GILDING, Paul. *The Great Disruption*. New York, Bloomsbury Press, 2011.

GODIN, Christian. *La Haine de la Nature*. Seyssel, Éditions Champ Vallon, 2012.

GOODLAND, Robert & ANHANG, Jeff, *Livestock and Climate Change. What if the key actors in climate change are cows pig and chickens?* *World Watch Magazine*, 22, 6, November-December 2009, pp. 10–19. Worldwatch Institute.

HADOT, Pierre. "L'homme antique et la nature" (1989). *Études de philosophie ancienne*. Paris, Belles Lettres, 1998, pp. 307–318.

_____. *Le voile d'Isis. Essai sur l'histoire de l'idée de nature*. Paris, Gallimard, 2004.

HUNTER, Robert, *Thermageddon. Countdown to 2030*. New York, 2002.

JAMIESON, Dale. *Reason in a Dark Time: Why the Struggle Against Climate Change Failed and What It Means for Our Future*. Oxford University Press, 2014.

JONAS, Hans. *The Imperative of Responsibility. In search of an Ethics for the Technological Age* (1979). The University of Chicago Press, 1984.

JOSE, Coleen, WALL, Kim, HINZEL, Jan Hendrik, "This dome in the Pacific houses tons of radioactive waste – and it's leaking". *The Guardian*, 3/VII/2015.

Le TREUT, Hervé & SOMMERVILLE, Richard (coordinating lead authors), "Historical Overview of Climate Change Science". IPCC Fourth Assessment Report (AR4) 2007.

LORIUS, Claude & CARPENTIER, Laurent. *Voyage dans l'Anthropocène: cette nouvelle ère dont nous sommes les héros*. Arles, Actes Sud, 2010.

LYMBERY, Philip & OAKESHOTT, Isabel, *Farmaggedon: the true cost of cheap meat*. London, BLOOMSBURY, 2014.

MACDONALD, Mia, "Food Security and Equity in a Climate-Constrained World". *State of the World 2012. Moving Toward Sustainable Prosperity*, WorldWatch, 2012.

MARSH, George Perkins. *The Earth as Modified by Human Action* (1874). A last revision of *Man and Nature* (1864). New York, 1907.

MEIRELLES F, João. "Você já comeu a Amazônia hoje?", Instituto Peabiru, Belém, 2005.

_____. "É possível superar a herança da ditadura brasileira (1964-1985) e controlar o desmatamento na Amazônia? Não, enquanto a pecuária bovina prosseguir como principal vetor de desmatamento". *Boletim do Museu Par. Emílio Goeldi*, 9, 1, 2014, pp. 219–241.

MONASTERSKY, Richard, "First atomic blast proposed as start of Anthropocene". *Nature*, 16/I/2015.

MONBIOT, George, "Overgrowth". *The Guardian*, 16/XII/2014.

MORA, Camilo *et al.*, "The projected timing of climate departure from recent variability". *Nature*, 502, 10/X/2013, pp. 183–187.

PIMENTEL, David, "Eight Meaty Facts about Animal Food". *Cornell Chronicle*, 7/VIII/1997.

PLASS, Gilbert, "Ther Carbon Dioxide Theory of Climatic Change". *Tellus*, 1956, pp. 140–154.

_____, "Carbon Dioxide and Climate". *Scientific American*, July, 1959, pp. 41–47.

REVELLE, Roger & SUESS HANS E. "Carbon Dioxide Exchange between Atmosphere and Ocean and the Question of an Increase of Atmospheric CO_2 During the Past Decades". *Tellus*, 9, 1957, pp. 18–27.

REVKIN, Andrew C., "Confronting the Anthropocene". *The New York Times*, 11/V/2011.

SANDERS, Robert, "Scientists uncover evidence of impending tipping point for Earth". *Berkeley News*, 6/VI/2012.

SERRES, Michel, *Éclaircissements. Entretiens avec Bruno Latour*. Paris, F. Bourin, 1992.

SIDEL, Victor W. & LEVY, Barry S., "The Threat of Nuclear War" (chapter in Ugo Bardi's book, 2014.

STEFFEN, Will; GRINEVALD, Jacques; CRUTZEN, Paul J. & MCNEILL, John. "The Anthropocene: conceptual and historical perspectives". *Philosophical Transactions of the Royal Society*, 369, 2011, pp. 842–867.

STEFFEN, Will *et al.*, "Planetary boundaries: Guiding human development on a changing planet". *Science*, 15/I/2015.

STEFFEN, Will *et al.*, "The trajectory of the Anthropocene: The Great Acceleration". *The Anthropocene Review*, 2015, 2(1), pp. 81–98.

SYVITSKI, James. "Anthropocene: an epoch of our making".*Global Change*, 78, March 2012, pp. 12–15.

TEILHARD DE CHARDIN, PIERRE, "L'Hominisation" (1923); "Le coeur de la matière" (1950). Republished in *Autobiographie spirituelle*. Paris, 1976.

THORNTON, Alex, "This is how many animals we eat each year". *World Economic Forum*, 8/II/2019.

TOYNBEE, Arnold J., *Mankind and Mother Earth*. Oxford University Press, 1976.

WALLACE-WELLS, David, "The Unhabitable Earth". *New York Magazine*, 2017.

WALLACE-WELLS, David, *The Unhabitable Earth. Life After Warming*. New York, 2019.

WEARN, Oliver R. *et al.* "Extinction Debt and Windows of Conservation Opportunity in the Brazilian Amazon". *Science*, 337, 6.091, 13/VII/2012, pp. 228–232.

WHYTE, Chelsea, "Living on the veg". *New Scientist*, 27/I/2018, pp. 26–31, p. 30.

WULF, Andrea, *The Invention of Nature. Alexander von Humboldt's New World*. New York, Vintage Books, 2015.

YANG, Hong *et al.*, "Antibiotic Application and Resistance in Swine Production in China: Current Situation and Future Perspectives". *Frontiers in Veterinary Science*, 17/V/2019.

ZALASIEWICZ, Jan, *A Stratigraphical Basis for the Anthropocene*. London, Geological Society, 2014.

ZALASIEWICZ, Jan & Mat, "Battle Scars". *New Scientist*, 225, 3014, 2015, pp. 36–39.

Part II
Three Concentric Illusions

Chapter 13
The Illusion of a Sustainable Capitalism

The preceding 12 chapters have sought to offer an overall picture of the multiple crises that—together, through interacting dynamics—are propelling us toward socio-environmental collapse. It is from this background that we begin the second part of this book. We will develop the book's two central theses, laid out in the Introduction and repeated here. First thesis, the illusion that capitalism can become environmentally sustainable is the most misleading idea in contemporary political, social, and economic thought. Second thesis, this first illusion is nourished by a second and a third one. The second illusion, discussed in Chap. 14, is the tenacious belief—at one time reasonable but now definitely fallacious—that the more material and energetic surplus we are able to produce, the safer will be our existence. These two illusions are grounded in a third one, discussed in Chap. 15—the anthropocentric illusion.

That capitalism is incapable of reversing the tendency toward global environmental collapse—the thesis of this chapter—is something that should not be considered a thesis, but an elementary fact of reality, given the evidence for it. This fact is even admitted by an authority on global capitalism, Pascal Lamy. In an interview held in 2007, the former director general of Crédit Lyonnais and former director general of WTO stated[1]:

> Capitalism cannot satisfy us. It is a means that must remain in the service of human development. Not an end in itself. A single example: if we do not vigorously question the dynamic of capitalism, do you believe we will succeed in mastering climate change? (…) You have, moreover, events that come to corroborate the least bearable aspects of the model: either its intrinsic dysfunctions, such as the subprime crisis, or the phenomena that capitalism and its value system don't allow us to deal with—the most obvious of those being global warming.

In the same year (2007), a similar verdict was issued: "Climate change is a result of the greatest market failure the world has seen." This sentence is not from an

---

[1] See Pascal Lamy, "Capitalism Cannot Satisfy Us". Interview granted to Daniel Fortin and Mathieu Magnaudeix. *Challenges*, 6/XII/2007.

© Springer Nature Switzerland AG 2020
L. Marques, *Capitalism and Environmental Collapse*,
https://doi.org/10.1007/978-3-030-47527-7_13

anti-capitalist manifesto, but from the *Stern Review. The Economics of Climate Change* authored by Lord Nicholas Stern, former president of the British Academy, former Chief Economist and Senior Vice-President of the World Bank, second permanent secretary to Her Majesty Treasury, and professor at the London School of Economics and the Collège de France. If there is someone who no longer has illusions about the compatibility between capitalism and any concept of environmental sustainability, it is Yvo de Boer, former Executive Secretary of the United Nations Framework Convention on Climate Change (UNFCCC), who resigned after the failure of the 15th Convention of the Parties (COP15) in Copenhagen in 2009. A diplomat well versed in the intricacies of international climate negotiations and professionally attached to the weight of words, he clarified his position in 2013 in an interview with *Bloomberg Business*: "The only way that a 2015 agreement can achieve a 2°[C] goal is to shut down the whole global economy" (Jung et al. 2015). Not too long ago, this statement could summarize the content of this chapter. Today, meeting the 2 °C target no longer depends on a deal. It has become a socio-physical impossibility, as discussed in Chap. 7. What is now on the agenda is to divert ourselves from our current trajectory which is leading us to an average global warming of over 2 °C in the second quarter of the century and over 3 °C probably in the third quarter of the century. We will repeat: any average global warming above 3 °C is considered "catastrophic" (Xu and Ramanathan 2017), as it will lead to low latitudes becoming uninhabitable during the summer, the frequent flooding of urban infrastructure in coastal cities, and drastic decreases in agricultural productivity. It will also probably trigger even greater warming (>5 °C) which will have unfathomable impacts.

To minimize the destruction to the Earth system resulting from the predatory dynamics of global capitalism, regulatory frameworks have been implemented on a "deregulated" capitalism. The adjective is in quotation marks to underline its redundancy. In other words, would an economy operating within ecological frameworks still be capitalist? One might ask a more modest question: would an economy still be capitalist if it is capable of functioning under the ten key recommendations for a Global Action Plan proposed in 2014 by Lord Nicholas Stern and Felipe Calderón in their report, *Better Growth Better Climate*? The first part of this book presents ample evidence that capitalism is incompatible with the adoption of these ten proposals, each of which equates to an admission of capitalism's congenital environmental unsustainability:

(1) "Accelerate low-carbon transformation by integrating climate into core economic decision-making processes." In their investment decisions, corporations will not consider the impact of global warming whenever that impact conflicts with the raison d'être of the investment: the expectation of profitability in the shortest time possible. As long as the investment decision is an inalienable legal prerogative of those in charge of companies (private or state-controlled) and as long as this decision does not emanate from a democratic authority that is guided by science, this first recommendation will be ignored. This is a trivial observation based on overwhelming historical experience, and it is quite puzzling that it would still be necessary to mention it in the twenty-first century.

(2) "Enter into a strong, lasting and equitable international climate agreement."
This second recommendation brings to mind Antonio Gramsci's famous
adage: "History teaches, but it has no pupils." Stern and Calderón should
remember, in fact, a lesson that has been relentlessly repeated for over 40 years.
The first international resolution to reduce GHG emissions dates back to June
1979 (Rich 2018; Klein 2018; Mecklin 2018). In 1988, in Toronto, the World
Conference on the Changing Atmosphere took place. More than 340 participants from 46 countries attended it. The same virtuous resolution was promoted here, but this time the goals were well quantified. Participants agreed
that there should be a 20% cut in global CO_2 emissions by 2005 and, eventually, a 50% cut. Since then, international meetings and protocols have followed monotonously; their central theme has been the reduction of GHG
emissions. And yet, since 1990, CO_2 emissions have increased by over 63%,
as is well known. This pattern is repeated even after the Paris Agreement. The
simplest and irrefutable proof of this is that by 2018 global CO_2 emissions
were already 4.7% higher than in 2015. There is nothing ambitious about the
Paris Agreement, it is not legally binding, and it remains ignored by the world
of high finance. A report released in March 2020 by Rainforest Action
Network, BankTrack, Indigenous Environmental Network, Oil Change
International, Reclaim France, and Sierra Club, and endorsed by over 160
organizations around the world, showed that 35 global banks from Canada,
China, Europe, Japan, and the United States have together funneled USD 2.7
trillion into fossil fuels in the 4 years since the Paris Agreement was adopted
(2016–2019). "The massive scale at which global banks continue to pump billions of dollars into fossil fuels is flatly incompatible with a livable future,"
said rightfully Alison Kirsch (Corbet 2019). The Guardian, together with two
think tanks (InfluenceMap and ProxyInsight), has revealed that, since the Paris
Agreement, the world's three largest money managers (BlackRock, Vanguard,
and State Street) have built a combined USD 300 billion fossil fuel investment
portfolio (Greenfield 2019). Furthermore, the Paris Agreement has not been
ratified by many OPEC countries, has been abandoned by the United States,
and pledges to reduce GHG emissions and to transfer resources to the poorest
countries are not being observed by the signatory countries. Germany did not
reach its goal of a 40% CO_2 emission reduction (from its 1990 level) by 2020.
France's 2015–2018 carbon budget has not been met either. During this period,
its annual emissions decreased by only 1.1%, much less than planned. Brazil,
the world's seventh largest GHG emitter, saw an 8.9% increase in emissions in
2016 (compared to 2015), despite the worst economic recession in its recent
history. It is true that these emissions decreased by 2.3% in 2017 (2071
$GtCO_2$), compared to 2016 (2119 $GtCO_2$), but the surge in fires and the acceleration of deforestation in the Amazon and other Brazilian biomes in 2019
should completely reverse this very modest advance. Brazil's current president, Jair Bolsonaro, one of the most heinous promoters of ecocide of our
time, probably does not even know what commitments Brazil took on in the
Paris Agreement. In September 2019, Antonio Guterres coordinated a high-
level UN meeting with the goal of increasing the ambitions of the Paris

Agreement. All major economies of the planet failed to answer. The complete failure of the COP25 in Madrid to cope with the climate emergency has once again shown that James Hansen was right when he declared in 2015: "Promises like Paris don't mean much, it's wishful thinking. It's a hoax that governments have played on us since the 1990s." What still needs to be understood in order to conclude that the Paris Agreement is doomed to the same pathetic fate as the Kyoto Protocol?

(3) "Phase out subsidies for fossil fuels and agricultural inputs, and incentives for urban sprawl." In 2009, the G20 issued a solemn statement: "We pledge to phase out fossil fuel subsidies." In 2015, the G20 countries spent US\$ 452 billion on direct subsidies to fossil fuels (Bast et al. 2015), and there was no mention of this in the Paris Agreement. An IMF working paper (Coady et al. 2019) provides estimates of direct and indirect fossil fuel subsidies (defined as the gap between existing and efficient prices for 191 countries): "Globally, subsidies remained large at \$4.7 trillion (6.3% of global GDP) in 2015 and are projected at \$5.2 trillion (6.5% of GDP) in 2017." The Climate Accountability Institute keeps reporting on the responsibility of the big polluters. As already pointed out in the Introduction, 63% of global emissions occurring between 1751 and 2010 originated from the activities of 90 corporations in the fossil fuel and cement industries. Just 100 corporations have been responsible for 71% of global emissions since 1988, and just 20 of them are directly linked to more than 33% of all GHG emissions since the beginning of the Industrial Revolution. These corporations are not paying the huge costs imposed on our societies by the burning of fossil fuels. We are. As stated by Lord Nicholas Stern, the IMF estimate "shatters the myth that fossil fuels are cheap" (Carrington 2015). In 2016, G7 leaders again urged all countries to phase out fossil fuel subsidies by 2025. Three years later, no significant step has been taken in this direction. And this has not happened for the same reason as always, one that everyone is aware of: seven out of the top ten corporations in the world by revenue (according to the 2018 Fortune Global 500 list) are fossil fuel industries or are umbilically linked to them. Together, the revenues of these seven corporations amount to almost two trillion dollars. Not only do they control states, but the two largest of them—Sinopec Group and China National Petroleum—are state-owned enterprises and are an essential part of the Chinese state's power strategies.

(4) "Introduce strong, predictable carbon prices as part of good fiscal reform and good business practice." With their unwavering faith in the market, economists continue to believe in the carbon pricing myth, as if the energy transition—at the required scale and speed—could be induced through pricing mechanisms. Actually, in 2019, more than 25 national or subnational carbon tax systems have already been implemented or are scheduled to be implemented around the world. So far, there has been no observed impact of these initiatives on fossil fuel consumption. And even as carbon tax systems become more widespread and more aggressive, the market will always be able to adapt to them without significantly reducing fossil fuel consumption, simply because these

fuels are not commodities like any other. As seen in Chap. 5 (Fig. 5.1), the huge variations in oil price between 1990 and 2019 had very little impact on the almost constant increase in oil consumption.

(5) "Substantially reduce capital costs for low-carbon infrastructure investments." These costs have been reduced without influencing the growth of fossil fuel consumption, as seen above. All projections indicate that there will be no significant reduction in gas, oil, and coal consumption in the discernible future.

(6) "Scale up innovation in key low-carbon and climate resilient technologies, tripling public investment in clean energy R&D and removing barriers to entrepreneurship and creativity." With the exception, perhaps, of China and India, there is no global expectation of tripling the allocation of resources for such research. Instead, we see a slight reduction in these investments on a global scale between 2011 and 2017, as shown in Fig. 13.1.

(7) "Make connected and compact cities the preferred form of urban development, by encouraging better managed urban growth and prioritising investments in efficient and safe mass transit systems." As seen in Chap. 9, urban sprawl and chaos increase with the proliferation of carmakers, fossil and cement industries, intensive agriculture, urban solid waste, and other unprocessed waste, particularly in so-called "developing" countries where gigantic conurbations tend to be concentrated.

(8) "Stop deforestation of natural forests by 2030." As the Global Forest Watch and several other indicators show, deforestation continues to accelerate in

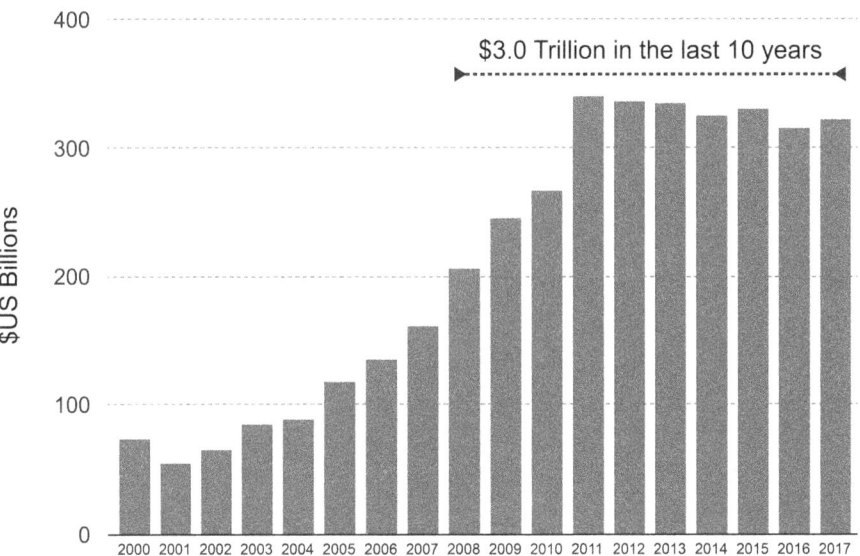

Fig. 13.1 Global investment in renewable energy supply (2000–2017) in US$ billions. [Source: IEA (2000–2016); Frankfurt School – UNEP Collaborating Centre for Climate & Sustainable Energy Finance (2017)] Note: the renewable category shown in this graph includes investments in electricity, transport, and heat

tropical and boreal forests on a global scale (see Chap. 2). Deforestation of tropical forests will cease simply because by 2030, as seen in Chap. 2, many of them will have been wiped out.

(9) "Restore at least 500 million hectares of lost or degraded forests and agricultural lands by 2030." Reforestation has been limited to little more than planting a few—usually exotic—species, ones that are considered inputs for industries. In addition, soils continue to be degraded and will remain so, as long as we maintain the two paradigms on which agribusiness is based on: (a) a commodity agriculture that is toxic-intensive and strongly export-oriented, with food self-sufficiency decreasing in a growing number of countries, and (b) a diet based on carnivorism, which is evidently unsustainable.

(10) "Accelerate the shift away from polluting coal-fired power generation, phasing out new unabated coal plants in developed economies immediately and in middle-income countries by 2025." As seen in Chap. 6, coal consumption rose again in 2017 after reaching a plateau and even a slight decrease over the previous 3 years. There are no signs of a significant, much less accelerated, decrease in the burning of coal for power generation on a global scale. Moreover, if the opening of hundreds of thousands of hydraulic fracturing oil and gas wells in more than 20 US states since 2005 has led to a reduction in coal use, it has not resulted in lower atmospheric GHG emissions. In Chap. 5, we refer to the work of Jeff Tollefson and colleagues ("Methane leaks erode green credentials of natural gas") published in 2013 in *Nature*. The results of this study have been confirmed by successive observations and measurements. The most recent of these is a study published in April 2016 by the Environment America Research and Policy Center, according to which, in 2014 alone, at least 5.3 billion pounds of methane have leaked from fracking wells (for gas extraction) in the United States. This is equivalent to the average emissions of 22 coal-fired thermoelectric plants in that year (Ridlington et al. 2016).

Obviously, evoking historical evidence would not be enough to demonstrate the structural unsustainability of capitalism, since such a system could change, as Lord Nicholas Stern and Felipe Calderón might argue, this being the very raison d'être of their document. It turns out, though, that globalized capitalism cannot change. More than a lesson from history, it is the logic of accumulation that can demonstrate the unfeasibility of the ten recommendations proposed in this document. The regulatory frameworks that the authors dream of are not within the aims of global capitalism and will never occupy a central position in its agenda. This chapter, thus, examines the two impossibilities of implementing regulatory frameworks capable of containing the tendency toward collapse within the realm of globalized capitalism:

(1) The self-regulation of economic agents induced by the presence of mechanisms that emanate from the market itself

(2) Regulation induced not only by market mechanisms but by agreements negotiated between businesses, the state, and civil society

13.1   The Capitalist Market Is Not Homeostatic

The idea of self-regulation does not apply to capitalism. It is not ruled by the principle of homeostasis, pertaining to the dynamics of optimizing the internal stability of an organism or system. Maurice Brown (1988) notes that this idea goes back to Adam Smith's "faith in the homeostatic properties of a perfectly competitive market economy." Even today, the belief that capitalism is self-regulating has the value of a maxim, being accepted by many scholars. An example of the use of this analogy comparing the mechanisms of the capitalist market to that of a living organism is found in Eduardo Giannetti (2013):

> [The market] "has a functioning logic equipped with surprising properties from the stand-point of productive and allocative efficiency. It is a homeostatic system governed by negative feedback. Every time the system becomes disturbed, it seeks to return to equilibrium."

The analogy between the market mechanism and that of a homeostatic system is a mistake. Since Claude Bernard's idea of milieu intérieur or the internal environment (Canguilhem 1968/1983)[2] and Walter Cannon's notion of homeostasis, we know that any influence disturbing the balance (deficits or excesses) of the vital functions in an organism or organic system triggers regulatory and compensatory activities that seek to neutralize this influence, resulting in the recovery of balance or, more precisely, in a new balance (allostasis). The maintenance of this efficient stability of the internal environment in its constant exchanges with the external environment is what guides the activity of every organism. Even though it is dependent on the external environment, even though it is, therefore, an "open" system, all the energies of an organism are ultimately centripetal: they are directed toward the survival, security, and reinforcement of the organism's centrality and stability, in short, of its own identity.

Now, regarding the basic functioning of the capitalist market, not only does it not work by negative feedback, but it is even opposed to the mechanism of homeostasis. This is because the fundamental force that drives the market is not the law of supply and demand, which operates within the scope of commodity circulation, but the law of capital accumulation, which operates within the scope of commodity production and is, by definition, expansive. The market can even force a cyclical crisis and less production, but expansion is the basic rule of the return on capital, in other words, of the physiology of capitalism.

This leads us to the second misconception of ascribing the attributes of homeostasis to the market: once it reaches its ideal size, every organism ceases to grow and goes onto the phase in which conservative adaptations prevail. This phenomenon does not occur in the capitalist market, which is driven by centrifugal forces (imposed by capital accumulation) toward unlimited growth. The ideal size of the

---

[2] See Claude Bernard, *Introduction à l'étude de la medicine expérimentale"* (1865): "La science antique n'a pu concevoir que le milieu extérieur; mais il faut, pour fonder la science biologique expérimentale, concevoir de plus un *milieu intérieur*. Je crois avoir le premier exprimé clairement cette idée". Cited by Georges Canguilhem, (1968/1983, p. 148).

capitalist market is, by definition, one that is infinite. Unlike an organism, if the capitalist market stops growing, it becomes unbalanced. If the age of capitalist growth is coming to an end, this is not due to a homeostatic virtue of the market, but to something extraneous to it: the physical limits of the biosphere's resilience. Since the 1970s, Ivan Illich (2003) has noticed that:

> Being open, human equilibrium is liable to change due to flexible but finite parameters; if men can change, they do so within certain limits. On the contrary, the dynamic of the industrial system underlies its instability: it is organized toward indefinite growth and an unlimited creation of new needs that quickly become mandatory in the industrial framework.

Another argument of Giannetti found in the same essay is, however, absolutely correct: "the pricing system, notwithstanding all of its surprising merits and properties, has one serious flaw: it does not give the right signs regarding the use of environmental resources." In this respect, André Lara Resende (2013) is adamant:

> Regarding the physical limits of the planet or the destruction of the environment caused by human action, relying on the market price system [...] makes no sense. Any student in a basic microeconomics course should know this.

13.1.1   The Inversion of Taxis

The only pricing operated by the market is that of the relationship between economic costs and profit rate. Capitalism cannot price its action on ecosystems because ecosystems are broader in space and time than the investment/profit cycle. In *Energy and Economic Myths* (1975), Nicholas Georgescu-Roegen essentially states the same thing: "the phenomenal domain covered by ecology is broader than that covered by economics." In this way, he continues, "economics will have to merge into ecology." Herman Daly (1990) formulates this thesis equally elegantly: "in its physical dimensions the economy is an open subsystem of the earth ecosystem, which is finite, nongrowing, and materially closed." In capitalism, the world is upside down: the physical environment is conceived of as a raw material, that is, as an open subsystem of the economic system; there is a reversal of taxis which results in an equally inverted hierarchy of the world, one that is incompatible with its sustainability. Therefore, the ability to subordinate economic goals to the environmental imperative is not within the sphere of capitalism.

13.2   Milton Friedman and Corporate Mentality

There is no moral judgment here. Capitalism is unsustainable not because corporate leaders are bad or unscrupulous men. It would be absurd to suppose that corporate owners, shareholders, and CEOs are people who lack a moral compass. Nothing allows us to affirm that there is less moral sense in business circles than there is in any other civil society environment, for example, in trade unions and universities and in religious, artistic, or sports groups. The problem is that, no matter how much they want to improve the ethical conduct of their corporations, managers *cannot* afford to subordinate their corporate goals to the environmental imperative.

To demonstrate this impossibility, we must start with a trivial example: money loses purchasing power due to inflation and has varying rates of purchasing power or profitability due to unequal market opportunities. To avoid depreciation or its use in disadvantageous conditions, every owner of a certain sum of money must choose the best exchange option in each moment. This applies to both the worker seeking to exchange his salary for as many goods as possible and the investor who chooses the most promising transactions or funds. In light of this elementary market reality, corporations must present the comparative advantages of one investment opportunity over others to their current or future investors and shareholders. If British Petroleum, for example, forgoes a potentially profitable investment because of its environmental impact, investors will have two alternatives: they will replace that decision-maker if they have the power to do so; or, if not, they will redirect their investments to other corporations or even other sectors of the economy that are more likely to be profitable.

Both those who offer and those who raise funds in the market are subordinate to this relentless rationality. It explains why corporations cannot self-regulate around variables other than profit maximization. They have minimal leeway to adopt what Seev Hirsch (2011) calls "enlightened self-interest," as this most often entails sacrificing investment opportunities, raising costs, losing competitiveness, or limiting profits in the short term. Both critics and supporters of capitalism agree on this. In 1876, Friedrich Engels wrote (as cited by Magdoff 2011):

> As individual capitalists are engaged in production and exchange for the sake of the immediate profit, only the nearest, most immediate results must first be taken into account. As long as the individual manufacturer or merchant sells a manufactured or purchased commodity with the usual coveted profit, he is satisfied and does not concern himself with what afterwards becomes of the commodity and its purchasers. The same thing applies to the natural effects of the same actions.

This passage could have been undersigned by Milton Friedman (1912–2006), winner of the Nobel Prize for economics in 1976, adviser to Ronald Reagan, professor at the Chicago School of Economics, and, according to *The Economist*, "the most influential economist of the second half of the 20th century." Friedman justly classifies as immoral any initiative of a corporation manager aimed at mitigating environmental impacts if such initiative entails a decrease in profit. Asked in 2004 whether John Browne, then President of British Petroleum, had the right to take environmental measures that would divert BP from its optimal profit, Friedman replied (quoted in Magdoff and Bellamy Foster 2011):

> No... He can do it with his own money. [But] if he pursues those environmental interests in such a way as to run the corporation less effectively for its stockholders, then I think he's being immoral. He's an employee of the stockholders, however elevated his position may appear to be. As such, he has a very strong moral responsibility to them.

Friedman's answer is irrefutably logical. It defines a corporation's "moral responsibility" as the commitment of its governing bodies to its shareholders. This logic and conception of moral responsibility were defended by the *New Individualist Review*, on whose editorial board Friedman served.[3] For the same reason, Rex Tillerson, chief executive officer of ExxonMobil (2006–2017) and the US Secretary of State (from February 2017 to March 2018), was applauded in May 2015 at the Exxon Mobil Corporation's annual meeting of shareholders in Dallas by justifying his refusal to host climate change specialists and set greenhouse gas emission limits: "We choose not to lose money on purpose." Incidentally, a proposal at this meeting to set GHG emission limits obtained less than 10% of the votes (Koenig 2015). This same moral responsibility toward shareholders is illustrated in another case analyzed by *The Economist* in a 2012 report on rising global obesity levels: in 2010, a PepsiCo leader gave up making her products healthier, as shareholders started becoming outraged. And rightly so, Friedman would say, since shareholders only put their resources and trust in PepsiCo because it promised them the best expected market return. Frederic Ghys and Hanna Hinkkanen (2013) showed why "socially responsible investments" (SRI) are "just marketing," as they do not really differ from traditional portfolios. According to a financial investment expert, quoted in their report: "the bank would transgress its financial role as an asset manager when including environmental and social considerations in investment decisions for clients who had not directly requested it."

We know that in order to maintain a reasonable chance of not exceeding a global average warming of 2 °C above the pre-industrial period, we have a carbon budget of approximately 600 GtCO$_2$ (Anderson 2015; Figueres et al. 2017). According to Friedman's logic, for corporations to be a moral entity, in other words, to keep their stock prices high, thereby honoring their contracts and commitments to their shareholders, they must continue burning the coal, oil, and gas reserves controlled by them and by the state-owned corporations that live off the sale of these fuels. In an open letter to Christiana Figueres (UNFCCC Executive Secretary), written by Cameron Fenton (director of the Canadian Youth Climate Coalition) and signed by over 160 people and NGOs (2012), we read:

> All together, the global oil, coal and gas industries are planning to burn over five times that amount, roughly 2,795 Gigatonnes of carbon. Indeed, their share prices depend on exploiting these reserves. (...) Their business plan is incompatible with our survival. Frighteningly, there are also states, parties to the convention [UNFCCC], with the same plan.

---

[3] Founded by Ralph Raico, the *New Individualist Review. A Journal of Classical Liberal Thought* was published between 1961 and 1968. In the Introduction to its reprint, in 1981, Friedman declared that his articles "remain timely and relevant." See M. Friedman, "Introduction". In *New Individualist Review*, Indianapolis, Liberty Press, 1981, pp. ix-xiv.

13.3   Three Aspects of the Impossibility of a Sustainable Capitalism

The logic of the impossibility of a sustainable capitalism is concretely proven in numerous aspects of its *modus operandi*. Let us isolate three aspects of this impossibility.

But before that, it is best to start by giving voice not to pure and staunch liberals like Milton Friedman, but to those who believe that capitalism has nothing to fear from environmental regulation. Many of them reject a defensive stance and put themselves on the offensive, claiming that environmental sustainability and increased profits are not only compatible but reciprocally strengthen each other in a virtuous circle.

If I am not mistaken, the advocates of this thesis prefer the following argument: adopting innovative solutions to increase the efficiency of the input/product or product/waste ratio and improve environmental safety in the production process increases the company's competitiveness (as opposed to reducing it) because it is a value-generating process, be it in terms of risk management, brand image, and, finally, effective financial results. If this is true, then taking the lead and being at the forefront of economic processes with lower environmental impact and risk will ensure a better profitability than the average profit rate. I hope to not underestimate the literature on the business and sustainability binomial by saying that it limits itself to elaborating variations on this theme while offering several case studies on the direct relationship between sustainability and profitability. There are a growing number of economists and NGOs committed to encouraging companies to embrace this belief. They naturally render a tremendous service to society and to the companies themselves through their work. However, their success is limited by the three aspects that render an environmentally sustainable capitalism impossible, as stated in the title of this section.

(1) Decoupling and Circular Economy

Decoupling is the hope that eco-efficient technologies and production processes in industrialized countries with mature economies will enable the miracle of increased production and consumption with less pressure (or at least no corresponding increase in pressure) on ecosystems (Jöstrom and Östblom 2010). It is true that a greater efficiency in the production process may allow for relative decoupling, meaning that it enables a reduction in pressure per product or per unit of GDP. But it does not decrease this pressure in absolute terms, since the number of products does not cease to increase on a global scale. The mechanism known as the "Jevons paradox" or rebound effect describes how increasing demand for energy or natural resources always tends to offset the eco-efficiency gain of technological innovation. Thus, although energy efficiency per product has doubled or even tripled since 1950, this gain is offset by the expansion of production at a greater rate than the eco-efficiency gain.

The actions of institutions and business foundations that advocate for an eco-efficient and circular economy based on reverse engineering, recycling, reuse, and remanufacturing are certainly positive. We know, however, that there is no circular economy. No economy, let alone a global economy trapped in the paradigm of expansion, can evade the second law of thermodynamics, whose relationship with economics has been analyzed by Nicholas Georgescu-Roegen since the 1970s (1971, 1975 and 1995). Here we must state the obvious: even though the surplus energy supplied by oil and other fossil fuels in relation to the energy invested to obtain them is declining (for this declining EROI, see Chap. 5, Sect. 5.5), low-carbon renewable energies are not yet, and may never be, as efficient as oil. This means that the energy transition, while urgent and imperative, will further distance us from a circular economy. According to calculations by Dominique Guyonnet, "to provide one Kw/h of electricity through land-based wind energy requires about 10 times more reinforced concrete and steel and 20 times more copper and aluminum than a coal-fired thermal power plant" (Madeline 2016). The only way, therefore, to lessen the environmental impact of capitalism is to reduce, in absolute terms, the consumption of energy and goods by the richest 10% or 20% of the planet. This is incompatible with capitalism's basic mechanism of expansive functioning and with the worldview that it sells to society.

(2) The Law of Resources Pyramid

The increasing scarcity of certain inputs and the need to secure their large-scale and low-cost supply nullify the potential benefits of various green initiatives taken on by companies. These cannot, in fact, evade the law of the resources pyramid, described by Richard Heinberg (2007):

> The capstone [of the pyramid] represents the easily and cheaply extracted portion of the resource; the next layer is the portion of the resource base that can be extracted with more difficulty and expense, and with worse environmental impacts; while the remaining bulk of the pyramid represents resources unlikely to be extracted under any realistic pricing scenario

This law of the resources pyramid can be stated in an even simpler form: in capitalism, the logic of capital accumulation and surplus, together with the growing scarcity of finite natural resources, necessarily exacerbates the negative environmental impact of economic activity.

(3) The Impossibility of Internalizing the Environmental Cost

What makes it specifically impossible for corporations to submit themselves to the environmental imperative is the impossibility of "internalizing" the costs of increasing environmental damage that they bring about. Methodologies to "price" nature are now multiplying. But whatever the methodology (always based on the assumption that the value of nature is reducible to a market price), the result is the same: it is impossible for corporations to internalize their environmental cost because the total value generated by their activity is often less than the monetary expression of the value of the natural heritage that was destroyed by that activity.[4] A report was prepared for The Economics of Ecosystems and Biodiversity (TEEB), titled Natural Capital at Risk. The top 100 externalities of business (2013) show that:

> The estimated cost of land use, water consumption, GHG emissions, air pollution, land and water pollution and waste for the world's primary sectors amounts to almost US$7.3 trillion. The analysis takes account of impacts under standard operating practices, but excludes the cost of, and risk from, low-probability, high-impact catastrophic events. (…) This equates to 13% of global economic output in 2009. Risk to business overall would be higher if all upstream sector impacts were included.

13.4 Regulation by a Mixed Mechanism

Let us now examine the second general impossibility of sustainable capitalism mentioned in the beginning of this chapter: sustainability achieved through regulatory frameworks negotiated between organized sectors of civil society, on the one hand, and states and corporations, or state–corporations, on the other. Here we touch on the punctus dolens of all the problems discussed in this chapter and even in this book: the impossibility, at least up to now, of this second route stems from the lack of parity between the two parties, a necessary condition for an effective negotiation.

There is still a huge gap between science and societies' perception of reality. The latter continue to consider the issue of climate change and the decline of biodiversity as non-priority issues in their list of concerns and expectations. But this is changing very quickly. Increasingly aware of the bankrupt planet being bequeathed to them by the values and paradigms of globalized capitalism, society as a whole, led by youth, is beginning to mobilize around the idea of an alternative paradigm, characterized by the subordination of the economy to ecology. Their movements have been gaining much more momentum and global reach than had been foreseen until recently by even the most optimistic. It is true that, up to now, their protests and claims have not slowed GHG emissions, nor the decline in biodiversity, nor the pollution of soils, water, and air. But history, it never hurts to repeat, is unpredictable, and a sudden paradigm shift in civilization, capable of overcoming the imperative of economic growth and anthropocentrism, may be closer than ever before. In any case, the possibility of mitigating the ongoing collapse is dependent on the strengthening of these socio-environmental movements, society's greater awareness of the extreme severity of the current situation, and the ability to impose on state–corporations' policies that are congruent with the current state of urgency.

---

[4] On the concept of the economic value of nature and its measurement, as proposed, among others, by Pavan Sukhdev, see The Economics of Ecosystems and Biodiversity in Business and Enterprise and his videos available on YouTube.

13.4.1 The State and the Financial System

We must recognize that, on the other side of the battlefield, the constituted powers are increasingly united in defending themselves. We should not expect from the state initiatives that might lead corporations toward activities with low environmental impact. We saw in the Introduction that there appears to be a true transformation toward a new type of state that is the partner, creditor, and debtor of corporations: the state–corporation. Contrary to the 1929 crisis, which led to the New Deal in the United States and to a new role of the state in the international stage, the financial crisis unleashed in 2008 displayed the impotence of the state and the loss of its identity. Rather than regulating financial activity, governments embarked on the most comprehensive bailout operation of banks. Since September 2008, the bulk of US and European financial resources has been used to bail out the banking system and "calm the markets." As a July 2011 document from the US Government Accountability Office (GAO) shows, from December 1, 2007, to July 21, 2010, the Federal Reserve Bank (FED) had, through various emergency programs and other assistance provided directly to institutions facing liquidity strains, given out loans that amounted to US\$ 1139 trillion.[5] The momentum of the crisis has led banks to, more than ever, take control of the state and raid their resources. According to a GAO report on conflict of interest, requested by Senator Bernie Sanders and published by him on June 12, 2012[6]:

> During the financial crisis, at least 18 former and current directors from Federal Reserve Banks worked in banks and corporations that collectively received over US\$4 trillion in low-interest loans from the Federal Reserve.

According to information reported by *Bloomberg*, as of March 2009, the Federal Reserve had pledged US\$ 7.7 trillion in guarantees and credit limits to the North American financial system (Ivry et al. 2011). As shown in Table 8 of the GAO Report to Congressional Addressees cited above, between December 1, 2007, and July 21, 2010, funds from FED emergency programs were mobilized by 21 US and European banks in the form of not term-adjusted transactions, with an aggregate value of US\$ 16.115 trillion. As George Monbiot asked in 2011, "Why is it so easy to save the banks, but so hard to save the biosphere?" The question has an unambiguous answer: because saving banks and other corporations have become a primary function of states. According to Moody's Bank Ratings 2012, which assessed 7 banks in Germany in June 2012 and an additional 17 in July 2012 (plus seven in

---

[5] "Report to Congressional Addressees. Opportunities exist to Strengthen Policies and Processes for Managing Emergency Assistance", July 2011, GAO-11-696. United States Government Accountability Office. I thank Dr. Orice Williams Brown, director of GAO's Financial Markets and Community Investment, for kindly sharing this document with me. http://www.gao.gov/products/GAO-11-696.

[6] US Senator Bernard Sanders (I-Vt.), Washington, D.C., 12/VI/2012, "Jamie Dimon Is Not Alone". http://www.sanders.senate.gov/imo/media/doc/061212DimonIsNotAlone.pdf

The Netherlands), even the richest banks in Europe cannot manage their losses alone and cannot strategically survive without the state's safety net.

13.4.2   The Obsolescence of the Statesman

There is no longer any place in the state for the classic figure of the statesman. Voters complain about the increasing corruption of parties, the loss of values and principles, and politicians' venal attachment to state benefits. They also complain about managerial incompetence, disloyalty, or lack of leadership of heads of state who betray their ideological profiles and break the promises that galvanized their electoral victories. It has become common to compare yesterday's politicians with their successors in a way that is always disadvantageous to the latter: De Gaulle with Hollande or Macron, Churchill with Cameron or Boris Johnson, Franklin D. Roosevelt with Obama or Trump, Adenauer with Merkel, De Gasperi with Berlusconi, Renzi, or Giuseppe Conte, etc. But it would be absurd to suppose that societies have lost the ability to produce personalities equal to the great statesmen who led Western democracies at critical moments in their history. What has been lost is the power of the state as the quintessential place of power and of political representation.

13.4.3   Threats to the Democratic Tradition of Political Representation

The idea that political leaders are provisional bearers of a mandate granted to them by the governed, the idea, in short, of political representation, a cornerstone of the democratic tradition born in Athens and expanded by universal suffrage in the contemporary age, obviously continues to be the only legitimate form of power that the state can exercise and must always be furthered. Nevertheless, this legitimacy is critically endangered in our time. By deterritorializing power and shifting strategic decisions to the anonymous boards of corporations (whereby states and their resources are activated to finance and execute these decisions), the globalization of capitalism produces, along with the chronic indebtedness of national states, a progressive transformation in the historical meaning of state political power. With statesmen being unable to set the conduct for and impose boundaries on corporations, popular mandates are increasingly becoming the loci of spectacular ritualizations of power, and its dignitaries are increasingly masters in the art of gesticulation. The meaning of the term "representation" exercised by the representatives of popular votes is, thus, increasingly understood in its pantomimic sense.

13.4.4   State Indebtedness

The international financial network controls states, mainly through their debts. Total government debt reached US$ 59 trillion in 2016 and exceeded US$ 65 trillion in 2018, up from US$ 37 trillion just a decade ago (Cox 2019). This debt has ballooned since the 2008 financial crisis, reaching levels never seen before in peacetime (Stubbington 2019). The dramatic leaps in public debt over the past decade, in both advanced and (so-called) emerging economies, are described in Fig. 13.2, according to data from the Bank for International Settlements (BIS).

Based on data from 2016 to 2017, of the 153 nations listed by the IMF or the CIA World Factbook, 102 now have public debts that exceed 50% of their GDP, 32 have public debts that exceed 80% of their GDP, and 16 countries, including the United States, Japan, and Italy, have debts that exceed 100% of their GDP: "the world's major economies have debts on average of more than 70% of GDP, the highest level of the past 150 years, except for a spike around the second world war" (Stubbington 2019). François Morin (2015) shows how 28 large banks, resulting from successive merges spurred by globalization and the deregulation of the Reagan-Thatcher era, have a total balance of US$ 50.3 trillion. These major banks have been labelled global systemically important banks (G-SIBs) by the Financial Stability Board (FSB), created in 2009 during the G20 summit in London. According to François Morin, they have fraudulently colluded into an oligopoly that he equates to a world-wide hydra (De Filippis 2015):

> Public debt plagues every major country. The private and toxic debts were massively trans-ferred to states during the last financial crisis. This public over-indebtedness, linked exclu-sively to the crisis and to these banks, explains—while completely denying the causes of the crisis—the policies of rigor and austerity applied everywhere. [...] States are not only disciplined by markets but, above all, they are hostage to the world hydra.

Corporations manage European public debt through a vicious cycle: (1) Prohibited by their regulations and by the Lisbon Treaty from buying government bonds directly from insolvent states, the European Central Bank (ECB) should buy them

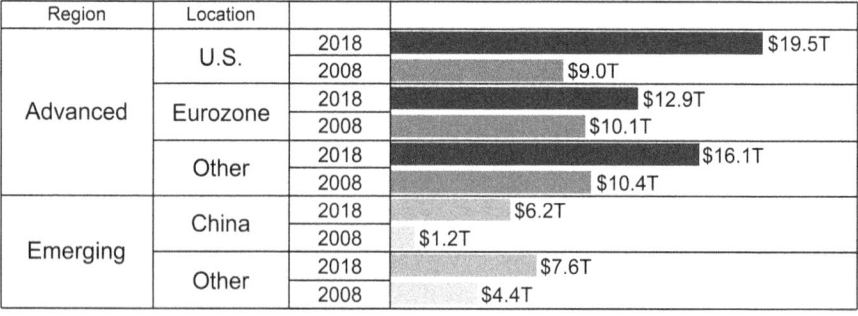

Region	Location		
Advanced	U.S.	2018	$19.5T
		2008	$9.0T
	Eurozone	2018	$12.9T
		2008	$10.1T
	Other	2018	$16.1T
		2008	$10.4T
Emerging	China	2018	$6.2T
		2008	$1.2T
	Other	2018	$7.6T
		2008	$4.4T

Fig. 13.2 Government debt in advanced and emerging economies (trillions of dollars) in 2008 and 2018. (Source: Jeff Cox, "Global debt is up 50% over the past decade, but S&P still says next crisis won't be as bad." *CNBC*, 12/III/2019, based on data from the Bank for International Settlements)

from banks in the secondary market, thus improving the balance sheet of these banks and avoiding the next systemic banking crisis. In addition, the ECB gives loans to banks at rates of 1% to 1.5%, obtaining "junk" bonds or high-risk government bonds as a guarantee.[7] (2) Recapitalized, banks lend "new" money to defaulting states so that they (3) avoid default and repay their creditors. (4) In this way, banks can continue to finance states at higher interest rates, since the state is poorly rated by credit rating agencies. To be able to pay off their debts, states (5) sacrifice their investments and public services for the imperative of reducing the budget deficit and public debt. The so-called austerity measures (6) weaken the economy and lower revenue, which (7) pushes states into default, completing the vicious cycle and taking it to the next level.

"Those in insolvency have to sell everything they have to pay their creditors," said Joseph Schlarmann, one of the leaders of the CDU, the party that leads Angela Merkel's coalition in Germany. This diktat led Greece to sell the island of Oxia in the Ionian Sea to Shaykh Hamad bin Khalifa al-Thani, the emir of Qatar, who bought it for a paltry sum of five million euros. Other Greek islands (out of a total of six thousand), such as Dolicha, were also put on sale. The natural, territorial, and cultural heritage of Mediterranean Europe is considered by creditors to be little more than a bankrupt estate. This type of negligence of Mediterranean civilizational heritage gave rise to bitter assessments done by Salvatore Settis (2002) and Silvia Dell'Orso (2002), this time on the Italian state's abandonment of its responsibilities toward the nation's extraordinary cultural memory. Formerly, the state guaranteed citizens the enjoyment of their heritage and devotion to their monuments through museums and the educational system. Having custody over and conserving this memory, the state was the nexus between the generations, and, through research, it also promoted the critical historical revision of this heritage.[8] Today, even when it does not simply sell this natural, territorial, or cultural heritage, the state–corporation denatures it by conceiving it as an input for tourism to be managed according to the industry's profitability motives.

---

[7] Between May 2010 and March 2011, the ECB bought 66 billion euros from bankers and other investors. In August 2011 alone, the ECB bought a further 36 billion euros (always in the secondary market and at a much higher price than the one traded in that market) of government bonds from Greece, Ireland, Portugal, Spain, and Italy. Not satisfied with this rescue operation, banks took the opportunity to buy more junk bonds in the secondary market at a rate of 42.5% of their face value on August 8, 2011 (and even lower later on), and to resell them to the ECB at 80% of this value. Eric Toussaint, "La BCE, fidèle serviteur des intérêts privés." Interview granted to CADTM, 16/IX/2011

[8] In "The Use and Abuse of History for Life." (1874), Nietzsche discusses the three meanings of history necessary for the man who lives in his own time: as a being that is active and has aspirations (monumental history), as a being that preserves and venerates (antiquarian history), and as a being that suffers and has a need for liberation (critical history). Precisely between the years of Nietzsche and the twentieth century, the social democratic state was the guarantor of these "uses" (*Nutzen*) of history that Nietzsche refers to.

13.4.5   Tax Evasion, the Huge "Black Hole"

The impoverishment of states comes, above all, from tax evasion. In 2000, an article published in the newspaper *Libération* estimated that approximately six trillion euros worth of resources were diverted to 65 tax havens, with an increase of 12% per year over the previous 3 years (1997–1999). According to a July 2012 report by economists from the Tax Justice Network (TJN):

> At least $21 trillion of unreported private financial wealth was owned by wealthy individuals via tax havens at the end of 2010. This sum is equivalent to the size of the United States and Japanese economies combined. There may be as much as $32 trillion of hidden financial assets held offshore by high net worth individuals (HNWIs), according to our report *The Price of Offshore Revisited* which is thought to be the most detailed and rigorous study ever made of financial assets held in offshore financial centres and secrecy structures. We consider these numbers to be conservative. This is only financial wealth and excludes a welter of real estate, yachts and other non-financial assets owned via offshore structures. (…) The number of the global super rich who have amassed a $21 trillion offshore fortune is fewer than 10 million people. Of these, less than 100,000 people worldwide own $9.8 trillion of wealth held offshore. (…) This at a time when governments around the world are starved for resources.

In 2008, Edouard Chambost, an expert on the subject, stated that "55% of international trade or 35% of financial flows pass through tax havens."[9] Gabriel Zucman (2014) of the London School of Economics estimates that states lose US$ 190 billion every year to tax evasion. What the tax havens and financial opacity show is the impotence of the states and, at the same time, its complicity in relation to the power of corporations. According to Thomas Piketty (2016):

> Unfortunately, in this area there is a huge gap between the triumphant declarations of governments and the reality of what they actually do. (…) The truth is that almost nothing has been done since the crisis in 2008. In some ways, things have even got worse. (…)

Thus, another vicious cycle, complementary to the one described above, is set up between corporations, investors, and big fortunes. These (1) divert a considerable portion of the taxes that are due to tax havens, and, through the banks that receive these funds, (2) they provide loans to states at high interest rates. Such interest (3) further puts states at the mercy of creditors. From *de jure* creditors of corporations, states become their chronic debtors, which, finally, (4) fuels the ideology that social democracy is unfeasible because it generates leviathan states that are deficit-prone and wasteful.

And, as if this vicious cycle was not enough, part of the state's revenue is directed to subsidize or finance—through the public treasury, public "development" banks, and tax exemptions—agribusiness, the auto industry, big mining and fossil fuel projects, the military–industrial complex, and other segments with a high concentration of corporate capital that have a deadly environmental impact. The current deterioration of states' financial health is comparable only to the situation at the end

---

[9] Quoted in *La Tribune*, 16/X/2008, p. 38

of World War II, when public finances had been destroyed. The difference, however, is that the destruction of the biosphere eliminates the prospect of a new cycle of economic growth, like the one that occurred in 1947–1973.

13.4.6   What to Expect from States?

In this context, what efforts can we still expect from states regarding the environmental regulation of corporate activity? Can the Brazilian state be expected— even before the election of Jair Bolsonaro—to implement an active energy transition policy and/or a policy to protect the country's forests? What to expect from the United States with its debt of US$ 16 trillion in December 2012 which then jumped to US$ 19 trillion in February 2016 and is expected to exceed US$ 23 trillion by 2020? In 2018, Trump signed a US$ 1.2 trillion plan to revise the entire US nuclear arsenal and authorized a new nuclear warhead for the first time in 34 years (Hennigan 2018). To sustain the country's military–industrial corporate complex, one of the most polluting and environmentally unsustainable, its "defense" budget must be the third item in the national budget. Dwight Eisenhower had already warned about this in his famous farewell address to the nation in 1961:

> This conjunction of an immense military establishment and a large arms industry is new in the American experience. The total influence—economic, political, even spiritual—is felt in every city, every Statehouse, every office of the Federal government. We recognize the imperative need for this development. Yet, we must not fail to comprehend its grave implications.

What Eisenhower called in 1961 the military–industrial complex has completely taken over the United States and is now best known as the military–industrial–congressional complex (MICC). The civil and military arms industry lobby maintains control over Congress, with the latter approving funds that are not even required by the military, such as financing the production of Abrams battle tanks which the army says it does not want, since the current fleet of 2400 units is only about 3 years old (Lardner 2013). Similarly, in France, although the Centre International de Recherche sur le Cancer (CIRC) has been claiming since 1988 that diesel is carcinogenic, the French government, followed by Germany, continues to subsidize some models of diesel engines, making France the country with the highest percentage (61%) of diesel-powered vehicles in the world. In conclusion, can we expect states to impose efficient environmental controls on the network of corporations that control them? Notwithstanding timid advances, the answer is essentially a negative one.

13.5   A Super-entity. The Greatest Level of Inequality in Human History

When comparing corporate vs. government revenues, it becomes very clear that corporate power is greater than that of states. According to the 2017 *Fortune Global 500*, revenues of the world's 500 largest corporations amounted to US$ 28 trillion, the equivalent of 37% of the world's GDP in 2015. As revealed by the NGO Global Justice Now (2016), "69 of the world's top economic entities are corporations rather than countries in 2015," and 10 of top 28 economic entities are corporations. "When looking at the top 200 economic entities, the figures are even more extreme, with 153 being corporations."[10] Nick Dearden, director of Global Justice Now, declared: "The vast wealth and power of corporations is at the heart of so many of the world's problems – like inequality and climate change. The drive for short-term profits today seems to trump basic human rights for millions of people on the planet."[11]

Even more important than the power of an isolated corporation is the obscure and highly concentrated power within the corporate network itself. This network is controlled by a caste that is barely visible and that is impervious to the pressures of governing parties and societies. Their investment decisions define the destinies of the world economy and, therefore, of humanity. This is what research on the architecture of the international ownership network, conducted by Stefania Vitali, James B. Glattfelder, and Stefano Battiston (2011) from the Eidgenössische Technische Hochschule (ETH) in Zurich, has shown. Based on a list of 43,060 transnational corporations (TNCs), taken from a sample of about 30 million economic actors in 194 countries contained in the Orbis 2007 database, the authors discovered that "nearly 4/10 of the control over the economic value of TNCs in the world is held, via a complicated web of ownership relations, by a group of 147 TNCs in the core, which has almost full control over itself." These 147 conglomerates occupy the center of a tentacular power structure, as these authors further elaborate:

> We find that transnational corporations form a giant bow-tie structure and that a large portion of control flows to a small tightly-knit core of financial institutions. This core can be seen as an economic "super-entity" that raises new important issues both for researchers and policy makers. (…) About 3/4 of the ownership of firms in the core remains in the hands of firms of the core itself. In other words, this is a tightly-knit group of corporations that cumulatively hold the majority share of each other.

---

[10] See Global Justice Now, "Corporations vs governments revenues: 2015 data." http://www.globaljustice.org.uk/sites/default/files/files/resources/corporations_vs_governments_final.pdf.

[11] "10 biggest corporations make more money than most countries in the world combined." *Global Justice Now*, 12/IX/2016

13.5.1   An Emerging Subspecies of Homo sapiens: The UHNWI

The concentration of so much economic power in the hands of a numerically insignificant caste is unprecedented in human history. By combining data from the Crédit Suisse Global Wealth Report pyramid (already presented in the Introduction, see Fig. 1.1) with the Wealth-X and UBS World Ultra Wealth Report 2014 and the two Oxfam International reports (2014 and 2015), we can examine this trend in more detail. As seen in the Introduction, in 2017, at the top of the global wealth pyramid, 0.7% of adults, or 36 million individuals, owned 45.9% of the world's wealth (US$ 128.7 trillion). Penetrating into the vertex of this global asset pyramid, we see that in this group of 36 million (with assets worth over US$ 1 million), there are 211,275 billionaires—the *ultra-high net-worth individuals* (UHNWI)—corresponding to 0.004% of adult humanity. Their assets alone total US$ 29.7 trillion, assets that, moreover, increased by 7% in 2014 compared to the previous year (64% of UHNWI are in North America and Europe, while 22% are in Asia).

Now, with the aid of a magnifying lens provided by two listings (from *Forbes Magazine* and the *Bloomberg Billionaires Index*), we will move up to examine the most exclusive stratum of this UHNWI club. In 2013, *Forbes Magazine* listed 1426 billionaires amassing US$ 5.4 trillion, which is equivalent to the GDP of Japan (the world's third largest GDP). *Forbes Magazine* 2018 lists 2208 billionaires with US$ 9.1 trillion, an 18% increase in wealth compared to the previous annual survey. The *Bloomberg Billionaires Index* contains an even more stratospheric list: the world's 300 richest individuals had a wealth of US$ 3.7 trillion in December 31, 2013. These 300 people became even richer throughout 2013, adding another US$ 524 billion to their net worth. From the top of this nanopyramid, situated at the extreme end of the Crédit Suisse pyramid, Table 13.1 allows us to contemplate the general picture of human inequality in the current phase of capitalism.

Table 13.1 The nanopyramid at the extreme end of the global wealth pyramid

Number of UHNWI	Assets owned by each UHNWI
98.700	More than 50 million dollars
33,900	More than 100 million dollars
3100	More than 500 million dollars
2208	More than 1 billion dollars
300	More than 12 billion dollars
85	More than 20 billion dollars
20	More than 60 billion dollars

Sources: *The Crédit Suisse Global Wealth Report 2013*; *Wealth-X and UBS World Ultra Wealth Report 2014*; Fuentes-Nieva, Galasso (2014, p. 3), in Oxfam International, *Working for the few* (2014); Oxfam International, *Wealth: Having It All and Wanting More* (2015); and *Forbes Magazine* 2018

Table   13.2 Number   of
individuals whose combined
wealth equals that of the
poorest half of humanity,
2010–2017 (Oxfam)

2010	388
2014	85
2015	62
2017	8

In 2014, Oxfam showed that the 85 richest individuals on the planet had a combined wealth of more than US\$ 1.7 trillion, which is equivalent to the wealth held by 3.5 billion people, the poorest half of humanity. The concentration of these assets continues to grow at a staggering rate, as shown in the Table 13.2 which shows the declining number of individuals whose combined wealth equals that of the bottom 3.6 billion people (the poorest half of humanity which is becoming increasingly poorer).

According to the Oxfam report, *An Economy for the 1%* (2016),

> The wealth of the richest 62 people has risen by 45% in the five years since 2010 – that's an increase of more than half a trillion dollars (\$542bn), to \$1.76 trillion. Meanwhile, the wealth of the bottom half fell by just over a trillion dollars in the same period – a drop of 38%.

Oxfam's 2015 report stated that in 2014, the richest 1% owned 48% of global wealth, leaving just 52% to be shared between the other 99% of adults on the planet:

> Almost all of that 52% is owned by those included in the richest 20%, leaving just 5.5% for the remaining 80% of people in the world. If this trend continues of an increasing wealth share to the richest, the top 1% will have more wealth than the remaining 99% of people in just two years.

This prognosis was confirmed in the following year. The 2016 report (*An Economy for the 1%*) states that "the global inequality crisis is reaching new extremes. The richest 1% now have more wealth than the rest of the world combined." Bill Gates's fortune, estimated at US\$ 78.5 billion (*Bloomberg*), is greater than the GDP of 66% of the world's countries. In present-day Russia, 110 people own 35% of the country's wealth.[12] Another way to understand this extreme concentration of wealth is to look at the large financial holding companies. Seven of the largest US financial holding companies (JPMorgan Chase, Bank of America, Citigroup, Wells Fargo, Goldman Sachs, MetLife, and Morgan Stanley) have more than US\$ 10 trillion in consolidated assets, representing 70.1% of all financial assets in the country (Avraham et al. 2012).[13] Also according to Oxfam, among the countries for which data is available, Brazil is the one with the greatest concentration of wealth in the hands of the richest 1%. Six billionaires in the country hold assets equivalent to that of the poorest half of the Brazilian people, who saw their share of national wealth further reduced from 2.7% to 2% (Oxfam 2017).

---

[12] "Band of brothers". *The Economist*, 22–28/XI/2014, p. 77. Review of Karen Dawisha, *Putin's Kleptocracy: Who owns Russia?* New York, 2014

[13] See also Mark Thoma, "How To (Maybe) End Too Big to Fail." *The Economist's View*, 25/II/2013

This emerging subspecies—the 0.004% of the human species known by the acronym UHNWI—owns the planet. Their economic and political power is greater than that of those who have a popular mandate in national governments. Their domination is also ideological, since economic policies are formulated—and evaluated by most opinion leaders—to benefit the business strategies of this caste. Its power surpasses—in scale, reach, and depth and in a way that is transversal and tentacular—everything that the most powerful rulers in the history of pre-capitalist societies could ever have conceived of or had reason to desire. Additionally, the power of these corporations is infinitely disproportionate to their social function of job creation. In 2009, the 100 largest among them employed 13.5 million people, that is, only 0.4% of the world's economically active population, estimated by the International Labour Organization at 3.210 billion potential workers. Who in such circumstances can still uphold the historical validity of the so-called Kuznets curve? As is well known, Simon Kuznets (1962) claimed, in the 1950s, that as an economy develops, market forces first increase and then decrease economic inequality. In reality, all the actions of this plutocratic super-entity are guided toward a single motto: defend and increase their wealth. Their interests are, therefore, incompatible with the conservation of the biophysical parameters that still support life on our planet.

13.6 "Degrowth Is Not Symmetrical to Growth"

Since the 1960s, evidence of the incompatibility between capitalism and the biophysical parameters that enable life on Earth has been recognized by experts from various disciplines. Some Marxist scholars (or those in close dialogue with Marx) belonging to two generations, from Murray Bookchin (1921–2006) and André Gorz (1923–2007) to a range of post-1960s scholars, such as John Bellamy Foster, Fred Magdoff, Brett Clark, Richard York, David Harvey, Michael Löwy, and Enrico Leff, clearly understand that the current historical situation is essentially characterized by an opposition between capitalism and conservation of the biosphere. In the preface to a book that is emblematic of this position (*The Ecological Rift. Capitalism's War on the Earth*, 2010), John Bellamy Foster, Brett Clark, and Richard York write[14]: "a deep chasm has opened up in the metabolic relation between human beings and nature—a metabolism that is the basis of life itself. The source of this unparalleled crisis is the capitalist society in which we live."

But capitalism is perceived as an environmentally unsustainable socioeconomic system also by those for whom Marx is not a central intellectual reference. Pascal Lamy and Yvo de Boer were already mentioned above. For them there is an unmistakable causal link between capitalism and the exacerbation of environmental crises. In fact, there are a number of non-Marxist thinkers who subscribe to this

---

[14] See also Foster, "Capitalism and Degrowth: An Impossibility Theorem". *Monthly Review*, 62, 8, 2011.

evidence. Two generations of pioneer thinkers, born between the beginning of the century and the interwar period, have laid the foundation for the understanding that capitalist accumulation is disrupting the climate system, depleting the planet's mineral, water, and biological resources, and provoking multiple disruptions in ecosystems and a collapse of biodiversity. We will point out the names of great economists, such as Kenneth E. Boulding, Nicholas Georgescu-Roegen, and Herman Daly; geographers like René Dumont (1973); key philosophers on ecology, such as Michel Serres and Arne Naess; theistic philosophers, such as Hans Jonas and Jacques Ellul; biologists, such as Rachel Carson and Paul and Anne Ehrlich (Tobias 2013); or a Christian-educated polymath and ecologist, Ivan Illich. Inspired by the writings of these thinkers who shaped the critical ecological thinking of the second half of the twentieth century, studies on socio-environmental crises are growing today. These studies share an awareness that the imperative of economic growth is increasingly threatening the maintenance of an organized society. Included here are authors of diverse ideological backgrounds and range of expertise, from Claude Lévi-Strauss, Edgar Morin (2007), Cornelius Castoriadis, and Richard Heinberg to Vittorio Hösle, Serge Latouche, and Hervé Kempf, who has a book suggestively titled, *Pour Sauver la Planète, Sortez du Capitalisme* [To Save the Planet, Leave Capitalism] (2009). In *A User's Guide to the Crisis of Civilization: And How to Save It* (2010), Nafeez Mosaddek Ahmed hit the bull's eye: "Global crises cannot be solved solely by such minor or even major policy reforms – but only by drastic reconfiguration of the system *itself.*" More recently, Hicham-Stéphane Afeissa (2007, 2012) has been working extensively on an overview of deep ecological thinking in philosophy, especially in the twentieth century.

13.6.1   The Idea of Managed Degrowth

The idea of managed degrowth, which implicitly or explicitly unites the names mentioned above, appears to be the most significant proposal today, perhaps the only one, that would be effective in creating a viable society. Any decreased human impact on the Earth system obviously requires abandoning the meat-based food system and the fossil fuel-based energy system. Moreover, the idea of degrowth rests on three assumptions, which, if not properly understood, would make degrowth seem absurd.

The first assumption is that economic downturn, far from being an option, is an inexorable trend. Precisely because we are depleting the planet's biodiversity and destabilizing the environmental coordinates that have prevailed in the Holocene, global economic growth rates are already declining when compared to the 1945–1973 average, as Gail Tverberg has shown (see Introduction, Sect. 1.7. The Phoenix that Turned into a Chicken). According to the World Bank, in the 2013–2017 period, the average growth of the global economy was 2.5%; this is a 0.9% decrease from the average a decade ago.[15] This tendency toward declining growth is inescapable. The current pandemic (SARS-CoV-2) will only accelerate this process of decline, as the economy will find it increasingly difficult to recover from its next crises. Conscious that the developmental illusion is pushing the services rendered by the biosphere toward collapse, supporters of degrowth realize that a managed degrowth would be the only way to prevent economic and socio-environmental collapse, one that will be more brutal and deadly the longer it is denied or underestimated.

The second assumption is that administered degrowth is essentially anti-capitalist. The idea of degrowth within the framework of capitalism was rightly defined by John Bellamy Foster (2011) as an impossibility theorem. Finally, the third assumption is that managed degrowth is not simply about a quantitative reduction in GDP. First of all, it means qualitatively redefining the objectives of the economic system which should be about adapting human societies to the limits of the biosphere and of natural resources. Obviously, this adaptation implies investments in places and in countries that lack basic infrastructure and, in general, an economic growth that is crucial for the transition to energy and transport systems that have a lower environmental impact. But these are localized investments that are oriented toward reducing environmental impacts (sanitary infrastructure, abandoning the use of firewood, public transportation, etc.); it is never about growth for the sake of growth.

Serge Latouche devoted almost all of his work to the question of degrowth (2004, 2006, 2007, 2012 and 2014). He explains (2014) the link between degrowth and overcoming capitalism: "The degrowth movement is revolutionary and anti-capitalist (and even anti-utilitarian) and its program is fundamentally a political one." Degrowth, as the same author insists, is the project of building an alternative to the current growth society: "this alternative has nothing to do with recession and crisis … There is nothing worse than a growth society without growth. [...] Degrowth is not symmetrical to growth." The most poignant formulation on the incompatibility between capitalism and sustainability comes from the theories of two economists, developed before the emergence of the concepts of sustainability and decay: (1) the theory by Nicholas Georgescu-Roegen in 1971 on the increasing generation of entropy from economic activity and, a fortiori, from an economy based on the paradigm of expansion and (2) the theory, developed in 1966 by Kenneth E. Boulding in *The Economics of the Coming Spaceship Earth*, on the need to overcome an open economy (cowboy economy) toward a closed economy (spaceman economy).[16] We

---

[15] World Bank, *Global Economic Prospects. Broad-Based Upturn, but for How Long?* January, 2018, p. xv

[16] Here is the meaning of the *cowboy economy* and *spaceman economy* metaphors: "For the sake of picturesqueness, I am tempted to call the open economy the 'cowboy economy,' the cowboy being symbolic of the illimitable plains and also associated with reckless, exploitative, romantic, and violent behavior, which is characteristic of open societies. The closed economy of the future might similarly be called the 'spaceman' economy, in which the earth has become a single spaceship, without unlimited reservoirs of anything, either for extraction or for pollution, and in which, there-

will return to this last theory in the next chapter. For now, it suffices to cite a central passage from this text in which Boulding shows that for the capitalist economy, production and consumption are seen as a commodity whereas in the economy toward which we should aspire—the closed economy or spaceman economy—what matters is minimizing throughput, namely, the rate of transformation (operated in the economic system) of raw materials into products and pollution. This means minimizing both production and consumption, which clearly conflicts with the capitalist view of the economic process:

> The difference between the two types of economy becomes most apparent in the attitude towards consumption. In the cowboy economy, consumption is regarded as a good thing and production likewise; and the success of the economy is measured by the amount of the throughput from the "factors of production," a part of which, at any rate, is extracted from the reservoirs of raw materials and noneconomic objects, and another part of which is output into the reservoirs of pollution. If there are infinite reservoirs from which material can be obtained and into which effluvia can be deposited, then the throughput is at least a plausible measure of the success of the economy. (…) By contrast, in the spaceman economy, throughput is by no means a desideratum, and is indeed to be regarded as something to be minimized rather than maximized. The essential measure of the success of the economy is not production and consumption at all, but the nature, extent, quality, and complexity of the total capital stock, including in this the state of the human bodies and minds included in the system. In the spaceman economy, what we are primarily concerned with is stock maintenance, and any technological change which results in the maintenance of a given total stock with a lessened throughput (that is, less production and consumption) is clearly a gain. This idea that both production and consumption are bad things rather than good things is very strange to economists, who have been obsessed with the income-flow concepts to the exclusion, almost, of capital-stock concepts.

The fundamental unsustainability of capitalism is demonstrated not only through the arguments of Georgescu-Roegen and Boulding but also by Herman Daly's impossibility theorem (1990). He claims that the impossibility—obvious, but not necessarily accepted in its consequences—of an economy based on the expanded reproduction of capital in a limited environment occupies a position in economic theory that is equivalent to the fundamental impossibilities of physics:

> Impossibility statements are the very foundation of science. It is impossible to: travel faster than the speed of light; create or destroy matter-energy; build a perpetual motion machine, etc. By respecting impossibility theorems we avoid wasting resources on projects that are bound to fail. Therefore economists should be very interested in impossibility theorems, especially the one to be demonstrated here, namely that it is impossible for the world economy to grow its way out of poverty and environmental degradation. In other words, sustainable growth is impossible.

---

fore, man must find his place in a cyclical ecological system."

13.7   Conclusion

What delays a broader acceptance of this set of reflections are not arguments in favor of capitalism, but the mantra of the absence of alternatives to it. Such is the hypnotic power of this mantra that even the scholars most aware of the links between environmental crisis and economic activity cling to the absurd idea of a "sustainable capitalism," an expression that Herman Daly aptly called "a bad oxymoron—self-contradictory as prose, and unevocative as poetry." It turns out that it is possible, and more than ever necessary, to overcome the idea that the failure of socialism has left society with no other alternative but to surrender to capitalism. Human thinking is not binary, and the unfeasibility of the socialist experience in the twentieth century does not ipso facto imply the viability of capitalism.

The 2012 OECD report, *Looking to 2060: Long-term global growth prospects*, published by Asa Johansson and colleagues (2012), states that: "the long-term scenario provides a relatively benign long-term outlook for the global economy." But in order for this scenario to prove itself benign, the report mentions, at the end of the Introduction, the factors that are being ignored in its forecast: "Indeed, a number of other factors are also ignored, including the possibility of disorderly debt defaults, trade disruptions and possible bottlenecks to growth due to an unsustainable use of natural resources and services from the environment." OECD economists still live in an enchanted kingdom in which economic forecasts can afford to ignore "possible [environmental] bottlenecks." These bottlenecks must indeed be ignored because recognizing them, not as "possible" but as inevitable, would compel these economists to review the assumption—now frankly absurd—on which their knowledge is based, namely, that the environment is a relatively abundant and stable factor of production, a fact and not a problem, therefore, for economic forecasts.

In short, capitalism is not an environmentally sustainable socioeconomic system if the establishment of regulatory frameworks capable of bringing it back to sustainability is left to the market. This is because the market is, at best, able to optimize the short-term cost–benefit relationship in the allocation of resources, but not the conservation of these resources. As Kim Stanley Robinson's formula rightly sums up: Adam Smith's "invisible hand never picks up the check" (as cited by Naomi Oreskes, in Conway, Oreskes 2014, p. 93).

Capitalism could perhaps approach sustainability if its regulation was driven by a mixed mechanism in which the state and civil society were strong enough to offset the power of the global corporate network with its four strongest binding points: Big Oil, Big Mining, Big Food, and Big Bank. This is no longer the case because the state–corporations of our days have no interest in confronting the corporate network, and if they did, they would no longer have the means to do so. Therefore, the immense task—of replacing the power of this global network with another model of society capable of combining local economy with an effective global and democratic political governance—lies on the shoulders of civil society, on its social, labor, and political organizations. Whether we will be able to accomplish this task is still an open question. It presupposes, first of all, awakening from the illusion that capitalism can become sustainable and renouncing our fascination with consumerism and with the age-old psychological constant: more surplus = more safety, to be discussed in the next chapter.

References

AFEISSA, Hicham-Stéphane (ed.), *Éthique de l'environment. Nature, valeur, respect.* Paris, Vrin, 2007.

———, *Portraits de philosophes en écologistes.* Paris, Éditions Dehors, 2012.

AHMED, Nafeez Mosaddek. *A User's Guide to the Crisis of Civilization: And How to Save it.* Pluto Press, 2010.

AVRAHAM, D., SELVAGGI, P. & VICKERY, J., "A Structural View of U.S. Bank Holding Companies". *FRBNY Economic Policy Review*, July 2012.

BAST, Elizabeth et al., *Empty promises. G20 subsidies to oil, gas and coal production.* Oil Change International. November 2015.

BELLAMY FOSTER, John; CLARK, Brett & YORK, Richard, *The ecological rift. Capitalism's war on the Earth.* New York, Monthly Review Press, 2010

BOULDING, Kenneth E., "The economics of the coming Spaceship Earth" (1966). In Jarrett, H. (ed.), *Environmental Quality in a Growing Economy*, pp. 3–14. Baltimore, Resources for the Future. Johns Hopkins University Press, 1966.

BROWN, Maurice, *Adam Smith's Economics: Its Place in the Development of Economic Thought.* London, Routledge, 1988.

CANGUILHEM, Georges, *Écrits sur la médecine.* Paris, PUF, 1989.

———, "Théorie et technique de l'expérimentation chez Claude Bernard". *Études d'Histoire et de Philosophie des Sciences* (1968). Paris: Vrin, 1983.

CARRINGTON, Damian, "Fossil fuels subsidised by $10m a minute". *The Guardian,* 18/V/2015.

COADY, David et al., "Global Fossil Fuel Subsidies Remain Large: An Update Based on Country-Level Estimates". IMF Working Paper, May 2019.

CONWAY, Eric M. & ORESKES, Naomi, *The Collapse of Western Civilization. A View from the Future.* Columbia University Press, 2014.

CORBET, Jessica, "Green Groups Call Out Big Banks for Pouring Billions Into Fossil Fuel Industry". *Common Dreams,* 20/III/2019.

DALY, Herman E. "Sustainable Growth. An Impossibility Theorem" (1990). *In*: DALY, Herman E. & TOWNSEND, Kenneth (eds.). *Valuing the Earth: Economics, Ecology, Ethics.* Cambridge (Mass.), MIT Press, 1993, pp. 267–285.

DELL'ORSO, Silvia. *Altro che musei. La questione dei beni culturali in Italia.* Rome/Bari, Laterza, 2002.

DE FILIPPIS, Vittorio, "François Morin: L'oligopole bancaire s'est transformé en hydre dévastatrice pour l'économie mondiale". *Libération,* 22/VII/2015.

DUMONT, René, *L'Utopie ou la mort!* Paris, Seuil, 1973.

GEORGESCU-ROEGEN, Nicholas. *The Entropy Law and the Economic Process.* Havard University Press, 1971.

———. "Energy and Economic Myths". *Southern Economic Journal*, 41, January 1975, pp. 347–381.

———. *La Décroissance* (1979). Paris, Sang de la Terre, 1995.

GHYS, Frederic & HINKKANEN, Hanna, "Searching for Socially Responsible Investments. Mission Impossible?". *The Guardian,* 2/VII/2013.

GIANNETTI, Eduardo, "A crise ambiental e a economia de mercado". In *Novo Contrato Social. Propostas para esta geração e para as futuras.* Instituto Ethos, 2013, pp. 69–75.

GREENFIELD, Patrick, "World's top three asset managers oversee $300bn fossil fuel investments". *The Guardian*, 12/X/2019.

HEINBERG, Richard, *Peak Everything: Waking Up to the Century of Declines*. Gabriola Island, New Society Publishers, 2007.

HENNIGAN, W. J., "Donald Trump Is Playing a Dangerous Game of Nuclear Poker". *Time*, 1/II/2018.

HIRSCH, S., "Making globalization moral?" *Transnational Corporations*, 20, 3, 2011, pp. 87–93.

ILLICH, Ivan. *La convivialité. Oeuvres completes*, vol. I. Paris, 2003, pp. 451–580.

IVRY, B., KEOUN, B. & KUNTZ, P., "Secret Fed Loans Gave Banks $ 13 Billion Undisclosed to Congress". *Bloomberg*, 27/XI/2011.

JÖSTROM, M. & ÖSTBLOM, G. "Decoupling waste generation from economic growth – A CGE analysis of the Swedish case". *Ecological Economics*, 69, 7, 15/V/2010, pp. 1545–1552.

JUNG, Alexander et al., "Warming world: is capitalism destroying our planet". *Spiegel Online International*, 25/II/2015.

KLEIN, Naomi, "Capitalism killed our climate momentum, not 'human nature'". *The Intercept*, 3/VIII/2018.

KOENIG, David, "Exxon Shareholders to Vote on Climate Change, Fracking". ABC*News*, 27/V/2015.

LARDNER, Richard, "Army says no to more tanks, but Congress insists". *Associated Press*, 29/IV/2013

LATOUCHE, Serge. *Survivre au développement. De la décolonisation de l'imaginaire économique à la construction d'une société alternative*. Paris, Mille et Une Nuits, 2004.

———. Le pari *de la décroissance*. Paris, Fayard, 2006.

———. Petit *traité de la décroissance sereine*. Paris, Fayard, 2007.

———. Bon pour *la casse*. Paris, Les liens qui libèrent, 2012.

———. *Itinérance. Du tiers-mondisme à la décroissance* (2014). *L'economia è una menzogna*. Turin, 2014.

MADELINE, Béatrice, « La ruée vers les métaux ». *Le Monde*, 12/IX/2016.

MAGDOFF, Fred. "Ecological Civilization". *Monthly Review*, 62, 8, 2011.

MAGDOFF, Fred & BELLAMY FOSTER, John. *What Every Environmentalist Needs to Know About Capitalism*. New York, Monthly Review Press, 2011.

MECKLIN, John, "Losing Earth" lost sight of some climate change villains, but not the scope of the problem, *The Bulletin of Atomic Scientists*, 6/VIII/2018.

MORIN, Edgar. *Vers l'abîme?*. Paris, Éditions de l'Herne, 2007

MORIN, François. *L'hydre mondiale. L'oligopole bancaire*. Montreal, Lux éditeur, 2015.

OXFAM, *Working for the few*, London, 2014

———, *Wealth: Having it all and wanting more*, London, 2015.

———, *A distância que nos une. Um retrato da desigualdade brasileira*. Oxfam Brasil, 2017.

———. "Panama Papers: Act now. Don't wait for another crisis". *The Guardian*, 10/IV/2016.

RICH, Nathaniel, 'Losing Earth: The decade we almost stopped climate change", *The New York Times*, 1/VIII/2018.

RESENDE, André Lara. *Os limites do possível. A economia além da conjuntura*. São Paulo, 2013.

RIDLINGTON, Elizabeth, NORMAN, Kim & RICHARDSON, Rachel, "Fracking by the Numbers. The Damage to Our Water, Land and Climate from a Decade of Dirty Drilling". Environment America Research & Policy Center, 2016.

SETTIS, Salvatore. *Italia S.p.A. L'assalto al patrimonio culturale*. Turin, Einaudi, 2002.

STERN, Nicholas & CALDERÓN, Felipe, *Better Growth Better Climate. The New Climate Economy Report. The Synthesis Report*, 2014.

STUBBINGTON, Tommy, "Global debt surges to highest level in peacetime". *Financial Times*, 25/IX/2019.

TOBIAS, Michael C., "The Ehrlich Factor: A Brief History of the Fate of Humanity, With Dr. Paul R. Ehrlich". *Bloomberg*, 2013.

VITALI, S., GLATTFELDER, J.B. & BATTISTON, S., "The Network of Global Corporate Control". Eidgenössische Technische Hochschule Zürich (ETH), *PLoS ONE,* 26/X/2011.

XU, Yangyang & RAMANATHAN, Veerabhadran, "Well below 2°C: Mitigation strategies for avoiding dangerous to catastrophic climate changes". *PNAS,* 14/IX/2017

ZUCMAN, Gabriel, "Taxing across Borders: Tracking Personal Wealth and Corporate Profits". *Journal of Economic Perspectives,* 28, 4, 2014, pp. 121–148.

Chapter 14
More Surplus = Less Security

Even at the risk of reiterating what is already well-known, it is useful to go over the reasons for capitalism's historical success. Capitalism has triumphed everywhere in the Contemporary Age because it has been able to offer European society and then the overwhelming majority of humanity (*volente nolente*) the most effective response—or at least what appeared to be so until the mid-twentieth century—to perennial problems of scarcity and hostility of nature, including threats from the human species itself. This response consisted in the generalization of a mode of production capable of (1) making profit maximization (through the continuous accumulation of surplus at the lowest possible cost) the raison d'être of economic activity (2) and redistributing part of the surplus produced in the form of income and wages.

Despite the extreme concentration of wealth present in the most recent forms of global capitalism, the complementarity between increase in surplus and distribution of this surplus was fundamental to the mechanism of capital and income accumulation in the twentieth century. Broadly speaking, until the 1980s, everything in capitalism—even (or especially) the cyclical crises, social crises, and wars—eventually led to an increase in surplus and to its social distribution in some form or another. Capitalism is a mode of production in which class struggle (i.e., all pressure for redistribution of income for the benefit of the non-owners of capital) implies, sooner or later, an increase in consumption, which successively feeds (through demand) the mechanism of accumulation. Even though the participation of non-owners of capital in the appropriation of surplus may be proportionally smaller than the increase in that surplus, the share of this surplus distributed to non-owners of capital has (at least until recently) been increasing in absolute terms.

In this way, capitalism is better than the preceding modes of production at fulfilling the following axiom: the greater (1) the accumulation of material, energetic, and informational surplus and (2) the redistribution of that surplus, the greater the security of the groups that benefit from it—even if very unequally—in the face of the perennial problems of scarcity and of the natural and human hostility mentioned above. Capitalism is not, therefore, an accidental system, initiated by chance and

© Springer Nature Switzerland AG 2020

L. Marques, *Capitalism and Environmental Collapse*,
https://doi.org/10.1007/978-3-030-47527-7_14

imposed from the outside by a tiny group of entrepreneurs on European societies and then on humanity. It is the most recent historical form of social organization, arrived at by man's search to diminish his existential precariousness.

14.1   Conviction of the Conquerors, Seduction of the Conquered

Of course, industrial capitalism would not have consolidated its worldwide expansion without an indisputable military superiority. But its victory would not have been lasting if the conquerors had not had the conviction, alongside their arms and profit, that their technology was capable of ensuring a superior civilization (through greater surplus production). The ideologists of eighteenth and nineteenth-century capitalism believed (and continue to believe in the twenty-first century) in the superiority and goodness of their socioeconomic system with the same intensity as the peoples of the Hijra have believed in the superiority, goodness, and universality of their religion since the seventh century, the Crusaders of the twelfth and thirteenth centuries, and the Jesuits of the sixteenth and seventeenth centuries. The mission of yesterday's ideologues was to bring salvation to the souls of the "pagans" through revealed religion. In the eyes of today's ideologues, this civilizing mission consists of the no less noble mission of exporting the religion of surplus accumulation. The first and most fundamental critics of the capitalist system—Marx and Engels—were also the ones who most enthusiastically made this connection between capitalist expansion and the civilizing process, not hesitating to state in the *Communist Manifesto*:

> The bourgeoisie has subjected the country to the rule of the towns. It has created enormous cities, has greatly increased the urban population as compared with the rural, and has thus rescued a considerable part of the population from the idiocy of rural life. Just as it has made the country dependent on the towns, so it has made barbarian and semi-barbarian countries dependent on the civilised ones, nations of peasants on nations of bourgeois, the East on the West.

The counterpart of this conviction on the part of the conquerors of yesterday and today is that the conquered, in addition to feeling impotence and terror, are seduced to a certain degree by the capitalist promise of a continuous increase in surplus and the consequent greater absolute appropriation of that surplus. And this seems to explain why even those who only possess a very residual quantity of that surplus (as seen in the Introduction, 56.6% of adult humanity owns only 1.8% of global wealth) saw and still persist in seeing capitalism as a good or as a minor evil which is, in any case, inevitable.

Not only the lower classes of Western society, but numerous non-Western societies in the nineteenth-century were militarily controlled, economically dispossessed, and then seduced by capitalism. In addition to gunpowder and steel, a key weapon in this conquest was the ability of capital to divest of meaning the symbolic forms of social life and practices of storing wealth in "archaic" societies, replacing them with the mechanism of accumulation as an end in itself (i.e., as a self-legitimizing goal). Again, it is in another passage of the *Communist Manifesto* that we find the most unmistakable praise of capitalism's ability to dissolve highly stratified and sophisticated civilizational structures, such as those of feudal Europe, China, Japan, India, and Southeast Asia. Formerly, the symbolic and ritual value of these societies' power structures made them relatively impervious to the imperative of capital accumulation. But, Marx and Engels state that:

> The bourgeoisie, by the rapid improvement of all instruments of production, by the immensely facilitated means of communication, draws all, even the most barbarian, nations into civilization. The cheap prices of commodities are the heavy artillery with which it batters down all Chinese walls, with which it forces the barbarians' intensely obstinate hatred of foreigners to capitulate.

14.2   From Ceiling Effect to the Principle of Infinite Accumulation

In fact, in pre-capitalist societies, not strongly monetized and not fully market-driven, the accumulation and concentration of wealth in the hands of an elite were, even when enormous, more symbolic than quantitative. The greatest limiting factor of this accumulation and concentration of wealth was the fact that it became dysfunctional after a certain point. The ruling castes of pre-capitalist societies could never accumulate more than a certain limit of land, palaces, herds, grains, treasures, sumptuous objects, copper or iron mines, gold coins, peasants, soldiers, and slaves. Past a certain limit, the capacity for accumulation hit a type of ceiling, beyond which it would no longer deliver effective material or symbolic power to its holders. On the contrary, beyond certain limits—geopolitical, military, financial, administrative, ethnic, religious, etc.—accumulation could become too costly, counterproductive, and ultimately disruptive and self-destructive (see item 14.5 Predominance of Centripetal Forces in Mediterranean Antiquity). Like the ideal size of a given organism in the history of evolution, the accumulation and concentration of wealth (as well as the perimeter of empires and the security of their borders) follow these criteria and limits of functionality. Disrespecting them was often a factor leading to imbalance and decline.

"Accumulate, accumulate! That is Moses and the prophets! (…) Accumulation for accumulation's sake, production for production's sake: by this formula classical economy expressed the historical mission of the bourgeoisie." With these incisive sentences, Marx (1867/1887, I, VII, chap. 24, p. 412) defined the new autonomy of the continuous expansion of surplus. Marx is probably the first to understand that the accumulation of industrial capital is not, in principle (at least in the capitalist's mind), subject to the physical limits of natural resources, a notion proposed by Malthus and Ricardo around the beginning of the nineteenth century. As Thomas Piketty (2013, p. 27) asserts:

> Capital can, therefore, potentially accumulate without limit. Indeed, his [Marx's] main conclusion is what we may call the 'principle of infinite accumulation,' that is, the inevitable tendency of capital to accumulate and to concentrate in infinite proportions, without a natural limit.

In *Capital*, Marx clearly describes how the centralization of capital completes the work of accumulation by enabling industrial capitalists to extend the scale of their operations (1867/1887, I, VII, chapter 25, p. 436). But, evidently, he could not yet glimpse another potent mechanism for the acceleration of accumulation: the subsequent final dematerialization of real values and of currency, as well as the extreme abstraction of financial market contracts in contemporary capitalism. With this recent phenomenon, the last remnants of this pre-capitalist ceiling effect of wealth accumulation and concentration have been definitively eliminated.

Free from the gravity of matter, that is, from real products, goods, and services and, finally, even from concrete "money," accumulation has reached stratospheric abstractions. Today, the magnitude of what is at stake in the world economy is no longer measured solely by its assets, that is, by its real goods and services, nor by present monetary values (bank deposits and the money supply), but also, and above all, by derivative contracts, which, by definition, are of no value in themselves, *but derive their value from the assumption that the economy should continue to grow*, a condition for the possibility of appreciation (above inflation) in an asset's market price: a property, the stock of a publicly traded company, a commodity on the Chicago stock exchange, the current and future interest rate, a currency's variation in exchange rate, etc. Thus, while global GDP in 2018 was around US$ 86 trillion, the Bank for International Settlements in Basel calculates that the notional value of outstanding OTC (over-the-counter) contracts derivatives reached US$ 640 trillion at end June 2019. In the United States, five megabanks monopolize round 95% of these contracts.

14.3   The Primitive Character of the Monetary Accumulation Drive

Separated from the sphere of human experience and without a measure that is common to human needs (however we define these), the progression of these notional amounts acquires its own ontological dynamic and density which, in the consciousness of economic agents, are almost unlinked to the real economic activity that, in principle, serves as their parameter. The monetary or post-monetary measure in which the value of such "wealth" is virtually expressed—a stock, a future right to buy or sell it for a set amount or an insurance against possible future devaluation of that stock—is the ultimate and superior form of labor alienation and commodity fetishism. Although still moving in the society of concrete commodities and belonging to the "material world," Marx's ideas on the alienation of labor and on the commodity fetish in the capitalist mode of production remains, if I am not mistaken, unsurpassed in regard to understanding the mechanism by which labor and its social value are estranged from commodities and, ultimately, from its general equivalent, money.

But Freud's contribution is no less important. Wealth expressed in coin, while taking on the form of an unlimited spiral and sublimating itself in post-currency, hides its transfigured anal drive. In "Character and anal erotism" (1908, p. 174), for example, Freud sheds light on the fact that the relations between the seemingly disparate complexes of interest in money and interest in defecation turn out to be very close:

> Whatever archaic modes of thought have predominated or persist–in the ancient civilizations, in myths, fairy tales and superstitions, in unconscious thinking, in dreams and in neuroses–money is brought into the most intimate relationship with dirt. We know that the gold which the devil gives his paramours turns into excrement after his departure, and the devil is certainly nothing else than the personification of the repressed unconscious instinctual life. We also know about the superstitions which connects the finding of treasure with defaecation and everyone is familiar with the figure of the 'shitter of ducats [*Dukatenscheisser*]. Indeed, even according to ancient Babylonian doctrine, gold is 'the faeces of Hell [*Mammon = ilu manman*]. (…) The original interest in defaecation is, as we know, destined to be extinguished in later years. In those years the interest in money makes its appearance as a new interest which has been absent in childhood. This makes it easier for the earlier impulsion, which is in process of losing its aim, to be carried over to the newly emerging aim.

The almost infinite accumulation of money–capital of which Piketty speaks, following in Marx's footprints, has nothing to do with the biological dimension of accumulation that was functional until recently for the human species (as it is for so many other species), since it effectively produced more security. Rather, it is propelled in the monetized and post-monetized economy by the very primitive psychic drive of child anal eroticism redirected to a new object: interest in money. No matter how much this inevitable tendency of money–capital to accumulate and concentrate in infinite proportions is sublimated today in its abstract contractual forms (thus expressing the most advanced historical form of capitalism), we understand with Freud that its underlying irrational drive remains its most powerful force.

14.4   Living Space of the Species and the Age of the Finite World

It is not necessary to insist on the obvious debt that material progress owes the accumulative drive. But it is not obvious that, from a certain level of surplus accumulation—a level we surpassed sometime in the second half of the twentieth century—the cumulative drive begins to threaten the existential security of societies by "producing" a type of nature that is more hostile, merciless, and mean to mankind than the one that once threatened us. In *The End of Nature* (1989), Bill McKibben summarizes the price of this polar opposite: "we have built a new Earth; it is not as nice as the old one."

In the Introduction (item 1.7 The Phoenix that Turned into a Chicken), I drew attention to a new law of capitalism that will be discussed here: scarcity and/or pollution of natural resources, climate change, and other environmental imbalances will increasingly be the decisive variables in determining the profit rate of capital. This law is really just the logical conclusion of the concept of *Lebensraum*, of "living space," coined by Friedrich Ratzel in 1897. It is a concept born within biogeography and anthropogeography, shortly before degenerating into imperialist rhetoric. Living space is the geographical area in which a particular species develops its habitat. As stated by Kenneth E. Boulding in *The Economics of Coming Spaceship Earth* (1966), with the passage from the "illimitable plane" of yore to the "closed sphere" of the contemporary world, the potential for expansion of the human species' living space has been exhausted:

> Over a very large part of the time that man has been on earth, there has been something like a frontier. That is, there was always some place else to go when things got too difficult, either by reason of the deterioration of the natural environment or a deterioration of the social structure in places where people happened to live. The image of the frontier is probably one of the oldest images of mankind, and it is not surprising that we find it hard to get rid of. (…) It was not until the Second World War and the development of the air age that the global nature of the planet really entered the popular imagination. Even now we are very far from having made the moral, political, and psychological adjustments which are implied in this transition from the illimitable plane to the closed sphere.

The image of an unlimited planet, open to economic exploitation (Boulding's "cowboy economy"), was celebrated throughout the slow development of habitable areas which began when the *Homo sapiens* left Africa maybe over 250,000 years ago. This psychological image of a world conceived of as an unlimited series of frontiers to be discovered still prevailed in the history of capitalism from its earliest beginnings to the mid-twentieth century. When faced with the problem of depletion of its immediate habitat (a scarcity of raw materials, natural resources, and consumer markets), capitalism has been able to maintain or increase its average rate of profit by using its centrifugal energies, namely, the military and commercial expansion of its borders. Schumpeter (1942, p. 107) teaches us that "we must not confuse geographical frontiers with economic ones." This is true, but it is also true that, in the history of capitalism, the crossing of geographical frontiers was one of the precursors for the crossing of economic frontiers.

The last two centuries saw the last three major waves of structural globalization, understood as leaps in the relationship between the value of global trade and that of global production: 1830–1885, 1905–1914, and after 1945 (Chase-Dunn et al. 2000). This third wave, which is now exhausting itself, seems to be the last. As early as 1931, sensing the advent of this new historical situation for mankind, Paul Valéry coined one of his incisive formulas: *Le temps du monde fini commence* ("The Age of the Finite World is Beginning"). In 1966, Kenneth Boulding returns to this point in the text cited above with the image of a circumscribed planet, the *Spaceship Earth*. In 2008, it is no longer the poet, nor the economist, but the geophysicist, André Lebeau, who states at the beginning of his *L'enfermement planétaire* (2008):

"The encounter of humanity with the limits of the planet is an unprecedented phenomenon in the history of the species."

The last major repositories of biological diversity—in the Arctic, the Amazon, Africa, and tropical Asia—are the extreme, ultimate Rubicons that the centrifugal logic of accumulation is crossing. And since capitalism cannot conceive of any logic other than a centrifugal one, it is left with science fiction as an escape valve. Thus, we observe a clear transformation in the history of this genre, as is reflected or elaborated in American cinema. Space exploration is no longer driven by a benign scientific and adventurous spirit reminiscent of the nineteenth-century tradition, as in *Star Trek* (1966). Rather, from the 1970s, it assumes an invariably malignant economic rationality. Space conquest is now directed toward the appropriation of resources—mineral, energy, and other—by militarized corporations, such as in *Alien* (1979), *Blade Runner* (1982), *Total Recall* (1990), *Avatar* (2009), and the TV series, *Terra Nova* (2011) or *The Expanse* (2015).

Returning to reality and to the only planet we can live on, exhaustion and disorganization make up the binary language of contemporary capitalism. At bay in its planetary living space and unable to colonize outer space, the society of growth begins to deplete its *living time*; that is, it begins to draw on the resources belonging to future generations. This ecological debt is quantified by the *Global Footprint Network*, according to which, with each year that goes by, the *Earth Overshoot Day* (the day that human consumption exceeds the capacity for renewal of planetary natural resources for that year), occurs earlier in the calendar. In 2013, humanity had consumed all the renewable natural resources available for that year by August 19; in 2019, that day was passed on July 29. Between 1993 and 2019 we went backwards about a month a decade. The *Living Planet Report* 2018, a WWF publication released every 2 years since 1989, warns that "over the past 50 years our Ecological Footprint—one measure of our consumption of natural resources—has increased by about 190%." This accelerating structural deficit is developed by Paul Gilding in his book, *The Great Disruption* (2011), as follows: we are "the first generation that, rather than sacrificing ourselves for our children's future, are sacrificing our children's future for ourselves." Indeed, a new report from the medical journal, *The Lancet* (Watts et al. 2019), warns that "the life of every child born today will be profoundly affected by climate change." According to a Native American saying, "We have not inherited the earth from our fathers, we are borrowing it from our children." This saying, much remembered today, was butchered with the natives themselves by the devastating expansion of capitalism, especially in the last 70 years. As early as 1992, Severn Cullis-Suzuki was well aware (and she was then only 12 years old) that her parents' generation would not give back the Earth they had borrowed from her when she delivered her famous speech at the closing session of the Earth Summit plenary (Eco-92): "Coming up here today, I have no hidden agenda. I am fighting for my future. Losing my future is not like losing an election, or a few points on the stock market." Ten years later, Robert Hunter (2002), co-founder of Greenpeace in 1971 and its first president, still seemed to hear the echo of those words:

Those of us living now in the industrialized countries should be thankful that the barrier of time appears to be firmly locked, because if they could ever break through, the people of the future (...) will surely come back to strangle us, the ancestors from hell, in our sleep, for having squandered the Earth's legacy in a handful of generations.

Also for Isabelle Stenger (2009/15), the present generations of adults will be held responsible for the ongoing environmental collapse:

I belong to a generation that will perhaps be the most hated in human memory, the generation that "knew" but did nothing or did too little (changing our lightbulbs, sorting our rubbish, riding bicycles...). But it is also a generation that will avoid the worst – we will already be dead.

14.5   The Predominance of Centripetal Forces in Mediterranean Antiquity

The sixteenth century is a game changer in an extremely abridged historical view of the logic that goes from the equation more surplus = more security to the equation more surplus = less security (i.e., from Antiquity to the Contemporary Age). As is well-known, the scientific and technological revolutions of the seventeenth and eighteenth centuries created the historical conditions that paved the way toward industrial capitalism. But we must return to the sixteenth century to see the birth of this fundamental mutation. Of course, those who speak of European expansion speak of the great Iberian voyages to such an extent that the two have become almost synonymous expressions. It is not a question of rediscussing the geographic, economic, geopolitical, and religious factors that are usually evoked to explain the genesis of the historical process of the great navigations. It is a question of understanding how the impulse of European man toward the mastery of new spaces and new landscapes is part of a larger civilizational discontinuity, separating a world dominated by centripetal forces, the Mediterranean world created by Antiquity, from a world dominated by the centrifugal forces ultimately generating contemporary capitalism—the North-Atlantic world (now about to become North-Pacific). André Chastel (1971) refers to this civilizational discontinuity between the Mediterranean world and the North Atlantic world in clear terms:

Industrial civilization, fundamentally northern and Atlantic, would crush to its last articulations the Mediterranean systems that had asserted themselves in the previous Age; the eloquence and formalism of the southern world were no longer necessary.

This discontinuity gains even more evidence when examined in light of the tension between centripetal and centrifugal forces that guide ancient Mediterranean and modern European societies.

Undoubtedly, like the civilizations of the Modern Age and other ancient civilizations, the civilizations of Mediterranean Antiquity contained centrifugal forces with expansionist and destructive dynamics. In his 1866 essay, *Du sentiment de la nature dans les sociétés modernes* (much earlier, thus, than the preoccupation of Toynbee and later historians with the collapse of ancient civilizations), the great French geographer and anarchist, Élisée Reclus, warned of this historical constant, present since Antiquity:

> In the history of mankind, among the causes of the disappearance of so many successive civilizations, is the brutal violence inflicted by most nations on the nurturing land. They cut down the forests, let the springs run out and rivers overflow, deteriorated the climates, surrounded the cities with marshy and pestilential zones.

Of course, the ancient world could not determine its environmental crises in the same way we have done since the second postwar period, firstly because they did not even have the term "environment" in their vocabulary. As Lukas Thommen (2009/2012) wisely reminds us: "Neither ancient Greek nor Latin had words for many of the concepts familiar to us today in connection with environmental issues; the word environment itself heads the list." This did not prevent concern from arising around the destruction of forests in the Mediterranean, particularly in Greece and Phoenicia, to stick to the two most well-documented examples. The cry of Critias (111) in Plato's homonymous dialogue about deforestation and erosion of Attica is much cited (translated by Benjamin Jowett, 2018):

> The consequence is, that in comparison of what then was, there are remaining only the bones of the wasted body, as they may be called, as in the case of small islands, all the richer and softer parts of the soil having fallen away, and the mere skeleton of the land being left. But in the primitive state of the country, its mountains were high hills covered with soil, and the plains (…) were full of rich earth, and there was abundance of wood in the mountains.

Alongside him, Pliny the Elder describes—not without alarm—the technique used for gold extraction by the *ruina montium* (*H.N., 33, 21*), the razing of mountains in the Iberian Peninsula, as *opera vicerit Gigantum*, that is, as a feat that surpasses the labors of the Giants who had heaped a number of mountains in their war with the gods. However, Pliny does not forget to point out that such practices had been banned in Italy, even though Italy was as rich or richer in gold reserves than the regions of Asturias, Galicia, and Lusitania.

That said, it is necessary to underline a fundamental difference between the ancient world and the one born from the sixteenth century. The ancient world seeks to understand itself from myths of origin which are concentric and centripetal. An example in this regard is the cult of omphalos in various Mediterranean sanctuaries, such as the Cretan temple linked to the birth of Zeus; the one of Phlius in the Peloponnese, mentioned by Pausanias; the one at Delphi, which has a Roman copy in the Delphi Archaeological Museum; and the *Catholicon* in the church of the Holy Sepulchre in Jerusalem, among others. The modern world, by contrast, is guided by myths of the future, which are expansive and centrifugal. In the modern world, man's self-image shifts from the center to the frontier, and the myth of overcoming limits (in every sense of the term) is the founding idea of his pride and identity. Reflecting, in his autobiography (1576), on the natural events that weave his own extraordinary destiny, Girolamo Cardano (1576/1932) applauds the fact that he is born in the century that surpassed Antiquity (chapter 41): "Among the extraordinary, though quite natural circumstances of my life, the first and most unusual is that

I was born in this century in which the whole world became known; whereas the ancients were familiar with but little more than a third part of it." This passage is emblematic of the *topos* that equates expanding the frontiers of the known world with overcoming Antiquity, an idea tirelessly repeated in the sixteenth century from the likes of Egidio da Viterbo to Giorgio Vasari, Paolo Giovio, and others.

The sixteenth century expansionist ideal was already in the antipodes of the ancient world. Although ancient Mediterranean civilizations often devastated the regions surrounding their sea, the geographical, geopolitical, and technological difficulties in expanding beyond them led the very idea of expansion to be seen as alien—if not negative—to the dominant values of these civilizations. Indeed, well-known contingencies set the brakes and limits on the tendency of these peoples to move away from themselves and to put their forces at the service of expansion. The centrifugal forces that led to the expansion of the Macedonian kingdom into Central Asia were only sporadic and short-lived. Since the middle of the third century BC, the new Parthian Empire had restored Iranian traditions, rebuilding the barrier that would seal the Roman Empire from any aspiration of having a lasting influence beyond the Roman province of Syria. The civilization which, since the time of Johann Gustav Droysen, has been referred to as Hellenistic eventually became concentrated on the eastern shores of the Mediterranean between Athens, Alexandria, Antioch, Pergamon, Ephesus, and other Aegean port cities in Asia Minor (Miletus, Izmir (Smyrna), Halicarnassus, etc.). It is true that the Romans were able to extend their power over the northern regions of western Europe, but the cornerstone of their expansion was control of the natural resources of western Europe, from the Atlantic to the Danube; their contact with the Germanic tribes, mentioned almost exclusively by Tacitus, had little influence on the Hellenistic conception of the Roman world. The "alien wisdom," to put it in the words of Arnaldo Momigliano (1975/1980), remains relatively marginal to this world, as well as being mediated by the Greek filter. In fact, throughout the Julio-Claudian, Flavian, and Antonine dynasties, with the exception of the campaigns of Claudio and Trajan, there is a neutralization of centrifugal forces by centripetal ones. Physical and political–military barriers induced the Roman Empire to stabilize the Imperial *limes*, to concentrate its ambitions and its identity on the *Mare Nostrum* and its surroundings, and to invest, especially with Hadrian (117–138), on its centripetal forces, that is, in a reconversion to the Rome-Athens axis (Calandra 1996). Hadrian's disinterest in the expansion of the Imperial *limes* only prolongs what Floro (1981) calls in the 130s a secular *inertia Caesarum*:

> A Caesare Augusto in saeculum nostrum haud multo minus anni ducenti, quibus inertia Caesarum quasi consenuit atque decoxit.

> From Augustus to our time not much less than two hundred years passed, during which time the inertia of the Caesars caused, as it were, [the empire] to grow old and nearly reduced it to nothing.

This predominance of the centripetal over the centrifugal was not shaped, however, only by "the inertia of the Caesars" and "external" barriers, such as geography or the military resistance of the Germans, Sarmatians, Parthians, etc. Mediterranean Antiquity created virtue out of necessity by adopting the Pythagorean and Socratic teachings as the paradigm of every aspiration to wisdom: the ontological primacy of unity over plurality, of center over periphery, of the measured over the unmeasured, and of self-knowledge over knowledge of the other, thus making centripetal forces prevail over their most primitive and spasmodic centrifugal drives.

This appreciation occurs throughout the whole mental arc of Mediterranean life, beginning with its cosmology. Unlike the modern conception of an infinite cosmos or the current Standard Cosmological Model which postulates an expanding universe, the ancient universe is finite, and the very concept of infinity is only potential, that is, negative, at least until Plotinus. "In the Greek tradition," as Pierre Aubenque teaches (1972/1999), "a being appeared to be more perfect the more determined it was; the infinite, to which something can always be added, was therefore a sign of incompleteness and lack." The Greek, Platonic, Aristotelian, Stoic, or Christian universe is fixed, spatially concentric, and palingenetic. Here, the principle of multiplicity, the dyad, is subordinated to and absorbed by the principle of unity. The architecture of Greek cosmology, from Plato to Plotinus, has as its foundation the claim of the metaphysical preexistence of unity over plurality. In Book V, 1 of the *Enneads*, Plotinus recaps the story of this superiority of the One over the dyad, of Being over becoming, from Anaxagoras and Parmenides to Plato. And in book VI, 9, he concludes, "It is by virtue of unity that beings are beings." The unmovable energy of the center desires nothing and seeks nothing outside of itself, but it draws to itself all centrifugal energy, an attraction that constitutes and reconstitutes the world forever: "it rests by changing" (*metaballon anapauetai*, translation by Malcolm Crowe), writes Heraclitus, quoted by Plotinus (IV, 8, 1) at the other end of the historical arch of ancient philosophy. For Plotinus, the principle of evil is the dissemination and forgetting of origin and center (V, I):

> The evil that has overtaken them [the souls, *psukhâi*] has its source in self-will, in the entry into the sphere of process, and in the primal differentiation with the desire for self-ownership. They conceived a pleasure in this freedom and largely indulged their own motion; thus they were hurried down the wrong path and in the end, drifting further and further, they came to lose even the thought of their origin in the Divine. A child wrenched young from home and brought up during many years at a distance will fail in knowledge of its father and of itself: the souls, in the same way, no longer discern either the divinity or their own nature; ignorance of their rank brings self-depreciation; they misplace their respect, honoring everything more than themselves; all their awe and admiration is for the alien, and, clinging to this, they have broken apart, as far as a soul may, and they make light of what they have deserted; their regard for the mundane and their disregard of themselves bring about their utter ignoring of the divine. Admiring pursuit of the external is a confession of inferiority (translated by Stephen MacKenna and B. S. Page).

This centrifugal movement of diversion is one in which the hypostases, among them the souls, originating from the One (*En*), depart in emanation (*proodos*). It is in opposition to the return to the One in reversion (*epistrophe*), a movement of return to the center, as Plotinus reiterates (VI, 9, 7–8) (translated by Giuseppe Faggin 2000):

> A child distraught will not recognise its father; to find ourselves is to know our source.

(...) Every soul that knows its history is aware, also, that its movement, unthwarted, is
not that of an outgoing line; its natural course may be likened to that in which a circle turns
not upon some external but on its own centre, the point to which it owes its rise. The soul's
movement will be about its source.

The movement of reversion toward the center and the One was well analyzed by
Pierre Hadot (2002) and Pierre Aubenque (1972/1989). They point out that such a
movement should not be understood as a relationship of mechanical inversion
between going and returning or descending and climbing. Aubenque writes:

For, as in the Platonic cave allegory, here there is not a pre-existing topography: it is the
walk (*cheminement*) that precedes and constitutes the path. Reversion must be represented
as the act by which the flow recedes, 'remembers' its source, and, in this act, is fixed. (...)
Reversion [*epistrophe*] thus structures emanation [*proodos*] and makes it a constitutive
element.

Expansive emanation, recollection of one's origin, and return to the origin do not
occur, therefore, in time, in succession, but are logical moments of the formation of
the world. In the psychological sense, reversion is already present in the fundamental enlightenment experience of the philosopher who withdraws from the sensible
world to gain access to the intelligible world (Plato *Republic*, 518c), or in the case
of the philosopher who gains intuition or ecstasy in the One, as with Plotinus
(according to what his biographer, Porphyry, writes about him). For both Plato and
Plotinus, this experience refers to the essence of the process of knowledge, which is
not exploration *ad extra*, but recollection.

Similarly, in the *forma urbis* of the Roman civilization, the center is inaugural
and, therefore, is itself a representation of the universe. The center of the Roman
city is surrounded by the *pomoerium*, a *termini* (boundary) line arranged short of the
fortifications themselves. Inside this space, the *urbs* affirms the holiness and sacred-
ness of its soil, as well as its privilege over *ager* (the space of cemeteries and *imperium militiae*). The *sulcus primigenius*, traced by the plow, which delimits the space
of the city at the moment of its foundation, protects it from the harmful influences
of the exterior. And at its center, marked by the intersection of its orthogonal axes
(*cardo* and *decumanus*), chunks of earth from the original homeland brought by the
founders are placed in a cavity, together with the household gods (*Di penates*), and
the sacrificial and propitiating objects during the augural foundation rites of the
Roman colonies. This cavity opened at the moment of its foundation was the *mundus*, a chamber covered by a vault and that was a miniature reproduction of the three
constitutive parts of the universe. André Magdelain (1990, p. 187) summarizes
this well:

The *mundus* has three parts: above, there is a vault, under it a *sacrum Cereris* (...), that is,
probably a chamber dedicated to Ceres, and finally an *inferior pars veluti consecrata dis
Manibus* (Fest. 144 L), one should note, an infernal well (...). The three elements are
undoubtedly a miniature representation of the universe, bringing together the sky, the earth,
represented by Ceres, and the mouth of hell.

The prevalence of centripetal conceptions of the world is favored by the imbrica-
tion (until saturation) of Greek culture in Etruscan–Latin culture, food

self-sufficiency, the mild climate, and the "closed" form of the Mediterranean Sea, whose extent is, as Paul Valéry (1933/1960) recalls, appropriate for the "scale of man's primitive means." The thought, the collective imaginary, and the fundamental artistic forms of Greco-Roman culture conform to this notion. Tragedy was, as is well-known, a warning against the urge to venture outside the boundary of the human condition, against the crime of excessiveness (*hubris*), punished by the reparative justice of Nemesis. In the speech in which Artabanus warns his nephew (Xerxes) not to leave his empire to conquer the Greeks, Herodotus (VII, 10) stresses how the gods punish those who exceed bounds:

> You see how the god smites with his thunderbolt creatures of greatness more than common, nor suffers them to display their pride, but such as are little move him not to anger; and you see how it is ever on the tallest buildings and trees that his bolts fall; for it is heaven's way to bring low all things of surpassing bigness. Thus a numerous host is destroyed by one that is lesser, the god of his jealousy sending panic fear or thunderbolt among them, whereby they do unworthily perish; for the god suffers pride in none but himself (translated by A. D. Godley).

Also, Polybius (VI, 3–4) attributes to *pleonexia* (the will to have or to be more than is necessary) the motor principle of his theory of anacyclosis, a theory on the cycle of successive erosions and the overthrowing of constitutions (*politeiai*). To avoid such erosion, a mixed constitutional regime was needed, one that was capable of containing— thanks to a system of counterweights—each member of the body politic within the limits of his/her powers.

Epic poems expressed this system of offsetting centrifugal energies by centripetal ones. In the *Iliad*, it is not a centrifugal force that leads the Greeks to the Trojan expedition, but a centripetal countervailing force: the need to restore order, transgressed by Paris. Ulysses' journey, the most paradigmatic ancient epic, does not discuss the boldness of venturing outside the confines of the world itself, but narrates, on the contrary, his return to Ithaca, to his own kingdom and home. From the fabulous travels that Ulysses recounts in the Court of Alcinous and from similar narratives, such as those of Ctesias in India and of Jambulus in the "Great Sea," Lucian of Samosata will make, in *A True Story*, a parody illustrating the Greek lack of interest in effective exploitation of the non-Mediterranean world. The journey that the epic poem narrates is, in short, less exploratory than initiatory and confirmatory. The hero's journey— Ulysses, Hercules, Theseus, Jason, Aeneas, or Psyche— has no purpose but a return to origin, to the world in which the hero is finally recognized and recognizes himself. In this way, the experience of the epic hero can be summed up in the Delphic–Socratic[1] maxim, inscribed, according to Pausanias (10.24.1), in the *pronaos* of the Temple of Apollo in Delphi: *gnōthi seautón* or *nosce te ipsum*, that is, know yourself. As Lara Nicolini (2005) notes, all the trials of Lucius and Psyche in the *Metamorphoses* of Apuleius stem from their *nimia curiositas*, an "excessive curiosity," defined as an inability to perceive the contradiction

---

[1] This maxim is indeed equally Socratic. Plato quotes it no less than six times in his Dialogues: Charmides (164D), Protagoras (343B), Phaedrus (229E), Philebus (48C), Alcibiades I (124A, 129A, 132C), and Laws (II.923A).

between the centrifugal knowledge of many things (*multiscius*) and the centripetal worship of wisdom (*prudentia*). To confuse wisdom with the exploration of domains other than one's own is the hallmark of the loss of the Golden Age. In *Metamorphoses* (I, 89–90), Ovid characterizes the Golden Age as one in which "No pine tree felled in the mountains had yet reached the flowing waves to travel to other lands: human beings only knew their own shores" (*Montibus in liquidas pinus descenderat undas/ Nullaque mortales praeter sua litora norant*). Even the journey of Aeneas is not exploratory, but it is a rite of revival of Troy in Latium, assured by the *translatio* of tutelary gods of Aeneas in Latin lands. The subject of Virgil's poem is not the trip itself but this transfer of Aeneas's gods, and, therefore, they will appear without fail in Roman coins and statuettes representing his escape from Troy. Like the coloniza-tion of the Mediterranean by the Greeks starting in the eighth century BC, Aeneas's *translatio* has little or nothing to do with the idea of expanding limits and cannot be understood as the result of an expansionist force, such as in the exploration and conquest of other continents by European societies, especially since the second half of the fifteenth century. The myth of Rome as a restoration of Troy, narrated by Virgil, is a poetic-political operation designed to remedy the inferiority complex of the Romans in relation to the greater antiquity of the Greeks (Momigliano 1988): the Romans did not consider themselves superior to the Greeks for being more modern than them (like New World Americans in relation to Europeans), but for being older, as attested to by their alleged Trojan origins.

Despite the attempts of Herodotus and Tacitus to understand cultures outside their civilizational coordinates, for the ancient Mediterranean man there is no pos-sible civilization far from the shores of his sea. As he moves away from his shores, the world becomes uninhabited or inhabited by monstrous peoples and animals; this is when it does not regress to more primitive forms. In a passage from his book (XXXIV, 3–4), Polybius conveys Pitheas of Massalia's lost account (*On the Ocean*) of his sea voyage to Thule on the northern border of the European continent (occur-ring at around 325 BC):

> […] those regions in which there was no longer any proper land nor sea nor air, but a sort of mixture of all three of the consistency of a jelly-fish in which one can neither walk nor sail, holding everything together, so to speak. He says he himself saw this jellyfish-like substance but the rest he derives from hearsay.

14.5.1   The Spatial Limit as a Sign of Wisdom: The Pillars of Hercules

This Western spatial limit of the old world was established by the pillars that Hercules had built above the promontories of the strait of Gibraltar after robbing the Cattle of Geryon (the giant who lived on the Island of Erytheia) of the mythic Hesperides in the far west of the Mediterranean. This was the tenth of Hercules' twelve labors and marked the westward extent of his travels. According to Strabo (*Geographia* 3.5.5.), these were "the pillars which Pindar calls the 'gates of Gades' when he asserts that they are the farthermost limits reached by Heracles." The columns of Hercules were in Antiquity a *Nec plus ultra*, a landmark to be understood through two meanings. First of all, they were a *Non Terrae Plus Ultra*, that is, a cosmographic warning, a space mark not to be passed, since Hercules himself, the hero par excellence, had not passed it. But they also expressed a philosophical and moral dimension, for they marked the Delphic–Socratic precept, "nothing in excess" (*mêdén ágan*). This precept is underlined by Pindar in the final verses of the third Olympian Ode (in praise of Theron's victory at Olympia):

> If water is best and gold is the most honored of all possessions, so now Theron reaches the farthest point by his own native excellence; he touches the pillars of Heracles. Beyond that the wise cannot set foot; nor can the unskilled set foot beyond that. I will not pursue it; I would be a fool (translated by Diane Arnson Svarlien).

The double cosmographic and moral–philosophical dimension of this *Nec plus ultra* established by Hercules conveyed a constellation of central notions of Greek and Latin thought. These central notions can be condensed into the concepts of *sophrosyne*, *phronesis*, or *prudentia*, urging the ancient man to preserve (or restore when transgressed) the ideas of origin and measure.

The essence of these notions was man's acceptance of his lot in existence, an essence expressed in both divine law (*themis*) and human law (*nomos*), as well as in philosophical, religious, and legal figures in the Greco-Roman world. Let us briefly mention some of them, starting with the myth of the two jars on the doorsil of Zeus, which Homer recalls in the pathetic speech of Achilles to Priam, king of Troy:

> You must have iron courage: sit now upon this seat, and for all our grief we will hide our sorrows in our hearts, for weeping will not avail us. The immortals know no care, yet the lot they spin for man is full of sorrow; on the floor of Jove's palace there stand two urns, the one filled with evil gifts, and the other with good ones. He for whom Jove the lord of thunder mixes the gifts he sends, will meet now with good and now with evil fortune; but he to whom Jove sends none but evil gifts will be pointed at by the finger of scorn, the hand of famine will pursue him to the ends of the world, and he will go up and down the face of the earth, respected neither by gods nor men (translated by Samuel Butler).

The same idea is expressed in *psychostasis*, in the ominous balance of Hermes *Psychopompos*, which we see painted in Greek vases from at least the sixth century BC. It also returns in Themis, the abstract deity, whose union with Zeus—according to Hesiod (*Theog.* 901–906)—will give birth to the Horae and the Parcae, sisters who regulate man's (good and bad) fate, as well as in *Tyké* and *Nemesis*, the pre-Olympic deities, present in both the emblems of cities and in funerary iconography, who administer divine justice and the destiny of men and cities and who guaranteed the distributive justice of the gods. In *Works and Days* (v. 40), Hesiod exclaims against those who intend to transgress the ideal of moderation, the veneration of limit, and the acceptance of distributive justice: "Fools! They know not how much more the half is than the whole."

In Rome, the notion of limit and prudence was embodied in the *Terminalia* in honor of *Terminus*, the primordial and irremovable patron deity of the boundary between properties (instituted by Numa Pompilius), whose image was worshiped in the temple of the Roman central deity, Jupiter Optimus Maximus. These notions are reflected in the most incisive among the Latin legal maxims: *Suum cuique tribuere*, to render to every man his due. They are also found, above all, in the most terrible of myths, that of Erysichthon, king of Thessaly, transmitted by various poets, from Callimachus (HymnVI) to Ovid (*Met*, VIII, 738–878): driven by his recklessness and greed to destroy the sacred forest, his crime is punished by an insatiable hunger that leads him to devour himself. Finally (but the examples could multiply), they ring in Pliny's condemnation (*H.N.*, 33.1) of the desecration of the earth by the lust that gold and other minerals from the earth stir in men:

> *Quam innocens, quam beata, immo uero etiam delicata esset uita, si nihil aliunde quam supra terras concupisceret, breuiterque, nisi quod secum est!*

> How innocent, how happy, how truly delightful even would life be, if we were to desire nothing but what is to be found upon the face of the earth; in a word, nothing but what is provided ready to our hands!

14.5.1.1   Dante

Dante did not know the abovementioned final verses of the third Olympian Ode of Pindar on the insanity of venturing beyond the columns of Hercules, but he reiterates them in his poem (Inferno 26) when he conveys the narrative of Ulysses who was condemned to the circle of deceitful advisers. Transformed into flame, the king of Ithaca laments how he misguided his companions, urging them to sail beyond the pillars of Hercules:

> *Io e i compagni eravam vecchi e tardi*
> *Quando venimmo a quella fosse stretta,*
> *Ov'Ercole segnò li suoi riguardi,*
> *Acciò che l'uom più oltre non si metta*

> Were I and my companions, when we came
> To the strait pass, where Hercules ordain'd
> The bound'ries not to be o'erstepp'ed by man.
> (translated by Henry Francis Cary)

Dante concludes the episode by narrating how the fatal whirlwind surrounding the ship of Ulysses and his companions emerges just as the heroes rejoice at the sight of Purgatory Island, which they mistakenly think is an island of the new world and of their salvation. The euphoria for this careless attempt to expand the ancient cosmos is, for Dante, excessive; it is a desecration of the place where divine justice is exercised, concealed from men. Almost two millennia after Pindar, the columns of Hercules still condensed the lesson of the philosophical, moral, and religious superiority of centripetal forces over centrifugal ones.

14.5.1.2   From Dante's Cosmography to Petrarch's Psychology

It is not surprising that Ulysses' self-reproach in the *Divine Comedy* translates mutatis mutandis into Petrarch's crisis of consciousness—modern man's first crisis of consciousness—in the face of the temptation and vanity of another spatial transgression: climbing Mount Ventoux, supposedly in 1336.[2] In an epistle to Dionigio of Borgo San Sepolcro, Petrarch thus begins this moral fable, already masterfully commented on by Jacob Burckhardt in *The Civilization of Renaissance in Italy* (1860):

> *Altissimum regionis huius montem, quem non immerito Ventosum vocant, hodierno die, sola vivendi insigni loci altitudinem cupiditate ductus, ascendi*

> Today I ascended the highest mountain in this region, which, not without cause, they call the Windy Peak [Ventoux]. Nothing but the desire to see its conspicuous height was the reason for this undertaking.

As he reaches the summit of his exhausting climb, Petrarch, opening by chance the volume of the *Confessions* that Dionigio had given him and that he had been carrying like a *vade mecum* (a pocket companion), is warned by Augustine:

> Where I fixed my eyes first, it was written: "And men go to admire the high mountains, the vast floods of the sea, the huge streams of the rivers, the circumference of the ocean and the revolutions of the stars - and desert themselves" [X.8.15]. I was stunned, I confess. I bade my brother, who wanted to hear more, not to molest me, and closed the book, angry with myself that I still admired earthly things. Long since I ought to have learned, even from pagan philosophers, that "nothing is admirable besides the mind; compared to its greatness nothing is great" [Seneca, Epistle 8.5].

Not only Seneca, but so many other "pagan philosophers" converge on this central aspect of ancient wisdom, well formulated once again by Seneca, when he advises Lucilius (*Ep.* 2.6):

> *Non discurris nec locorum mutationibus inquietaris. Aegri animi ista iactatio est: primum argumentum compositae mentis existimo posse consistere et secum morari.*

> You do not run hither and thither and distract yourself by changing your abode; for such restlessness is the sign of a disordered spirit. The primary indication, to my thinking, of a well-ordered mind is a man's ability to remain in one place and linger in his own company (translated by Richard M. Gummere).

---

[2] "Ad Dyonisium de Burgo Sancti Sepulcri ordinis sancti augustini et sacre pagine professorem, de curis propriis". *Familiarum rerum libri*, IV, 1, in *Opere*, Florence, Sansoni, 1992, pp. 385–392. For an alternative dating of the letter, in 1353, see Dotti (1987, pp. 38–39).

14.6   The Emblem of Charles V and the Affirmation
of Centrifugal Forces

Throughout the fifteenth century, the columns of Hercules lost their significance as a geographical landmark. But this loss sharpened their symbolic dimension, on the one hand, and reversed their former meaning, on the other. This happened in October 1516 when Luigi Marliano forged the motto, *Plus Oultre*, for the future Emperor Charles V. This *impresa* [emblem], in its French form, still remains on the ceiling of the Alhambra palace in Granada, but in 1518 the insignia would come to be fixed in a German form, *Noch Weiterer,* in Hans Weidtiz's portrait of Charles V, or in its most common Latin form, *Plus Ultra*. At times, the *Plus Ultra* insignia is shown only by the icon of the two columns, as in the coin minted for the coronation of Charles V in Bologna in 1530, or in a famous 1556 etching of the victories of Charles V where the two columns of Hercules serve as a canopy to the emperor's throne.

It is unlikely, as Earl Rosenthal (1971) argues, that in 1516, Charles V's emblem alluded to the colonies of New Spain. It translated the general and abstract idea of a new conception of *virtù* which consisted, henceforth, in the primacy of impulse over self-restraint. This idea henceforth defied, refused, and openly opposed the tradition of *sophrosyne, phronesis,* and *prudentia*. This centrifugal drive could easily evade the accusation of arrogance since its alibi was the holy ambition to universalize Christianity, as in an effort to compensate for the losses suffered in the eastern Mediterranean. But it could not triumph over the pagans without sacrificing for itself at least two of the cardinal Platonic virtues embraced by Christian theology: prudence and temperance.

Of course, origin myths will not disappear in the sixteenth century and beyond. Lisbon was founded by the king of Ithaca (as evoked in Fernando Pessoa's poem *Ulysses*); according to Edmund Spencer's, *The Faerie Queene* (1596), the Tudors had their remote origins in King Arthur and, beyond him, in Troy ("For noble Britons sprung from Troians bold"); several Italian lineages until the seventeenth century prided themselves on the fact that their *capostipite* was some character of Greco-Latin mythology. But all this would inexorably become only a matter of genealogy and poetry. In his entry "Histoire" (1756) for the *Encyclopédie*, Voltaire mocks these myths of origin and considers that "all peoples' origins are clearly fables." It is the myths of the future that, henceforth, begin to operate.

If the Modern Age, which opens with the Spanish-Habsburg domination of Italy from the third decade of the sixteenth century, can be rightly called "modern," it is because it affirms this centrifugal imperative. In the Iberian epic, nothing remains of the restorative meaning of an order that had been transgressed, typical of the ancient epic. The story of *The Lusiads* of Camoes does not evoke the redressing of a transgression like that of Paris, a homecoming like that of Ulysses, or a myth of origin such as in the foundation of Rome. What the poem sings right away in its celebrated opening verses is, in fact, the overcoming of Antiquity:

Cessem do sábio grego e do Troiano
As navegações grandes que fizeram;

Cale-se de Alexandre e de Trajano
A fama das vitórias que tiveram;
Que eu canto o peito ilustre Lusitano,
A quem Neptuno e Marte obedeceram.
Cesse tudo o que a Musa antiga canta,
Que outro valor mais alto se alevanta.

Tell no more tales about the subtle Greek,
The Trojan refugee on epic seas,
The Roman Trajan or grand Alexander
With all their Asiatic victories.
Look to the men who made Neptune and Mars
Obey: I sing the daring Portuguese.
Have done with all the Ancient Muses prize.
A higher code of honor's on the rise.
(translated by A.Z. Foreman)

Camonian heroes no longer heed the reproaches against maritime expansion launched from the beach by "a reverend figure" (*velho d'aspeito venerando*) who personifies the populace of Portugal (Book IV):

Ó glória de mandar, ó vã cobiça
Desta vaidade a quem chamamos Fama!
Ó fraudulento gosto, que se atiça
Cua aura popular, que honra se chama!

frantic thirst of honour and of fame,
The crowd's blind tribute, a fallacious name;
What stings, what plagues, what secret scourges curs'd,
Torment those bosoms where thy pride is nurs'd!
(translated by William Julius Mickle)

The verses from this reverend figure are the last echo of the now abandoned Delphic-Socratic warning. It is not by chance that the old man compares Vasco da Gama and his men to Phaethon and Icarus, guilty of the *hubris* of a *plus ultra,* punished by the gods. In the middle of the next century, Pascal (*Pensées*, 139) will renew this ancient *topos*, believing to have found the source of all human unhappiness in the urge to abandon one's home:

> *J'ai découvert que tout le malheur des hommes vient d'une seule chose, qui est de ne savoir pas demeurer en repos dans une chambre. Un homme qui a assez de bien pour vivre, s'il savait demeurer chez soi avec plaisir, n'en sortirait pas pour aller sur la mer ou au siège d'une place.*

> I have discovered that all the unhappiness of men arises from one single fact, that they cannot stay quietly in their own chamber. A man who has enough to live on, if he knew how to stay with pleasure at home, would not leave it to go to sea or to besiege a town.

It is tempting to see here a critique of the frontispiece of the *Instauratio magna*, published in 1620, in which Sir Francis Bacon returns to the image of the columns of Hercules. This time, however, he does so to identify the columns with the science of Aristotle that the *Novum organum scientiarum* is surpassing by launching itself into the open sea of the future. To make this allegory even more explicit, Bacon writes below it, always in the future tense: *Multi pertransibunt & augebitur scientia*

(Daniel 12:4), which means "Many will travel [beyond the Pillars of Hercules] and knowledge will be increased." In his *Meditations Sacrae* (1597), Francis Bacon would coin an aphorism that would know a huge fortune: *ipsa scientia potestas est*, i.e., knowledge itself is power. This is repeated almost *ipsis verbis*, for example, by Thomas Hobbes in *De homine* (1658). It is the essence of the myth of the future: the transformation of knowledge into an operation, into a technical appropriation of the cause-effect relationship, whereby increased scientific knowledge will result in a correlative increase in power over nature. In "Aphorisms concerning the interpretation of nature" (I, 3), Bacon expresses this better than anyone else: "Human knowledge and human power meet at a point; for where the cause isn't known the effect can't be produced."

As is well-known and was once again reiterated in the beginning of this section, the scientific and technological revolutions of the seventeenth and eighteenth centuries are the historical conditions for the possibility of capitalism. Indeed, economic activity cannot, technically and mentally, aim at increasing capital reproduction until knowledge itself is aimed at indefinitely increasing the power to operate nature. The birth of capitalism is also, and perhaps above all, the result of an epistemological operation. With capitalism, knowledge could no longer be *reflexive* and will henceforth be operational knowledge, "know-how"; this is why technology will eventually become the defining instance of human specificity. In a footnote in *Capital* (I, IV, Chapter 13), Marx distinguishes, with fine irony, the ancient man from the bourgeois man of his century:

> Strictly, Aristotle's definition is that man is by nature a town-citizen. This is quite as characteristic of ancient classical society as [Benjamin] Franklin's definition of man, as a tool-making animal, is characteristic of Yankeedom.

In this context, there is no longer any place for the values to which the columns of Hercules: Greco-Latin *sophrosyne, phronesis* and *prudentia*. At the dawn of industrial capitalism, the virtue of prudence, central to classical tradition, loses much of its value. In a letter to La Harpe dated March 31, 1775, Voltaire considered it *une sotte vertue* (a foolish virtue) and "Kant banished it from morality because its imperative was only hypothetical" (Aubenque 1963). Prudence evidently could not stand out in the value scale of this "tool-making animal."

14.7   Technolatry, Manifest Destiny, and Dystopia

About two and a half centuries after Benjamin Franklin, Ray Kurzweil (2005) proposed a more updated definition of the meaning of being human: "Being human means being part of a civilization that seeks to extend its boundaries." A similar definition of Western mentality was proposed, not without some irony, by Lévi-Strauss (1989/2013): "Western thinking is centrifugal" (*La pensée occidentale est centrifuge*). In fact, the modern ideal of the *Plus ultra* will flow into the typically North American myth of the continual and necessary conquest of "new frontiers."

This postponement into the future rather than existence in the present and this defi-nition by the boundary rather than the center and by deficiency rather than fullness are the touchstone of the modern and, even more, of the contemporary *Weltanschauung*. Observing the obsolescence of Plotinus, Giuseppe Faggin (2000) characterizes this spirit well in the conclusion of his meditation on the Alexandrian philosopher's wisdom:

> It is inevitable that we will be tempted at this point to put Plotinian thought in the coordi-nates of our time (…). We should note that contemporary man knows no craving for returns, neither historical returns, nor metaphysical ones. (…) His emblem is speed, that is, fleeing from himself, surpassing himself, always going beyond, that is, refusing to turn back and to stop. The Plotinian One is to him an empty abstraction. He knows, if at all, future projects of unity and dreams of them on the technocratic level.

It would be just as childish to criticize our time as to defend it. The technology and rationality that preside over it do not require defense lawyers; such is the evidence with which its benefits are imposed. Since the *machinisme* of the seventeenth cen-tury (Schuhl 1947), man's ability to think and conceive the world as a mechanism and then as a thermodynamic system allowed him to understand nature as a system of forces which he was able to operate to his own advantage. Heidegger understands this well in his essay on the question of technology (1953/1954):

> Modern physics is not experimental physics because it applies apparatus to the questioning of nature. The reverse is true. Because physics, indeed already as pure theory, sets nature up to exhibit itself as a coherence of forces calculable in advance, it orders its experiments precisely for the purpose of asking whether and how nature reports itself when set up in this way.

The risks, however, increase with the increase in technology. Several scientists and philosophers, such as Martin Rees (2003), Julie Wakefield (2012), Wayt Gibbs and Christine Soares (2012), and Toby Ord (2020) highlighted the most likely risks: a pandemic triggered by a biotechnology slippage, a cyberattack capable of caus-ing—intentionally or unintentionally—a computer blackout (or a nuclear war), the hypothetical future scenario of an ecophagy produced by nanotechnology (gray goo), or the risks involved in artificial intelligence (AI) research in what is already proclaimed as the dawn of augmented humanity, the era in which humans transcend biology (Kurweil 2005). Nick Bostrom (2014) sums up the potential dangers of self-improving AI:

> We cannot blithely assume that a superintelligence with the final goal of calculating the decimals of pi (or making paperclips, or counting grains of sand) would limit its activities in such a way as not to infringe on human interests. An agent with such a final goal would have a convergent instrumental reason, in many situations, to acquire an unlimited amount of physical resources and, if possible, to eliminate potential threats to itself and its goal system. Human beings might constitute potential threats; they certainly constitute physical resources.

The rapid evolution of contemporary technology evokes primal fears of a cybernetic Golem. Gérard Lebrun (2006) rightly warns that a criticism of technology should not result in technophobia or, at the limit, a technology-breaking movement not far from the nineteenth century fetishism of the Luddites against the machine itself.

Indeed, the neo-Luddite group in Jack Plagen and Wally Pfister's *Transcendence* (2014) provides a recent example of a satanization of the machine that goes back at least to the terrible image of Blake's *dark Satanic Mills*. But Lebrun seems more attentive to technophobia than to technolatry, which clings to technology as an amulet for warding off evil.

14.7.1   Manifest Destiny

Technolatry is a religion in which God is *Homo faber* and his high clergy is an elite of scientists devoted to their "mission" of saving men from their deviation and enabling them to fulfill their "manifest destiny." Technolatry has been well analyzed and illustrated in the books of Clive Hamilton (2010), Déborah Danowski, and Eduardo Viveiros de Castro (2014). Between scary and amusing pages, they guide us through this world of *condottieri* who are dazzled by the technologies of the future. Hamilton evokes, for example, Edward Teller and his *protégé*, Lowell Wood, of the Lawrence Livermore National Laboratory in San Francisco, who consider themselves visionaries. A father of the H Bomb in 1951, Teller inspired the "Star Wars" (Strategic Defense Initiative) of the Reagan-era and, above all, the character of Dr. Strangelove in Stanley Kubrick's homonymous 1964 film. Lowell Wood, a member of the Hoover Institute, was for years a Pentagon top weaponeer, even chairing the electromagnetic pulse (EMP) Commission aimed at assessing the effects of the high-altitude explosion of a nuclear bomb generating EMP on the network and electronic equipment of the targeted territory.

Such macabre developments in technology are legitimized in the "Manifest Destiny" theology, this mixture of dollar and Bible that emerged in the United States in the 1840s in the historical context of the war against Mexico, the destruction of forests, and the genocide of indigenous peoples. In 1872, John Gast depicted it in *American Progress* (Los Angeles, Museum of the American West), a horrible painting; it is a devotional image disseminated in engravings as a grotesque and perhaps unconscious echo of Eugène Delacroix's *La Liberté guidant le peuple* (1830). In 1893, Frederick Jackson Turner publishes *The Significance of the Frontier in American History*, an essay republished in 1921 as the first chapter of his classic, *The Frontier in American History*. His central thesis is that westward expansion had reached, in the 1880s, the closing of the American frontier, with profound social and economic consequences. But this momentum should not cease and would probably continue beyond the Pacific border, driven by the expansive character of American life:

> Since the days when the fleet of Columbus sailed into the waters of the New World, America has been another name for opportunity, and the people of the United States have taken their tone from the incessant expansion which has not only been open but has even been forced upon them. He would be a rash prophet who should assert that the expansive character of American life has now entirely ceased. Movement has been its dominant fact, and, unless

this training has no effect upon a people, the American energy will continually demand a wider field for its exercise.

In fact, in the twentieth century, the theology of "manifest destiny" was updated to justify US interventionism in the world, from Hawaii to the Philippines. In 1920, in his annual message to Congress, Woodrow Wilson declared: "It is surely the manifest destiny of the United States to lead in the attempt to make this spirit [of democracy] prevail." Following the invasion of Iraq, carried out in the same spirit, Lowell Wood returned to this expression in 2004, in the context of man's colonization of Mars: "It is the manifest destiny of the human race!", he told a meeting of the Mars Society (as cited by Hamilton 2010). Danowski and Viveiros de Castro analyze, with equal severity, technology and its theology of manifest destiny:

> But there are those who look with enthusiasm to the prospect of the loss of the world, seeing it as a mere elimination of provisional scaffolding, a support structure no longer needed by humans because they understand that the end of the world, as the end of a nonhuman or antihuman 'Nature,' will take place in the form of the fulfillment of our manifest destiny.

In this context, they evoke the idea of the emergence of a "Singularity," a process that would lead to a fusion between human biology and technology, "creating a superior form of machine consciousness, which will remain at the service of human design (…) Death, to whom we owe the very idea of necessity, will finally become optional." Looking at the shallow reading of Nietzsche that guides the book by Ted Nordhaus and Michael Shellenberg (2007), founders and mentors of the *Breakthrough Institute*, a mecca of technology, Danowski and Castro Nurseries are particularly incisive:

> The authors thus imagine a mind-blowing concubinage between Nietzsche and Pollyana from whose abominable copulation would emerge a monstrous daughter, an ecopolitical Barbie whom we might call the Gratitude of the Rich (p. 68).

14.7.2 From Dominion to Self-Control: The Audacity of Prudence

After World War II, industrial societies begin to face the fact that the continual enhancement of their "tools" begins to work against them. This was noted by Hans Jonas (1979/1984) who, in *The Imperative of Responsibility*, underlines precisely the ominous side of the Baconian ideal:

> The danger derives from the excessive dimensions of the scientific-technological-industrial civilization. What we could call the Baconian program–namely, to aim knowledge at power over nature, and to utilize power over nature for the improvement of the human lot–lacked in its capitalist execution from the outset the rationality as well as the justice with which it could have been conjoined.

In its "capitalist execution," to use the language of Hans Jonas, the rationalist tradition boils down to an acritical, non-reflexive, and purely utilitarian form of instrumental reason. This was well defined by Max Horkheimer in his essay *The End of*

Reason (1941) "as the optimum adaptation of means to ends, thinking as an energy saving operation." No contemporary philosopher seems to be as accurate and radical in his analysis of instrumental reason as Herbert Marcuse. In his critique of Max Weber (1968), Marcuse reveals how instrumental reason conceals its ideological domination:

> The very concept of technical reason is perhaps ideological. Not only the application of technology but technology itself is domination (of nature and men)–methodical, scientific, calculated, calculating control. Specific purposes and interests of domination are not foisted upon technology 'subsequently' and from the outside; they enter the very construction of the technical apparatus. Technology is always a historical-social *project*: in it is projected what a society and its ruling interests intend to do with men and things. Such a 'purpose' of domination is 'substantive' and to this extent belongs to the very form of technical reason.

In the same year of 1968 when everything seemed possible, Jürgen Habermas quoted this same passage in a text titled "Technology and Science as 'Ideology'" and dedicated it to Herbert Marcuse on his seventieth birthday. Whatever their perspectives, neither Habermas nor Marcuse can be accused of technophobia since they do not advocate for a refusal of the technological sphere as such in favor of a naive retreat into the sphere of nature or religion. Both, like Hans Jonas, advocate for a more or less radical critique (but not a refusal) of technology. Thanks to this criticism and to the constant increase in awareness of the responsibility of instrumental reason in the ongoing socio-environmental collapse, the recollection of the Greco-Latin lesson of the prudent self-management of ingenuity rings ever more strongly. In order to survive the trap of his own power, contemporary man begins to realize that nothing requires greater audacity than prudence. He begins to understand that breaking the philosophical and ideological blockade of accumulation and instrumental reason means acquiring a supreme power: the power over his own power. In Antiquity, this was considered the greatest demonstration of greatness. In *Natural History* (XXXV, 87), Pliny clearly expresses this essential *topos* of ancient wisdom, saying that Alexander the Great was *magnus animo, maior imperio sui*, that is, great in spirit, but even greater for the empire he had conquered over himself. As Michel Serres writes (1990/1992): "Why must we henceforth try to dominate our dominion? Because being disorderly, overstepping its goal, and counterproductive, pure dominion turns against itself" (p. 61).

If the Delphic–Socratic ideal of measurement, self-control, and knowledge inspired the thought and art of Antiquity, the most penetrating and impactful literary and cinematographic pieces of the contemporary world since the first postwar period will be warnings and reactions precisely toward the absence of this ideal. We will not trace the origins of these warnings and reactions which date back to the nineteenth century. Suffice it to remember that the great icons of the 1920s and 1960s will be dystopian analyses of the world created by instrumental reason and shaped by machines or by the manipulation of psychic mechanisms: Fritz Lang's *Metropolis* (1926), Chaplin's *Modern Times* (1933–1936), Aldous Huxley's *Brave New World* (1932), Orwell's *1984* (1949), Stanley Kubrick's diptych, *Dr. Strangelove* (1964), and *2001-A Space Odyssey (1968)*, among others.

From the 1970s, anticipatory fictions, literary or film, some of the latter by such inventive designers such as Syd Mead, took on a form not foreseen by Aldous Huxley and his contemporaries in the 1930s, who projected a dark but aseptic future. These new fictions, not coincidentally called *cyberpunk*, are characterized by hyper-urban societies dominated by "megacorporations," represented by "a combination of low life and high tech," in the words of William Gibson, who coined the term "megacorporation." In cinema, the "classic" of the cyberpunk genre is Ridley Scott's *Blade Runner* (1982), inspired by a Philip Dick novel of 1968.

The counterpart of the golden years of capitalism (1945–1973) are the dystopias of Ray Bradbury, John Brunner, Philip Dick, and William Gibson. As a logical development of them, we witness, successively, the revival of another subgenre of science fiction called post-apocalyptic, in which the world we know is destroyed, not by an external agent, but by the very machinery of technology, through a nuclear cataclysm, the emergence of technological singularity, or an environmental collapse. This subgenre has enjoyed great success, from *The Terminator* (1984) by James Cameron and *Matrix* (1999) by brothers Andy and Larry Wachowski, to Cormac McCarthy's novel, *The Road* (2006), and Andrew Stanton's *WALL-E* (2008), previously mentioned in Chap. 4. *The Lorax* (2012) is another animation that follows the same line of thought on environmental collapse. It is based on Dr. Seuss's book set in Thneedville, an equally post-apocalyptic city made of plastic, where its innocent inhabitants live cloistered, no longer in a spaceship, but in a wall that separates them from a lifeless world.

A common denominator to all this fictional production (of which I mention a few examples without any intention of making an inventory) is, of course, the critique of technology, which, as is routinely repeated today, assumes a form equivalent to that of *hubris* in the economy of the Greek tragedy.[3] Now, before moving to the last chapter, it is essential to avoid the misunderstanding of Jacques Ellul (1954) and others who attribute the environmental crisis to technology as if this had become, finally, an originating instance. Technology is the objectification of a competence that is inherent to all species. It seems impossible, moreover, to surgically separate its benevolent side from its ominous one. However, its progress is essential today more than ever, as the imperative transition to a low-impact society will require accelerated technological innovation. Putting human ingenuity at the service of decreasing anthropic pressure on the biosphere—rather than blindly tying it to an anachronistic and dysfunctional accumulative drive—is the inescapable question defining a new agenda and a new political-ideological spectrum, which are inconceivable as long as the illusion of the unlimited persists.

---

[3] On the analogy between man"s evil action over nature and tragic *hubris*, see, for example, Argullol (2004, pp. 151–156), but the analogy has become commonplace.

References

ARGULLOL, R. "Vers un humanisme polycentrique". *Diogène*, 206, April–June 2004, pp. 151–156.

AUBENQUE, Pierre. *La prudence chez Aristote*. Paris, PUF, 1963.

———. "Plotin et le néoplatonisme". *In*: CHÂTELET, F. *La philosophie païenne* (1972). Paris, Hachette, 1999, pp. 228–246.

BOSTROM, Nick, *Superintelligence. Paths, Dangers, Strategies*. Oxford University Press, 2014.

BOULDING, Kenneth E., "The economics of the coming Spaceship Earth" (1966). In Jarrett, H. (ed.), *Environmental Quality in a Growing Economy*, pp. 3–14. Baltimore, Resources for the Future. Johns Hopkins University Press.

CALANDRA, E. *Oltre la Grecia. Alle origini del filellenismo di Adriano*. Perugia, 1996.

CARDANO, Girolamo, *De propria vita* (1576), Italian translation, *L'autobiografia di Gerolamo Cardano*, Milan, La Famiglia Meneghina Editrice, 1932.

CHASE-DUNN, Christopher; KAWANO, Yukio & BREWER, Benjamin D. "Trade Globalization since 1795: Waves of Integration in the World-System". *American Sociological Review*, 65, 1, 2000, pp. 77–95.

CHASTEL, André, *Leçon inaugurale fait le Mercredi 20 janvier 1971*. Collège de France. Chaire d'Art et de Civilisation de la Renaissance en Italie, 1971.

DANOWSKI, Deborah & VIVEIROS DE CASTRO, Eduardo. *Há mundo por vir? Ensaio sobre os medos e os fins*. Rio de Janeiro, 2014.

ELLUL, Jacques. *La Technique, ou l'enjeu du siècle*. Paris, Armand Colin, 1954.

FAGGIN, Giuseppe. "Introduzione" a Plotino, *Enneadi*. Milan, Bompiani, 2000, pp. xvii–xxxiv.

FLORO, *Epitome di Storia Romana* (130c.–138c.), Milan, Rusconi, 1981.

FREUD, Sigmund, "Character and anal erotism" (1908), in S. Freud, The Standard Edition of the Complete Psychological Works, translated by James Strachey, London, 1959, pp. 167–176.

GIBBS, W. Wayt & SOARES, Christine, "Preparing for a Pandemic". In, "How it all ends". *Scientific American,* 2012.

GILDING, Paul. *The Great Disruption*. New York, Bloomsbury Press, 2011.

HADOT, Pierre. "Conversion". *Exercices spirituels et philosophie antique*. Paris, Albin Michel, 2002.

HAMILTON, Clive. *Requiem for a Species. Why We Resist the Truth About Climate Change,* Earthscan, 2010.

HEIDEGGER, Martin, "Die Frage nach der Technik" (1953). In: *Vorträge und Aufsätze*. Neske: Pfullingen, 1954. Translated by William Lovitt, "The Question Concerning Technology", in M. Heidgger: *Basic Writings*. New York, Harper & Row, 1977, pp. 287–317.

HORKHEIMER, Max, "The End of Reason". *Studies in Philosophy and Social Science*, 9, 1941, pp. 366–388.

HUNTER, Robert, *Thermageddon. Countdown to 2030*. New York, 2002.

KURZWEIL, Ray, *The singularity is near: When humans transcend biology*. Penguin, 2005.

LEBEAU, André, *L'enfermement planétaire*, Paris, Gallimard, 2008

LEBRUN, Gérard, "Sobre a tecnofobia". *A filosofia e sua história*. São Paulo, 2006, pp. 481–508.

LÉVI-STRAUSS, Claude, "Tout à l'envers" (1989), in *Nous sommes tous des cannibales*, Paris, Seuil, 2013.

MAGDELAIN, André. "Le pomerium archaïque et le mundus". *Jus imperium auctoritas. Études de droit romain*. Rome, Publications de l'École Française de Rome, 1990, pp. 155–191.

MARCUSE, Herbert, "Industrialization and Capitalism in the Work of Max Weber". *Negations: Essays in Critical Theory* (1968), MayFlyBooks/Ephemera, 2009.

MARX, Karl. *Das Kapital* (1867). English translation (1887) from the 4th German edition by Samuel Moore & Edward Aveling, edited by Friedrich Engels. Moscow, Progress Publishers.

McKIBBEN, Bill, *The end of nature*. New York, Anchor Books, 1989.

MOMIGLIANO, Arnaldo. *Alien wisdom. The limits of Hellenization* (1975). Italian translation *Saggezza straniera. L'Ellenismo e le altre culture* (1975). Turin, Einaudi, 1980.

————. "La leggenda di Enea nella storia di Roma fino ad Augusto". *Saggi di Storia della Religione Romana*. Brescia, Morcelliana, 1988, pp. 171–183.

NICOLINI, Lara, Introduzione a Apuleio, *Le Metamorfosi*, Roma, BUR, 2005

NORDHAUS, Ted & SHELLENBERGER, Michael. *Break Through: From the Death of Environmentalism to the Politics of Possibility*. Boston, Houghton Miflin, 2007.

ORD, Tobby, *The Precipice. Existential Risk and the Future of Humanity*. New York, 2020.

PIKETTY, Thomas. *Le capital au 21ᵉ siècle*. Paris, Seuil, 2013.

PLATO, *Critias* translated by Benjamin Jowett, Global Grey, 2018.

PLOTINUS, Enneads. Paris, Belles Lettres, 1931 (translated by Émile Bréhier); Milan, Bompiani, 2000, p. 793 (translated by Giuseppe Faggin).

REES, Martin, *Our final century. Will civilization survive the twenty-first century?*. London, Arrow Books, 2003.

ROSENTHAL, Earl. "Plus Ultra, Non plus Ultra, and the Columnar Device of Emperor Charles V". *Journal of the Warburg and Courtauld Institutes*, 34, 1971, pp. 204–228.

SCHUHL, Pierre-Maxime. *Machinisme et philosophie*. Paris, PUF, 1947.

SCHUMPETER, Joseph A. *Can capitalism survive? Creative destruction and the future of the global economy* (1942). New York, HarperCollins, 1976.

SERRES, Michel. *Le Contrat naturel*. Paris, Éditions François Bourin, 1990; Flammarion, 1992.

STENGERS, Isabelle, "Au temps des catastrophes. Résister à la barbarie qui vient" (2009), English translation, "In Catastrophic Times: Resisting the Coming Barbarism". In Tom Cohen & Claire Colebrook (ed.), *Critical Climate Change*, Open Humanities Press & Meson press, 2015.

THOMMEN, Lukas. *An Environmental History of Ancient Greece and Rome* (2009). Cambridge University Press, 2012.

VALÉRY, Paul. *Regards sur le monde actuel* (1945), "Avant-propos" (1931). *Oeuvres*, vol. II. Paris, Gallimard, 1960, pp. 913–928.

————. "Le Centre Universitaire Méditerranéen" (1933). *Regards sur le monde actuel* (1945). *Oeuvres*, vol. II. Paris, Gallimard, 1960, pp. 1128–1145.

WAKEFIELD, Julie, "Doom and Gloom by 2100". In, "How it all ends". *Scientific American*, 2012.

Chapter 15
The Anthropocentric Illusion

In the previous two chapters, we have seen how much belief in the possibility of a "sustainable capitalism" is illusory. We then saw how this illusion rests on another one, rooted in primitive phases of the psyche and primitive species' behavior: the illusion that our security continues to be directly proportional to the increase in surplus. In this chapter, we place these two illusions within the scope of a third one (of a metaphysical and religious nature), which generates, sustains, and encompasses them: the anthropocentric illusion.

The term anthropocentrism has at least two very different meanings, and it is important to specify them in order to avoid misunderstandings. In one sense, anthropocentrism is the inescapable logical prison of the principle of identity. In 1899, Sully Prudhomme pointed it out by affirming the mental constant that consists in "conceiving all activity in the external world from one's own [activity], as revealed by its conscience." In this sense, Serge Moscovici (1968, p. 22) is correct when he makes the following claim about anthropocentrism: "All our models of nature are anthropocentric, in one way or another." At the limit, there can only be for man a *human* history of nature in such a way that the very title of Moscovici's work—*Essai de l'Histoire Humaine de la Nature*—is redundant, so to speak. Carl Friedrich von Weizsäcker's statement in his *The History of Nature* (1951) must be understood in this same way: "We know nature only through the medium of human experience" (as cited by Moscovici 1968, p. 40). Such affirmations are, evidently, tautological, since they are limited to affirming that man's conception of nature cannot be but the human conception of nature. Nicholas Georgescu-Roegen (1971) refutes the criticism that thermodynamics bears the mark of anthropomorphism (a term that he uses to convey the sense of anthropocentrism), remembering that "the idea that man can think of nature in wholly nonanthropomorphic terms is a patent contradiction in terms" (p. 227). Later, the author is even more assertive about this logical impossibility of human thought being able to "go out" of itself:

> The natural scientist came to realize that, as Louis de Broglie put it, he is in a continuous hand-to-hand battle with nature. And being a man, he cannot possibly describe nature

© Springer Nature Switzerland AG 2020
L. Marques, *Capitalism and Environmental Collapse*,
https://doi.org/10.1007/978-3-030-47527-7_15

otherwise than in terms 'adapted to our mentality'. (…) By the very nature of its actor, every intellectual endeavor of man is and will never cease to be human (p. 342).

Indeed, the point of view from which we perceive the world *cannot* but be anthropocentric precisely because this point of view is ours, in the same way that a dog's world view *cannot* not be "canine centric." We think as if we were the wind or a mountain only as a poetic metaphor, to remember the beautiful images used by Aldo Leopold (1949) when he titled one of the sections of his book, "On Top (Thinking Like a Mountain)." This meaning of anthropocentrism undoubtedly has the value of an epistemological caveat in reminding us that we are enclosed in the logical principle of identity, but it does not say anything specific about what the term anthropocentrism designates *within* this principle of identity.

Let us open two philosophy dictionaries at random. The first defines anthropocentrism as: "any thought orientation that puts man at the center of reality and considers the good of humanity as the ultimate cause of all things.[1]" The second reiterates the first: "Anthropocentrism refers to a doctrine that places man at the center of the world. [...] Anthropocentrism further enunciates the idea that all things in the universe (minerals, vegetables, animals) are subordinate to man.[2]" Therefore, anthropocentrism is not restricted to the principle of identity, since identity and the presumption of superiority are not synonymous notions. It is one thing to admit that we are stuck with the human point of view; another, quite different thing, is to claim that this point of view enjoys the privilege of centrality, superiority, and an ultimate purpose, capable of relegating others to subordinate, peripheral, and instrumental positions. What interests us here, of course, is to analyze anthropocentrism as a presumption of centrality, superiority, and of ultimate purpose.

15.1 Three Historical Emphases of the Anthropocentric Presumption

It is impossible to deconstruct anthropocentrism into distinct elements, since it refers to a mix of beliefs that reciprocally interpenetrate, engender, and imply each other. For the purpose of a descriptive exercise only, it is useful to detect three inclinations that are more recurrent and that were inherited from Antiquity:

1. The cosmotheological and teleological presumption which sees man as the mediating center and the purpose of the cosmos
2. The biological presumption which asserts a superiority and a radical discontinuity of man in the web of other forms of life

---

[1] *Enciclopedia Filosofica*. Fondazione Centro Studi Filosofici di Gallarate. Milan, Bompiani, 2006, vol. I, *ad vocem*

[2] See André Jacob (dir.), *Encyclopédie Philosophique Universelle*, vol. II – *Les Notions Philosophiques*, under the direction of Sylvain Auroux, Paris, PUF, 1990, p. 105.

3. The ecological presumption, based on the belief that man generally adapts his *habitat* to his ends, unlike other species which, generally, adapt to him

The biological and ecological inclinations, which are most closely related (since the third is a particular case of the second), are the ones that interest us here the most, but it is important to realize that they take root in the first inclination (especially with regard to their teleological aspect), which is why it is important to discuss it firstly.

15.1.1   The Cosmotheological and Teleological Emphases

In its origin, this first inclination comes to the fore through the analogy between microcosm and macrocosm. The cosmos and the human body would be governed by the same structures, proportions, and harmonies. The human body would be, therefore, a microcosm, that is, a kind of epitome of the cosmic order. In Greece, the term microcosm and the idea behind it seem to go back to a period between Pythagoras and Democritus (Conger 1922). In any case, it can be found in Democritus (frg. B 34), according to an uncertain passage from the *Prolegomena to Aristotle* (38, 14) by David of Nerken, a fifth-century AD Armenian philosopher (Diels-Kranz/Reale 2006, pp. 1364–1365):

> And just as in the Universe we see beings that, like the gods, only govern, beings that at the same time govern and are governed, like men (these are, in fact, governed by the gods and govern the animals without language / *alogon zoion*), and in short, beings that are only governed, like animals without language, so we also observe in man—which, according to Democritus, is a *microcosm*—that same distribution [mind, heart, and passions].

The inextricable interpenetration between the cosmotheological emphasis and the biological emphasis is already evident in this passage. In both the history of visual arts and the history of ideas, this idea of man as microcosm, as an image of the cosmos, would become commonplace. It returns in Polykleitos canon and in Vitruvius' *Homo quadratus* and then applied both to architectural proportions and to the human body, as in Villard de Honnecourt and Leonardo da Vinci, among others. In Vitruvius (*De architectura decem libri*, III, 1), we read:

> [...] *non potest aedes nulla sine symmetria atque proportione rationem habere compositionis, nisi uti ad hominis bene figurati membrorum habuerit exactam rationem.*

> [...] no building can be said to be well designed which wants symmetry and proportion. In truth they are as necessary to the beauty of a building as to that of a well formed human figure,

The microcosm/macrocosm analogy would also be seen in several other systems of proportion of the human body expressed in the canons representing the body of the crucified Christ, namely, in the history of painting in the twelfth and thirteenth centuries (Sandberg-Vavalà 1929; Battisti 1967). In the context of the history of ideas, the correlation between microcosm and macrocosm radiates from Plato's *Timaeus*.

Calcidius (thanks to whom the Latin Middle Ages partially knew this dialogue) writes in his extensive commentary on its translation (c. 320) that man was called *mundum brevem* by the ancients (*veteribus*).[3] More than Plato, it is Neoplatonism that will appropriate this correlation, from the *Hermetica*, Boethius (477–524), and Maximus the Confessor (c. 580–662) to Bernardus Silvestris (1998), author in the mid-twelfth century of a prosimetrum that became famous (titled *Cosmographia. De mundi universitate sive megacosmus et microcosmos*). In any case, in 1486, in his *Oration on the Dignity of Man* (*Oratio de hominis dignitate*), Giovanni Pico della Mirandola inventories the traditional ways of defining man as a center, mediator, and purpose of the created cosmos before advancing his own considerations[4]:

> I was not satisfied by the many reasons that have been advanced for the preeminence of human nature: that man is the intermediary between creatures, that he is the intimate of the gods above him, as he is the lord of the beings beneath him; that, by the acuteness of his senses, the inquiry of his reason, and the light of his intelligence, he is the interpreter of nature, set midway between unchanging timelessness and the flux of time; that he is the living union and (as the Persians say), the marriage hymn of the world, and, by David's testimony, but little lower than the angels.

Judeo-Christian anthropology could not but reinforce the cosmotheological aspect of anthropocentrism already present in the classical tradition, for example, in Ovid (*Met.* I, 76–79) which accepts, as equally possible, two hypotheses to explain the origin of man: "either the creator god, source of a better world, seeded it from the divine (*siue diuino semine fecit / Ille opifex rerum, mundi melioris origo*), or Prometheus molded them into an image of the all-controlling gods" (*Finxit in effigiem moderantum cuncta deorum*). In the Elohist source of *Genesis* (Gn, 1, 26), Elohim created man on the sixth day as a coronation of all creation, infusing his own forms in him as a mark of this prerogative:

> Then Elohim said, 'Let us make humans in our image, in our likeness. Let them rule the fish in the sea, the birds in the sky, the domestic animals all over the earth, and all the animals that crawl on the earth'.

As in the fragment referring to Democritus, mentioned above, the relationship between the cosmotheological privilege of man and his right to rule over other forms of life in the created world is also inextricable. If it is true that Christianity was able to avoid a literal interpretation of these two verses (the heresy of the Anthropomorphites, according to which God would have defined his own form in this process, never seriously threatened Christian exegesis), it is no less true that Christian doctrine never denied this divine predilection for man, His exceptional love for him, expressed by His care in the Yahwist source of *Genesis* (Gn 2,7) through inoculating him with a *nefesh* (or *ruah* in Gn, 6, 17), that is, an exclusive

---

[3] Chapter 25: "*Unde opinor hominem, inquit, mundum brevem adpellatum.*" See Van Winden (1965, p. 133).

[4] *Oratio*, op. cit., ed. cit. p. 103: *esse hominem creaturarum internuntium, superis familiarem, regem inferiorum; sensuum perspicacia, rationis indagine, intelligentiae lumine, naturae interpretem; stabilis aevi et fluxi temporis interstitium, et (quod Persae dicunt) mundi copulam, immo hymenaeum, ab angelis, teste Davide, paulo deminutum.*

"breath" of immortality; this was, in short, God's will that a kind of continuous theophany be expressed through the presence of man. So much so that, since the Apologists, the dogma of incarnation, the essence and raison d'être of this religion, would confirm the passage of *Genesis* mentioned above. It is, therefore, out of love for man and in the form of man, in the human figure of Jesus, that the antinomy between creator and creature, between the intelligible and the sensible, and between unity and multiplicity, is resolved and overcome again.

In his pedagogical best seller of 1609, reissued in 1638, *The Wisdom of the Ancients*,[5] Francis Bacon still dwells, and perhaps for the last time, on the *topos* of man as a microcosm when analyzing the myth of Prometheus (chapter XXVI):

> Man seems to be the thing in which the whole world centres, with respect to final causes; so that if he were away, all other things would stray and fluctuate, without end or intention, or become perfectly disjointed, and out of frame; for all things are made subservient to man, and he receives use and benefit from them all. (…) And it is not without reason added, that the mass of matter whereof man was formed, should be mixed up with particles taken from different animals, and wrought in with the clay, because it is certain, that of all things in the universe, man is the most compounded and recompounded body; so that the ancients, not improperly, styled him a Microcosm, or little world within himself.

15.1.2   The Biological Emphasis

As shown in the lines above, the biological emphasis on anthropocentric beliefs is only a logical development of the cosmotheological and teleological presumption. It consists in placing man at the apex of the chain of life and, at the same time, placing him in discontinuity with it. Arthur O. Lovejoy (1936/1957, p. 186) crudely defined this presumption as "one of the most curious monuments of human imbecility." As might be expected, the classic model of anthropocentrism becomes fixed in Western thought starting with Plato and Aristotle. In *Timaeus* (90e-92), Plato elaborates a zoogony based on the idea of *Urtier*—the prototype of the male human animal—and of his chain of being according to a curve of transmutations:

> Of the men who came into the world, those who were cowards or led unrighteous lives may with reason be supposed to have changed into the nature of women in the second generation.[…] Thus were created women and the female sex in general. But the race of birds was created out of innocent light-minded men, who, although their minds were directed toward heaven, imagined, in their simplicity, that the clearest demonstration of the things above was to be obtained by sight; these were remodelled and transformed into birds, and they grew feathers instead of hair. The race of wild pedestrian animals, again, came from those who had no philosophy in any of their thoughts, and never considered at all about the nature of the heavens, because they had ceased to use the courses of the head, but followed the guidance of those parts of the soul which are in the breast. In consequence of these habits of theirs they had their front-legs and their heads resting upon the earth to which they were drawn by natural affinity; and the crowns of their heads were elongated and of all sorts of shapes, into which the courses of the soul were crushed by reason of disuse. And this was

---

[5] About the success of the book, see Malherbe (2011, pp. 131–134).

the reason why they were created quadrupeds and polypods: God gave the more senseless of them the more support that they might be more attracted to the earth. And the most foolish of them, who trail their bodies entirely upon the ground and have no longer any need of feet, he made without feet to crawl upon the earth. The fourth class were the inhabitants of the water: these were made out of the most entirely senseless and ignorant of all, whom the transformers did not think any longer worthy of pure respiration, because they possessed a soul which was made impure by all sorts of transgression; and instead of the subtle and pure medium of air, they gave them the deep and muddy sea to be their element of respiration; and hence arose the race of fishes and oysters, and other aquatic animals, which have received the most remote habitations as a punishment of their outlandish ignorance. These are the laws by which animals pass into one another, now, as ever, changing as they lose or gain wisdom and folly (translated by Benjamin Jowett).

With Aristotle (*Politics*, 1.5, 1254b7), the hierarchy of beings, starting with the human male, is consolidated and becomes fixed:

Tame animals have a better nature than wild, and all tame animals are better off when they are ruled by man; for then they are preserved. Again, the male is by nature superior, and the female inferior; and the one rules, and the other is ruled; this principle, of necessity, extends to all mankind (translated by Benjamin Jowett).

Further on (*Pol.* 1.8, 1256b115), he not only calcifies this hierarchy, but makes man, by nature, the raison d'être of nonhumans:

We may infer that, after the birth of animals, plants exist for their sake, and that the other animals exist for the sake of man, the tame for use and food, the wild, if not all, at least the greater part of them, for food, and for the provision of clothing and various instruments. Now if nature makes nothing incomplete, and nothing in vain, the inference must be that she has made all animals and plants for the sake of man (translated by Benjamin Jowett).

For Aristotle there is, therefore, no right for the non-man. Hence, there is no need for an ethics of the free-human-male in relation to other creatures, as we read equally clearly in the *Nicomachean Ethics*, VIII, 11, 6: "there is no friendship nor justice towards lifeless things. But neither is there friendship towards a horse or an ox, not to a slave qua slave. For there is nothing common to the two parties" (translated by W.D. Ross).

In Cicero, anthropocentrism maintains these extreme forms and, from a modern lens, is almost comical, as in this passage from *De Natura Deorum* (II, lxiii): "What other end do sheep serve except that of clothing men with their wool, when it has been prepared and woven? (*Quid enim oves aliud afferunt, nisi ut earum villis confectis atque contextis homines vestiantur?*). And the examples multiply: the ox's neck is made for the yoke, the dog's nose to help the hunter, etc. Likewise, narratives of colossal killings of animals brought from Africa and Asia to animate the games are commonplace in the biographies of emperors, from Suetonius to the collection, *Scriptores historiae augustae*. "Augustus had 3,500 animals killed in twenty-six venationes. At the dedication of the Colosseum under Titus, 9,000 were destroyed in 100 days, and Trajan's conquest over Dacia was celebrated by the slaughter of 11,000 wild animals" (Hughes 1975, p. 102). These were, Cicero would say, the raison d'être of these animals.

Scholasticism just continues this tradition. In *Summa contra gentiles* (III, chapter 22,7), in discussing the concept of Providence, St. Thomas Aquinas will operate a sort of Christian synthesis between man's cosmotheological and biological prerogatives:

> The ultimate end of the whole process of generation is the human soul, and matter tends toward it as toward an ultimate form. So, elements exist for the sake of mixed bodies; these latter exist for the sake of living bodies, among which plants exist for animals, and animals for men. Therefore, man is the end of the whole order of generation.

Based on this anthropology, at the same time ancient and Judeo-Christian, Giannozzo Manetti (1396–1459) can claim—in his *De dignitate et excellentia hominis* (1453) in a passage cited by Kraye (1988, p. 110)—that man owns the world. This is, moreover, a simple paraphrasing of another section of Cícero's *De Natura Deorum*: "Ours are the lands, the fields, the prairies, the hills [...] ours are the oxen, the bulls, the camels [...] ours, the seas and all the fish" (*Nostre sunt terre, nostri agri, nostri campi, nostri montes [...] nostri boves, nostri tauri, nostri cameli [...] nostra maria, nostri omnes pisces*). Likewise, at the other end of the Renaissance parable, Francesco Buonamici, in his *De moto libri I* (1591), will repeat once again the *leit-motiv* that nature exists for man (Kraye 1988, p. 310): "The elements serve man, [...] men are also given, for their health, many plants and stones and medicinal metals" (*Homini elementa serviunt [...] multis etiam plantis lapidibusque atque metallis medicae vires datae sunt ad unam eius salutem*).

15.1.3   From Descartes to Kant: Anthropocentrism as Discontinuity

In reality, with the end of geocentrism, the liquidation of heavenly hierarchies by the theologians of the Reformation (Mason 1957/1971), and the growing crisis of the theory of general correspondence between the four moods, the four elements, the four seasons, and the four states of matter (cold, hot, dry, humid), the figure of the human as a microcosm lost its cosmological argument and, with it, much of its power of persuasion. As is well-known, Pascal expresses better than anyone else the religious distress caused by this loss.

Reduced to the idea of the soul as a divine gift and a fundamental and distinctive essence of man, anthropocentrism would undergo an essential mutation from then on. Starting with Descartes, anthropocentrism will no longer defend the speculative relationship between microcosm and macrocosm, that is, a continuity between man and nature (man being the metaphysical or religious quintessence of the universe), but a radical discontinuity: the inaugural ontological distinction between *res cogitans*, an exclusive characteristic of man, and nonhuman (*res extensa*).

Departing from the speculative philosophy of the schools, according to which all animals would be endowed with a vegetative and sensitive soul, leaving only the rational soul to man, Descartes reduces the nonhuman to bodies whose only attribute will be extension. This allows them to be conceived of in terms of pure mechanical and measurable forces. The *locus classicus* of this modern anthropocentric conception of nature is found in the sixth and last part of Descartes' *Discourse on the Method of Rightly Conducting One's Reason and Seeking Truth in the Sciences* (1637):

> For they–these scientific notions of mine–showed me that we can get knowledge that would be very useful in life, and that in place of the speculative philosophy taught in the schools we might find a practical philosophy through which, knowing the power and the actions of fire, water, air, the stars, the heavens and all the other bodies in our environment as clearly as we know the various crafts of our artisans, we could (like artisans) put these bodies to use in all the appropriate ways, and thus make ourselves the masters and (as it were) owners of nature (translated by Jonathan Bennett).

From this inaugural distinction between *res cogitans* and *res extensa* which turns men into masters and owners of nature, it follows that animals, since they have no soul, must be understood as purely mechanical entities. In the well-known letter to the Marquess of Newcastle, of November 23, 1646, Descartes hurls animals in the Tartarus of automatism without hesitation[6]:

> I cannot share the opinion of Montaigne and others who attribute understanding and thought to animals. (…) I know that animals do many things better than we do, but this does not surprise me. It can even be used to prove they act naturally and mechanically, like a clock which tells the time better than our judgement does.

Despite the criticisms of Rousseau (1754) and Voltaire in their article "Animals" (*Bêtes*) for the *Encyclopédie*, this reduction of the animal to a kind of automaton devoid of sensitivity and conscience (as it had been previously devoid of its own purpose) establishes the terrain on which modern anthropocentrism thrives, formulated by Kant in paragraph 86 of the *Critique of Judgment*[7]:

> The commonest Understanding, if it thinks over the presence of things in the world, and the existence of the world itself, cannot forbear from the judgement that all the various creatures (…) would be for nothing, if there were not also men (rational beings in general). Without men the whole creation would be a mere waste, in vain, and without final purpose.

Leonel Ribeiro dos Santos (2012) [8] examines, in Kant's work, the nuances of the expression that man is *Herr der Natur, Meister über die Natur* (Lord of nature,

---

[6] *Oeuvres et lettres*, Paris, Gallimard, Pléiade, 1953, p. 1256. See also Descartes's letter to Henri Morus, of February 5, 1649, *loc. cit.*, p. 1312.

[7] Cf. Immanuel Kant, *Kritik der Urteilskraft* (1790, Ak V,442): *Es ist ein Urtheil, dessen sich selbst der gemeinste Verstand nicht entschlagen kann, wenn er über das Dasein der Dinge in der Welt und die Existenz der Welt selbst nachdenkt: dass nämlich alle die mannigfaltigen Geschöpfe, von wie grosser Kunsteinrichtung und wie mannigfaltigem zweckmässig auf einandere bezogenen Zusammenhange sie auch sein mögen, ja selbst das Ganze so vieler Systeme derselben, die wir unrichtiger Weise Welten nennen, zu nichts da sein würden, wenn es in ihnen nicht Menschen (vernünftige Wesen überhaupt) gäbe; d. i. dass ohne den Menschen die ganze Schöpfung eine blosse Wüste, um sonst und ohne Endzweck sein würde.*

[8] I thank the author for the courtesy of making me aware of his book, which I use as a guide in the discussion of Kant's anthropocentrism.

Master of the nature), an expression that is already present in the *Critique of Pure Reason*. Kant conceives the man–nature relationship in a way not substantially different from the one expressed in Buffon's *Époques de la nature* (1780), for whom nature "fertilized" by man is superior to brute nature, as seen in the beginning of Chap. 11. As Ribeiro dos Santos stresses, in some aspects Kant returns to the old and medieval idea of man:

> [...] suspended between two worlds, an amphibious being, placed in a 'middle situation' (*Mittelstand*) equally remote from the extremes, located in the 'dangerous middle point' (*gefährliche Zwischenpunkt*) or in the 'dangerous middle way' (*in der gefährlichen Mittelstrasse*) of the hierarchy of beings. For this very reason, he is the being of mediation, the isthmus, the *copula mundi*, the *terminus medius* (p. 144).

That being said, for Kant, man, the recipient of moral law, is the only one capable of giving meaning to things and the only being endowed with reason. This gives him, and only him, an inalienable purpose of his own. If Cartesian philosophy started from the separation between *res cogitans* and *res extensa*, for Kant, the latter (nature) is so far removed from philosophical reflection that, as Ribeiro dos Santos also emphasizes (p. 145), it is entirely absorbed by the first pole:

> the first separation that Kant proposes to us is not between men and things, but rather in man himself, between Human and Humanity (between man, as a sensitive and rational physical being, and man as a person or moral rational being), between *homo phaenomenon* and *homo noumenon*.

Man no longer affirms his humanity, therefore, negatively, in opposition to nonhumans, but in the tension between his empirical being and his transcendental being. This form of Kant's "hyper-anthropocentrism" guarantees the final removal of nature from the field of reflection on man as a moral being. And, inevitably, we find in it a certain similitude with the initial distinction established by Hegel (in his *Lectures on Aesthetics*) between the beauty of art as beauty born of the spirit and the beauty of nature, which is surgically expelled from the philosophical reflection on the beautiful. In her famous work, *The Human Condition* (1958), Hanna Arendt undertook perhaps the first ecological criticism of Kantian anthropocentrism, which can be summarized in this expression: Kant's operation that establishes man as the "supreme end" permits him "to degrade nature and the world into mere means, robbing both of their independent dignity."

15.1.4   The Ecological Presumption

The thesis that man, lacking the physical endowments of other animals (as Descartes reminds the Marquess of Newcastle in his aforementioned letter), is able to surpass them by virtue of his mental capacity is complemented by the conviction that man, precisely because of these gifts, has an adaptive relationship with his habitat that is predominantly active, whereas other animals have a predominantly passive adaptive relationship with their respective habitats. Such a conviction is as old as the myth of

Epimetheus. According to this myth, the brother of Prometheus forgot about man when he distributed all the physical attributes to other animals. This was corrected by Prometheus who, thus granted fire and the arts to man. However, in Aeschylus, Prometheus already appears as *philanthropos*, since "all arts (*tékhnai*) of men come from Prometheus" (Pucci 2005); the myth does not appear in the form of a pure contrast between Prometheus and his brother, Epimetheus, except in the narrative that Plato attributes to Protagoras in the homonymous youth dialogue (320–323):

> Once upon a time there were gods only, and no mortal creatures. But when the time came that these also should be created, the gods fashioned them out of earth and fire and various mixtures of both elements in the interior of the earth; and when they were about to bring them into the light of day, they ordered Prometheus and Epimetheus to equip them, and to distribute to them severally their proper qualities. Epimetheus said to Prometheus: 'Let me distribute, and do you inspect.' (…) Thus did Epimetheus, who, not being very wise, forgot that he had distributed among the brute animals all the qualities which he had to give,—and when he came to man, who was still unprovided, he was terribly perplexed. Now while he was in this perplexity, Prometheus came to inspect the distribution, and he found that the other animals were suitably furnished, but that man alone was naked and shoeless, and had neither bed nor arms of defense. (…) Prometheus, not knowing how he could devise his salvation, stole the mechanical arts of Hephaestus and Athene, and fire with them (they could neither have been acquired nor used without fire), and gave them to man. (…) Now man, having a share of the divine attributes, was at first the only one of the animals who had any gods, because he alone was of their kindred.

The contrast between human ingenuity, shared with the gods, and the laws of nature will, as we know, sees a prodigious posterity. This contrast is found entirely, for example, in the *Oration on the Dignity of Man* (*Oratio de hominis dignitate*, 1486), already mentioned, in which Giovanni Pico della Mirandola (2018) creates the discourse of a god that places man above the laws of nature[9]:

> Neither a fixed abode, nor a face of thine own, nor any endowment particular to thee have We given thee, O Adam, so that by thine own judgement and decision thou mayst have and possess whatever place, whatever form, whatever gifts thou dost desire. The nature of all other creatures is defined and restricted within the laws which We have laid down. Thou, impeded by no restrictions, according to thine own free will, in whose hand We have placed thee, shalt define thyself (translated by Charles Glenn Wallis).

This contrast between man and nature is reflected in both branches of the origin myth of human culture, in Hesiod's Golden Age and in the lost Paradise of *Genesis*, its correlative in the Old Testament. In a distant beginning, humans enjoyed divine benevolence and a status close to that of divinity. But a cataclysm threw them into a condition close to that of animals, a condition from which they would be able to raise themselves only gradually, thanks to their *tékhnai*, their *artes*, until they

---

[9] *Oratio Ioannis Pici Mirandulani Concordia Comitis* (1486), ed. and translation by Eugenio Garin, Florence, Vallecchi, 1942, p. 107: *Nec certam sedem, nec propriam faciem, nec munus ullum peculiare tibi dedimus, o Adam, ut quam sedem, quam faciem, quae munera tute optaveris, ea, pro voto, pro tua sententia, habeas et possideas. Definita ceteris natura intra praescriptas a nobis leges coercetur. Tu, nullis angustiis coercitus, pro tuo arbitrio, in cuius manu te posui, tibi illam praefinies. Medium te mundi posui, ut circumspiceres inde commodius quicquid est in mundo.*

reached the state of civilization. Thus, human history would be essentially the story of the progressive differentiation of humans from the condition imposed on other animals and the story of mankind's increasing capacity to know, control, and use the laws of nature to its own advantage, if not (in a future promised only to us) to redesign them. The epitome of this conception can be seen in Bacon's program of a science whose mission is "to put [nature] in constraint (...) bound into service, hounded in her wanderings and put on the rack and tortured for her secrets."

Two scientific manifestos of the twentieth century provide, perhaps, the most vivid evidence of the ecological assumption of a Baconian matrix. *Transhumanism*, signed by Sir Julian Huxley in 1957, is the first of them. It is an ardent defense and profession of faith in "man's responsibility and destiny—to be an agent for the rest of the world in the job of realizing its inherent potentialities as fully as possible."

> It is as if man had been suddenly appointed managing director of the biggest business of all, the business of evolution —appointed without being asked if he wanted it, and without proper warning and preparation. What is more, he can't refuse the job. Whether he wants to or not, whether he is conscious of what he is doing or not, he is in point of fact determining the future direction of evolution on this earth. That is his inescapable destiny, and the sooner he realizes it and starts believing in it, the better for all concerned.

This great naturalist would not fail to realize, soon after, how disastrously man was managing his "business." Thus, in 1961, he would participate in the creation of the WWF, and in 1972, he would co-sign Edward Goldsmith's and Robert Allen's *Blueprint for Survival*, a manifesto that is the antithesis of the ecological presumption.

The second scientific manifesto, the "Heidelberg Appeal," from 1992, will be the last link in the long history of the ecological presumption of anthropocentrism. It will be discussed below (see below item 15.4 The Great Mental Block).

15.2   The Fourth "Severe Blow"

In "A Difficulty in the Path of Psychoanalysis" (1917), Freud observes that "the universal narcissism of men, their self-love, has up to the present suffered three severe blows (*drei schwere Kränkungen*) from the researches of Science:" Copernican heliocentrism, Darwinian evolution, and the discovery of the preeminence of will or of the unconscious over intellectual representations, something proposed by Schopenhauer and by psychoanalysis itself.[10] These three discoveries shook anthropocentrism by removing three presumptions from the *Homo sapiens*:

---

[10] Freud is the first to recognize and honor, in this same text and in others, Schopenhauer's precedence over some of his fundamental ideas. In his *Autobiographical Study* (1925), he writes: "The large extent to which psycho-analysis coincides with the philosophy of Schopenhauer — not only did he assert the dominance of the emotions and the supreme importance of sexuality but he was even aware of the mechanism of repression — is not to be traced to my acquaintance with his teaching. I read Schopenhauer very late in my life."

the presumption of cosmological centrality, the presumption of biological exceptionality, and the presumption of consciousness as the founding instance of the subject. The environmental crises that have been piling up since the second half of the twentieth century add a fourth severe blow to anthropocentrism, this time to the ecological presumption, outlined above. According to this presumption, man—thanks to his mental skills—can adapt his habitat to his own needs rather than himself adapting to it, thus escaping the condition to which other animals are subject.

15.2.1   The Limit Between Adaptation and Counter-Adaptation

It is true that, since the advent of agriculture, societies have increasingly intervened in their environment to submit it to their strategies for boosting energy, production, and consumption. But this prehistoric and historical intervention until the nineteenth century has no common ground with the contemporary crises that threaten our global society. This is because there is a limit to the habitats' adaptation to humans. Beyond this limit, it becomes counterproductive; in short, it becomes something that could be called a counter-adaptation, since the resulting habitat will probably be more unfavorable to humans than the previous one. When humans exceed this limit, when their adaptive interventions destroy biodiversity, alter the chemical equilibrium of the atmosphere, soil, and water, polluting them, these interventions tend to trigger disruption mechanisms in the biosphere that lead it to other equilibrium points which will, most likely, be more hostile to humans. This leads to an even greater effort to manipulate the environmental coordinates, an effort that, in turn, leads nature to re-equilibrium points that are, most likely, even more problematic for humans (and for other species). This spiraling dynamic ends up creating a bigger transformation in the relationship between man and his planetary habitat: the relationship between man and nature ceases being reciprocally adaptive to become a reciprocally destructive interaction. In essence, the transition from reciprocally adaptive dynamics to reciprocally destructive ones had already been understood by Friedrich Engels, who expresses it in the image of an illusory victory (as cited by Magdoff and Bellamy Foster 2011):

> Let us not, however, flatter ourselves overmuch on account of our human conquest over nature. For each such conquest takes its revenge on us. Each of them, it is true, has in the first place the consequences on which we counted, but in the second and third places it has quite different, unforeseen effects which only too often cancel out the first.

Finally, the contemporary observer will say: there will reach a point in which the illusion of maximum hegemony of human technology over nature is gradually, or catastrophically, converted into its opposite, that is, into a maximum hegemony of nature over man (or without man).

15.2.2 Counterproductivity: a Defining Feature of Our Time

This dynamic, similar to that of a sorcerer's apprentice, this destructive interaction between humans and nature, has been emerging more clearly in the last half century in the form of a general historical constant: beyond a certain degree of interference of economic activity in ecosystem balances, the more man tries to submit nature to his law, the more she submits him to her law. This effect may be defined as that of a backfire from a gun with too large a caliber, that is, a reverse effect, one that is negative and often completely contrary to that which was expected. Ivan Illich (2003) formulated this phenomenon well: "When an activity powered by instruments exceeds a limit defined by the *ad hoc* scale, it immediately returns against its goal, and then threatens to destroy the entire social body." Jean-Pierre Dupuy (2002), who is one of those who introduced Illich's thinking in France, assimilates this "Illichian" concept of "counterproductivity" in his reflection on the catastrophe, but it is Michel Serres (1990/1992, p. 61) who expresses it perfectly: "through our mastery, we have become so much and so little masters of the Earth that it once again threatens to master us in turn." In fact, the image that man has of his action in the world begins to change and even to reverse itself. He begins to realize that the more he imagines himself to be master of nature, the more the image of his disastrous ability to exhaust it, disorganize it, and turn it against himself is revealed to him. In its most general expression, this phenomenon has already been extensively discussed in Chap. 14: the more global capitalism produces surplus, the more the destruction of the environment reveals the impossibility of improving, through this path, the quality of life. This law was confirmed by a study coordinated by Ida Kubiszewski and colleagues (2013), as well as by Robert Costanza and colleagues (2014). These authors compared the gross domestic product (GDP) per capita and the genuine progress indicator (GPI) per capita of 17 countries comprising just over half the global population. GPI is an index capable of quantifying and integrating several variables of human well-being (Talberth et al. 2007). The measure of both indicators "were highly correlated from 1950 until about 1978, when they moved apart as environmental and social costs began to outweigh the benefits of increased GDP." The divorce between these two curves after 1978 is well evidenced by Fig. 15.1.

The examples of "counterproductivity" accumulate, taking different forms. The more industrial man boasts of his technological capacity to increase surplus, the more garbage he produces, and the more his inability to clean himself of his own waste (like a baby) is revealed. As seen in Chap. 4 (4.1. Sewage and Municipal Solid Waste, see Fig. 4.1), if the current trend is maintained, production of waste and intoxication (caused by industries) of humans and of the biosphere will increase incessantly throughout the twenty-first century (Hoornweg et al. 2013; Stromberg 2013), with waste generated per capita per day ranging currently between 0.11 kg and 4.54 kg (World Bank).

The more plastic industrial man throws into the ocean, the greater the likelihood that particles of these polymers will accumulate in the food chain, and he will end up intoxicating himself, as seen in Chap. 11 (Sect. 11.5 Plastic Kills and Travels

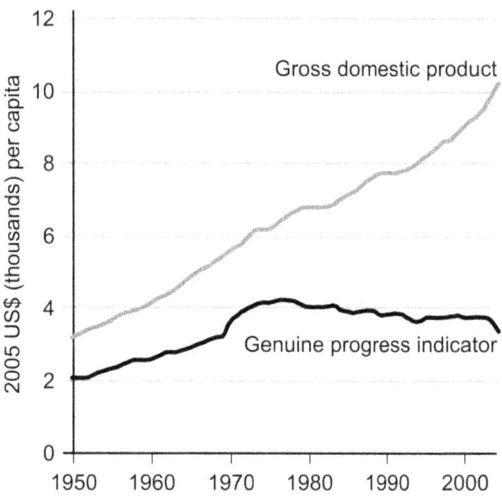

Fig. 15.1 Evolution of GDP and GPI in 17 countries between 1950 and 2013 (per capita in thousands of dollars in 2005). Based on Ida Kubiszewski et al. "Beyond GDP: Measuring and Achieving Global Genuine Progress." *Ecological Economics*, 93, 2013, pp. 57–68 and Robert Costanza et al., "Time to Leave GDP Behind." *Nature*, 7483, 505, 16/I/2014, pp. 283–285

through the Food Web) and described in an increasing number of scientific papers (Van Cauwenberghe and Janssen 2018; Schwabl et al. 2018). The more industrial man "enriches" the soil with industrial fertilizers, the more it becomes impoverished, and the more it also impoverishes maritime biodiversity and the sea's potential to feed humans. The more industrial man believes he must defend his agriculture with "pesticides," the more offensive the attacked species become, and the more brutal the doses of these pesticides must become, or the less abundant become pollinators, who, due to their irreplaceable pollination service, could increase harvests. This is demonstrated by a study published by Lucas Garibaldi and colleagues (2013), whose results are thus presented by Lawrence D. Harder, one of its authors[11]: "Paradoxically, most common approaches to increase agricultural efficiency, such as cultivation of all available land and the use of pesticides, reduce the abundance and variety of wild insects that could increase production of these crops." Between 1965 and 2010, global per capita energy consumption increased by over 50%, that is, from less than 50 gigajoules to around 75 gigajoules (Tverberg 2012). This immense increase in less than half a century had an opposite effect than what was intended, because the more the contemporary man yearns to hold (or believes he is about to hold) the key that will give him access to nature's almost infinite energy resources, the more he is threatened by energy scarcity or by the destructive effects of its abundance. The more the ways of accessing primary energy become sophisticated, the more energy is needed to obtain the same amount of energy and to try to "manage" the disorder in nature and in society caused by the extreme processes of obtaining and spending this energy.

---

[11]As cited in *Red Orbit*. http://www.redorbit.com/news/science/1112794650/wild-bee-loss-affecting-crop-pollination-030113/

15.3 Human Health Backfires

The more man industrializes his food and the more he turns it into processed food and fast food, the less nutritious and less healthy it becomes. On the carcinogenicity of the consumption of red meat and processed meat,[12] the World Health Organization (2015) states: "Red meat was classified as Group 2A, probably carcinogenic to humans" and "processed meat was classified as Group 1, carcinogenic to humans (see also Chap. 12, Sect. 12.4 Hypobiosphere)". The more agribusiness makes use of genetic "improvement" to manipulate agricultural products, introducing genes from other species, the more it exposes the population to imponderable risks, with no real advantages in relation to organic agriculture.

The more the contemporary man boasts of penetrating more deeply the laws of life's behavior, the more this behavior is, ultimately, hostile to him. This is affirmed in the *Global Burden of Disease Report* (GBD 2010), a survey involving almost 500 scientists in 50 countries and considered by its publisher Richard Horton (2012), editor of *The Lancet*, "the most comprehensive assessment of human health in the history of medicine." The conclusions of this research, which compared data from 2010 with data from 1970 and 1990, were summarized as follows: "we now have a grip on some common infectious diseases, which has saved millions of children from early deaths. Collectively, however, we are spending more of our lives living in poor health and with disability." (Hamzelou 2012). Another statement of the researchers is also well formulated[13]: "Globally, health advances present most people with a devastating irony: avoid premature death but live longer and sicker."

The most notable example of this, the research points out, is the epidemic increase in noninfectious and chronic diseases, such as obesity and obesity-related diseases. In addition to these paradoxical effects, we might add the proliferation of musculoskeletal, respiratory, allergic, and autoimmune diseases. As pointed out by Douglas Kerr (2007), "the prevalence of autoimmune diseases like systemic lupus erythematosus or lupus, multiple sclerosis, and type 1 diabetes is on the rise." With regard to syndromes involving allergies and respiratory problems, due at least in part to the increase in air pollution, a work published by Ruby Pawankar in the *World Allergy Organization Journal* (2014) shows an epidemic picture:

> The prevalence of allergic diseases and asthma are increasing worldwide, particularly in low and middle income countries. Moreover, the complexity and severity of allergic diseases, including asthma, continue to increase especially in children and young adults, who are bearing the greatest burden of these trends. (…) Allergic diseases include life-threatening anaphylaxis, food allergies, certain forms of asthma, rhinitis, conjunctivitis, angioedema, urticaria, eczema, eosinophilic disorders, including eosinophilic esophagitis, and drug and

---

[12] According to the World Health Organization, red meat refers to all mammalian muscle meat, including beef, veal, pork, lamb, mutton, horse, and goat. Examples of processed meat include hot dogs (frankfurters), ham, sausages, corned beef, and biltong or beef jerky as well as canned meat and meat-based preparations and sauces.

[13] Cf. "Global Burden of Disease: Massive shifts reshape the health landscape worldwide." Institute of Health Metrics and Evaluation (IHME), University of Washington, Washington D.C., 2010.

insect allergies. (…) For instance, asthma prevalence is rising in several high as well as low and middle income countries and the prevalence and impact of allergic diseases continue to grow. (…) According to the World Health Organization, the number of patients having asthma is 300 million and with the rising trends it is expected to increase to 400 million, by 2025.

According to *The State of Food Security and Nutrition in the World 2017*, from FAO, "childhood overweight and obesity are increasing in most regions, and in all regions for adults. In 2016, 41 million children under 5 years of age were overweight." The USA is becoming a nation of obese people, as shown by a large survey conducted by Ward J. Zachary and colleagues (2019):

The findings from our approach suggest with high predictive accuracy that by 2030 nearly 1 in 2 adults will have obesity (48.9%; 95% confidence interval [CI], 47.7 to 50.1), and the prevalence will be higher than 50% in 29 states and not below 35% in any state. Nearly 1 in 4 adults is projected to have severe obesity by 2030 (24.2%; 95% CI, 22.9 to 25.5), and the prevalence will be higher than 25% in 25 states. We predict that, nationally, severe obesity is likely to become the most common body-mass index (BMI) category among women (27.6%; 95% CI, 26.1 to 29.2), non-Hispanic black adults (31.7%; 95% CI, 29.9 to 33.4), and low-income adults (31.7%; 95% CI, 30.2 to 33.2).

While this trend is particularly acute in the USA, it is widespread. In fact, for the first time in human history and on a global scale, overweightness, obesity, and severe obesity represent a greater public health problem than hunger and chronic malnutrition, problems that are, moreover, again on the rise in the 2016–2018 triennium. Noninfectious diseases include heart disease, high blood pressure, and diabetes. Between 1980 and 2014, in a time span of just 35 years, the number of diabetic adults almost quadrupled, from 108 million to 422 million, that is, from 4.8% to 8.5% of the world's adult population, according to the WHO's *Global Report on Diabetes* (2016). In 2030, diabetes will be the seventh leading cause of mortality in the world. Until recently, type 2 diabetes used to occur nearly entirely among adults, but now it occurs in children too. There is a clear negative return effect also with the use of noncaloric artificial sweeteners, which are increasingly consumed on a global scale. Alison Abbot (2014) warns of the fact that "the artificial sweeteners that are widely seen as a way to combat obesity and diabetes could, in part, be contributing to the global epidemic of these conditions." Abbot mentions a paper by Jotham Suez and colleagues (2014) whose "results link noncaloric artificial sweeteners (NAS) consumption, dysbiosis[14] and metabolic abnormalities." The authors call for a reassessment of massive NAS usage. More recently, the review published by Adekunle Sanyaolu and colleagues (2018) comes to the same conclusions regarding the five artificial sweeteners most commonly used today (aspartame, saccharin, acesulfame potassium, neotame, and sucralose): "despite the widespread consumption of artificial sweeteners by lean, overweight and obese individuals alike, obesity and diabetes continue to dramatically rise." Aspartame, for instance, "has been linked to the

---

[14] Dysbiosis is a term for a microbial imbalance or maladaptation on or inside the body, such as an impaired microbiota.

exacerbation of diabetes, headache, seizures, depression, arthritis and other medical conditions."

Let us also remember that in Chap. 7 (Sect. 7.4 Suffering and Greater Lethality Due to the Current Warming), we mentioned the increasing burden of insect-borne, tick-borne, and fungi-borne (serious and sometimes lethal) diseases which are increasing at low latitudes and spreading toward higher latitudes, thanks to milder winters and longer and warmer summers. In fact, "the geographical distributions of the arboviruses dengue, yellow fever, chikungunya and Zika have expanded, causing severe disease outbreaks in many urban populations" (Kraemer et al. 2019). According to these authors, the estimated rate of spread of *Ae. aegypti* across the USA and of *Ae. albopictus* across Europe occurred at least 60 km per year since 1990. Both *Ae. aegypti* and *Ae. albopictus* will continue to expand beyond their current distributions, reaching as far north as Chicago, Germany, and Shanghai by 2050.

What follows is a summary of five other cases, undoubtedly the most serious ones, illustrating threats and damages to human health resulting from the ecological presumption of anthropocentrism. Such aggravations occur well above the rate of demographic growth and improvement in diagnosis.

(1) Antimicrobial Resistance and "The End of Modern Medicine"

Antimicrobial resistance (AMR), which started in the 1970s among Gram-negative bacteria (Aghapour et al. 2019), is a didactic example of global counterproductivity. The Global Burden of Disease Study 2017 (WHO 2017) once again drew attention to increased mortality from diseases and disorders linked to antibiotic resistance: "Since 2007, there have been rapid increases in emerging diseases and disorders due to antibiotic use or resistance." In May 2016, Jim O'Neill, coordinator of a comprehensive report on the global growth of bacterial resistance to antibiotics, estimated that:

> [...] without policies to stop the worrying spread of AMR (Antimicrobial resistance), today's already large 700,000 deaths every year would become an extremely disturbing ten million every year, more people than currently die from cancer. Indeed, even at the current rates, it is fair to assume that over one million people will have died from AMR since I started this Review in the summer of 2014.

This resistance grows all the more rapidly the more the pharmaceutical industry induces doctors and veterinarians to routinely administer antibiotics to humans and farm animals (in the latter case preventively). Sally Davies, England's Chief Medical Officer (the highest public office in health policy-making in that country), warns that the ongoing generalization of this resistance will mean "the end of the modern medicine."[15] For her, one in three prescriptions for antibiotics in her country is probably unnecessary.

But the even bigger problem is the prescription of antibiotics to animals raised for our carnivorous diet. These animals receive daily doses of antibiotics to stimulate

---

[15] "Chief medical officer warns antibiotic resistance could signal 'end of modern medicine'". *The Pharmaceutical Journal*, 17/X/2017.

growth and disease prevention, that is, to reduce the high risks of infection typical of confinement and unhealthy conditions to which agribusiness subjects them to. It is true that a new European legislation is restricting this use, and recent data from the European Medicines Agency (European Surveillance of Veterinary Antimicrobial Consumption, 2018) shows that sales of antibiotics for use in food-producing animals decreased by more than 20% across the European Union between 2011 and 2016. In the USA, sales of antibiotics for use in animals increased 23% between 2009 and 2014.[16] Although they have also declined significantly in recent years, as shown by an FDA report, such sales remain very high: "livestock sales of antibiotics in 2018 still represent almost two-thirds [64%] of sales of these drugs, or nearly 12.3 million pounds for animal use vs. approximately 7 million pounds for human use," and 92% of these drugs are distributed in feed or water to animals that are not necessarily sick (Kar and Wallinga 2018). Thomas P. Van Boeckel and colleagues (2015) show that[17]:

> These practices contribute to the spread of drug-resistant pathogens in both livestock and humans, posing a significant public health threat. We present the first global map (228 countries) of antibiotic consumption in livestock and conservatively estimate the total consumption in 2010 at 63,151 tons. We project that antimicrobial consumption will rise by 67% by 2030, and nearly double in Brazil, Russia, India, China, and South Africa. This rise is likely to be driven by the growth in consumer demand for livestock products in middle-income countries and a shift to large-scale farms where antimicrobials are used routinely.

The estimated application of antibiotics in animal production in China accounted for about 84,240 tons in 2013 (Yang et al. 2019). Among antibiotics, one of the most used is colistin, an antibiotic for treatment of most multidrug-resistant Gram-negative bacteria. In 2016, the Chinese government banned colistin for animal use after the MCR-1 gene, which confers resistance to colistin, was detected in food samples. But from October 2015 to October 2016, China exported nearly 100 tons of colistin-premixed animal feed, supplements and additives to India alone (Khullar 2019). Introduced in 1959, colistin was sparsely used until a few years ago because of its toxicity to the kidneys, which explains the low rate of bacterial resistance that developed against it. Because it is cheap and because it is an alternative to other antibiotics already ineffective against various bacterial infections, it started to be used massively in animals. The result is that colistin-resistant bacteria are now developing not only in China but also in Europe. As Maryn MacKenna (2015) states:

> The rapid dissemination of previous resistance mechanisms (e.g., NDM-1) indicates that, with the advent of transmissible colistin resistance, progression of *Enterobacteriaceae* from extensive drug resistance to pan-drug resistance is inevitable and will ultimately become global.

The issue is not restricted to resistance to colistin. Sara Reardon (2015) describes the recent increase in bacterial resistance to five classes of antibiotics. The title of

---

[16] Cf. Danielle Nierenberg, Food Tank https://mail.google.com/mail/u/0/#inbox/1464dada2ffb475a
[17] See also Mackenzie (2015).

her article, "Spread of antibiotic-resistance gene does not spell bacterial apocalypse—yet," gives a clear indication of what is to come. Reardon, in fact, warns:

> It is only a matter of time, however, before some kinds of infections may not be treatable
> with any of our current antibiotics. The U.S. Food and Drug Administration has approved
> about half a dozen new antibiotics in the past two years, and about 30 more are in the pipe-
> line. But most are similar to existing drugs and may not work any better. The most recently
> discovered class of antibiotics, lipopeptides, was identified in the late 1980s.

An example of the selection of super-resistant bacteria is seen in the so-called
carbapenem-resistant Enterobacteriaceae (CRE), a family of more than 70 bacteria
that have progressively acquired resistance to one of these five classes of antibiotics,
the carbapenems. According to Thomas Frieden, director of the Centers for Disease
Control and Prevention (CDC), these bacteria pose a triple threat: they are resistant
to virtually all antibiotics, even the most potent; they kill half of the patients if they
cause blood infections; and they can transfer their antibiotic resistance to other bacteria in the same family, making them potentially untreatable as well. Despite two
dozen new antibiotics launched since 2000 and a global market of 40 billion dollars,
the WHO data shows that "in the past four years, more than 1,200 cases have been
registered in French hospitals. A simple *Escherichia coli* or *Staphylococcus aureus*
can kill: 25,000 people a year succumb to these infections in Europe and 23,000 in
the United States" (Hecketsweiler 2014).

In 2015, the WHO launched the Global Antimicrobial Surveillance System
(GLASS), whose first report, released in January 2018, shows the occurrence of
antibiotic resistance in more than 500,000 people suspected of bacterial infections
in 22 countries with available data, especially *Escherichia coli*, *Klebsiella pneumoniae*, *Staphylococcus aureus*, *Streptococcus pneumoniae*, and *Salmonella*.[18]

Tuberculosis

Tuberculosis (TB) is one of the top 10 causes of death worldwide and remains the
leading killer from a single infectious agent, with ten million people falling ill with
TB in 2018 (see WHO, *Global Tuberculosis Report* 2019). An upsurge in new
strains of tuberculosis is observed by multidrug-resistant tuberculosis or MDR-TB,
which are resistant to treatment with at least two of the most powerful first-line anti-
TB medications, isoniazid and rifampin. In 2018, there were about half a million
new cases of rifampicin-resistant TB (of which 78% had multidrug-resistant TB).
Globally, 3.4% of new TB cases and 18% of previously treated cases were multidrug-
resistant TB or rifampicin-resistant TB (MDR/RR-TB). In extreme cases, some
forms of TB, called extensively drug-resistant tuberculosis or XDR-TB, are also
resistant to second-line antibiotics (at least one of the fluoroquinolones and one of
the injectable agents used in MDR-TB treatment regimens). The average proportion
of MDR-TB new cases with XDR-TB was 6.2% in 2018. Globally, 11,403 patients
with XDR-TB were enrolled in treatment in 78 countries and territories, a 16%

---

[18]Cf. WHO, "High levels of antibiotic resistance found worldwide, new data shows." 29/I/2018;
"WHO releases its first report on global antibiotic resistance." Center for Infectious Disease
Research and Policy (CIDRAP), University of Minnesota, 29/I/2018.

increase compared to 2017. Globally, the latest available data (2018) show a treatment success rate of 85% for drug-susceptible TB, 56% for MDR-TB, and only 39% for XDR-TB. According to a report published by the All-Party Parliamentary Group on Global Tuberculosis of the UK Parliament (2014), in a scenario where an additional 40% of all cases of TB are resistant to first-line drugs, leading to a doubling of the infection rate, 75 million additional people will lose their lives over the next 35 years (2015–2050) as a result of tuberculosis and MDR-TB. This is a figure equivalent to the number of fatal victims of the 1918–1919 H1N1 virus pandemic, commonly known as the Spanish flu.

(2) **The Increasing Cancer Burden**

Globally, 1 in 6 deaths is due to cancer, and the cancer burden continues to grow. The *WHO Report on Cancer 2020* states:

> In 2018, 18.1 million people around the world had cancer, and 9.6 million died from the disease. By 2040, those figures will nearly double [29.4 million people], with the greatest increase in Low and Middle Income Countries (LMIC), where more than two thirds of the world's cancers will occur. (…) In absolute terms, the burden is highest in countries with high and very high Human Development Index (HDIs), although, in relative terms, the increases will be proportionally greater in countries with low and medium HDIs.

According to the International Agency for Research on Cancer (IARC/WHO), *World Cancer Report 2014* (Stewart and Wild 2014), the incidence of cancer had increased from 12.7 million in 2008 to 14.1 million in 2012. In 2018, it reached 18.1 new cases. Incident cases of cancer all ages increased dramatically between 2008 and 2018 (+44%). In 2040, they are expected to increase by 133%, compared to 2008. The previous IARC/WHO *World Cancer Report 2003* (Stewart and Kleihues 2003) estimated that cancer rates were set to increase at an alarming rate, from ten million new cases globally in 2000 to 15 million in 2020. But in 2012, there were already 12.6 million new cases of cancer (Stewart and Wild 2014), and in 2018, 18.1 million new cases. A study published in 2012 in *Lancet Oncology*, coordinated by Freddie Bray, from the International Agency for Research on Cancer (IARC), stated that "the incidence of cancer is expected to increase by more than 75% by the year 2030 in developed countries, and over 90% in developing nations" (as cited by Rattue 2012).

Environmental Causes of Cancer   The increase in the incidence of several types of cancer is caused, at least in part, by lifestyles that are typical in the contemporary world (a sedentary lifestyle, eating processed foods, etc.) and by the increasing human exposure to carcinogenic substances created by the chemical industry. "It has been estimated that about 40% of risk factors are attributed to environmental and lifestyle conditions which can be preventable in both China or in other developed countries" (Feng et al. 2019). About 1000 chemical compounds are classified as endocrine-disrupting chemicals (EDCs) due to their ability to mimic or block the action of natural hormones. The International Agency for Research on Cancer classifies environmental factors into five hazard groups. As pointed out by Catherine Vincent (2012),

Since 1971, more than 900 chemical, physical or biological agents have been classified as such, of which more than 100 have been identified as *carcinogenic* (group 1), and more than 300 as *probably carcinogenic,* or *possibly carcinogenic* to humans (groups 2A and 2B).

In the USA, under the 1976 Toxic Substances Control Act (TSCA), the Environmental Protection Agency (EPA) had inventoried approximately 62,000 industrial chemicals between 1978 and 1982. This law required that the industry submit to the EPA every new chemical using a process called Premanufacture Notification (PMN). From 1982 to 2012, the agency added 22,000 new chemicals to this inventory, which now contains around 84,000 chemical substances that may possibly be commercialized.[19] Organisms, societies, and the planetary biota, in general, are immersed today in an industrial chemical soup of many tens of thousands of industrial chemical compounds whose direct effects are unknown, not to mention the effects of the combined interactions between them. The chemical industry is one of the most oligopolized industrial sectors of global capitalism and, in reality, one of the most damaging to living organisms. Its power to remain above any health control is reflected in UNEP estimates: the global volume of sales of chemical substances has increased from US$ 171 billion in 1975 to US$ 4.1 trillion in 2013. As seen in Chap. 4, many of these compounds are known to be cancerogenic: glyphosate, bisphenol-A (BPA), phthalates, persistent organic pollutants (POPs), and volatile organic compounds (VOCs), such as Mesitylene or 1,3,5-trimethylbenzene (a derivative of benzene) or xylene and aliphatic compounds, linked to the exploration of shale gas and oil and the burning of electronic waste. Furthermore, cancer is also caused by the irradiation of UV-B and UV-C rays, especially in the southernmost parts of the planet where the ozone's protective layer (in the stratosphere) has been destroyed by ozone-depleting chemicals (ODCs), especially chlorofluorocarbons. The responsibility of industrially processed meat in increasing mortality from heart disease and cancer has also been verified. This is categorically stated by Sabine Rohrmann (who led a team of 47 researchers) in a paper published on March 7, 2013, in *BMC Medicine*: "The results of our analysis support a moderate positive association between processed meat consumption and mortality, in particular due to cardiovascular diseases, but also to cancer."

Lung Cancer and China Globally, lung cancer leads the incidence of new cancer cases in 2018, with 2,093,876 new cases (12.3%), followed very closely by breast cancer (2,088,849). In addition, lung cancer represents the majority of all cancer death modalities, accounting for 1,761,007 deaths in 2018 alone (18.6% of cancer deaths). Here, the responsibility of the cigarette industry has long been established, but there are also important contributors, including the auto industry. Despite the denial of corporations whose campaigns to discredit science date back to at least the

---

[19] See "Identifying and Reducing Environmental Health Risks of Chemicals in Our Society: Workshop Summary." (Roundtable on Environmental Health Services, Research and Medicine: Board on Population Health and Public Health Practice. Institute of Medicine. Washington (DC) National Academic Press, 2/X/2014.

1960s, intoxication by gases released when burning gasoline and diesel has been confirmed. On June 12, 2012, scientists from IARC/WHO issued a document presenting the unanimous conclusion that "exhaust gases from diesel engines are one of the causes of lung cancer."

But along with cigarettes and the combustion engine, many other toxic substances suspended in the atmosphere most likely have an effect; these explain why the upward curve of China's accelerated industrialization has led to a corresponding increase in lung cancer. "Cancer is the leading cause of death in China. (…) An estimated 4.3 million new cancer cases and 2.9 million new cancer deaths occurred in China in 2018" (Feng et al. 2019). In 2018, the most commonly diagnosed cancers in Chinese males was lung cancer (21.9% of total new cases and 26,4% of cancer-related deaths), with lung cancer being the second among women, after breast cancer (Feng et al. 2019; Chen 2016). In October 2011, the *China Daily* published statistics from the Beijing Institute for Cancer Research of the Peking University Cancer Hospital. In Beijing, instances of lung cancer rose 56% between 2000 and 2009. In 2010, about 105 people were diagnosed with cancer every day, with the disease responsible for one in four deaths in this city. "Pollution and unhealthy lifestyles are the primary causes for the high cancer rate," stated Li Pingping, co-author of the report, adding that the fast pace of life in the city, as well as excessive pressure, disrupt people's hormones and increases the cancer risk (Yingqi 2012).

(3) **Decline in Male Fertility**

Chemical pollution is the hypothesis—ever more recurrent, more plausible, and even more likely—of the etiology of various pathologies, dysfunctions, and neurobehavioral disorders that have been affecting our societies. One of these dysfunctions now being studied more is the decline in male fertility. Bisphenol-A (BPA) has been associated with reduced fetal testosterone, which increases the likelihood of cancer. It has also been linked to decreased fertility in people exposed to this substance in utero.

Since the 1990s, the first consistent reports have emerged showing that the quantity and quality of sperm had started to decline. A team of Danish researchers coordinated by Elisabeth Carlsen (1992) concluded that "there has been a genuine decline in semen quality over the past 50 years. As male fertility is to some extent correlated with sperm count the results may reflect an overall reduction in male fertility." Although the causes (which must be multiple) of this decline are not yet certain, the work converges toward endocrine disruptors, which are chemicals that, at certain doses, can interfere with hormonal systems. Regarding the decline in male fertility, Joëlle Le Moal, epidemiologist at the Institut national de Veille Sanitaire (InVS), France, says, for example, that "the hypothesis of endocrine disorders is strong, given the globally diffused chemicals in the environment to which the population is exposed by every possible route, whether through food or air" (Benkimoun 2012). More recently, a meta-analysis with data from 42,935 men, signed by eight scientists and coordinated by Hagai Levine (2017), confirmed the consistency of previous studies, established new and more reliable results, and further strengthened the environmental hypothesis:

> We report a significant overall decline in both sperm concentration [SC, 10^6/mL] and total sperm count (TSC) in samples collected between 1973 and 2011. Declines were significant only in studies from North America, Europe, Australia (and New Zealand), where they were most pronounced among men unselected by fertility. In this latter group, SC declined 52.4% (-1.4% per year) and TSC 59.3% (-1.6% per year) over the study period. (…) There was no sign of 'leveling off' of the decline, when analyses were restricted to studies with sample collection in 1996–2011. (…) While the current study is not designed to provide direct information on the causes of the observed declines, sperm count has been plausibly associated with multiple environmental and lifestyle influences, both prenatally and in adult life. In particular, endocrine disruption from chemical exposures or maternal smoking during critical windows of male reproductive development may play a role in prenatal life, while lifestyle changes and exposure to pesticides may play a role in adult life.

On the other hand, in data collected in South American, Asian, and African countries, there were no significant trends in total sperm count (TSC) and in sperm concentration (SC) of sperm in semen. This discrepancy can be explained for two reasons. The first, discussed by the authors, is a more incomplete and less temporally comprehensive knowledge of the data, given the much smaller and more recent number of studies on male infertility carried out in these countries.[20] The second reason, not mentioned in this study, seems, in my opinion, to play at least as important a role as the first: the lesser exposure of the populations of the so-called developing countries to these chemical compounds, whose impacts on fertility sometimes have already occurred in the prenatal care period and can only be observed many years later. In fact, in the 1970s–1990s, these societies were even less urbanized, their agriculture was less intensely subject to industrial fertilizers and pesticides, and, above all, their daily contact with chemicals used on a large scale in common products in industrialized countries was much smaller. If this second reason is correct, the rapid urbanization of these countries in the past 25 years and their more recent exposure as producers and/or consumers of these intoxicating products should be reflected in future declines in their fertility rates.

The decline in male fertility seen in industrialized countries has not (yet) condemned males to terminal scenarios, such as the one imagined in the novel *The Children of Men* by P. D. James (1992) and the homonymous film by Alfonso Cuarón (2006), in which they portray a society in 2027, immersed in the chaos generated by an absolute collapse in human fertility, among other factors. If we refer to the fertility thresholds defined by D. S. Guzick et al. (2001) and Patela et al. (2018), these countries are still a long way from a collective state of male infertility or subfertility. But sperm concentrations in the "Unselected Western" men had dropped by more than half (52.4%) in just under 40 years (from 99 million sperm per ml in 1973 to 47.1 million/ml in 2011); this is a rate of decline of 1.4% per year in this group. Therefore, in 2018, if this rate is maintained, this number should already be around

---

[20] Levine et al. (2017, p. 8): "As in prior analyses, we saw no significant declines for studies from South America, Asia and Africa, which may, in part be accounted for by limited statistical power and an absence of studies in unselected men from these countries prior to 1985."

42.6 million/mL. And in the next 40 years, it should drop again by half, that is, to 24 million/mL, placing these countries already very close to the male subfertility zone. Consequently, if this trend is maintained, male infertility (at least in industrialized countries) could stop being science fiction in the next half century.

The problem, however, does not lie in the future, but in the present. According to Hagai Levine and colleagues, in the review mentioned above, "the high proportion of men from western countries with concentration below 40 million/ml is particularly concerning given the evidence that sperm concentration below this threshold is associated with a decreased monthly probability of conception" (p. 9). In addition, this decline implies health problems not pertaining to reproduction. For example, reduced levels of sperm in the semen are predictors of several pathological conditions and are correlated with a higher incidence of cancer in the testicles and with genital malformations, such as cryptorchidism and hypospadias (see Chap. 4, Sect. 4.2 Plastic, the Throwaway Lifestyle), in addition to delayed puberty and lower total testosterone levels, problems whose incidence have also been increasing in percentage in the last decades.

(4) **Endocrine and Neurobehavioral Disorders**

A review published in 2017 in the *Global Pediatric Health* magazine based on observations of two cohorts of pregnant mothers and their children in China's Tongliang province since 2002 and 2005 concludes that children with less prenatal exposure to polycyclic aromatic hydrocarbons (PAHs) and to lead and mercury had benefits in their neuronal development (Kalia et al. 2017). There is nothing surprising about this result. The interference of industrial chemicals in the endocrine system, including fetal development, neurobehavioral changes, and decreased fertility, was denounced as early as 1996 by Theo Colborn, Dianne Dumanoski, and John Peterson Myers in a book that became famous at the time: *Our Stolen Future*. More recently, this interference has also been examined in the etiology of the so-called autism syndrome disorders (ASDs), first described by Leo Kanner in 1943. The editorial in the journal *Environmental Health Perspectives* (25/IV/ 2012), titled "A Research Strategy to Discover the Environmental Causes of Autism and Neurodevelopmental Disabilities" (Landrigan et al. 2012), notes that:

> Autism, attention deficit/hyperactivity disorder (ADHD), mental retardation, dyslexia, and other biologically based disorders of brain development affect between 400,000 and 600,000 of the four million children born in the United States each year. (…) Prospective studies (…) have linked autistic behaviors with prenatal exposures to the organophosphate insecticide chlorpyrifos and also with prenatal exposures to phthalates. Additional prospective studies have linked loss of cognition (IQ), dyslexia, and ADHD to lead, methylmercury, organophosphate insecticides, organo-chlorine insecticides, polychlorinated biphenyls, arsenic, manganese, polycyclic aromatic hydrocarbons, bisphenol A, brominated flame retardants, and perfluorinated compounds. Toxic chemicals likely cause injury to the developing human brain either through direct toxicity or inter-actions with the genome. An expert committee convened by the U.S. National Academy of Sciences (NAS) estimated that 3% of neuro-behavioral disorders are caused directly by toxic environmental exposures and that another 25% are caused by inter-actions between environmental factors, defined broadly, and inherited susceptibilities. (…) A major unanswered question is whether there are still undiscovered environmental causes of autism or other neuro-developmental

disorders (NDDs) among the thousands of chemicals currently in wide use in the United States. (...) The U.S. Environmental Protection Agency has identified 3000 "high production volume" (HPV)[21] chemicals that are in widest use and thus pose greatest potential for human exposure. These HPV chemicals are used today in millions of consumer products. Children and pregnant women are exposed extensively to them, and CDC surveys detect quantifiable levels of nearly 200 HPV chemicals in the bodies of virtually all Americans, including pregnant women.

Janie Shelton and colleagues (2014) showed that proximity (within 1.5 km) to organophosphates at some point during gestation was associated with a 60% increased risk of autism spectrum disorders (ASDs). "Children of mothers residing near pyrethroid insecticide applications just before conception or during third trimester were at greater risk for both ASD and developmental delay." Ondine S. von Ehrenstein and colleagues (2019) reached similar results:

> Findings suggest that an offspring's risk of autism spectrum disorder increases following prenatal exposure to ambient pesticides within 2000 m of their mother's residence during pregnancy, compared with offspring of women from the same agricultural region without such exposure. Infant exposure could further increase risks for autism spectrum disorder with comorbid intellectual disability.

As Karen Weintraub had already well understood in 2011, "shifting diagnoses and heightened awareness explain only part of the apparent rise in autism." The effort of the medical establishment to discard environmental factors in favor of the hypothesis of genetic factors in the genesis of this epidemic is well reflected in the fact that in the first decade of the twenty-first century, "the US federal government has spent about US$ 1 billion researching the genetics of autism and only about $40 million on studies of possible environmental factors" (Weintraub 2011).

That said, the autism prevalence curve in the USA is exponential and typical of the spread of an epidemic. In 1975, only one in five thousand children born in the USA suffered from autism. In 1995, according to data from the US Centers for Disease Control and Prevention (CDC), for every 500 children born in that country, one was a victim of this syndrome (including deep autism and the so-called Rett and Asperger syndromes). This proportion increased in 2000 to one child in 150 and in 2006 to one child in 110; in 2012 (from 2010 data), it was one child in 88 and in 2014 (from 2012 data), one child in 68! Figure 15.2 shows this progression.

In 2018, the Centers for Disease Control and Prevention (CDC) released its biennial update of autism's estimated prevalence among the US's children, based on an analysis of 2014 medical records and educational records of 8-year-old children from 11 monitoring sites across the country (Baio et al. 2018). The 2014 data guiding the 2018 assessment shows a 15% increase over 2012:

> For 2014, the overall prevalence of autism spectrum disorder (ASD) among the 11 Autism and Developmental Disabilities Monitoring (ADDM) sites was 16.8 per 1000 (one in 59) children aged 8 years. Overall ASD prevalence estimates varied among sites, from 13.1–29.3 per 1000 children aged 8 years. (...) Among the nine sites with sufficient data on

---

[21] High production volume or HPV are chemical substances produced or imported into the USA in quantities of 500 tons or more per year.

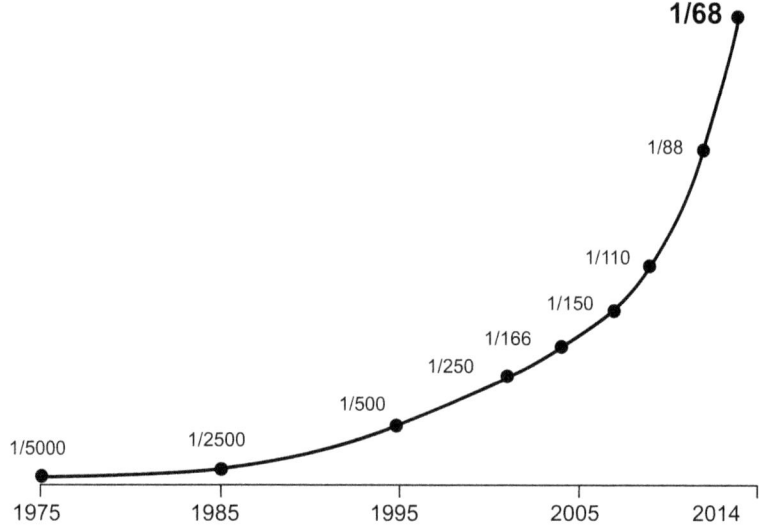

Fig. 15.2 Evolution of Diagnoses of Deep autism, Rett and Asperger Syndromes in the USA. Source: based on Barbara Demeneix, *Losing Our Minds. How Environmental Pollution Impairs Human Intelligence and Mental Health*, Oxford University Press, 2014, according to Karen Weintraub, "The Prevalence Puzzle: Autism Counts". *Nature*, 479, 2/XI/2011, pp. 22–24 and data from the US Centers for Disease Control and Prevention (CDC)

intellectual ability, 31% of children with ASD were classified in the range of intellectual disability (intelligence quotient [IQ] <70), 25% were in the borderline range (IQ 71–85), and 44% had IQ scores in the average to above average range (i.e., IQ >85).

According to Autism Speaks Chief Science Officer Thomas Frazier, "the new national prevalence estimate of 1 in 59 still reflects a significant undercount of autism's true prevalence among our children." In fact, a survey by Michael Kogan and colleagues (2018) most recently showed a prevalence of ASD in 1 out of 40 children assessed in the USA. However, such estimates are based upon parent reports and should not be compared to the clinical data offered by CDC biennial assessments.

In her book, *Losing Our Minds: How Environmental Pollution Impairs Human Intelligence and Mental Health* (2014), Barbara Demeneix, who studies the thyroid hormone signaling that allows the human brain to develop and function normally, writes that the incidence of autism spectrum disorders in the USA in 2014 occurred in one in 56 boys with a marked increase in incidence in the twenty-first century. The author goes to the heart of the ecological presumption and of counterproductivity (the most obvious consequence of this presumption), by warning that:

Failure to correct future injustices is also a failure to recognize that human intelligence and ingenuity have been responsible for the creation and production of a vast array of untested and potentially dangerous substances. It should be possible for human intelligence and ingenuity to find the means to control and eliminate their nefarious consequences. If not, then future generation could well find themselves lacking the intelligence and ingenuity ever to do so.

National surveys of children's health (NSCH) between 2003 and 2016 in the USA show, moreover, an almost equally alarming rate of attention deficit hyperactivity disorder (ADHD). The estimated number of US children and adolescents who have ever had a diagnosis of ADHD was 4.4 million in 2003. This number rose to 6.1 million in 2016 (9.4%). This number includes 388,000 children aged 2–5 years; four million children aged 6–11 years and three million children aged 12–17 years. According to a national 2016 parent survey, six in ten children with ADHD had at least one other mental, emotional, or behavioral disorder, as shown in Fig. 15.3.

(5) DSM-5 and Depressions

The assumption of the big pharmaceutical corporations that they can "handle" the electrochemistry of the brain generates many other negative effects, creating or exacerbating situations of dependence, depression, and suicide—deliberate or by overdose. In fact, the more the pharmaceutical industry induces medicine to use drugs acting on the central nervous system to treat even mild depressions, the greater the gap between research on brain circuitry and the diagnoses categorized in mental illness classification manuals. A petition signed by 14,000 mental health professionals calls for "a critical change" in the last edition (2012) of the *Diagnostic and Statistical Manual of Mental Disorders*, known as DSM-5, published by the American Psychiatric Association (APA). The perplexity is so widespread that "some critics argue that it's time to rip up the manual and start again – with wider input" (Aldhous 2012). As stated by Allen Frances, who chaired the task force that produced the DSM-4 (1994), there is clear pressure from the pharmaceutical industry to expand the spectrum of what are considered to be medicable mental illnesses: "DSM-5 opens up the possibility that millions and millions of people currently considered normal will be diagnosed as having a mental disorder and will receive medication and stigma that they don't need" (as cited by Brauser 2012). According to Marcia Angell, Professor at Harvard University, of the 170 experts who contrib-

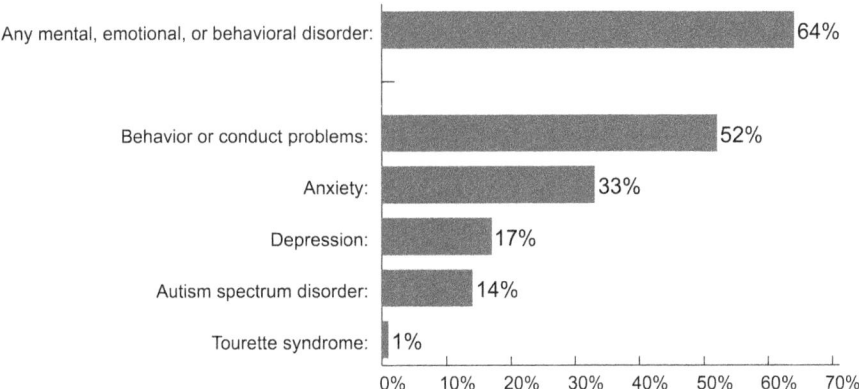

Fig. 15.3 Percentage of children and adolescents with ADHD and another disorder. (Source: Data and Statistics about ADHD. Centers for Disease Control and Prevention (CDC)). 15/X/2019 https://www.cdc.gov/ncbddd/adhd/data.html

uted to writing the fourth edition of the *Diagnostic and Statistical Manual of Mental Disorders* (DSM-4), published by the *American Psychiatric Association*, 95 had financial ties to pharmaceutical corporations. As for the authors of the chapters on schizophrenia and mood disorders, 100% of them had ties to this industry (Angell 2004).

Since the 1950s, with the emergence of the so-called tricyclic antidepressants and especially since prozac (1988), the first of a class of drugs known as selective serotonin reuptake inhibitors (SSRIs), the medicalization of depression has become one of the most striking phenomena of our day. The number of people enrolled as mentally ill in Supplemental Security Income (SSI) or Social Security Disability Insurance (SSDI) in the USA between 1987 (prozac's launch date) and 2007 increased two and a half times, going from one in 184 to one in 76. Among children, there was an increase of 35 times, so that mental illness became the main cause of disability in this age group (Angell 2011).

In seeking to eliminate depression, which would mean eradicating a fundamental dimension of the human condition, the pharmaceutical industry has only contributed to generalizing it. In fact, the question of depression has been omnipresent in the history of philosophical reflection since classical antiquity. The Hippocratic Corpus already addressed the issue of depression when dealing with "mania" or "melancholia" (black bile) and in the famous *Problem XXX,1*, Aristotle raises once again the question: "Why is it that all those men who have achieved exceptional things (*perittoi*), whether in philosophy, in politics, in poetry, or in the arts, are clearly melancholic?" (Pigeaud 1988). Depression (or what was then called melancholy, *Weltschmerz* or *spleen*), a typical temperament trait of the philosopher or artist, was an introspective affection of fruitful sadness in the mind or "soul," an object of recurring reflection and self-reflection since the advent of the Modern Age, found in famous masterpieces by Ficino, Dürer, Michelangelo, Robert Burton, Blake, Füssli, Jean Paul, Leopardi, Baudelaire, Dostoevsky, Freud, etc. (Wittkower 1963; Klibansky, Panofsky & Saxl 1964; Préaud 1982; Hersant 2005; Clair 2005).

Our days have made depressive syndromes almost always unfruitful and, in any case, typical of the "common man." Based on data from the Global Burden Disease Study 2017, the World Health Organization (2019) states that depression is a leading cause of disability worldwide, with more than 264 million people of all ages suffering from this disease (1 in 29 people). Between 1990 and 2007, the number of all-age years lived with disability (YLDs) attributed to depressive disorders increased by 33.4% (31% to 35.8%), becoming the third leading cause of all-age YLDs in 2007: "Three causes (low back pain, headache disorders, and depressive disorders) have prevailed as leading causes of non-fatal health loss for nearly three decades, while diabetes has emerged as the fourth leading cause of disability globally." Evelyn Bromet and colleagues (2011) presented data on the prevalence, impairment, and demographic correlates of depression from 18 high and low- to middle-income countries in the World Mental Health Survey Initiative. They stress the World Health Organization projection that by 2020 depression will be the second leading cause of disability worldwide.

As could be expected, the authors of this survey find that major depressive episodes are "strongly linked to social conditions." This epidemic is undoubtedly due to several factors, many of which are socioeconomic in nature. These factors include unemployment, stress at work, and insecurity regarding the maintenance of employment, resulting from the new normality of global capitalism: lower rates of economic growth, bankruptcy of the *Welfare State*, dismantling of the social achievements of the past, and the progressive obsolescence of human productive activity by automation. We cannot underestimate, on the other hand, factors that are psychic in nature, such as the growing existential malaise in consumer society or the impact, still little known, of social networks on the sociability of young people.

But two other additional factors must be remembered; the first is exposure to air pollution or particulate matter (PM). Isobel Braithwaite and colleagues (2019) published a systematic review and a meta-analysis on the emerging evidence supporting a possible etiological link between exposure to particulate matter and depression. According to the authors:

> Air pollution contains many individual pollutants, including particulate matter (PM), gaseous pollutants and metallic and organic compounds. In this systematic review, we have focused on PM—itself a complex, heterogeneous mixture—because it is responsible for the largest proportion of air pollution's physical health impacts and there is mounting evidence for tenable mechanisms by which it might affect risk of multiple mental health outcomes. Inflammation involving the central nervous system (CNS; neuroinflammation), has been implicated as having an important role in the pathophysiology of both depression and psychosis. Hypothalamo–pituitary–adrenal (HPA) axis dysregulation has also been implicated in depression. The plausibility of these as potential etiological pathways between PM exposure and such outcomes is backed up by evidence from animal and human studies into the pathophysiological effects of PM exposure. For example, PM exposure has been shown to be associated with markers of neuroinflammation such as glial activation and oxidative stress in humans and rodents; with changes in brain structure, as shown in humans and animals; and with increased stress hormone (cortisol) production. There is also emerging evidence to suggest that PM may adversely affect cognitive development, cognitive performance, and dementia risk as well as stress and psychological well-being. Taken together, these findings demonstrate the wide-ranging impacts of PM on brain health and functioning and may support the hypothesis of an association with clinically relevant mental health outcomes.

Finally, the second supplementary factor is the aforementioned over-medicalization, especially since the 1990s. This is a result of pharmaceutical corporations' commercial strategies, as well as epistemological impasses on criteria for categorization and diagnosis. For the pharmaceutical industry, this epidemic of depression (with the potential to become one of the largest in human history) has had very positive impacts on its sales volume over the past 30 years. Currently, around 40 antidepressants are available, and they are among the most commonly prescribed drugs in many Western countries. Between 2000 and 2015, prescriptions increased in all 29 countries surveyed by the OECD, on average doubling (Wilson 2018). Australia is the second largest consumer of antidepressants in the OECD, coming only behind Iceland which has a consumption rate of 106 doses per day for every thousand people. In the UK, the use of antidepressants increased by 234% in 10 years up to 2002 (Moncrieff 2004). Between 2006 and 2016, the number of medical prescriptions for antidepressants increased from 31 million to 64.7 million (Wilson 2018). In the USA, a 2005 study reports that in the realm of the non-institutionalized civilian population (i.e., civilians over 16 years of age not admitted to criminal or health institutions), 11% of women and 5% of men took antidepressants in 2002 (Stagnitti 2005). In 2011, 10% of Americans over the age of six took antidepressants. In Brazil, the sale of antidepressants grew 44.8% between 2006 and 2009. Brazil is part of a group of countries classified as "pharma-emerging" markets (Brazil, Russia, India, Korea, Mexico, and Turkey). Together, they account for 50% of the global growth of the drug market (Guimarães 2009). And the global outlook for this business is the best possible. Allied Market Research recently published a report, titled, *"Antidepressant Drugs Market by Depressive Disorder. (...) Global Opportunity Analysis and Industry Forecast, 2017–2023."* According to this report, the global antidepressant drugs market accounted for US$ 14.11 billion in 2017 and is expected to reach US$ 15.98 billion by 2023, registering a compound annual growth rate (CAGR) of 2.1% during the forecast period.

The consequences are well-known. According to Datasus data, between 1996 and 2012, there was a 705% increase in deaths related to depression in Brazil and a significant increase in the number of suicides (Cambrioli 2014). In the USA, more than nine million adults reported suicidal thoughts in 2013 and more than one million attempted suicide (Locklear 2016). David Healy, Professor of Psychiatry at Bangor University (Wales), writes in his blog: "Adverse drug events are now the fourth leading cause of death in hospitals. (…) In mental health for instance drug-induced problems are the leading cause of death."

In addition to their harms, Irving Kirsch, a psychologist at Harvard Medical School, demonstrated that antidepressants are little more than a formidable source of profit with no real benefit for most patients with mild depressions. He showed that people on antidepressants were only a little more likely to get better than those on placebos. "It is an extremely small effect size," he says, as cited by Clare Wilson (2018) who explains:

> "Depression is usually assessed using a questionnaire that gives a number on the Hamilton Depression Scale between 0 and 52, rising with severity. (…) Those who took the drugs showed an average reduction on the Hamilton scale that was only about two points greater than that of those taking the placebo tablets.

In addition to the aforementioned depressive syndromes and other neurobehavioral illnesses and disorders, there is a rise in the diagnosis of neuroses, obsessions, anxieties, posttraumatic stress disorder (PTSD), phobias, panic disorder, bipolar disorder, insomnia, occupational burnout, chronic fatigue syndrome (CFS), computer vision syndrome (CVS), etc. The phenomena of tolerance, dependence, and side effects of psychotropic drugs, including suicide and homicidal violence, are also aggravated in the same proportion. David Healy revealed that the GlaxoSmithKline (GSK) laboratory hid clinical trials proving that Deroxat, an antidepressant, was responsible for aggressive behavior in 25% of the patients tested. Michèle Rivasi (2015), an MEP, reports the suspected impact of agomelatine (a drug developed by Servier laboratories) on the behavior of Andreas Lubitz, the copilot who, in March 2015, caused a Germanwings plane to crash (the medical magazine *Prescrire* evaluated in January 2015 that this substance causes suicidal and aggressive behaviors). According to Marcia Angell (2008), mentioned above, antipsychotics such as Risperdal, Zyprexa, and Seroquel are now more sold in the USA than drugs to control cholesterol, which were, until recently, best sellers in that country's pharmacies.

In general terms, we have observed that the more contemporary man claims to be the only rational being in the biosphere, the more he reveals his psychic vulnerability and irrationality; also, the more his IQ decreases statistically (a phenomenon called the Flynn effect in reverse), his behavior becomes dominated by fanaticisms, elementary drives for territoriality, and flight or aggression in the face of the "unknown." All this occurs through endemic or epidemic manifestations in the twentieth and twenty-first centuries: social phobias, superstitious behavior, religious obscurantism, sectarian hatreds, racism, xenophobia, genocides, ethnic and religious "cleansings," militarism, scientific research focused on military technology, etc. This calls for a general conclusion on the value of biological and ecological assumptions of anthropocentrism: the more contemporary man reveres and takes pride in his uniqueness (in nature), the more the evidence accumulates that such a presumption results from the tautological operation of putting himself at the top of a scale of values defined based on himself.

15.4   The Great Mental Block

In 1992, two documents stimulated by ECO-92 and signed by fundamental segments of the scientific community collided: the *Heildelberg Appeal* and the *World Scientists' Warning to Humanity*. The first, organized by Michel Salomon, a French dermatologist, was published on June 1, 1992, in the *Wall Street Journal*. The document was signed by 46 eminent scientists and intellectuals and, subsequently, by another 4000, including 72 Nobel Prize winners. These supporters expressed, in its purest form, the ecological presumption of anthropocentrism discussed above:

> We want to make our full contribution to the preservation of our common heritage, the Earth. We are, however, worried at the dawn of the twenty-first century, at the emergence of an irrational ideology which is opposed to scientific and industrial progress and impedes economic and social development. We contend that a Natural State, sometimes idealized by movements with a tendency to look toward the past, does not exist and has probably never existed since man's first appearance in the biosphere, insofar as humanity has always progressed by increasingly harnessing Nature to its needs and not the reverse.

This appeal was confronted in November 1992 by another one titled, as stated, *World Scientists' Warning to Humanity*. Sponsored by MIT's Union of Concerned Scientists, it was written by Henry W. Kendall (1926–1999, Nobel Prize for Physics, 1990) and signed by 1700 scientists, including most winners of the Nobel Prize in various fields of science. His conception of the relationship between man and nature had nothing of the former document's anthropocentric presumption:

> Human beings and the natural world are on a collision course. Human activities inflict harsh and often irreversible damage on the environment and on critical resources. If not checked, many of our current practices put at serious risk the future that we wish for human society and the plant and animal kingdoms, and may so alter the living world that it will be unable to sustain life in the manner that we know. Fundamental changes are urgent if we are to avoid the collision our present course will bring about.

The curious thing is that 49 out of the 72 scientists who won the Nobel Prize and who signed this second document also signed the *Heidelberg Appeal*, without apparently paying attention to the fact that the theses of the two documents were mutually exclusive. It is possible that in 1992 this was not yet evident. The historical setback of almost 30 years and, above all, the worsening of the situation and the increasing evidence of the tendency toward environmental collapse make the antagonism between the two documents unequivocal today. No one disagrees with the constant claim in the *Heidelberg Appeal* that "humanity has always progressed by increasingly harnessing Nature to its needs and not the reverse." Such is precisely the historical basis on which its multimillennial ecological presumption, discussed in the previous item, is built. But the supporters of the first document still believe that science has the mission of allowing this constant of the human past to persist in the present and in the future. The supporters of the second document, on the other hand, no longer understand science according to the Baconian paradigm, that is, as a growing power, supposedly benign and unstoppable, of "harnessing" nature for human purposes. Because of their perception that "human beings and the natural world are on a collision course" and because of their condemnation of the socioeconomic system that keeps us on this course, the supporters of the second document are today at the forefront of critical thinking.

From the "Scientific Consensus" (2013) to "Climate Emergency" (2019)

It is tempting to hypothesize that this ecological presumption of anthropocentric science is beginning to be overcome, and an important sign of this are the various collective calls from the scientific community that have been occurring since 2013. Initiated by Anthony Barnosky and Elizabeth Hadly (see also 2015), both from Stanford University, the first call represents a very important step forward, since it is motivated by a new sense of urgency and by the perception of the need to definitively overcome the ecological presumption of anthropocentrism. This is the *Scientific Consensus on Maintaining Humanity's Life Support Systems in the twenty-first Century: Information for Policy Makers*. When presented to the governor of California in May 2013, it had already been signed by 522 leading scientists from 41 countries. In 2016, it had been signed by more than 1300 scientists and researchers in general but also by members of NGOs, students, and the general public in more than 60 countries. The "scientific consensus" calls on society to guide all economic activity and all political action by the urgent imperative to minimize the transformation of the Earth's remaining ecosystems:

Earth is rapidly approaching a tipping point. Human impacts are causing alarming levels of harm to our planet. As scientists who study the interaction of people with the rest of the biosphere using a wide range of approaches, we agree that the evidence that humans are damaging their ecological life-supporting systems is overwhelming. We further agree that, based on the best scientific information available, human quality of life will suffer substantial degradation by the year 2050, if we continue on our current path.

In 2017, taking stock of the 25 years since ECO-92 and the launch of the *World Scientists' Warning to Humanity*, William J. Ripple, Christopher Wolf, Mauro Galetti, Thomas M Newsome, Mohammed Alamgir, Eileen Crist, Mahmoud I Mahmoud, and William F. Laurance launched a manifesto, co-signed by 15,364 scientists and intellectuals from 184 countries, titled *"World Scientists' Warning to Humanity: A Second Notice."* This document, published in November 2017 in the journal *BioScience*, reiterates the warnings made by the 1992 document and cautions that:

Since 1992, with the exception of stabilizing the stratospheric ozone layer, humanity has failed to make sufficient progress in generally solving these foreseen environmental challenges, and alarmingly, most of them are getting far worse. Especially troubling is the current trajectory of potentially catastrophic climate change due to rising GHGs from burning fossil fuels, deforestation, and agricultural production— particularly from farming ruminants for meat consumption. Moreover, we have unleashed a mass extinction event, the sixth in roughly 540 million years, wherein many current life forms could be annihilated or at least committed to extinction by the end of this century.

On September 9, 2018, the newspaper *Libération* published the *SOS from 700 French Scientists*: we are already fully engaged in the "climate future"; this is the main message from these scientists. Finally, on November 5, 2019, *BioScience* published a document titled *World Scientists' Warning of a Climate Emergency* which was signed by more than 11 thousand scientists from around the world:

Scientists have a moral obligation to clearly warn humanity of any catastrophic threat and to "tell it like it is." On the basis of this obligation (…) we declare clearly and unequivocally that planet Earth is facing a climate emergency. (…) An immense increase of scale in endeavors to conserve our biosphere is needed to avoid untold suffering due to the climate crisis.

Distant from the old Baconian conception of science, these four documents (2013, 2017, 2018, and 2019) provide society and its leaders with the current state of the man–nature relationship; based on this, the meaning and the priorities of the economy and of politics should be redefined. As this is clearly not happening, some scientists are increasingly giving in to the idea that the response to the climate emergency must rest on the triad: mitigation, adaptation, and restoration. There are several strategies of climate restoration: enhancing phytoplankton growth, increasing the weathering of minerals, developing devices that remove CO_2 directly from the air, etc. The common goal of these different technologies is to decrease atmospheric CO_2 concentrations from the current 415 ppm to no more than 350 ppm. It is not a matter of discussing the pros and cons of geoengineering and bioengineering but of admitting that technologies for removing CO_2 from the atmosphere at the required scale simply do not exist yet and possibly never will. Dominic Lenzi and colleagues

(2018) show that, to keep global warming at 2 °C above the pre-industrial period, it would be necessary, in addition to radically reducing emissions, to artificially remove 12 Gt CO_2 per year from the atmosphere every year until 2100. To put it in context, this means removing more CO_2 than China emits today annually (10 Gt CO_2). It also means removing more CO_2 than the entire average land carbon uptake over the past 15 years (11 Gt CO_2 per year). It is, therefore, not surprising that an editorial in *Nature* (21/II/2018) is titled: "Why current negative-emissions strategies remain 'magical thinking.'" Tim Lenton thinks similarly: "This [geoengineering] is complete science fiction. We ought to stop talking about it" (as cited by Battersby 2013).

This way of understanding the relationship between man and nature can, in fact, lead to results similar to those of Bouvard and Pécuchet, the comical characters in Flaubert's last novel. It is opposed to the critical and self-critical spirit, essential to science, an unreasonable reaction that consists, in a word, of persisting in the illusion that we can "harness Nature," to remember the expression used in the *Heidelberg Appeal*. Attempts to interfere on a large scale in infinitely complex systems based on theoretical models and specific experimental practices—on the whole understood as scientific management—will, in all probability, end up adding to the anthropic pressure on the Earth system. In fact, everything begs us to suppose that no technology will be able to cool the Arctic and, in general, to "manage" the planet's climate, just as no technology will be able to penetrate and manipulate the fundamental mechanisms of life, of consciousness, of aesthetic emotion, of matter, and of the universe. Such mechanisms are still fundamentally inaccessible to knowledge and will remain so for a long time, perhaps forever. Nature is always more complex and stranger than we imagine it to be. As John Haldane (1928) had already warned, "my own suspicion is that the Universe is not only queerer than we suppose, but queerer than we *can* suppose."

In fact, the only geoengineering that is known, effective, and safe is to try to drastically reduce anthropic interference on ecosystems, hoping that they can still, in the medium and long term, return to the efficiency previously shaped by evolution. This means not only stopping deforestation but reforesting on a large scale (which has nothing to do with planting stocks of cellulose), abandoning fossil fuels, returning rivers to their normal flow, reversing the tendency toward carnivorism, allowing the oceans to replenish themselves, encouraging policies of demographic rationality, guaranteeing women's sexual and reproductive freedom, refusing the intoxicating monoculture of agribusiness that reduces food to commodities, opting for agriculture that is organic and local, and recognizing the insanity of consumerism and its inevitable counterpart: physical and mental asphyxiation by garbage. But to be feasible, this genre of geoengineering supposes abandoning the growth paradigm and the paradigm of the expanded reproduction of capital as the purposes of economic activity. It supposes a deepening of democracy. It supposes, above all, understanding that we can no longer think of the man–nature relationship in anthropocentric terms, as the scientists who subscribed to the *Heidelberg Appeal* still persisted in doing so in 1992.

15.5 Conclusion

The great mental block that we are victims to is the illusion that, as in the past, we will continue to "grow" in the future, since there would be no limits to our "manifest destiny," to our exceptionality in the web of life, and to nature's adaptation to ingenuity and to human demands for "more." To fuel this illusion, there are the spectacular rates of increase in production and consumption of energy per capita of the last two centuries, followed by lower mortality rates under the age of 5 (which never cease to regress), and increased access of different population groups to education, information, and medical assistance (benefits that are, of course, not extendable to "peripheral" peoples, extinct or degraded by the steamroller effect of capitalist expansion).

This growth and these improvements have essentially been, even for its beneficiaries, a kind of *short-term optimization* of the allocation of "resources and services provided" by nature. There has not yet been a generalized perception that these positive indicators, brought about by science and technology, were achieved at an excessive real cost when we account for the erosion of the biosphere, an erosion until recently overshadowed by the exciting spectacle of technical and scientific progress. And because this real cost, in addition to being excessive, is increasing, because the conquest was made at the expense of the planet's biological diversity and the stability of the Holocene's climate system, these positive short-term indicators are turning into their respective opposites. In summary, what appears little by little in the counterproductive effects mentioned above is that the unrestricted potentiation of energy, production, and consumption (which essentially characterizes the techno-scientific reason enthroned by capitalism) does not suppress or even diminish the adversities of nature that capitalism proclaims to be able to combat: scarcity, climatic rigors, disease, and human aggressiveness. Rather, it only progressively transforms these adversities *into worse forms* of scarcity, climatic rigors, diseases, and human aggressiveness.

From Johann Spies and Marlowe in the end of the sixteenth century to Goethe, Chamisso, Murnau, Thomas Mann, and Valéry in the 1940s, we never cease to return to Faust. It is not, of course, always the same Faust, because our conception of the meaning of his pact was inverted between Spies and Goethe, to again be inverted between Goethe and Thomas Mann. But these inversions only advanced the fact that in his desire to overcome limits, Faust synthesizes—for better or for worse—the modern and contemporary conception of man's destiny (Bianquis 1955). No one is allowed to predict the limits of that destiny, if there are limits. But the best science we have today warns us, at the top of its lungs, that if we do not learn to reconcile ourselves with the very notion of limits these limits will, very soon, be those at the bottom of an abyss, not in outer space.

References

ABBOT, Alison, "Sugar substitutes linked to obesity". *Nature*, 513, 290, 18/IX/2014, p. 290

AGHAPOUR, Zahra *et al.*, "Molecular mechanisms related to colistin resistance in Enterobacteriaceae". *Infection and drug resistance*, 12, 2019, pp. 965–975.

ALDHOUS, Peter, "Forget labels, target faulty wiring to help mental illness". *New Scientist*, 15/XII/2012.

ANGELL, Marcia, *The Truth About the Drug Companies: How They Deceive Us and What to Do About It*. London, Random House, 2004.

———, "Industry-Sponsored Clinical Research: A Broken System". *The Journal of the American Medical Association*, 300, 9, 2008, pp. 1069–1071.

———, "The Epidemic of Mental Illness. Why?" *The New York Review of Books*, 23/VI/2011.

ARENDT, Hanna, *The Human Condition*, Chicago University Press, 1958.

BAIO, Jon *et al.*, "Prevalence of Autism Spectrum Disorder Among Children Aged 8 Years — Autism and Developmental Disabilities Monitoring Network, 11 Sites, United States, 2014". Centers for Disease Control and Prevention, 27/IV/2018.

BARNOSKY, Anthony & Hadly, Elizabeth. *End Game: Tipping Point for Planet Earth?* London, HarperCollins, 2015.

BATTERSBY, Stephen, "Can geoengineering avert climate chaos?". *New Scientist*, 22/IX/2013.

BATTISTI, Eugenio. *Il Crocifisso di Cimabue in Santa Croce*. Milão, 1967.

BENKIMOUN, Paul, "Chute spectaculaire de la qualité du sperme". *Le Monde*, 6/XII/2012.

BIANQUIS, Geneviève. *Faust à travers quatre siècles*. Paris, Aubier Montaigne, 1955.

BRAITHWAITE, Isobel *et al.*, "Air Pollution (Particulate Matter) Exposure and Associations with Depression, Anxiety, Bipolar, Psychosis and Suicide Risk: A Systematic Review and Meta-Analysis". *Environmental Health Perspectives*, 127, 12, 18/XII/2019.

BRAUSER, Debora, "Experts React to DSM-5 Approval". *Medscape Medical News*, 3/XII/2012.

BROMET, E. *et al.*, "Cross-national epidemiology of DSM-IV major depressive episode". *BMC Medicine*, 9, 90, 2011.

CAMBRIOLI, Fabiana, "Mortes por depressão crescem 705%". *O Estado de São Paulo*, 17/VIII/2014.

CARLSEN, Elisabeth *et al.*, "Evidence for decreasing quality of semen during past 50 years". *BMJ*, 305, September, 1992.

CHEN, Wanquing, "Cancer statistics in China". *Cancer Journal of Clinicians*, 25/I/2016.

CLAIR, Jean, *Mélancolie. Génie et folie en Occident*, Paris, 2005.

CONGER, George Perrigo, *Theories of Macrocosms and Microcosms in the History of Philosophy*. New York, Columbia University Press, 1922.

COSTANZA, Robert *et al.* "Time to leave GDP behind". *Nature*, 7.483, 505, 16/I/2014, pp. 283–285.

DEMENEIX, Barbara, *Losing Our Minds. How Environmental Pollution Impairs Human Intelligence and Mental Health*, Oxford University Press, 2014.

DIELS, Hermann & KRANZ (ed.), *Die Fragmente de Vorsokratiker*; REALE, Giovanni, *I Presocratici*, Democritus, frag. B 34, Milan, Bompiani, 2006.

DUPUY, Jean Pierre. *Pour un catastrophisme éclairé. Quand l'impossible est certain*. Paris, Éditions du Seuil, 2002.

EHRENSTEIN, Ondine S. von *et al.*, "Prenatal and infant exposure to ambient pesticides and autisym spectrum disorder in children: population based case-control study". *BMJ*, 364, 2019.

FENG, Rui-Mei *et al.*, "Current cancer situation in China: good or bad news from the 2018 Global Cancer Statistics?" *Cancer Communications*, 39, 2019.

GARIBALDI, Lucas A. *et al.* "Wild Pollinators Enhance Fruit Set of Crops Regardless of Honey Bee Abundance". *Science*, 28/II/2013.

GEORGESCU-ROEGEN, Nicholas. *The Entropy Law and the Economic Process*. Harvard University Press, 1971.

GUIMARÃES, Lígia, "Venda de antidepressivos no Brasil cresce 44,8% em 4 anos, diz pesquisa". *Globo.com*, 26/XII/2009.

GUZICK, D. S. *et al.*, "Sperm morphology, motility, and concentration in fertile and infertile men". *New England Journal of Medicine*, 2001, pp. 1388–1393.

HALDANE, John Burdon. *Possible Worlds and Other Papers*. New York, Harper, 1928.

HAMZELOU, Jessica, "Global health report card". *New Scientist*, 22–29/XII/2012.

HECKETSWEILER, Chloé, "Le 'business model' cassé des antibiotiques". *Le Monde*, 18/XI/2014.

HERSANT, Yves, *Mélancolies. De l'Antiquité au XXe siècle*. Paris, Bouquins, 2005.

HOORNWEG, Daniel; BHADA-TATA, Perinaz & KENNEDY, Chris. "Waste production must peak this century". *Nature*, 502, 31/X/2013, pp. 615–617.

HORTON, Richard (ed.) *Global Burden of Disease Report* (GBD 2010), *The Lancet* 2012.

HUGHES, J. Donald, *Ecology in Ancient Civilizations*, University of New Mexico Press, 1975.

ILLICH, Ivan. *La convivialité. Oeuvres completes*, vol. I. Paris, 2003, pp. 451–580.

KALIA, Vrinda, PERERA, Frederica & TANG, Deliang. "Environmental Pollutants and Neurodevelopment: Review of Benefits From Closure of a Coal-Burning Power Plant in Tongliang, China". *Global Pediatric Health*, 4, 31/VII/2017.

KAR, Avinash & WALLINGA, David, "Livestock Antibiotic Sales See Big Drop, but Remain High". National Resources Defense Council (NRDC), 18/XII/2018.

KERR, Douglas, Preface to Donna Jackson Nakazawa, *The Autoimmune Epidemic*, Touchstone/Simon and Schuster, 2007.

R. KLIBANSKY, R., PANOFSKY, E., SAXL, F., *Saturn and Melancholy. Studies in the History of Natural Philosophy Religion and Art*, 1964.

KHULLAR, Bhavya, "Why colistin should be banned in food-animal production sector in India". *Down to Earth*, 9/VII/2019.

KOGAN, Michael D. *et al.* "The Prevalence of Parent-Reported Autism Spectrum Disorder Among US Children". *Pediatrics*, December 2018.

KRAEMER, Moritz U.G., REINER Jr., Robert C. & GOLDING, Nick, "Past and future spread of the arbovirus *Aedes aegypti* and *Aedes albopictus*". *Nature microbiology*, 4, 4/III/2019, pp. 854–863.

KRAYE, Jill, "Moral Philosophy", in Charles B. Schmitt, *The Cambridge History of Renaissance Philosophy*. Cambridge University Press, 1988, pp. 301–386.

KUBISZEWSKI, Ida *et al.* "Beyond GDP: Measuring and achieving global genuine progress". *Ecological Economics*, 93, 2013, pp. 57–68.

LANDRIGAN, Philip J.; LAMBERTINI, Luca & BIRNBAUM, Linda S. "A Research Strategy to Discover the Environmental Causes of Autism and Neurodevelopmental Disabilities" (Editorial). *Environmental Health Perspectives*, 25/IV/2012.

LENZI, Dominic *et al.*, "Don't deploy negative emissions technologies without ethical analysis". *Nature*, 19/IX/2018.

LEOPOLD, Aldo, *A Sand County Almanac*. Oxford University Press, 1949.

LEVINE, Hagai *et al.*, "Temporal trends in sperm count: a systematic review and meta-regression analysis". *Human Reproduction Update*, 2017, pp. 1–14.

LOCKLEAR, Mallory, "Drug quickly quells suicidal thoughts". *New Scientist*, 6/II/2016, p. 12.

LOVEJOY, Arthur O., *The Great Chain of Being. A Study of the History of an Idea* (1936). Harvard University Press, 1957.

MacKENNA, Maryn, "Apocalypse Pig: The Last Antibiotic Begins to Fail". *National Geographic*, 21/XI/2015.

MACKENZIE, Debora, "Pork chop... with side of superbugs". *New Scientist*, 28/III/2015.

MAGDOFF, Fred & BELLAMY FOSTER, John. *What Every Environmentalist Needs to Know About Capitalism*. New York, Monthly Review Press, 2011.

MALHERBE, Michel, *La philosophie de Francis Bacon*, Paris, Vrin, 2011.

MASON, F. S., "Scienza e religione nell'Inghilterra del XVII secolo". *In*: HILL, Christopher. *Saggi sulla rivoluzione inglese* (1957). Milan, Feltrinelli, 1971, pp. 283–303.

MONCRIEFF, Joanna, "Psychiatric drug promotion and the politics of neoliberalism" (Editorial). *British Journal of Psychiatry*, 188, 2004, pp. 301–302.

MOSCOVICI, Serge, *Essai de l'histoire humaine de la nature*. Paris, Flammarion, 1968.

O'NEILL, Jim (coord.), *Tackling Drug-Resistant Infections Globally: Final Report and Recommendations. The Review on Antimicrobial Resistance*. May 2016.

PATELA, Amir S. *et al.*, "Prediction of male infertility by the World Health Organization laboratory manual for assessment of semen analysis: A systematic review", *Arab Journal of Urology*, 16, 1/III/2018.

PAWANKAR, Ruby, "Allergic diseases and asthma: a global public health concern and a call to action". *World Allergy Organization Journal*, 7, 2014.

PICO DELLA MIRANDOLA, Giovanni, *Oratio de hominis dignitate*. English edition, Oration on the dignity of man, edited by Sebastian Michael, based on the translation by C. Glenn Wallis, London, 2018.

PIGEAUD, Jackie, *Aristote. L'Homme de génie et la mélancolie. Problème XXX,1*. Paris, Éditions Rivages, 1988.

PRÉAUD, Maxime, *Mélancolies*. Paris: Herscher, 1982.

PUCCI, Pietro, "Prométhée, de Hésiode à Platon". *Communications*, 78, 2005, pp. 51–70.

RATTUE, Grace, "Cancer Rates Expected To Increase 75% By 2030" *Medical New Today*, 1/VI/2012.

REARDON, Sara, "Spread of antibiotic-resistance gene does not spell bacterial apocalypse – yet". *Nature*, 21/XII/2015.

RIBEIRO DOS SANTOS, Leonel, *Regresso a Kant. Ética, estética, filosofia política*. Lisbon, 2012.

RIVASI, Michèle, "Ces scandaleux antidépresseurs". *Le Monde*, 5/VI/2015.

ROHRMANN, Sabine *et al.*, "Meat consumption and mortality. Results from the European Prospective Investigation into Cancer and Nutrition", *BMC Medicine*, 7/III/2013.

SANDBERG-VAVALÀ, Evelyn. *La croce dipinta italiana e l'iconografia della Passione*. Verona, 1929.

SANYAOLU, Adekunle *et al.*, "Artificial sweeteners and their association with Diabetes: A review". *Journal of Public Health Catalog*, I, 2018.

SCHWABL, Philipp *et al.*, "Assessment of microplastic concentrations in human stool". *United European Gastroenterology (UEG)*, 23/X/2018.

SERRES, Michel. *Le Contrat naturel*. Paris, Éditions François Bourin, 1990; Flammarion, 1992.

SHELTON, Janie *et al.*, "Neurodevelopmental Disorders and Prenatal Residential Proximity to Agricultural Pesticides: The CHARGE Study". *Environmental Health Perspective*, 122, 10, October 2014.

SILVESTRIS, Bernardus, *Cosmographie. De mundi universitate sive megacosmus et microcosmos*. French translation, introduction and notes by M. Lemoine, Paris, Cerf, 1998.

STAGNITTI, M. N., "Antidepressant Use in the U.S. Civilian Noninstitutionalized Population, 2002" *Medical Expenditure Panel Service*, 9/V/2005.

STEWART, Bernard W. & KLEIHUES, Paul (eds.), *World Cancer Report*, IARC/WHO, Lyon, 2003.

STEWART, Bernard W. & WILD, Christopher P. (eds.), *World Cancer Report*, IARC/WHO, 2014.

STROMBERG, Joseph, "When Will We Hit Peak Garbage?" *Smithsonian Institution*, Washington, 30/X/2013.

SUEZ, Jotham *et al.*, "Artificial sweeteners induce glucose intolerance by altering the gut microbiota". *Nature*, 17/IX/2014.

SULLY PRUDHOMME, "Anthropomorphisme et causes finales". *Revue scientifique*, 4 March 1899.

TALBERTH, John; COBB, Clifford & SLATTERY, Noah. "The Genuine Progress Indicator 2006. A Tool for Sustainable Development", February 2007.

TVERBERG, Gail, "World Energy Consumption Since 1820 in Charts". *Our finite world*, 12/III/2012.

VINCENT, Catherine, "Les gaz émis par les moteurs diesel reconnus comme cancérogènes". *Le Monde*, 13/VI/2012.

VAN BOECKEL, Thomas P. *et al.* "Global trends in antimicrobial use in food animals". *PNAS*, 112, 18, 5/V/2015.

VAN CAUWENBERGHE, L. & JANSSEN, C., "Microplastics in bivalves cultured for human consumption". *Environmental Pollution*, 193, 2018, pp. 65–70.

VAN WINDEN, J. M. C. *Calcidius On Matter. His Doctrines and Sources. A Chapter in the History of Platonism*. Leiden, Brill, 1965.

WEINTRAUB, Karen, "The prevalence puzzle: autism counts". *Nature*, 479, 2/XI/2011, pp. 22–24.

WEIZSÄCKER, Carl Friedrich Freiheer von, *The History of Nature*, Routledge & Kegan Paul, 1951.

WHO (World Health Organization), "Q&A on the carcinogenicity of the consumption of red meat and processed meat". October 2015.

WHO (World Health Organization), *The Global Burden of Disease Study*, 2017.

WHO (World Health Organization), *Global Antimicrobial Resistance Surveillance System (GLASS) Report: early implementation 2016–2017*. Geneva, 2017.

WHO (World Health Organization), *Global Tuberculosis Report 2019*.

———, *Report on Cancer. Setting Priorities, Investing Wisely, and Providing Care for All.* Geneva, 2020.

WILSON, Clare, "Nobody can agree about antidepressants. Here's what you need to know". *New Scientist*, 2/X/2018.

WITTKOWER, R. & M. *Born under Saturn. The Character and Conduct of Artists: A Documented History from Antiquity to the French Revolution*. London, Weidenfeld & Nicolson, 1963.

YANG, Hong *et al.*, "Antibiotic Application and Resistance in Swine Production in China: Current Situation and Future Perspectives". *Frontiers in Veterinary Science*, 17/V/2019.

YINGQI, Cheng, "Beijing residents face rising cancer threat" *China Daily*. 12/X/2012.

ZACHARY, J. Ward *et al.*, "Projected U.S. State-Level Prevalence of Adult Obesity and Severe Obesity". *The New England Journal of Medicine*, 19/XII/2019.

Chapter 16
Conclusion: From the Social Contract to the Natural Contract

To conclude, let us return for a moment to the great documents and scientific manifestos of the last 50 years. Although signed by scientists who have very different backgrounds and written in very different contexts, they all share the perception that the second quarter of the twenty-first century will be one of substantial degradation, if not an environmental collapse, of our global society. In 1972, the manifesto *A Blueprint for Survival*, signed by over 30 leading scientists, stated:

> The principal defect of the industrial way of life with its ethos of expansion is that it is not sustainable. Its termination within the lifetime of someone born today is inevitable—unless it continues to be sustained for a while longer by an entrenched minority at the cost of imposing great suffering on the rest of mankind.

For the generation born in the 1970s, their years of life will come to an end precisely during the second quarter of the twenty-first century. Twenty years later, something very similar was stated in the document, *Scientists' Warning to Humanity*, discussed in the previous chapter. For its 1700 signatories, "a great change in our stewardship of the Earth and the life on it is required, if vast human misery is to be avoided." And the time remaining for this "great change" was clearly fixed: "no more than one or a few decades remain before the chance to avert the threats we now confront will be lost and the prospects for humanity immeasurably diminished." Once again, the time horizon for the catastrophe was fixed to at most a "few decades," that is, to the second quarter of the twenty-first century. In 2013, the document "Scientific Consensus on Maintaining Humanity's Life Support Systems in the 21st Century," signed by more than 1300 scientists, researchers, members of NGOs, students, and the general public, in more than 60 countries, reiterated the same deadlines[1]:

> Earth is rapidly approaching a tipping point. Human impacts are causing alarming levels of harm to our planet. As scientists who study the interaction of people with the rest of the biosphere using a wide range of approaches, we agree that the evidence that humans are damaging their ecological life-support systems is overwhelming. We further agree that,

---

[1] See also Barnosky et al. (2012) and Barnosky and Hadly (2015), Chap. 10: "End Game."

© Springer Nature Switzerland AG 2020
L. Marques, *Capitalism and Environmental Collapse*,
https://doi.org/10.1007/978-3-030-47527-7_16

based on the best scientific information available, *human quality of life will suffer substantial degradation by the year 2050 if we continue on our current path.*

Twenty-five years after the 1992 manifesto, William Ripple et al. (2017) wrote the *World Scientists' Warning to Humanity: A Second Notice*, signed by 15,364 scientists from 184 countries. The document underlined the imminence of a catastrophe: "Soon it will be too late to shift course away from our failing trajectory, and time is running out." In 2019, William Ripple and colleagues signed another document, this time with a specific focus on the current imbalances in the climate system: *World Scientists' Warning of a Climate Emergency*. In this document, it is no longer a question of alerting to imminent crises, but to the fact that the climate crisis has become an emergency and is already upon us:

> We declare, with more than 11,000 scientist signatories from around the world, clearly and unequivocally that planet Earth is facing a climate emergency. (…) The climate crisis has arrived and is accelerating faster than most scientists expected. It is more severe than anticipated, threatening natural ecosystems and the fate of humanity.

The same perception that the future has arrived was affirmed in a manifesto published on September 9, 2018, by the French newspaper *Libération*, "SOS of 700 Scientists," a document that begins with this verdict: "We have already fully entered the 'climate future'" (*Nous sommes d'ores et déjà pleinement entrés dans le 'futur climatique'*). The developments and impacts of the "Great Acceleration" (Steffen et al. 2015) are occurring at an exponential rate; thus, if written in 2020, the 2013 "Scientific Consensus" would possibly say: "human quality of life *is already suffering* substantial degradation."

The deleterious effects of the Anthropocene—this new world made in our image and likeness—are already part of the present. Day by day, its features are seen in the news, in our daily perceptions, and, even more, in the signals and data emitted by nature, now accessible to the senses, in addition to being observed and analyzed according to different measurement strategies and scientific methodologies. And the more sophisticated the observations, analysis, and projections, the more unambiguous is the message they send out: we are now witnessing the first phases of an environmental collapse of unfathomable proportions because civilization's elementary way of functioning, which has become hegemonic since modern history began, puts increasing destructive pressure on the biosphere and on the balances of the climate system.

The capitalist economic system triumphed because it proved to be the most capable of accumulating wealth, creating surplus, and expanding. But precisely because it is distinguished by these qualities, precisely because these qualities define it, this economic system is a prisoner of them. It does not have the freedom to limit itself. In other words, in its accumulative mechanism, it is incapable of not destroying the biological wealth of our planet and the balance of the Earth system that permitted the rise of civilizations throughout the Holocene. Today, the choice is between disassembling this mechanism piece by piece or condemning ourselves, at best, to life conditions that are much more adverse than those that this civilization of accumulation has provided us with. Stephane Hessel and Edgar Morin (2011) are expeditious regarding the consequences of this choice: "our societies must now choose: meta-morphosis or death."

However unequivocal and well-founded it may be, this message is inaudible or still sounds to many as an exercise in futurology, with the lack of value that the term deserves. Others, it is true, no longer deny it the weight of reality, but almost all, and in particular those who have a decision-making voice, refuse to admit what this message implies: the need for an organized degrowth, which—we will repeat this as often as necessary—is not symmetrical to growth, but it does mean a qualitative redefinition of society's political and economic structure and priorities, with the objective of drastically reducing its impacts on the biosphere. The *incipit* to the Gospel of the overwhelming majority of politicians, economists, intellectuals, and opinion makers is "in the beginning there was growth." So any contestation of this belief is dismissed as ignorance of the basic market mechanisms that found and regulate the order of the world in a way that is more unalterable than universal gravitation. Everything would be possible, according to this consensus, except stop-ping the accumulation machine, because the maintenance, well-being, and even the happiness of societies depend on its functioning.

Numerous initiatives by civil society, states, and even some corporations have tried to make the world's capitalist order more "sustainable." As long as they tend to mitigate the destructiveness of human action on the biosphere, all these initiatives, even the most modest individual and local gestures, are precious. Our ability to delay and mitigate the environmental collapse outlined before us depends on them. But these initiatives only allow us to delay and mitigate the collapse, for it is not possible to *reverse* the tendency toward environmental collapse within the scope of capitalism, as the preceding pages were intended to demonstrate. There is no sus-tainable capitalism because there can be no sustainability (1) when the legal order guarantees that decisions about strategic investment flows emanate from the inter-ests of a small group of people, whether these be businessmen or bureaucrats, and (2) when the raison d'être of these investments is the increased remuneration/repro-duction of capital, whether private or owned by the state. Unsustainability is consti-tutive of capitalism. The effort to "educate" capitalism for sustainability is, therefore, what I have called in this book the most misleading illusion of contemporary thought because it has the most serious consequences.

There remains the alternative of overcoming capitalism. As everyone knows, the transformation of the socialist project (the most generous and rational legacy of the Enlightenment and of the nineteenth century) into the monstrosity to which "real" socialism in the twentieth century reduced it explains why the appeal to an alterna-tive society to capitalism has lost so much support. Socialism is a closed historical experience and its failure is not from yesterday. It was already clearly announced in the savage repression of the Tambov and Kronstadt rebellions of 1920 and 1921. China's turnaround, the fall of the Berlin Wall, and the implosion of the Soviet Union between 1987 and 1991 only sealed the bankruptcy of a historic project that never sought to overcome, *not even in an ideal horizon*, the principle of accumulation.

This final failure resulted in an exultant conviction by those on the right and in a widely shared resignation by those on the left that capitalism is not a transitory system, but something "second nature," the "natural" place toward which gravitates any civilization capable of accumulating surplus and, thus, satisfying the demand of increasing population contingents. With all its defects, it would be the most efficient resource allocator and the best system of social organization that humanity has been able to equip itself with since the so-called Neolithic revolution, given that it faithfully mirrors the contradictions of our own species, its character at the same time vulnerable and resourceful, individualistic and gregarious, aggressive and cooperative.

The crisis which began in 2008 is starting to produce some cracks in this conviction. The noise produced by the clamor between different political and economic policies on the best strategy to resume a new growth cycle smothers, almost completely, the argument of the environmental infeasibility of this growth. But this argument is starting to impose itself. Climate change, the signs of decline or depletion of natural resources, and the effects of counterproductivity caused by the anthropogenic imbalances of the biosphere, with all the suffering this entails, are beginning to leave the circumscribed domain of scientific magazines to make headlines in the mainstream press and even in the Davos agenda, now concerned with the inconvenience that this topic of the environment might cause to business. More importantly, consumer society is becoming more and more disenchanted in the face of capitalism's declining capacity to supply it with the drug that it has become dependent on, precisely consumption. In short, there is a diffused, but growing, "malaise in capitalism."

Moreover, the old left, which has absorbed the corporate worldview, according to which there is no prosperity for workers without economic growth, has been losing what remains of its social representativeness. A new left is beginning to emerge, identified with ideas such as environmental conservation, divestment in fossil fuels, managed degrowth, and eco-socialism, terms that tend to converge more and more. In this context, the final capitulation of socialist bureaucracies to the global market proved to be good because it freed critical thought from the weight of the legacy of these brutal regimes (to which it felt or was accused of being indebted to). The condemnation of the chaos to which corporations are leading the planet has taken on forms and discourses that no longer have anything to do with the fossilized socialist rhetoric.

Science and social movements are now converging for the first time. In addition to the scientific manifestos mentioned above, one should also remember the manifesto, "The greatest challenge in the history of mankind," published by *Le Monde* in 2018 and endorsed by more than 200 personalities, among them artists, philosophers, scientists, and journalists.[2] In the economic area, the campaign for the divestment of fossil fuels, *Keep it in the ground*, launched in 2011 by Bill McKibben, with the support of groups such as People & Planet and *The Guardian*, today includes

---

[2] Cf. "Le plus grand défi de l'histoire de l'humanité". *Le Monde*, 3/IX/2018.

around 1000 investment institutions that have obtained commitments to withdraw 8 trillion dollars from this sector (McKibben 2018). In France, four NGOs (Fondation pour la nature et l'homme, Notre affaire à tous, Greenpeace France, and Oxfam France) garnered immediate support with nearly two million signatures (Garric 2018) when they launched the *l'Affaire du siècle*, a lawsuit against the French state for "climate inaction," i.e., for not abiding by existing legislation. In Holland, a lawsuit initiated by the NGO Urgenda, in the name of 900 citizens, demanded that the Dutch state reduce its GHG emissions by 25% by 2020.[3] According to the Sabin Center for Climate Change Law at Columbia University, in August 2018, there were more than 1.110 cases registered, 888 in the USA, some of which were launched by children and youth with the support of scientists such as James Hansen, against governments and corporations accused of being criminally responsible for climate changes. Outside of the USA, there are 240 cases in countries such as Belgium, Colombia, Ireland, New Zealand, Portugal, Switzerland, Norway, the UK, and Uganda.[4] Between 2018 and 2019, the *Rise for Climate* convoked by the NGO 350. org took crowds of people to the streets, from Katmandu to Bogota, passing through Rio de Janeiro, Lagos, and San Francisco where they met with the Guardians of the Forest, a group of 20 indigenous peoples from 14 countries.

The mobilization of teenagers and youth is now a driving force for these movements. Along the same lines, the call for action by the *Extinction Rebellion* nonviolent, civil disobedience movement in the UK, launched in October 2018 with the *Rising Up!* Campaign, is another example of this political empowerment. Among the 100 eminent signatories of *Extinction Rebellion's* second manifesto are Noam Chomsky, Naomi Klein, Bill McKibben, William Ripple, and Vandana Shiva. Its central idea goes to the heart of the agenda of our time[5]:

> "Political leaders worldwide are failing to address the environmental crisis. If global corporate capitalism continues to drive the international economy, global catastrophe is inevitable. (…) We must collectively do whatever's necessary non-violently, to persuade politicians and business leaders to relinquish their complacency and denial"

These environmental campaigns online and in the streets are vital signs—even if they, unfortunately, are unable to influence decision-making at the moment—of a renewed critical energy.

Certainly, we do not yet know what form a postcapitalist society will take on. But there can no longer be uncertainty about the central fact of our time: there is an irreconcilable incompatibility between the globalized capitalism of today and the conservation of the coordinates of the Earth system that allow for the survival of our societies and of the majestic biodiversity of our planet. Defining and building an

---

[3] Cf. A. Neslen, "Dutch government appeals against court ruling over emissions cuts". *The Guardian*, 28/V/2018: "The Netherlands is 34th in the world when it comes to greenhouse gas emissions. But when it comes to per capita emissions the Netherlands ranks ninth—the highest of any EU country. A Dutch person emits twice as much as the global average and 1.5 times more than the average EU citizen."

[4] See Fred Pearce, "Polluter pays?". *New Scientist*, 18/VIII/2018, pp. 38–41.

[5] See "Extinction Rebellion," Wikipedia.

alternative socioeconomic logic to that of the corporations that govern humanity's economy, politics, and life remain the greatest challenge of contemporary political thought and practice. In fact, it is the biggest challenge in history. To face it, we must start a double, synchronous discussion on *how* to get *where* we want to go. But we must say that for the first time in the history of politics, the discussion about *how* should precede the discussion about *where* (paradoxically, tactics must precede strategy) because the strategic objective can no longer be defined beforehand by a "vanguard" group, but will be defined collectively as greater spaces for political participation are won.

16.1 Decentralization and Power Sharing

With regard, therefore, to how to move away from the worst-case scenarios of the ongoing collapse, everything that, in my view, can be said at the moment stems from the principle of decentralization. The term decentralization is ambiguous and needs to be clarified in order to avoid two mistakes, the first of a political nature and the second of an economic nature: (1) Already in *A Blueprint for Survival* (1972), Edward Goldsmith and the editorial board of *The Ecologist* magazine wrote: "Possibly the most radical change we propose in the creation of a new social system is decentralization." Exactly, but the meaning given in 1972 to the term decentralization had nothing to do with the meaning proposed here, almost 50 years later. In *A Blueprint for Survival*, decentralization was synonymous with local power, "localism." This is because in 1972 the need for a global political approach to global problems, such as the climate emergency that can only be tackled globally, was not yet clear. Today, on the other hand, the term decentralization requires the consolidation of a shared global political power. (2) The second mistake to be avoided is the hypocrisy of economic liberalism. Criticism of centralization is commonplace among those who defend liberal rhetoric, since, according to them, decentralization is synonymous with the market. But what this rhetoric hides is that real power—the decision on investments in strategic sectors of society—has never been more centralized. The state-corporation, in which the interests of the high technocracy of states and large corporations are merged, concentrates today a power never held or dreamed of by the despots of yore. Decentralization means precisely the dismantling of this extreme concentration of economic power in favor of effective democratic global governance. In a word, decentralization here means, at the same time, a dismantling of economic globalization and a deepening of shared global governance.

Those who speak about global governance must, of course, take into consideration the United Nations. Rising 75 years ago from the ashes of the League of Nations (1919–1939) and based on a 1945 reality, it is obvious that the UN has repeated the same mistakes as its predecessor. In a letter to Pierre Comert dated April 11, 1923, Albert Einstein justified his departure from the International Committee on Intellectual Cooperation, an agency of the League of Nations, for not believing that the latter could fulfill its mission: "By its silence and its actions, the League [of Nations] functions as a tool of those nations which, at this point of history, happen to be the dominant powers."[6]

Einstein's diagnosis applies, mutatis mutandis, to the current situation of the UN. Even its foremost and most elementary mission, that of guaranteeing peace, has never been so far from being fulfilled. From 1946 to 2007, the Correlates of War (COW) project at the University of Michigan[7] lists 238 wars, which, incidentally, have increased in number with each decade: 29 wars between 1946 and 1955; 33 between 1956 and 1965; 43 between 1966 and 1975; 40 between 1976 and 1985; 46 between 1986 and 1995; and 47 between 1996 and 2007. According to Jan and Mat Zalasiewicz (2015), about 50 armed conflicts were occurring in 2015. Just as the failure of the League of Nations, the failure of the UN and of the summit meetings sponsored by it also stem from the fact that they are an instance of international legitimation of the interests of "powers" making up its Security Council.

In its current form, the UN is the expression of the maximum concentration of power. But it is the only available structure to advance in the sphere of global governance and must, therefore, be supported and strengthened. Strengthening it means suppressing the veto power of its Security Council and submitting the latter to the General Assembly, from which a power, similar to that of a parliament, must emanate, one that is superior to the power of national sovereignties and imperialist plans. This is the only way to peacefully and rationally face problems that cannot be tackled only at the national level, such as the environmental crises and the global socio-environmental collapse that is underway. The ecocide that continues to be perpetrated against the oceans, against the planet's native vegetation covers, and against biodiversity in general must be subject to penalties that are effectively dissuasive. Some progress has been made in this direction, although still very slow. In 2002, governed by the document the Rome Statute of the International Criminal Court, the International Criminal Court (ICC) that sits in The Hague entered into force. There are currently 122 states that are signatories to the Rome Statute. As an intergovernmental organization, the ICC has jurisdiction to prosecute individuals for crimes of genocide, war crimes against humanity, and crimes of aggression. In 2012, Anja Gauger, Mai Pouye Rabatel-Fernel, Louise Kulbicki, Damien Short, and Polly Higgins published *The Ecocide Project: 'Ecocide is the Missing 5th Crime Against Peace.'* In 2016, the ICC recognized the crime of ecocide, following a joint call for the creation of an International and European Criminal Court for the Environment, which was launched in the European Parliament in 2014, along with the "Charte de Bruxelles." The ICC will start focusing "on crimes linked to environmental destruction, the illegal exploitation of natural resources and unlawful dispossession of land" (Arsenault 2016). There is an urgent need to equip this international arbitration body and give legal and coercive power to its verdicts.

---

[6] Quoted by Christophe David, Preface to Albert Einstein, Sigmund Freud, *Warum Krieg?* (1932), *Pourquoi la Guerre?* Paris, 2005, p. 11.

[7] See Correlates of War Project (University of Michigan) in <http://www.correlatesofwar.org/>

16.2   Neither Nation nor Empire: Overcoming the Notion of Absolute National Sovereignty

To effectively combat ecocide, however, it is essential to expand its meaning to include, for example, the exploration of fossil fuels. And above all, it is essential to overcome—mentally and legally—the notion of absolute national sovereignty. Sir Nicholas Stern stated in 2010:

> We will need new institutions. Indeed, I would argue that if John Maynard Keynes and Harry Dexter White were conducting a Bretton Woods Conference now instead of in 1944, they would have had three different institutions instead of World Bank, IMF, and WTO. We surely do need institutions for finance and for trade, but we now need one for the environment, a World Environmental Organisation.

A World Environmental Organization is a key initiative. But if the legal status of the nation-state is not limited, it will be just one more bureaucratic agency, unable to fulfill its historical function. Just 5 years ago, Christophe McGlade and Paul Ekins (2015) still believed that there was a 50% chance of limiting average global warming to 2 °C above the pre-industrial period, if, and only if, the impulse of states to increase the use of fossil fuels available in their territories were neutralized:

> Our results show that policy makers' instincts to exploit rapidly and completely their territorial fossil fuels are, in aggregate, inconsistent with their commitments to this temperature limit. Implementation of this policy commitment would also render unnecessary continued substantial expenditure on fossil fuel exploration, because any new discoveries could not lead to increased aggregate production.

Although the goal of limiting average global warming to 2 °C above the pre-industrial period is no longer possible (see Chaps. 7 and 8), it is necessary, precisely for this reason, to redouble efforts to limit this warming to the minimum amount still possible. The reasoning of Christophe McGlade and Paul Ekins, therefore, remains valid: the remaining fossil fuel reserves, considered to be strategic by the states that own them, must now be considered stranded assets. We must exchange their commercial value for the value of our survival as an organized society. The countries where these reserves are located can no longer claim the right to profit from them while driving us toward an environmental collapse. This, of course, applies both to countries with large reserves of gas, coal, and oil and to those that hold the last large forest reserves. These countries can no longer afford to devastate these forests to satisfy the greed of farmers and businessmen who also do not generate lasting wealth even for their fellow citizens. But how can we prevent these suicidal policies if not by limiting—through global governance—the power of national states which are in the process of becoming state-corporations?

Nationalism has nothing to do with belonging to a linguistic, cultural, and territorial community that shares a collective memory and a heritage of historical, artistic and religious experiences. As long as it is not hijacked by xenophobic campaigns, this feeling of belonging is much richer than the contingency of being part of a nation-state. The history of nationalism and imperialism (the nationalism of the most materially powerful countries) is, as we know, the history of the development of capitalism expressed in ideology. The recent extreme globalization of capitalism and the state's closer association with the international corporate network tend to make this nationalist ideology anachronistic. But nationalist ideology resists. And the more it resists, the more xenophobic, belligerent, and irrational it becomes. That said, the idea of the nation as the maximum expression of the authority exercised over a territory can no longer be something of our time. Just as the nation-state is inseparable from the historical development of capitalism, the state-corporation, in its contemporary expression, is inseparable from the combined crises of capitalism and of the environment of our days. The nation-state, in its historical or contemporary form, is the negation of the universality of man, of the *humanus qua humanus*, and of his cultural diversity. Nationalism should disappear with capitalism because it comes from the same matrix, one that is accumulative, militaristic, and expansionist. Paradoxically, it is the final globalization of capitalism itself that placed the urgent demand of a shared power on the agenda, for it is only in this world now unified by capital that the consciousness of a common political destiny and of a planetary community of interests can prosper. Furthermore, it is only in this world unified by capital that there can emerge the non-anthropocentric ideal of a planetary community of beings of nature (Afeissa 2010).

In addition to the failure of socialism, the other great lesson that the history of the last century contains is that nationalisms and their imperialist developments made this century probably the most violent in Western history. Imperialist, colonial, neocolonial, local, regional, or world wars, the most horrible wars of the twentieth century—of an unprecedented horror which we did not even believe ourselves capable of as a species—were invariably waged in the name of the nation. In both Kant's 1795 proposal that "the law of nations must be founded on a federation of free states" and Marx's exhortation in 1848 for an international union of workers, there is a critical perception of a bond of causality between nation and war. In this sense, Kant and Marx are more up-to-date than ever. In *Eclipse of Reason* (1947), Max Horkheimer notes that:

> The idea of the national community (*Volksgemeinshaft*), first set up as an idol, can eventually be maintained only by terror. This explains the tendency of liberalism to tilt over into fascism and of the intellectual and political representatives of liberalism to make their peace to its opposites.

For no other reason, Arnold Toynbee defined his *Study of History* in the 1920s and 1930s as a moral warning in the face of nationalism's tendency to create "Wars of Nationality, which began in the eighteenth century and are still the scourge of the twentieth."[8] In this game of continuous belligerence between nations, in which only corporations profit, the following type of declaration of war on the world launched by Theodore Roosevelt in 1899 is emblematic: "If we shrink from the hard contests where men must win at hazard of their lives and at the risk of all they hold dear, then the bolder and stronger peoples will pass us by, and will win for themselves the domination of the world." Just before the 1992 Earth Summit in Rio de Janeiro,

---

[8] See Arnold J. Toynbee, *A Study of History*, Oxford University Press, 1935, Vol. I, p. 147.

Roosevelt's warmongering nationalism was renewed in George H. W. Bush's insolent claim that "the American way of life is not up for negotiations. Period."

It turns out that nature is also no longer willing to negotiate with the American way of life (mecca for the way of life of contemporary societies), nor with its ideologues, much less with its diplomats and political representatives, as Ban Ki-moon wisely recalled at the end of Rio + 20 (see Introduction, Sect. 1.4). Nor is nature impressed by military power. More clearly than in the past, recent history teaches us that no nation or group of nations is capable of imposing its will on the world, even if it is endowed with undisputed military superiority. This is because, as already mentioned in the Introduction, the current global socio-environmental precariousness impacts all societies and makes the select club of the rich more vulnerable than ever to the collapse of the poor. Even societies with the most advanced technology, in fact, especially them, will not survive in a polluted world, which is on average more than 2 °C hotter than the pre-industrial period, threatened by extreme heat waves, with floods and extreme weather events that are more frequent and more devastating, in a world disorganized by massive migrations, without biodiversity, without the crucial services provided by forests and other terrestrial and aquatic ecosystems, without arable lands, with declining phytoplankton, and with cities submerged by oceans without fish.

16.3   A Power of Arbitration and Veto Emanating from Society

This shared globalization of political power will not survive without a radical reduction in inequality. The patterns of consumption of energy, goods, and services of 8.6% of the world's adult population who owned 85.6% of the world wealth in 2017 (as seen in the Introduction and in Chap. 13) are the main accelerators of environmental unsustainability and of the trend toward collapse. This pyramid of wealth is incompatible with an economy that respects the biosphere. A return to progressive fiscal policy, including the confiscatory taxation of fortunes, would eliminate the moral and socio-environmental aberration that humanity has arrived at in the twenty-first century. Additionally, eliminating this aberration is not "socialist." As Thomas Piketty (2013, Chap. 14) recalls: "When we look at the history of progressive taxation in the twentieth century, it is remarkable to see how Great Britain and the United States were extremely advanced, especially the United States, which invented the confiscatory tax on 'income and excessive fortunes.'"

Decentralizing world wealth, however, requires something much more effective than putting mechanisms of income distribution in place. It requires giving constituents back the power that their rulers have to a certain extent lost to corporations. The center of power is no longer in the state. The center of power is in the plutosphere, composed, as seen in Chap. 13 (Sect. 13.5 A Super-Entity: The Greatest Level of Inequality in Human History), by *ultra-high-net-worth individuals* (UHNWI), corresponding to 0.004% of adult humanity and by 147 conglomerates that hold economic control of the corporate network, closely associated with the high techno-bureaucracy of the state.

And here we get to the heart of a first definition of what it means to overcome capitalism. It is a question of refashioning the constitutional pact on a subnational, national, and international scale, so as to confer societies with arbitration and veto power in all decisions involving:

(1) Maintenance of the accumulation mechanism
(2) Increase in pollution and waste of resources
(3) Impact of economic activity on the climate system, natural resources, and eco-system balances beyond the limit recommended by the scientific consensus of international institutions belonging to the UN or accredited by its treaties and conventions

A new constitutional body would be charged with ensuring that these three strict clauses of the new constitutional pact are observed. This constitutional body would be elected also by universal suffrage and composed equally of:

(1) Representatives of civil society, including professional associations and NGOs
(2) Representatives of scientific institutions
(3) Representatives of future generations, charged with examining the possible future consequences of each decision
(4) Biologists, botanists, and animal protection societies capable of representing the interests of nonhuman species (Dobson 1998)

16.4   Citizens Must Summon Science to Understand Their Own Political Interest

In *Vers une démocratie écologique* (*Toward an Ecological Democracy* 2010), Dominique Bourg and Kerry Whiteside take on the task of imagining a new political structure capable of reconciling democracy with a power that is capable of preventing or, more realistically, mitigating the impacts of environmental collapse as much as possible. This new structure entails overcoming classic representative government based on the liberal individualistic tradition, one in which the individual makes the final judgment on his own interests. The authors remind us that (p. 10):

> The classical representative system assumes that I am the only judge of my condition. Who can, effectively, judge my well-being better than myself? Representatives must, therefore, regularly return to their voters to ensure that their policies are well-founded. Now, the complexity of environmental problems, the fact that they affect us indirectly or from afar, prevents us from assessing them for ourselves.

Indeed, because the environmental crises in which the contemporary world is sinking are not only local, but transnational; because they are systemic and cumulative and their cascading impacts are more likely to become "severe, pervasive and irreversible" (IPCC, AR5 2014)[9]; and, finally, because their feedback dynamics are taking us closer to nonlinear transitions that lead to another state of the Earth system, citizens no longer have the means that enable them to evaluate their own interests and, therefore, their political decisions.

Just as industrial capitalism meant the separation of the worker from the means of production, globalized capitalism in the era of the great environmental crises means the separation of the citizen from the "spontaneous" means of his own political judgment. This does not mean his political infantilization because obviously nobody can claim to be responsible for him. On the contrary. It means the greatest emancipation in the history of the establishment of citizenship: the inalienable right and the duty that we now have to integrate into our political judgment basic scientific information on human impacts on the biosphere, because, ignorant of science, the citizen can no longer understand today where his own political interest lies. Environmental crises pose problems which the democratic institutions of the past are no longer qualified to solve because science and politics, knowledge and interest, were still separate instances in the previous paradigm. They are no longer separate today. Therefore, there is a need for science to be directly present as a veto power in all strategic economic and political decisions. Inversely, there is also the need for science to overcome the adolescent Baconian claim of infinite expansion of the knowledge–power binomial and to take responsibility for alerting society and deepening knowledge of the dangers that hover over the biosphere today. Since Hans Jonas (1979, 1985), we began to realize that the greatest challenge for humans today is to philosophically manage the contradiction between the potential infinity of knowledge and the self-limitation of doing.

16.5   From the Negative to the Positive Principle: The Natural Contract

This new power of arbitration and veto capable of controlling the accumulative drive still refers, however, to a *negative* concept of overcoming capitalism, that is, the need to deny the logic of destruction. A second approach to what it means to overcome capitalism should make it possible to point to the *positive* principle on which we can reestablish the constitutional pact: the transition from the social contract to the natural contract. The natural contract is a term used in legal philosophy and attributed, in modern times, to Michel Serres, but that goes back, as he recalls, to Antiquity and its legacy: to the "pacts of nature" (*foedera naturae*)[10] of Lucretius

---

[9]Cf. IPCC 5/XI/2014, *Climate Change 2014. Synthesis Report*: "Continued emission of greenhouse gases will cause further warming and long-lasting changes in all components of the climate system, increasing the likelihood of severe, pervasive and irreversible impacts for people and ecosystems."

[10]For Lucretius, all animals, including humans, are part of these pacts or laws of nature (*foedera naturae*) that guarantee its functioning. See Lucretius, *De rerum natura*, I, 584–592; Droz-Vincent

and the *Cantico delle Creature* of S. Francisco de Assis, which was renewed in the encyclical *Laudato Si'* (2015), to remember only the most important paradigms.

The common objection to the possibility of establishing a contract between humans and nonhumans on the grounds that the latter cannot speak for themselves has long been refuted by Hans Jonas and many other thinkers (Regan 1988; Dobson 1998; Afeissa 2010, etc.). This objection falls apart when we consider that today's children and that the next generations cannot take the floor either and that, *precisely for this reason*, their status as full actors must be ensured at the negotiating table. The natural contract states that socioeconomic relations between humans can no longer be governed by the premise that nature is only the starting point of the production process. Nothing but the madness about which Lucretius speaks of (see Introduction, Sect. 1.9) can further support the immature belief that nature and other forms of life were "created for man" and that we are, therefore, their reason for being. We are a part of the web of the biosphere, and we closely depend on it not only for our survival but also for our material and spiritual happiness. As Michel Serres (1998, p. 15) rightly states, "I prefer to speak of nature not in its ordinary senses, but in the pure etymological sense, as it [nature] is in the process of being born, totally new for us, for our knowledge, and for our globalized acts." Serres refers to nature in its etymological sense specifically as *natura*, that is, "the action of giving birth."[11] One cannot kill that which gives birth merely for it to acquire the form of merchandise, profit, and garbage.

Michel Serres' proposal was accepted by an economist who is committed to thinking and acting in favor of environmental sustainability, namely, Ignacy Sachs (2009), who writes: "The social contract on which our society's governability is based must be complemented by a natural contract (Michel Serres)."[12] More than complementary to it, the natural contract must *establish* the social contract, establish its conditions of possibility. This means no longer reducing the *Homo sapiens* to *Homo oeconomicus*, a reduction typical of the Modern Age. It means refusing to see the bourgeois as the ideal type of man and, by extension, his utilitarian way of thinking as the prototype of thought. It is necessary to reject Schumpeter's premise, according to which "all logic is derived from the pattern of the economic decision or, to use a pet phrase of mine, that the economic pattern is the matrix of logic."[13] This phrase that Schumpeter indulges in is found everywhere in classical economic theory, in its developments in the nineteenth century and, not without irony, in Nietzsche: "Fixing prices, setting values, working out equivalents, exchanging— this preoccupied man's first thoughts to such a degree that in a certain sense it *constitutes* thought."[14]

(1996, pp. 191–211); Serres (1998); Takakijy (2013, p. 1).

[11] Nominal form of the verb *nascor*, to be born; "natura: action de faire naître." See Alfred Ernout, Antoine Meillet, *Dictionnaire étymologique de la langue latine* (1932), Paris, Klincksieck, 2001, p. 430.

[12] Cf. I. Sachs (2009, p. 49). This refers to the republication of a text written in 1998.

[13] Cf. Schumpeter (1942/1976, p. 118).

[14] Cf. Nietzsche, *Zur Genealogie der Moral. Eine Streitschrift* (1887). Zweite Abhandlung:

It matters little that the pattern of economic decision-making is at the root of logical thinking, even if this is nothing more than silly speculation. The important thing is to understand that what this matrix has provided us with in the past, it is taking away from us in the present. As noted by Vittorio Hösle (1991/2009, p. 57), it is crucial that the twenty-first century transitions from the economic paradigm to the ecological paradigm, that is, that the economy ceases to be the central element (*Zentralgebiet*) of civilization, such as it has been since the advent of the Modern Age, and give way to ecology. In other words, if we want to paraphrase Schumpeter, the ecological standard must, henceforth, be the matrix of logic.

This new natural contract implies not only overcoming the economic logic upon which the worldview of the modern man is based; it also means the parallel overcoming of classical democracy itself (based on an anthropocentric matrix) for the benefit of biocracy, which is nothing but a more comprehensive understanding of democracy. Biocracy was well defined by Terence Ball (Dobson and Eckersley 2006, p. 135):

> While biocracy certainly does not exclude human interests, neither does it—like liberal democracy, social democracy and people's democracy—place them at the apex of a hierarchical pyramid of moral considerability; rather, biocracy counts human interests as one set within a web of complexly interdependent interests.

Patterns of social depredation are built upon patterns of depredation of nature and vice versa. So that every democracy, when deepened, ends up implying a biocracy. Essentially, biocracy is the principle of government according to which nature as a whole and, in particular, the biosphere—the ensemble of human *and* nonhuman life—are *irreducible* to the self-propelling mechanism of capitalist accumulation.

The natural contract does not emanate only from legal philosophy; it is no longer just the last chapter in the long history of the universalization of legal subjects: from slaves to the foreigner, to the weakest, to women, to the elderly, to the physically challenged, to the sick, to children, and, finally, to sentient animals. It emanates from the urgent requirement to conserve what remains of the biosphere, *without which we are no longer only denying future sustainability, but the survival of today's society*.

16.6   Our Plan Is Survival

The italicized sentence above deserves to be emphasized, because it shows a crucial difference between the situations of man (and of his social aspirations) in the twentieth century and in the twenty-first century. The twentieth century taught us in a

---

"Schuld," "schlechtes Gewissen" und Verwandtes, Chap. 8: "Preise machen, Werthe abmessen, Äquivalente ausdenken, tauschen—das hat in einem solchen Maasse das aller erste Denken des Menschen präoccupirt, dass es in einem gewissen Sinne *das* Denken ist" (text from the Projekt Gutenberg).

harsh way to no longer accept the vanity and arrogance of those who believe that they can pull out of a hat the recipe for what is "best" for society, be it the philosopher, the scientist, the economist, the religious leader, the liberal leader, or Lenin's "vanguard" political party. Cornelius Castoriadis, as well as others, was aware of this when he stated in 1980 that a better society[15]:

> [...] will come out of the society itself or not. Recognizing this limit to political thought and action means not allowing oneself to redo the work of political philosophers of the past, taking the place of society and deciding, like Plato and even Aristotle, that such a musical scale is good for the education of young people, while that other one is bad and should, therefore, be banned from the polis.

These are words that should not be forgotten. But we must also not forget that the context which gave rise to them—the reflection on the big mistakes made by revolutionary projects of the twentieth century—is no longer relevant to our time. Discussing the "best" society and the mistakes or crimes committed in its name is no longer on today's agenda. What is on the agenda is only how to significantly mitigate the ongoing environmental collapse which jeopardizes the survival of *any* complex society. Today, the political program is to fight for a society capable of surviving, that is, capable, once more, of fitting into the biosphere.

To many, this may sound like excessive pessimism, defeatism, a renunciation of the moral demand for social justice, or an unacceptable minimalism. But it is the opposite! The paradox of our day is that the ruined idea of social revolution—expelled through the back door in the end of the twentieth century—returns today, albeit profoundly different, through the front door. This is because we will need to institute a natural contract to mitigate the approaching collapse and it will not be politically feasible, nor philosophically conceivable, without the most radical redefinition of our species' position in the web of life, a redefinition that amounts to an economic, social, and political revolution, one that is much more ambitious than all the preceding ones.

Those who consider remote the probabilities that this post-anthropocentric and, a fortiori, postcapitalist natural contract be signed have obvious reasons to do so. Such skepticism is easy, perhaps even inevitable, because everything takes us toward it, especially the fact that this contract is not limited to a mere delegation of power in the Hobbesian sense, to an international superstate or a committee of alleged experts on the scientific management of nature (bioengineering, geoengineering, etc.).

What this natural contract establishes is something more difficult. It is autonomy in the first sense of the term. It is the self-institution of a *nomos* that, far from the presumption of abolishing the social and ideological conflicts inherent in societies, reconciles us philosophically with the finitude of the biosphere so that we might conceive of ourselves and behave, in short, as a species among species. Walter

---

[15] Lecture given on February 27, 1980, to students from Louvain-la-Neuve on the topic "Anti-nuclear struggle, ecology and politics," in Cornelius Castoriadis & Daniel Cohn-Bendit, *De l'écologie à l'autonomie*. Lormont, Le bord de l'eau, 2014, p. 45.

Benjamin and Cornelius Castoriadis complement each other perfectly in formulating this program of a democratically organized and managed degrowth. The first one writes: "Marx said that revolutions are the locomotive of world history. But perhaps things are very different. It may be that revolutions are the act by which the human race travelling in the train applies the emergency brake."[16] "We need," writes Castoriadis, "not dominance, but a control over this desire for dominance, a self-limitation. Autonomy, moreover, means self-limitation. We need to eliminate this madness of expansion without limit".[17]

Such skepticism is easy, perhaps even inevitable, in short, because in spite of the scientific evidence that we are facing a collapse, societies have not shown themselves to be capable of pulling, nor even apparently willing to pull, the emergency brakes and claim such autonomy. But it is precisely because they have no more time and no other choice but to take an unprecedented leap toward self-overcoming that it is still possible to imagine it. History, as discussed multiple times in the Introduction, is unpredictable, and when situations are extreme and tensions reach the point of no return, the most unlikely solutions on the historical continuum can suddenly find their way.

The simple and unavoidable fact in our agenda is that contemporary man has no alternative but to attempt that which seems unrealistic today: to overcome capitalism, if by this we mean overcoming, at the same time, the insane accumulation mechanism and the philosophical poverty of anthropocentrism. Otherwise, humans should abandon their aspiration to live up to the title of *Homo sapiens* and conform to their current condition of *Homo exstinguens*, an ephemeral title, since the verb *exstinguo* in Latin also admits the reflective sense.

References

AFEISSA, Hicham-Stéphane, *La communauté des êtres de nature*, Éditions MF, 2010.
ARSENAULT, Chris, "International court to prosecute environmental crimes in major shift". *Reuters*, 15/IX/2016.
BACHOFEN, Blaise; ELBAZ, Sion & POIRIER, Nicolas. *Cornelius Castoriadis. Réinventer l'autonomie*. Paris, Éditions du Sandre, 2008.
BARNOSKY, Anthony *et al.* "Approaching a state shift in Earth's biosphere". *Nature* 486, 7/VI/2012, pp. 52–58.
BARNOSKY, Anthony & HADLY, Elizabeth. *End Game: Tipping Point for Planet Earth?* London, HarperCollins, 2015.
BALL, Terence, "Democracy". In, DOBSON, Andrew & ECKERSLEY, Robyn (eds.). *Political Theory and the Ecological Challenge*. Cambridge, University Press, 2006, pp. 131–146.

---

[16] See Michael Löwy, "The Revolution is the Emergency Brake. Walter Benjamin's political-ecological currency": "This is one of the preparatory notes to "On the Concept of History," which does not appear in the final versions of the document. The passage from Marx to which Benjamin refers appears in *The Civil War in France*: "Die Revolutionen sind die Lokomotiven der Geschichte" (online).

[17] Cf. Castoriadis, "Briser la clôture," in Bachofen; Elbaz & Poirier (2008, p. 282).

BOURG, Dominique & WHITESIDE, Kerry, *Vers une démocratie écologique*. Paris, Seuil, 2010.

DOBSON, Andrew, "Representative Democracy and the Environment". In, LAFFERTY, William (ed.), *Democracy and the Environment*, Edward Elgar, 1998, Chapter 7, pp. 124–139.

DROZ-VINCENT, G. "Les *foedera naturae* chez Lucrèce". In: LÉVY, C. (org.). *Le concept de nature à Rome*. Paris, 1996, pp. 191–211

GARRIC, Audrey, "Pétition pour le climat: un succès inédit". *Le Monde*, 28/XII/2018.

HESSEL, Stéphane & MORIN, Edgar. *Le chemin de l'espérance*. Paris, Fayard, 2011.

HORKHEIMER, Max. *Eclipse of Reason*. Oxford University Press, 1947.

HÖSLE, Vittorio. *Philosophie de la crise écologique* (1991). Paris, Payot, 2009.

JONAS, Hans. *The Imperative of Responsibility. In search of an Ethics for the Technological Age* (1979). The University of Chicago Press, 1984.

_____. *Sull'orlo dell'abisso. Conversazioni sul rapporto tra uomo e natura* (1985). Turin, Einaudi, 2000.

KANT, Immanuel, *Zum ewigen Frieden. Ein philosophischer Entwurf* (1795). English translation: *To perpetual peace: a philosophical sketch*, Hackett Publishing, 2003.

MARX, Karl & ENGELS, Friedrich, *The Communist Manifesto* (1848). <https://www.marxists.org/archive/marx/works/download/pdf/Manifesto.pdf>.

MCGLADE, Christophe & EKINS, Paul. "The geographical distribution of fossil fuels unused when limiting global warming to 2°C". *Nature*, 517, 8/I/2015, pp. 187–190.

MCKIBBEN, Bill, "At last, divestment is hitting the fossil fuel industry where it hurts". *The Guardian*, 16/XII/2018.

REGAN, Tom, *The case for Animal Rights*. Routledge, 1988.

RIPPLE, William J. *et al.*, "World Scientists' Warning to Humanity: A Second Notice". *BioScience*, 67, 12, December 2017, pp. 1026-1028.

RIPPLE, William J. *et al.*, "World Scientists' Warning of a Climate Emergency", *BioScience*, 70, 1, 5/XI/2019.

ROOSEVELT, Theodor, "The Strenuous Life" (1899). In *The Strenuous Life. Essays and Addresses*, New York, 1900.

SACHS, Ignacy. *Caminhos para o desenvolvimento sustentável* (1998). Rio de Janeiro, 2009.

SCHUMPETER, Joseph A. *Can capitalism survive? Creative destruction and the future of the global economy* (1942). New York, HarperCollins, 1976.

SERRES, Michel, *Retour au "Contrat naturel"*. Paris, Bibliothèque Nationale de France, 1998.

STEFFEN, Will *et al.*, "The trajectory of the Anthropocene: The Great Acceleration". *The Anthropocene Review*, 2015, 2(1), pp. 81–98.

STERN, Nicholas, *Managing climate change. Climate, growth and equitable development*, Paris, Collège de France, 2010.

TAKAKIJY, Laura Chason. *Lucretius, Pietas, and the* Foedera Naturae. Thesis. University of Texas at Austin, 2013.

ZALASIEWICZ, Jan & Mat, "Battle Scars". *New Scientist*, 28/III/2015, pp. 36–39

Index

CPSIA information can be obtained
at www.ICGtesting.com
Printed in the USA
BVHW091202021220
594679BV00006B/129